T0235162

Lecture Notes in Artificial Intelligence 10339

Subseries of Lecture Notes in Computer Science

More information about this series at http://www.springer.com/series/1244

David W. Aha · Jean Lieber (Eds.)

Case-Based Reasoning Research and Development

25th International Conference, ICCBR 2017 Trondheim, Norway, June 26–28, 2017 Proceedings

Springer

Editors
David W. Aha
Naval Research Laboratory
Navy Center for Applied Research in AI
Washington, DC
USA

Jean Lieber
LORIA/Bâtiment B/équipe Orpailleur
University of Lorraine
Vandœuvre-lès-Nancy Cedex
France

ISSN 0302-9743 ISSN 1611-3349 (electronic)
Lecture Notes in Artificial Intelligence
ISBN 978-3-319-61029-0 ISBN 978-3-319-61030-6 (eBook)
DOI 10.1007/978-3-319-61030-6

Library of Congress Control Number: 2017943841

LNCS Sublibrary: SL7 – Artificial Intelligence

Printed on acid-free paper

This Springer imprint is published by Springer Nature
The registered company is Springer International Publishing AG
The registered company address is: Gewerbestrasse 11, 6330 Cham, Switzerland

Preface

This volume contains the papers presented at the 25th International Conference on Case-Based Reasoning (ICCBR), held during June 26–28, 2017, in Trondheim, Norway. ICCBR is the premier annual meeting of the CBR research community. The theme of ICCBR 2017, "Analogy for Reuse," was highlighted in several events.

Prior events related to ICCBR also include the European Workshops and Conferences on CBR, which we also list here for completeness. These include: Kaiserslautern, Germany (1993); Chantilly, France (1994); Sesimbra, Portugal (1995); Lausanne, Switzerland (1996); Providence, USA (1997); Dublin, Ireland (1998); Seeon Monastery, Germany (1999); Trento, Italy (2000); Vancouver, Canada (2001); Aberdeen, UK (2002); Trondheim, Norway (2003); Madrid, Spain (2004); Chicago, USA (2005); Fethiye, Turkey (2006); Belfast, UK (2007); Trier, Germany (2008); Seattle, USA (2009); Alessandria, Italy (2010); Greenwich, UK (2011); Lyon, France (2012); Saratoga Springs, USA (2013); Cork, Ireland (2014); Frankfurt, Germany (2015); and Atlanta, Georgia (2016).

Each day of the conference began with an outstanding keynote presentation. Henri Prade described recent advances on analogical reasoning, Agnar Aamodt and Enric Plaza gave a joint presentation on the past, present, and future of CBR along with the history of artificial intelligence, and Mary Lou Maher talked about models of novelty and surprise in the context of their integration with CBR and recommender systems to encourage user curiosity. These proceedings contain an abstract of each of these presentations and an extended paper for two of them.

In addition to the first keynote talk, the conference's first day included workshops, a doctoral consortium, the Computer Cooking Contest (CCC), and ICCBR's first video competition. The themes of the three workshops were "CBR and Deep Learning" (the first of its kind), "Computational Analogy" and "Process-Oriented CBR." We thank the leaders of these workshops for their devotion to pursuing and encouraging further contributions to these topics.

During the following two days of the conference, 27 papers were presented as plenary oral or poster presentations; they were selected among 43 submissions. These papers, which are included in the proceedings, address many themes related to the theory and application of case-based reasoning. The authors of these papers represent 12 countries in Asia, Australasia, Europe, and North America.

It is our great pleasure to acknowledge the help of many people for their help in organizing this conference. First, Odd Erik Gunderson did a wonderful job bringing this conference to life and also managed the conference budget. Kerstin Bach performed the difficult tasks of publicity chair and sponsorship chair; we are thankful for what she has done, which included the development and maintenance of the conference website. Anders Kofod-Peterson and Antonio A. Sánchez Ruiz Granados were in charge of the workshops, from the call for workshop proposals to their organization at the conference. Stefania Montani and Jonathan Rubin organized the doctoral consortium, an event that is critical for passing the relay from one generation of researchers to the next. The CCC was organized by Nadia Najjar and David Wilson, following a tradition in this conference series since 2008. Amélie Cordier and Michael Floyd organized the Video Competition, which we hope is the first of many!

We are also grateful to the members of the Advisory Committee, who provided invaluable guidance when called upon, and to the members of the Program Committee (and additional reviewers), who thoughtfully assessed the submissions and did an excellent job in providing constructive feedback to the authors. Thanks also to the local support group at NTNU, led by Odd Erik Gundersen, for their help throughout the conference.

We also wish to thank Alexandra Coman for her great help during the submission phase with EasyChair and during the editing process.

We are very grateful for the support of our sponsors, including NTNU's Department of Computer and Information Science, the Research Council of Norway, and the new Telenor-NTNU AI Laboratory.

Last, but not least, we would like to thank all the authors who submitted papers to the conference and contributed to its high quality, as well as the conference attendees, who made this such an enjoyable event for everyone involved. Finally, it was a pleasure to return to Trondheim in 2017, the first example of location reuse in ICCBR's storied history.

June 2017

David W. Aha
Jean Lieber

Organization

Program Chairs

David W. Aha	Naval Research Laboratory, USA
Jean Lieber	Loria (UL, CNRS, Inria), France

Local Chair

Odd Erik Gundersen	Norwegian University of Science and Technology (NTNU), Norway and MazeMap AS, Norway

Workshop Chairs

Anders Kofod-Peterson	NTNU, Norway and Alexandra Instituttet, Denmark
Antonio A. Sánchez Ruiz	Universidad Complutense de Madrid, Spain

Sponsorship and Publicity Chair

Kerstin Bach	NTNU, Norway

Doctoral Consortium Chairs

Stefania Montani	Università del Piemonte Orientale, Italy
Jonathan Rubin	Philips Research North America, USA

Computer Cooking Contest Chairs

David Wilson	University of North Carolina at Charlotte, USA
Nadia Najjar	University of North Carolina at Charlotte, USA

Video Competition Chairs

Michael Floyd	Knexus Research Corporation, USA
Amélie Cordier	Hoomano, France

Advisory Committee

Agnar Aamodt	NTNU, Norway
Susan Craw	Robert Gordon University, UK
Ashok Goel	Georgia Institute of Technology, USA
Mehmet H. Göker	Salesforce, USA
Ramon Lopez De Mantaras	IIIA - CSIC, Spain
Luigi Portinale	Università del Piemonte Orientale, Italy

Program Committee

Klaus-Dieter Althoff	DFKI/University of Hildesheim, Germany
Kerstin Bach	NTNU, Norway
Ralph Bergmann	Universität Trier, Germany
Isabelle Bichindaritz	State University of New York at Oswego, USA
Derek Bridge	University College Cork, Ireland
Sutanu Chakraborti	Indian Institute of Technology, Madras, India
Alexandra Coman	NRC/Naval Research Laboratory, USA
Sarah Jane Delany	Dublin Institute of Technology, Ireland
Belén Díaz-Agudo	Universidad Complutense de Madrid, Spain
Michael Floyd	Knexus Research, USA
Eyke Hüllermeier	Universität Paderborn, Germany
Joseph Kendall-Morwick	University of Central Missouri, USA
Deepak Khemani	Indian Institute of Technology, Madras, India
Luc Lamontagne	Université Laval, Québec, Canada
David Leake	Indiana University, USA
Cindy Marling	Ohio University, USA
Stewart Massie	Robert Gordon University, UK
Mirjam Minor	Goethe-Universität, Frankfurt, Germany
Stefania Montani	Università del Piemonte Orientale, Italy
Héctor Muñoz-Avila	Lehigh University, USA
Emmanuel Nauer	Loria (UL, CNRS, Inria), France
Santiago Ontañon	Drexel University, USA
Pinar Øzturk	NTNU, Norway
Miltos Petridis	University of Brighton, UK
Enric Plaza	IIIA-CSIC, Spain
Juan A. Recio-García	Universidad Complutense de Madrid, Spain
Jonathan Rubin	Philips Research North America, USA
Antonio A. Sánchez-Ruiz	Universidad Complutense de Madrid, Spain
Barry Smyth	University College Dublin, Ireland
Frode Sørmo	Amazon Alexa, UK
Swaroop Vattam	MIT Lincoln Labs, USA
Ian Watson	University of Auckland, New Zealand
Rosina Weber	Drexel University, USA
David Wilson	University of North Carolina at Charlotte, USA
Nirmalie Wiratunga	Robert Gordon University, UK

Additional Reviewers

Viktor Ayzenshtadt	DFKI, Germany
Yoke Yie Chen	Robert Gordon University, UK
Jeremie Clos	Robert Gordon University, UK
Amélie Cordier	Hoomano, France

G. Devi	India Institute of Technology, Madras, India
Eric Kübler	Goethe-Universität, Germany
Alain Mille	LIRIS, Université de Lyon, CNRS, France
Rotem Stram	DFKI, Germany
Christian Zeyen	Universität Trier, Germany

Abstracts of Invited Papers

Case-Based Reasoning and the Upswing of AI (Extended Abstract)

Agnar Aamodt[1] and Enric Plaza[2]

[1] Department of Computer Science, NTNU, Trondheim, Norway
agnar@ntnu.no
[2] IIIA – CSIC, Campus U.A.B., Bellaterra, Catalonia
enric@iiia.csic.es

The history and evolution of AI has shaped Case-Based Reasoning (CBR) research and applications. We are currently living in an upswing of AI. To what degree does that mean an upswing of CBR as well? And what buttons should we push in order to increase the influence of CBR within the current AI summer, and beyond?

Artificial Intelligence as a scientific field was established at a Dartmouth College seminar in 1956, but ever since ancient times the idea of thinking as a formal and mechanistic process has occupied people's minds. After the firing of the starting shot in 56, the field has experienced both summers and winters, including two serious AI winters up to now. The causes behind these shifts in seasons have been subject to substantial discussion, and a compelling question of course is what to learn from this.

Case-Based Reasoning has had its own development history within the broader AI field. The grouping of AI methods into data-driven AI and knowledge-based AI [1] is also a familiar distinction in CBR. Recent trends in AI has clearly favoured the data-driven methods, and the well-known successes of Deep Neural Networks is a justification for that. But in order to widen the scope of AI methods, and be able to address a wider range of problems and applications, there are reasons to believe that a stronger knowledge-based influence will be needed in the years to come. Several authors have claimed we should look beyond the current upswing of AI, some have argued for methods inspired by human cognition, and others for a need to revitalize symbol-processing based on explicit knowledge representation. Pat Langley started the Cognitive Systems Movement [5], aimed at getting AI back to its roots of studying artefacts that explore the full range of human intelligence. The Artificial General Intelligence initiative addresses thinking machines with full human capabilities and beyond [3]. A focus on symbolic AI and knowledge representation issues has been strongly advocated by Hector Levesque [6], who also warns us to not to be blinded by short-term successes of particular methods.

Initiatives such as these are important to be aware of when we discuss future paths for CBR, and AI more generally. Moreover, the upswing of AI has created in the public, media, and decision makers a great confusion as to what AI is, where numerous concepts — AI, robots, ML, deep learning, and big data, together with the "smart" adjective before almost anything — are conflated and used interchangeably.

So, where is CBR in the overall AI landscape today? Does it live its own life alongside other main subareas, or are there sufficient similarities at the foundational level to group CBR with other methods? With a focus on machine learning, a division of the ML field into five "tribes" has been suggested [2], within which one such tribe is the "Analogizers", united by their reliance on similarity assessment as the basis for learning. It is a diverse tribe covering analogical reasoning, instance-based methods, and support vector machines. For each of the five tribes a unifying 'master algorithm' is proposed, and some people may fall off their chairs when kernel machines is assigned as the unifying method for this tribe. Anyway, views like this may trigger discussions that will lead to a better understanding of CBR in relation to other AI methods.

Given the growing interest in cognitive foundations of AI, we recall the notions of System 1 and System 2 in human cognition presented by Kahneman [4]. System 1 is a model of human memory capable of 'fast thinking', basically performing recognition of new inputs and responding intuitively, while System 2 models the deliberate, explicit reasoning performed by humans. An important issue to discuss is how they could be related to an integrated view of CBR encompassing both data-driven and knowledge-intensive processes.

All these considerations open up some important future challenges and opportunities for CBR, including: How to interpret the revitalized cognitive turn in the paradigm of CBR? How can data-driven CBR be competitive with current ML developments? Can CBR offer a new kind of synergy of data-driven and knowledge-intensive approaches for AI?

References

1. Barcelona Declaration for the Proper Development and Use of Artificial Intelligence in Europe: http://www.bdebate.org/sites/default/files/barcelona-declaration_v7-1-eng.pdf. Accessed 5 March 2017
2. Domingos, P.: The Master Algorithm: How the Quest for the Ultimate Learning Machine Will Remake Our World. Basic Books, New York (2015)
3. Goertzel, B.: Artificial general intelligence: concept, state of the art, and future prospects. J. Artifi. Gen. Intell. **5**(1), 1–48 (2014)
4. Kahneman, D.: Thinking, Fast and Slow. Penguin Books (2011)
5. Langley, P.: The cognitive systems paradigm. Adv. Cogn. Syst. J. **1**, 3–13 (2012)
6. Levesque, H.: On our best behavior. Artif. Intell. **212**(1), 27–35 (2014)

Encouraging Curiosity in Case-Based Reasoning and Recommender Systems

Mary Lou Maher and Kazjon Grace

University of North Carolina at Charlotte, Charlotte NC 28202, USA
{m.maher,k.grace}@uncc.edu

Abstract. A key benefit of case-based reasoning (CBR) and recommender systems is the use of past experience to guide the synthesis or selection of the best solution for a specific context or user. Typically, the solution presented to the user is based on a value system that privileges the closest match in a query and the solution that performs best when evaluated according to predefined requirements. In domains in which creativity is desirable or the user is engaged in a learning activity, there is a benefit to moving beyond the expected or "best match" and include results based on computational models of novelty and surprise. In our invited paper, we will describe models of novelty and surprise that are integrated with both CBR and Recommender Systems to encourage user curiosity.

Analogical Proportions and Analogical Reasoning – An Introduction

Henri Prade[1,2] and Gilles Richard[1]

[1] IRIT, Toulouse University, Toulouse, France
[2] QCIS, University of Technology, Sydney, Australia
{prade, richard}@irit.fr

Abstract. Analogical proportions are statements of the form "*a* is to *b* as *c* is to *d*". For more than a decade now, their formalization and use have been the focus of many researchers. In this talk we shall primarily focus on their modeling in logical settings, both in the Boolean and in the multiple-valued cases. This logical view makes clear that analogy is as much a matter of dissimilarity as a matter of similarity. Moreover analogical proportions emerge as being especially remarkable in the framework of logical proportions. The analogical proportion and seven other code independent logical proportions can be shown as being of particular interest. Besides, analogical proportions are at the basis of an inference mechanism that enables us to complete or create a fourth item from three other items. The relation with case-based reasoning and case-based decision making is emphasized. Potential applications and current developments are also discussed.

Contents

Invited Papers

Regular Papers

Invited Papers

Encouraging Curiosity in Case-Based Reasoning and Recommender Systems

Mary Lou Maher([✉]) and Kazjon Grace

University of North Carolina at Charlotte, Charlotte, NC 28202, USA
{m.maher,k.grace}@uncc.edu

Abstract. A key benefit of case-based reasoning (CBR) and recommender systems is the use of past experience to guide the synthesis or selection of the best solution for a specific context or user. Typically, the solution presented to the user is based on a value system that privileges the closest match in a query and the solution that performs best when evaluated according to predefined requirements. In domains in which creativity is desirable or the user is engaged in a learning activity, there is a benefit to moving beyond the expected or "best match" and include results based on computational models of novelty and surprise. In this paper, models of novelty and surprise are integrated with both CBR and Recommender Systems to encourage user curiosity.

Keywords: Curiosity · Case-based reasoning · Recommender systems

1 Why Encourage Curiosity?

Curiosity is a desire to learn or know something. With the increasing reliance on information available online and the use of AI to guide the user in both problem solving and their quest for information, there arises the potential to deliberately trigger the user's curiosity. Guiding our natural tendency to be curious is increasingly important as we are exposed to vast amounts of information and we often cope by paying attention to information that is familiar or matches our expectations and belief systems. In problem solving scenarios that reward creativity, encouraging curiosity can guide the user towards a broader range of potential solutions. In learning contexts, guiding the user towards material that expands their knowledge incrementally can assist in defining a personalized path through a vast amount of relevant material. In information retrieval and social media, encouraging curiosity will assist in broadening our exposure and reduce the information bubble effect. This paper focuses on encouraging curiosity in case-based reasoning and recommender systems with the exploration and application of computational models of novelty, value, expectation, and surprise.

While we claim that creativity and learning are areas of human reasoning that favor or reward the person in a curious state, typically problem solving and search has placed more emphasis on producing a result that is deemed correct rather than being creative. The significance of developing a computational model for encouraging specific curiosity raises the question of a reason for directing search or synthesis towards novelty. In some areas of problem solving, such as design and planning, creative solutions are more highly valued than solutions that are derivative. Studies of creativity

D.W. Aha and J. Lieber (Eds.): ICCBR 2017, LNAI 10339, pp. 3–15, 2017.
DOI: 10.1007/978-3-319-61030-6_1

in design have produced many explanations and models of human cognition during design. Schon's "reflective practitioner" account of design [1] suggests that specific curiosity is a key component of problem-framing. Suwa et al. identified unexpected discoveries as precursors to creative design [2], and found that unexpected discoveries lead to reflective reinterpretation of the current problem, which in turn leads to further unexpected discoveries. This reflective behavior suggests that surprise is one possible trigger for specific curiosity, and that it can lead users to reformulate their goals and their approach to a problem. In this paper we posit that computational approaches to triggering curiosity is important in areas related to user experience design, including encouraging curiosity towards diversity in diet and triggering curiosity in human learning experiences. More generally, by understanding how computational systems can encourage human curiosity, we can begin to address the broader issue of the role of computational systems in co-creation and innovation.

Case-based reasoning systems present or synthesize solutions to problems based on previous experiences that are most likely to satisfy the user's specified requirements. There are numerous alternatives to indexing, retrieving, and adapting cases, with a focus on responding to the user's predefined goals and evaluating results based on best fit. Incorporating computational models of expectation, novelty, and surprise in case-based reasoning systems allows the system to deliberately retrieve or synthesize alternatives that would stimulate curiosity and lead to creative solutions.

Recommender systems use various algorithms to filter the vast number of possible results to a query in order to increase the potential that the user will select one of the alternatives presented. A recommender system ranks the results of the information presented to the user according to a value system (e.g., according to the preferences of similar users). Incorporating computational models of expectation, novelty and surprise in recommender systems enables the system to deliberately present alternatives that stimulate the user's curiosity.

2 What Is Curiosity?

The concept of curiosity has been used to refer both to a trait and a state [3]. Curiosity-as-trait refers to an innate ability of a person, and individuals differ in how much curiosity they have. Curiosity-as-state refers to a motivational state of a person that causes the person to seek novel stimuli, and it varies within each person according to their context. Curiosity-as-state is malleable – a person is not restricted by their innate ability to be curious, and curiosity can be encouraged by external events or contexts. In this paper, we adopt the curiosity-as-state concept and consider ways in which cognitive systems can be extended to encourage curiosity. These proposed curiosity-stimulating systems would seek out and present novel stimuli during search, and support people in seeking novel solutions during problem solving and synthesis. Curiosity-as-state has been integrated into cognitive systems in the past, such as Saunders and Gero's computational model of curiosity in agents to guide creative behavior [4] and Merrick and Maher's model of curiosity in agents expressing motivational states as a guide to learning [5]. Our work extends this concept by proposing that a system emulates what would make its user curious, in order to encourage their curiosity [6–8].

To explore the role of computational systems in encouraging the state of curiosity we consider different ways in which curiosity can be experienced. Berlyne proposed that state curiosity can be considered along two dimensions [3]:

1. epistemic vs perceptual curiosity and
2. diversive vs specific curiosity.

In the first dimension, perceptual curiosity is the drive towards novel sensory stimuli and epistemic curiosity is the drive to acquire new knowledge. Along the second dimension, Berlyne describes diversive curiosity as unguided search for any new information or stimuli and specific curiosity is search for a novel solution to a specific problem or goal. Diversive curiosity is a good descriptor for the computational models in Saunders and Gero's agents [4] as well as Merrick and Maher's agents [5]. Diversive curiosity is an explanation of Schmidhuber's theory of creativity and intrinsic motivation [9]. Specific curiosity is the basis for reformulation of design goals and for stimulating the curiosity of a user in selecting surprising recipes in the Q-Chef system [8]. In this paper we explore specific curiosity and how it can be achieved with models of expectation, novelty and surprise in the context of CBR and recommender systems.

3 Computational Models of Novelty and Surprise

In this section we describe the basic principles for building computational models of novelty and surprise, as the key ingredients for encouraging curiosity in human search and problem solving. We also identify subtle differences between novelty and surprise. In most applications that employ novelty detectors, novelty is not a good thing. Outliers are either ignored in many data science applications, or they are an indication of error. In many applications, such as cybersecurity, control systems, or fraud monitoring, anomaly detection is an indicator that something bad has happened. In contrast, we are interested in novelty detection as a good thing.

Curiosity as a state can be triggered by novel stimuli. How something is novel depends on familiarity. An adaptation of the Wundt curve, shown in Fig. 1, illustrates how we can move from a state of boredom to curiosity when we experience something that has a certain threshold of novelty, but too much novelty can cause a negative response and fear. This suggests that novelty can be experienced along a spectrum of newness, and curiosity-as-state has a lower bound and upper bound in that spectrum.

While novelty as a concept is simply about the quality of something being new, the measurement of novelty relies on a representation of the events or experiences, and a metric for the degree of novelty. A generalized process for determining novelty has three stages:

1. Identify the source data as a set or series of data points or experiences.
2. Represent the source data in a common structure.
3. Measure novelty of a new data item within the structured representation.

Pimentel et al. provide a classification of novelty detection techniques in the following categories: distance-based, probabilistic, reconstruction-based, domain-based,

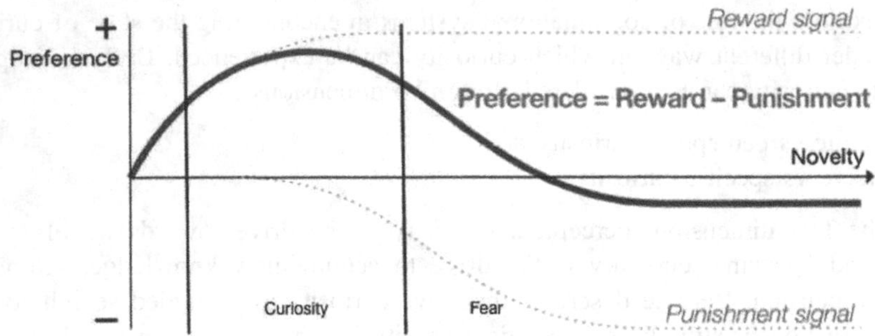

Fig. 1. An adaptation of the Wundt curve

and information theoretic [10]. Distance-based novelty uses concepts of nearest neighbour and clustering analysis. Probabilistic methods of novelty consider density and likelihood of occurrence. Reconstruction-based novelty involves training a regression model and recognizing abnormal data. Domain-based creates a representation of the domain by defining a boundary around the normal class. Information theoretic models of novelty use concepts such as entropy to measure the amount of new information.

The distance-based principle provides a range of values that measure the distance between the new event and previous experiences. This measurement assumes a structured representation on which distance can be measured. Grace et al. describe a process for measuring the novelty of mobile devices [11]: (1) each device is represented as a feature vector, (2) all existing device feature vectors are placed into clusters in order of their appearance on the market using a K-means algorithm, and (3) the novelty of a new device is represented by its distance to the mean value of the nearest cluster. By clustering the feature vectors of the designs, we can build a representation of how designs vary in a conceptual space, and then compare a new design to others of the same kind. Novelty (or similarity) is measured by the cluster distance and then normalized by the average similarity of designs within that cluster, reflecting that some clusters are tightly defined while others are broad. Our concept of average similarity of designs in a cluster is similar to the "Silhouette Score" introduced by Rousseeuw [13], which is a measure of how tightly grouped the data in the cluster are. In contrast to the Silhouette Score, we are interested in the distance between an individual data point and an existing cluster to identify designs that are not close to a cluster.

Novelty can capture how different a new event is from previous events, but it does not necessarily capture how surprising a new event is based on our experience of previous events. If a new event is following a trend, then the novelty is not unexpected. We claim that curiosity is triggered when our expectations are not met, not merely because something is novel. We define surprise as the degree to which the properties of a new event or experience violate expectations about those properties. With this definition, a surprising event is always novel, but a novel event is not always surprising.

The experience of surprise is mediated by how knowledgeable a person is about the new event or experience. If a person is not familiar with the space of possibilities, then everything is novel but nothing is surprising. We have modelled this with a confidence value based on the density of the data points in the clusters nearest the novel data point [14]. Expectations (and therefore surprise) are guided by temporal relationships as well: the experience of an event today that would be common in the past, for example a person wearing bell-bottom jeans, would be surprising.

Measuring surprise extends the notion of novelty by including a model of expectation. We decompose the process for measuring surprise into four canonical steps.

1. Collection of data as a series of events or experiences
2. Representation of source data in a common structure
3. Construction of a representation of expectation
4. Measure of unexpectedness of a new event or experience

The selection of a representation of the source data determines what properties play a role in setting expectations. This can be defined based on predictive models, or based on descriptive and performance requirements in a search or design process. Expectations are temporally and experientially contextualized, in that they are made in the context of a conceptual space and a set of past experiences.

Surprise can be measured as the inverse of the prior expected probability of observing a data item. This approach is also used by Macedo et al. [15] and Itti and Baldi [16]. When a design is observed, the degree of expectedness of each observed attribute of a design given its other attributes is calculated. Maher et al. use regression modelling to create expectations about future designs, and exclusively constructed expectations as a function of product release date [17]. Grace et al. extend the temporally-based expectation to consider the unexpectedness of an object to be the maximum of the unexpectedness of any of its parts in the given context [14]. The unexpectedness is calculated as the maximum value, not the mean, based on results from Reisenzein that the intensity of surprise increases directly with the unexpectedness of an event [18]. This is based on the intuition that a single highly unexpected aspect of the object is sufficient to elicit surprise, and that such unexpectedness would not be affected by the number of other surprising features.

4 PQE: A Model for Personalized Curiosity

We have developed PQE (pronounced pēk) for stimulating specific curiosity in design based on a process that has three components: a model of the user's preferences, a model of curiosity based on user behaviour and preferences, and a model that synthesises designs for the user, shown in Fig. 2. The PQE cycle iteratively *models* a user's curiosity in order to *stimulate* it. The user's feedback on designs refines the system's model of their preferences and familiarities, which are used to update a model of what will encourage the user towards the state of specific curiosity. The user model and the curiosity model are then used as a resource in a computational design system, providing the basis for evaluating designs for value (satisfies the user's requirements) and surprise (satisfies the user's specific curiosity). The resulting design is presented to the user, who can then refine the design and then give feedback on it to the system, starting the cycle anew. The system is capable of guiding the user towards surprising and preferred designs such that future generated designs share those traits. This is a form of meta-reasoning [19] common to designers: (re)formulation of design goals as a result of surprise [4]. Curiosity-triggered goals limit the search for new solutions in a space of possibilities to focus on what the system finds surprising, influencing the direction of synthesis [12, 20]. The PQE architecture is elaborated on in our other papers [6–8]. Here we describe it briefly and then focus on how it might be useful in CBR and recommender systems.

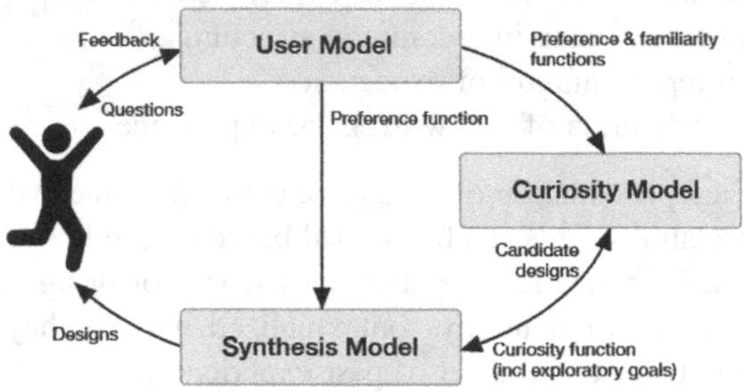

Fig. 2. PQE: a process for synthesis based on user preference and curiosity models [7].

5 Encouraging Curiosity in Case-Based Reasoning

We adapted the PQE model for encouraging diversity in a user's diet in a system we call Q-Chef. We hypothesize that users will be motivated to increase the diversity of their diet if they are presented with recipe suggestions designed to stimulate their curiosity. Changing food-related behavior is particularly challenging, requiring education, motivation, and regular supervision [21]. Nutritional knowledge has been called "necessary but not sufficient" [22] for changing food behavior, as much relevant information is poorly accessible, conflicting, or too abstract to be applicable directly to household food purchasing. Sustaining participant motivation has been suggested as a critical requirement of effective intervention in eating behavior [23], as motivation to eat healthy is a strong predictor of actual health eating behavior. These findings support the benefit of behavior change technologies that explicitly model and encourage the intrinsic curiosity motivations of users. The Q-Chef system is an interactive recipe generator that models user curiosity in response to their preferences, and then recommends recipes that will stimulate their interest in a more diverse diet.

Q-Chef builds on the PQE concepts of user model, curiosity model, and synthesis model. The user model incorporates feedback from the user about both their familiarity with recipes as well as their preference for those recipes. The preferences and familiarity are recorded separately in order to accurately model curiosity, as someone can be highly familiar with things that they greatly dislike, or vice versa. Familiarity ratings provide the basis for personalizing the curiosity model, in the same way that a preference rating provides the basis for personalizing the preference function. Familiarity is the system's estimate of which recipes the user has seen. Preference ratings are used for determining the value of a design and familiarity ratings are used to determine the novelty of the design.

The curiosity model uses the preference and familiarity functions to generate a model of user expectations based on novelty and to create exploratory goals. Curiosity is based on the Wundt curve model shown in Fig. 1, in which there is a peak level of novelty after which the stimulus becomes progressively more undesirable. What we learn from the Wundt curve is that the amount of difference from our knowledge and expectations is the relevant measure and this measure is not dependent on the content being measured. PQE's curiosity model builds on prior approaches to modeling diversive curiosity [3] and on related systems that model user knowledge in order to recommend things that lie just beyond it [24].

In Q-Chef the measure of surprise of a set of ingredients in a recipe is based on a probabilistic deep neural network, trained with a data set of existing recipes, which captures the expected likelihood for different ingredient combinations, coupled with algorithms for determining surprise and curiosity from those likelihoods. We model recipes as being composed of a set of features, where each feature is an ingredient in the recipe. The curiosity model evaluates recipes for surprising combinations, which consist of a surprising *feature* that is of low expected likelihood given the simultaneous presence in the recipe of a surprise *context*, a set of other features. Surprise can then be measured as the ratio of the rarity of the surprising feature in context to its rarity overall [16]. This is expressed as the ratio of the conditional probability of ingredient B given the context of ingredients A to the unconditional probability of the context A:

$$Surprise = \log_2 \frac{P(A \cap B)}{P(B)} - \log_2 P(A)$$

The recipe synthesis model uses value (the user's preference) and surprise (specific curiosity) functions as exploratory goals for generating recipe suggestions that are presented to the user. The synthesis model uses a case adaptation process. Cases are selected based on the surprising combinations identified by the curiosity model and they are completed using rules for extending the surprising combination to a complete set of ingredients. The objective of synthesis is to produce recipes that are in the user's *novelty sweet spot*, (i.e. are close to the user's preferred level of novelty, a parameter we will establish by experimentation). We use δ to refer to the user's preferred level of novelty, and it will be initially hand-tuned. When the synthesis model generates a recipe that has a combination of ingredients that is highly preferable and surprising, it is presented to the user as a suggested recipe.

Q-Chef has been framed as a case-based reasoner that integrates a model of surprise to encourage curiosity. Case retrieval is a process of selecting one or more cases that are close matches for the current problem. To encourage curiosity, we reconsider the retrieval process to search for ingredient combinations based on value (closest match) and surprise (unexpectedness). Case adaptation uses these ingredient combinations to search for and adapt cases to generate new recipes that will encourage curiosity.

The Q-chef CBR process consists of two complementary CBR cycles: problem framing and problem solving, as shown in Fig. 3. Both CBR cycles employ the same case base: the first cycle uses the case base to generate a new set of requirements that are surprising and the second cycle selects and adapts cases that match the surprising requirements. The model of surprise and similarity is based on the latent ingredient

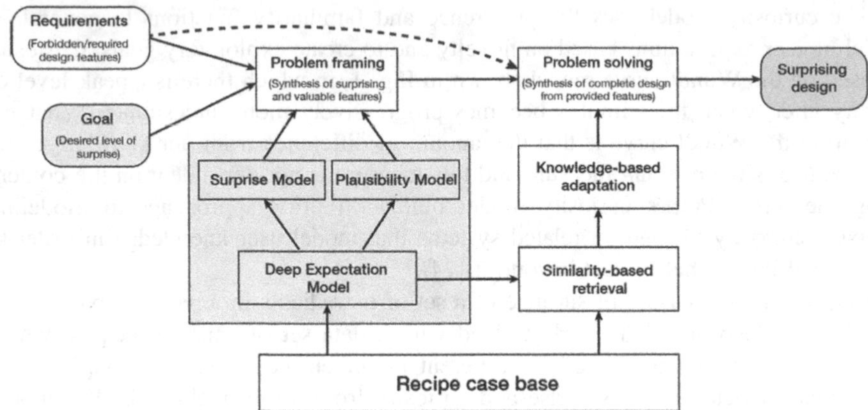

Fig. 3. Q-chef: a CBR system with problem framing to encourage curiosity [8].

association in the case base generated using a deep learning network trained over the case base of ingredient combinations.

Our implementation of specific curiosity generates a new set of ingredients as the output of problem framing, serving as a goal to generate recipes similar to the triggering surprise during problem solving. The set of ingredients selected as the goals for problem solving is based on a threshold for novelty and value.

We have implemented a sampling-based method to generate recipes from the probability distribution over ingredient combinations captured by the neural network used in expectation. As a stochastic generative deep network, the variational autoencoder learns a hidden vector of random variables that, when sampled, can be transformed into vectors of ingredients. These sampled vectors are drawn from the same distribution used by the model of curiosity, and can therefore be considered "plausible" recipes.

To simulate personalized curiosity we separated the dataset of $\sim$100 k recipes into two mutually exclusive sets, as an extreme example of two users with highly different familiarity within the design space of recipes. The split was made based on the presence of sugar (brown, plain, or icing/confectioner's) in the recipe, yielding $\sim$60,000 recipes without sugar (test user #1) and $\sim$40,000 recipes with sugar (test user #2). We compare these two users with a control: a hypothetical omniscient user who has knowledge of all recipes in the database (test user #3). We produce such exaggerated simulations of user profiles as a proof-of-concept of the Q-Chef model's ability to personalize. Specifically, that it is able to produce different expectations, construct different exploratory goals, and synthesize different recipes based on the user model.

We compared the most surprising recipes for each user, noting the level of surprise elicited in all conditions in each case. Among the most surprising recipes identified were "whisky bread pudding", "beef in a barrel" (which was served in a pineapple), and "bacon and onion muffins". Full results, including comparisons between what the sugar-only and no-sugar test cases found surprising, can be seen in Grace et al. (2016).

6 Encouraging Curiosity in Recommender Systems

We are adapting the PQE model for recommending a sequence of resources for open-ended learning tasks in a system we are calling Queue. We define an open-ended task as one in which the structure and pedagogical purpose is specified, but the sequence of the learning materials is up to the learner. Examples of open-ended learning tasks include literature reviews and research tasks, assuming that in each case the learner is given a broad theme or question and they define the exact scope of the work themselves. Queue models curiosity in the learner based on its representation of their preferences and knowledge, and then recommends a sequence of resources that will encourage curiosity. This kind of curiosity-based recommender system helps direct the learner's exploration of the learning materials towards resources that are sufficiently surprising as to stimulate curiosity, and yet sufficiently familiar to be approachable.

The four processes of the Queue system are shown in Fig. 4. The Learner Model is constructed from a combination of direct questions and feedback on suggested resources to generate a representation of the learner's *familiarity* with concepts (a vector function over the set of concepts describing the student's competence in each) and of their *interests* (another vector function over the set of concepts describing the student's preference for each). This model of the learner provides a basis for a cyclical interaction between expectation, composition and reflection. The expectation model is constructed as a probabilistic model of the content in the resources. The curiosity model measures the surprising content in each of the resources. Sequence composition is based on the surprising content and the learner's interests and familiarity, leading to reflection that updates the expectation model.

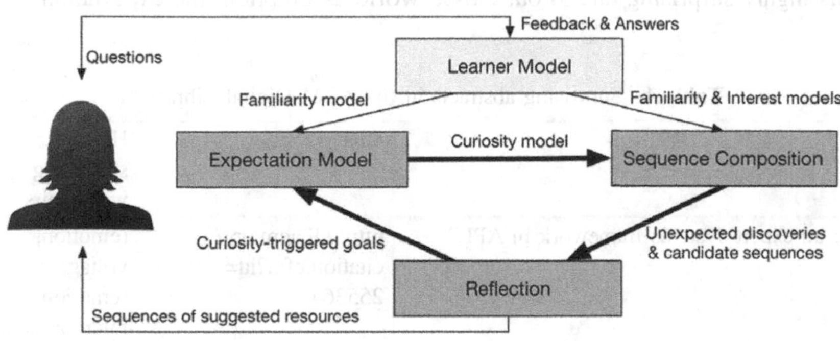

Fig. 4. Queue: a system for encouraging curiosity in learners.

In the sequence composition process, candidate sequences of resources are composed based on value and surprise. The measurement of value is determined using the learner model and the results of a typical recommender system. The measurement of surprise is based on the learner's expectations. We represent expectation with a probabilistic model constructed from the student's familiarities and interests with reference to the corpus of resources. This model captures what the learner expects to see in resources and is combined with curiosity-triggered exploratory goals. In the reflection process resources that are discovered to be highly unexpected according to the expectation model cause the formation of new goals that influence future search.

As a preliminary study of Queue we focus on combining a measure of surprise with existing algorithms for recommender systems for identifying resources that will be both familiar and surprising. Recommender systems are a core part of web search engines, e-commerce and social networks to help users find relevant information based on their preferences and interests [25]. Recommender systems address the problem of information overload by providing personalized information to the user. There are two primary algorithmic concepts used in current recommender systems: Content Based Filtering (CBF) [26] which utilizes the content of items in order to recommend new items with similar descriptions and Collaborative Filtering (CF) [27] which recommend items that other users with the same taste have rated positively. Due to limitations of lack of enough content in CBF and sparsity of data in CF, the use of hybrid techniques [28] is common.

We have gathered a dataset of research paper abstracts from the ACM Digital Library (ACMDL), and are exploring different models of expectation that capture what learners find surprising about academic literature. In this preliminary study we start with a model of expectation based on the unlikely co-occurrence of pairs of words, either adjacent words or any pair of words present in the abstract. This is similar to our model of expectation in Q-Chef in which we model expectations on the co-occurrence of ingredients. Our identification of surprising research papers is still in progress, but several interesting things can be seen in our preliminary results in Table 1, which were generated using an expectation model derived from the GloVe model of co-occurrence based word embeddings [29]. The most surprising abstract in the ACMDL is one that, through an apparent OCR error, contains the entire paper's body text. The third paper is rated as highly surprising due to our closed world assumption: the expectation model

Table 1. Surprising abstracts in the ACM Digital Library

Title	URL	Highly surprising word pairs
"XPL: an expert systems framework in APL"	http://dl.acm.org/citation.cfm?id=255364	(emotion, voltage) (emotion, microprocessor) (fun, voltage)
"Biomedical electronics serving as physical environmental and emotional watchdogs"	http://dl.acm.org/citation.cfm?id=2228362	(emotion, CMOS) (information, CMOS)
"EarSketch: a web-based environment for teaching introductory computer science through music remixing"	http://dl.acm.org/citation.cfm?id=2691869	(hop, curriculum) (hop, introductory) (hop, educator)

only knows the ACMDL, where the word "hop" is near-exclusively used in the context of routing and network infrastructure. That context does not overlap with papers on education, and thus the expectation model confidently predicts those words will not co-occur. The model is surprised by the existence of hip hop, much less that it might be used in CS education. We are in the process of developing models for personalizing and extending these results for application to educational recommendation.

7 Discussion and Future Work

Encouraging curiosity has enormous potential in our digital lives in which much of the information that is available to us is filtered in some way. We have developed a reliance on search engines, recommender systems, and AI-enabled problem solving because in addition to filtering, but they are also successful in providing focus to open ended tasks in a vast digital resource. The development of more capable recommender systems and case-based reasoning is measured by how well the results match the users' expectations and lead to the adoption of the recommendation. We challenge this assumed measurement and claim that in some circumstances it is appropriate to introduce surprising results to the user and go beyond their expectations.

In this paper we present basic principles of curiosity, novelty, and surprise and show how these can be incorporated as computational models in case-based reasoning and recommender systems. Our current application areas, encouraging dietary diversity and recommending sequences of learning materials, are a starting point for exploring computational models of curiosity. Our future work is to develop our models of expectation and surprise in both structured (e.g., recipes) and unstructured (e.g., research papers) representations and to pursue studies of how people recognize surprising content. The development of our computational models of surprise will be calibrated with studies of how and why people find research papers and recipes surprising.

Acknowledgements. These ideas have emerged from years of research in computational creativity and we acknowledge that our ideas have been influenced by the co-authors of our other papers and our PhD students, including Douglas H Fisher, Kate Brady, Katherine Merrick, Dave Wilson, Nadia Najjar, Mohammad Mahzoon, Maryam Mohseni, and Pegah Karimi. The authors acknowledge support from NSF IIS1618810 CompCog: RI: Small: Pique: A cognitive model of curiosity for personalizing sequences of learning resources.

References

1. Schön, D.A.: The Reflective Practitioner: How Professionals Think in Action, vol. 5126. Basic books, New York (1983)
2. Suwa, M., Gero, J.S., Purcell, T.: Unexpected discoveries and s-inventions of design requirements: a key to creative designs. In: Computational Models of Creative Design, Key Centre of Design Computing and Cognition, pp. 297–320. University of Sydney, Australia (1999)
3. Berlyne, D.E.: Exploration and curiosity. Science **153**, 25–33 (1966)

4. Saunders, R., Gero, J.S.: Artificial creativity: a synthetic approach to the study of creative behaviour. In: Computational and Cognitive Models of Creative Design V, University of Sydney, pp. 113–139 (2001)
5. Merrick, K., Maher, M.L.: Motivated learning from interesting events: adaptive, multitask learning agents for complex environments. Adapt. Behav. **17**(1), 7–27 (2009)
6. Grace, K., Maher, M.L.: Specific curiosity as a cause and consequence of transformational creativity. In: Proceedings of the Sixth International Conference on Computational Creativity June, p. 260 (2015)
7. Grace, K., Maher, M.L., Wilson, D., Najjar, N.: Personalised specific curiosity for computational design systems. In: Proceedings of Design Computing and Cognition (2016)
8. Grace, K., Maher, M.L., Wilson, D.C., Najjar, N.A.: Combining CBR and deep learning to generate surprising recipe designs. In: Goel, A., Díaz-Agudo, M.B., Roth-Berghofer, T. (eds.) ICCBR 2016. LNCS, vol. 9969, pp. 154–169. Springer, Cham (2016). doi:10.1007/978-3-319-47096-2_11
9. Schmidhuber, J.: Formal theory of creativity, fun, and intrinsic motivation (1990–2010). IEEE Trans. Auton. Mental Dev. **2**(3), 230–247 (2010)
10. Pimentel, M.A., Clifton, D.A., Clifton, L., Tarassenko, L.: A review of novelty detection. Signal Process. **99**, 215–249 (2014)
11. Grace, K., Maher, M.L., Fisher, D., Brady, K.: Modeling expectation for evaluating surprise in design creativity. In: Gero, John S., Hanna, S. (eds.) Design Computing and Cognition 2014, pp. 189–206. Springer, Cham (2015). doi:10.1007/978-3-319-14956-1_11
12. Grace, K., Maher, M.L.: Surprise-triggered reformulation of design goals. In: AAAI 2016 (2016)
13. Rousseeuw, P.J.: Silhouettes: a graphical aid to the interpretation and validation of cluster analysis. Comput. Appl. Math. **20**, 53–65 (1987)
14. Grace, K., Maher, M.L. Fisher, D.H., Brady, K.: Data-intensive prediction of creative designs using on novelty, value and surprise. Int. J. Des. Creativity Innov. (2014)
15. Macedo, L., Cardoso, A., Reisenzein, R., Lorini, E.: Artificial surprise. In: Handbook of Research on Synthetic Emotions and Sociable Robotics: New Applications in Affective Computing and Artificial Intelligence. IGI Global, Hershey (2009)
16. Itti, L., Baldi, P.: A surprising theory of attention. In: IEEE Workshop on Applied Imagery and Pattern Recognition, pp. 45–54 (2004)
17. Maher, M.L., Brady, K., Fisher, D.H.: Computational models of surprise in evaluating creative design. In: Proceedings of The Fourth International Conference on Computational Creativity, University of Sydney, pp. 147–151 (2013)
18. Reisenzein, R.: The subjective experience of surprise. In: The Message Within: The Role of Subjective Experience in Social Cognition and Behavior, pp. 262–279, (2000)
19. Cox, M.T., Raja, A.: Metareasoning: Thinking About Thinking. MIT Press, Cambridge (2011)
20. Grace, K., Maher, M.L.: Surprise and reformulation as meta-cognitive processes in creative design. In: Proceedings of the Third Annual Conference on Advances in Cognitive Systems (2015)
21. Barker, M., Swift, J.A.: The application of psychological theory to nutrition behaviour change. Proc. Nutr. Soc. **68**(02), 205–209 (2009)
22. Worsley, A.: Nutrition knowledge and food consumption: can nutrition knowledge change food behaviour? Asia Pac. J. Clin. Nutr. **11**(s3), S579–S585 (2002)
23. Renner, B., Schwarzer, R.: The motivation to eat a healthy diet: how intenders and nonintenders differ in terms of risk perception, outcome expectancies, self-efficacy, and nutrition behavior. Polish Psychol. Bull. **36**(1), 7–15 (2005). SRC - GoogleScholar FG - 0

24. Adamopoulos, P., Tuzhilin, A.: On unexpectedness in recommender systems: or how to better expect the unexpected. ACM Trans. Intell. Syst. Technol. (TIST) **5**(4), 54 (2014)
25. Schafer, J.B., Konstan, J., Riedl, J.: Recommender systems in e-commerce. In: Proceedings of the 1st ACM Conference on Electronic Commerce. ACM (1999)
26. Pazzani, M.J., Billsus, D.: Content-based recommendation systems. In: The Adaptive Web, pp. 325–341. Springer Berlin Heidelberg (2007)
27. Schafer, J.B., Frankowski, D., Herlocker, J., Sen, S.: Collaborative filtering recommender systems. In: The Adaptive Web, pp. 291–324 (2007)
28. Burke, R.: Hybrid recommender systems: survey and experiments. User Model. User-Adap. Inter. **12**(4), 331–370 (2002)
29. Pennington, J., Socher, R., Manning, C.D.: Glove: global vectors for word representation. EMNLP **14**, 1532–1543 (2014)

Analogical Proportions and Analogical Reasoning - An Introduction

Henri Prade[1,2](✉) and Gilles Richard[1]

[1] IRIT, Toulouse University, Toulouse, France
{prade,richard}@irit.fr
[2] QCIS, University of Technology, Sydney, Australia

Abstract. Analogical proportions are statements of the form "*a* is to *b* as *c* is to *d*". For more than a decade now, their formalization and use have raised the interest of a number of researchers. In this talk we shall primarily focus on their modeling in logical settings, both in the Boolean and in the multiple-valued cases. This logical view makes clear that analogy is as much a matter of dissimilarity as a matter of similarity. Moreover analogical proportions emerge as being especially remarkable in the framework of logical proportions. The analogical proportion and seven other code independent logical proportions can be shown as being of particular interest. Besides, analogical proportions are at the basis of an inference mechanism which enables us to complete or create a fourth item from three other items. The relation with case-based reasoning and case-based decision is emphasized. Potential applications and current developments are also discussed.

1 Introduction

Analogical reasoning has been regarded for a long time as a fruitful, creative way of drawing conclusions, or of explaining states of fact, even if this form of reasoning does not present the guarantees of validity offered by deductive reasoning. As such, it has been extensively studied in particular by philosophers, psychologists and computer scientists. We can cite for instance [12,14,15,34] where the power of analogical reasoning is emphasized. See also [28] for a computationally oriented survey of current trends.

Roughly speaking, the idea of analogy is to establish a parallel between two situations [11,36] on the basis of which, one tentatively concludes that what is true in the first situation may also be true in the second one. When the two situations refer to apparently unrelated domains, the parallel may be especially rich. Just think of the well known example of the Bohr's model of atom where electrons circle around the kernel, which is analogically linked to the model of planets running around the sun.

Closely related to analogical reasoning is the idea of analogical proportions, i.e., statements of the form "*a* is to *b* as *c* is to *d*", which dates back to Aristotle (in Western world). This establishes a parallel, here between the pair (a, b) and the pair (c, d) [13]. Case-based reasoning seems also to obey a reasoning pattern of

D.W. Aha and J. Lieber (Eds.): ICCBR 2017, LNAI 10339, pp. 16–32, 2017.
DOI: 10.1007/978-3-319-61030-6_2

the same kind. Indeed it establishes a collection of parallels between known cases referring to a pair (< *problem*$_i$ >, < *solution*$_i$ >) and a new < *problem*$_0$ >, for which one may think of a < *solution*$_0$ > that is all the more similar to < *solution*$_i$ > as < *problem*$_0$ > is similar to < *problem*$_i$ >. This suggests that case-based reasoning is a particular instance of analogical reasoning. Still we shall see that analogical proportion-based inference significantly departs from case-based reasoning.

This survey paper is structured as follows. First the notion of logical proportion is recalled in Sect. 2 before focusing on analogical proportion in Sect. 3 and other homogeneous proportions in Sect. 4, in a Boolean setting. The extension of analogical proportions to multiple-valued settings is briefly presented in Sect. 5. The analogical inference machinery is discussed in Sect. 6, applications are reviewed in Sect. 7; relations and differences with case-based reasoning are addressed in Sect. 8.

2 Boolean Logical Proportions

Proportions in mathematics state the identity of relations between two ordered pairs of entities, say (a, b) and (c, d). Thus, the geometric proportion corresponds to the equality of two ratios, i.e., $a/b = c/d$, while the arithmetic proportion compares two pairs of numbers in terms of their differences, i.e., $a - b = c - d$. In these equalities, which emphasize the symmetric role of the pairs (a, b) and (c, d), geometric or arithmetic ratios have an implicit comparative flavor, and the proportions express the invariance of the ratios. Note that by cross-product for geometric proportion, or by addition for the arithmetic one, the two proportions are respectively equivalent to $ad = bc$ and to $a+d = b+c$, which makes clear that b and c, or a and d, can be permuted without changing the validity of the proportion. Moreover, mathematical proportions are at the basis of reasoning procedures that enable us to "extrapolate" the fourth value knowing three of the four quantities. Indeed, assuming that d is unknown, one can deduce $d = c \times b/a$ in the first case, which corresponds to the well-known "rule of three", or $d = c + (b - a)$ in the second case. Besides, continuous proportions where $b = c$ are directly related to the idea of averaging, since taking $b = c$ as the unknown respectively yields the geometric mean $(ad)^{1/2}$ and the arithmetic mean $(a + d)/2$.

Generally speaking, the idea of proportion is a matter of comparison of comparisons, as suggested by the statement "a is to b as c is to d". In the Boolean setting there are four comparison indicators. On the one hand there are two similarity indicators, namely a positive one $a \wedge b$ and a negative one $\neg a \wedge \neg b$, and on the other hand two dissimilarity indicators $\neg a \wedge b$ and $a \wedge \neg b$. Logical proportions [27,29] connect four Boolean variables through a conjunction of two equivalences between similarity or dissimilarity indicators pertaining respectively to two pairs (a, b) and (c, d). More formally

Definition 1. *A logical proportion $T(a, b, c, d)$ is the conjunction of two equivalences between indicators for (a, b) on one side and indicators for (c, d) on the other side.*

For instance, $((a \wedge \neg b) \equiv (c \wedge \neg d)) \wedge ((a \wedge b) \equiv (c \wedge d))$ is a logical proportion, expressing that "a differs from b as c differs from d" and that "a is similar to b as c is similar to d". It has been established that there are 120 syntactically and semantically distinct logical equivalences. All these proportions share a remarkable property: they are true for exactly 6 patterns of values of $abcd$ among 2^4 possible values. Thus, the above example is true for 0000, 1111, 1010, 0101, 0001, and 0100. The interested reader is referred to [27,29] for thorough studies of the different types of logical proportions. In the following, we only consider those satisfying the *code independent* property. This property expresses that there should be no distinction when encoding information positively or negatively. In other words, encoding truth (resp. falsity) with 1 or with 0 (resp. with 0 and 1) is just a matter of convention, and should not impact the final result. Thus we should have the following entailment between the two logical expressions $T(a, b, c, d)$ and $T(\neg a, \neg b, \neg c, \neg d)$, i.e., $T(a, b, c, d) \Rightarrow T(\neg a, \neg b, \neg c, \neg d)$.

It has been established [27] that there only exist eight logical proportions that satisfy the above property. Indeed from a structural viewpoint, note that a proportion is built up with a pair of equivalences between indicators chosen among $4 \times 4 = 16$ equivalences. So, to ensure code independency, the only way to proceed is to first choose an equivalence then to pair it with its counterpart where every literal is negated: for instance $a \wedge b \equiv \neg c \wedge d$ should be paired with $\neg a \wedge \neg b \equiv c \wedge \neg d$ in order to get a code independent proportion. This simple reasoning shows that we have indeed $16/2 = 8$ possibilities.

The 8 code independent proportions split into 4 *homogeneous* proportions that are symmetrical (one can exchange (a, b) with (c, d)) and 4 *heterogeneous* ones that are not symmetrical. Homogeneity here refers to the fact that in the expression of the proportions, both equivalences link indicators of the same kind (similarity or dissimilarity), while in the case of heterogeneous proportions they link indicators of opposite kinds. This explains why there are four *homogeneous* and four *heterogeneous* logical proportions. In the following section, we focus our attention on one especially remarkable code independent logical proportion, the *analogical proportion*, reviewing the 7 others in the next section. Note also that the first example of logical proportion given above after Definition 1 is symmetrical, but not code independent.

3 Boolean Analogical Proportion

The analogical proportion "a is to b as c is to d" more formally states that a differs from b as c differs from d and b differs from a as d differs from c". This is logically expressed as [24] by the quaternary connective Ana:

$$\text{Ana}(a, b, c, d) \triangleq ((a \wedge \neg b) \equiv (c \wedge \neg d)) \wedge ((\neg a \wedge b) \equiv (\neg c \wedge d)) \tag{1}$$

Note that this logical expression of an analogical proportion only uses dissimilarity indicators, and does not mix a dissimilarity indicator and a similarity indicator as in the first example of logical expression we gave. In some sense

Table 1. Boolean patterns making Analogy true

a	b	c	d
0	0	0	0
1	1	1	1
0	0	1	1
1	1	0	0
0	1	0	1
1	0	1	0

Analogy is first a matter of (controlled) dissimilarity. Table 1 exhibits the 6 patterns for which $\text{Ana}(a, b, c, d)$, also traditionally denoted $a : b :: c : d$, is true. It can be easily checked on this table that the analogical proportion is indeed independent with respect to the positive or negative encoding of properties. Moreover, one can also see that the logical expression of $a : b :: c : d$ satisfies the key properties of an analogical proportion, namely

- reflexivity: $a : b :: a : b$
- symmetry: $a : b :: c : d \Rightarrow c : d :: a : b$
- central permutation: $a : b :: c : d \Rightarrow a : c :: b : d$

Consequently it also satisfies $a : a :: b : b$, and the external permutation $a : b :: c : d \Rightarrow d : b :: c : a$. Note also that these properties clearly hold for numerical proportions ($\frac{a}{b} = \frac{a}{b}$; $\frac{a}{b} = \frac{c}{d} \Rightarrow \frac{c}{d} = \frac{a}{b}$, and $\frac{a}{b} = \frac{c}{d} \Rightarrow \frac{a}{c} = \frac{b}{d}$). Table 1 is not the only Boolean model satisfying the three above postulates, but it is the minimal one. See [32] for this result and also for the justification of the 6 patterns in Table 1 in terms of (minimal) Kolmogorov complexity. Moreover, with this definition, the analogical proportion is transitive in the following sense:

$$(a : b :: c : d) \wedge (c : d :: e : f) \Rightarrow a : b :: e : f$$

Besides, note also that analogical proportion holds for the three following generic patterns: $s : s :: s : s$, $s : s :: t : t$ and $s : t :: s : t$ where s and t are distinct values, which is the basis for the extension of the definition of the analogical proportion to nominal values. The above Boolean logic view of analogical proportion agrees with other previous proposals aiming at formalizing the idea of analogical proportion in various algebraic settings (including set-theoretic definitions of analogical proportions) [19,22,35]; see [24,27] for details. Moreover, it is also worth noticing that the constraint $a - b = c - d$ defining arithmetic proportions, when restricted to $\{0, 1\}$, validates the same 6 patterns as in Table 1, although $a - b \in \{-1, 0, 1\}$ in this case. This arithmetic proportion view of analogical proportion is the one advocated by [33] between numerical vectors representing words in high-dimensional spaces.

Representing objects with a single Boolean value is not generally sufficient and we have to consider situations where items are represented by *vectors* of

Table 2. Pairing pairs (a, b) and (c, d)

	$\mathcal{A}_1$	...	$\mathcal{A}_{i-1}$	$\mathcal{A}_i$	...	$\mathcal{A}_{j-1}$	$\mathcal{A}_j$	...	$\mathcal{A}_{k-1}$	$\mathcal{A}_k$	...	$\mathcal{A}_{r-1}$	$\mathcal{A}_r$	...	$\mathcal{A}_{s-1}$	$\mathcal{A}_s$	...	$\mathcal{A}_n$
$\boldsymbol{a}$	1	...	1	0	...	0	1	...	1	0	...	0	1	...	1	0	...	0
$\boldsymbol{b}$	1	...	1	0	...	0	1	...	1	0	...	0	0	...	0	1	...	1
$\boldsymbol{c}$	1	...	1	0	...	0	0	...	0	1	...	1	1	...	1	0	...	0
$\boldsymbol{d}$	1	...	1	0	...	0	0	...	0	1	...	1	0	...	0	1	...	1

Boolean values, each component being the value of a binary attribute. A simple extension of the previous definition to Boolean vectors in $\mathbb{B}^n$ of the form $\boldsymbol{a} = (a_1, ..., a_n)$ can be done as follows:

$$\boldsymbol{a} : \boldsymbol{b} :: \boldsymbol{c} : \boldsymbol{d} \text{ iff } \forall i \in [1, n], a_i : b_i :: c_i : d_i$$

Obviously, all the basic properties (symmetry, central permutation) still hold for vectors. In that respect it is important to notice that the four vectors are of the same nature, since they refer to the same set of features. Then symmetry just means that comparing the results of the comparisons of the two vectors inside each pair of vectors $(\boldsymbol{a}, \boldsymbol{b})$ and $(\boldsymbol{c}, \boldsymbol{d})$ does not depend on the ordering of the two pairs. Thus the repeated applications of symmetry followed by central permutation yield 8 equivalent forms of the analogical proportion: $(\boldsymbol{a} : \boldsymbol{b} :: \boldsymbol{c} : \boldsymbol{d}) = (\boldsymbol{c} : \boldsymbol{d} :: \boldsymbol{a} : \boldsymbol{b}) = (\boldsymbol{c} : \boldsymbol{a} :: \boldsymbol{d} : \boldsymbol{b}) = (\boldsymbol{d} : \boldsymbol{b} :: \boldsymbol{c} : \boldsymbol{a}) = (\boldsymbol{d} : \boldsymbol{cvb} : \boldsymbol{a}) = (\boldsymbol{b} : \boldsymbol{a} :: \boldsymbol{d} : \boldsymbol{c}) = (\boldsymbol{b} : \boldsymbol{d} :: \boldsymbol{a} : \boldsymbol{c}) = (\boldsymbol{a} : \boldsymbol{c} :: \boldsymbol{b} : \boldsymbol{d})$. Table 2 pictures the situation, where the components of the vectors have been suitably reordered in such a way that the attributes for which one of the 6 patterns characterizing the analogical proportion is observed, have been gathered, e.g., attributes $\mathcal{A}_1$ to $\mathcal{A}_{i-1}$ exhibits the pattern 1111. In the general case, some of the patterns may be absent.

This table shows that building the analogical proportion $\boldsymbol{a} : \boldsymbol{b} :: \boldsymbol{c} : \boldsymbol{d}$ is a matter of pairing the pair $(\boldsymbol{a}, \boldsymbol{b})$ with the pair $(\boldsymbol{c}, \boldsymbol{d})$. More precisely, on attributes $\mathcal{A}_1$ to $\mathcal{A}_{j-1}$, the four vectors are equal; on attributes $\mathcal{A}_j$ to $\mathcal{A}_{r-1}$, $\boldsymbol{a} = \boldsymbol{b}$ and $\boldsymbol{c} = \boldsymbol{d}$, but $(\boldsymbol{a}, \boldsymbol{b}) \neq (\boldsymbol{c}, \boldsymbol{d})$. In other words, on attributes $\mathcal{A}_1$ to $\mathcal{A}_{r-1}$ $\boldsymbol{a}$ and $\boldsymbol{b}$ agree and $\boldsymbol{c}$ and $\boldsymbol{d}$ agree as well. This contrasts with attributes $\mathcal{A}_r$ to $\mathcal{A}_n$, for which we can see that $\boldsymbol{a}$ differs from $\boldsymbol{b}$ as $\boldsymbol{c}$ differs from $\boldsymbol{d}$ (and vice-versa). We recognize the meaning of the formal definition of the analogical proportion.

Table 3. Pairing (a, d) and (b, c)

	$\mathcal{A}_1$	...	$\mathcal{A}_{i-1}$	$\mathcal{A}_i$	...	$\mathcal{A}_{j-1}$	$\mathcal{A}_j$	...	$\mathcal{A}_{k-1}$	$\mathcal{A}_k$	...	$\mathcal{A}_{r-1}$	$\mathcal{A}_r$	...	$\mathcal{A}_{s-1}$	$\mathcal{A}_s$	...	$\mathcal{A}_n$
$\boldsymbol{a}$	1	...	1	0	...	0	1	...	1	0	...	0	1	...	1	0	...	0
$\boldsymbol{d}$	1	...	1	0	...	0	0	...	0	1	...	1	0	...	0	1	...	1
$\boldsymbol{b}$	1	...	1	0	...	0	1	...	1	0	...	0	0	...	0	1	...	1
$\boldsymbol{c}$	1	...	1	0	...	0	0	...	0	1	...	1	1	...	1	0	...	0

Let us now pair the vectors differently, namely considering pair $(\boldsymbol{a}, \boldsymbol{d})$ and pair $(\boldsymbol{b}, \boldsymbol{c})$, as in Table 3. First, we can see that $\boldsymbol{a} : \boldsymbol{d} :: \boldsymbol{b} : \boldsymbol{c}$ *does not hold* due to attributes $\mathcal{A}_s$ to $\mathcal{A}_n$. Obviously, we continue to have $\boldsymbol{a} = \boldsymbol{b} = \boldsymbol{c} = \boldsymbol{d}$ for attributes $\mathcal{A}_1$ to $\mathcal{A}_{j-1}$, while on the rest of the attributes the values inside each pair differ (in four different ways). Then the following definition of the analogical proportion [24], logically equivalent to Eq. 1, should not come as a surprise:

$$a : b :: c : d = ((a \wedge d) \equiv (b \wedge c)) \wedge ((\neg a \wedge \neg d) \equiv (\neg b \wedge \neg c)) \tag{2}$$

or equivalently

$$a : b :: c : d = ((a \wedge d) \equiv (b \wedge c)) \wedge ((a \vee d) \equiv (b \vee c)) \tag{3}$$

Expression (3) can be viewed as the logical counterpart of a well-known property of geometrical proportions: the product of the means is equal to the product of the extremes. Interestingly enough, Piaget [25] pp. 35–37) named *logical proportion* any logical expression between four propositional formulas a, b, c, d for which (3) is true. Apparently, and strangely enough, Piaget never related this expression to the idea of analogy.

4 Seven Other Remarkable Logical Proportions

As said in Sect. 2, there are 7 other code independent logical proportions. We start with two of them that are closely related to analogical proportion, before considering the last of the 4 homogeneous proportions, and finally the four heterogeneous proportions.

4.1 Two Proportions Associated with Analogy

2 other homogeneous logical proportions are closely related to analogical proportion:

- *reverse analogy*: $\text{Rev}(a, b, c, d) \triangleq ((\neg a \wedge b) \equiv (c \wedge \neg d)) \wedge ((a \wedge \neg b) \equiv (\neg c \wedge d))$
  It reverses analogy into "b is to a as c is to d". In fact $\text{Rev}(a, b, c, d) \Leftrightarrow \text{Ana}(b, a, c, d)$.
- *paralogy*: $\text{Par}(a, b, c, d) \triangleq ((a \wedge b) \equiv (c \wedge d)) \wedge ((\neg a \wedge \neg b) \equiv (\neg c \wedge \neg d))$
  It expresses that what a and b have in common (positively or negatively), c and d have it also, and conversely. Up to a permutation, we recognize an expression similar to the expression (2) of $a : b :: c : d$. It can be checked that $\text{Par}(a, b, c, d)) \Leftrightarrow \text{Ana}(c, b, a, d)$.

In columns 2 and 3 of Table 4, we give the 6 patterns that make true Rev and Par, together with Ana in column 1.

A geometric illustration of the three proportions Ana, Par, Rev can be given. Indeed given 3 points $\boldsymbol{a}$, $\boldsymbol{b}$, $\boldsymbol{c}$ of the real plan $\mathbb{R}^2$, one can always find a point $\boldsymbol{d}$ such that *abdc* is a parallelogram (see Fig. 1). In fact, from 3 non aligned points, one can build 3 distinct parallelograms. See Fig. 1 where the index of d refers to

Table 4. Boolean patterns making true Analogy, Reverse analogy, Paralogy, Inverse paralogy

Ana				Rev				Par				Inv			
0	0	0	0	0	0	0	0	0	0	0	0	1	1	0	0
1	1	1	1	1	1	1	1	1	1	1	1	0	0	1	1
0	0	1	1	0	0	1	1	1	0	0	1	1	0	0	1
1	1	0	0	1	1	0	0	0	1	1	0	0	1	1	0
0	1	0	1	0	1	1	0	0	1	0	1	0	1	0	1
1	0	1	0	1	0	0	1	1	0	1	0	1	0	1	0

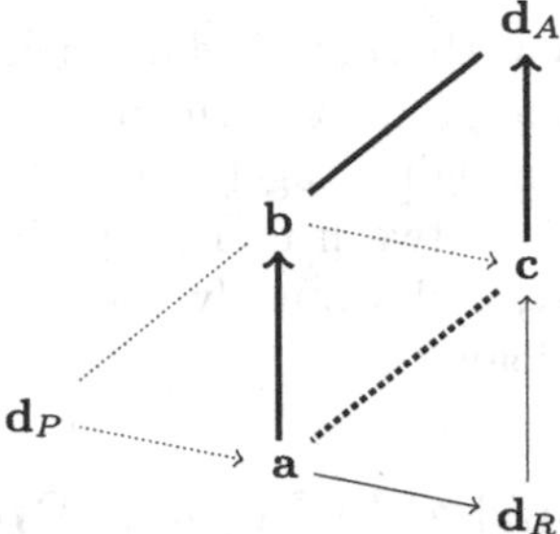

Fig. 1. Three parallelograms

the proportion that generates it from $(\boldsymbol{a}, \boldsymbol{b}, \boldsymbol{c})$. In Fig. 1, we have used different types of lines (with different width, dotted or not, arrows or not) to try to help visualizing the 3 parallelograms. Indeed if $(\boldsymbol{a}, \boldsymbol{b}, \boldsymbol{c}$ are respectively represented by coordinates $(0,0)$, $(0,1)$, and $(1,0)$, d_A is $(1,1)$ and $(0,0) : (0,1) :: (1,0) : (1,1)$ holds componentwise. It can be seen that geometrically, in $\mathbb{R}^2$, this corresponds to the equality of vectors $\overrightarrow{\boldsymbol{ab}}$ and $\overrightarrow{\boldsymbol{cd}_A}$. Applying the permutations linking Ana, Par, and Rev, we observe that $\overrightarrow{\boldsymbol{ab}} = \overrightarrow{\boldsymbol{d}_R\boldsymbol{c}}$ and $\overrightarrow{\boldsymbol{ad}_P} = \overrightarrow{\boldsymbol{cb}}$ as expected since $\boldsymbol{a} : \boldsymbol{b} v \boldsymbol{d}_R : \boldsymbol{c}$ iff $\boldsymbol{a}, \boldsymbol{b}, \boldsymbol{c}, \boldsymbol{d}_R$ make a reverse analogy, while $\boldsymbol{a} : \boldsymbol{d}_P :: \boldsymbol{c} : \boldsymbol{b}$ iff $\boldsymbol{a}, \boldsymbol{b}, \boldsymbol{c}, \boldsymbol{d}_P$ make a paralogy. Moreover, if we remember that $\boldsymbol{a} : \boldsymbol{b} :: \boldsymbol{c} : \boldsymbol{d}_A$ holds if and only if $\boldsymbol{a} - \boldsymbol{b} = \boldsymbol{c} - \boldsymbol{d}_A$ componentwise, which is equivalent in $\mathbb{R}$ to $\boldsymbol{d}_A = -\boldsymbol{a} + \boldsymbol{b} + \boldsymbol{c}$, we can easily deduce using permutations that $\boldsymbol{d}_R = \boldsymbol{a} - \boldsymbol{b} + \boldsymbol{c}$ and $\boldsymbol{d}_P = \boldsymbol{a} + \boldsymbol{b} - \boldsymbol{c}$. It yields $\boldsymbol{d}_R = (1, -1)$ and $\boldsymbol{d}_P = (-1, 1)$, which indeed corresponds to the coordinates of $\boldsymbol{d}_R$ and $\boldsymbol{d}_P$ in $\mathbb{R}^2$.

4.2 Inverse Paralogy

By switching the positive and the negative similarity indicators pertaining to the pair (c, d) in the definition of the paralogy, we obtain a new homogeneous logical proportion called *inverse paralogy*. Namely its expression is given by

$$\text{Inv}(a, b, c, d) \triangleq ((a \wedge b) \equiv (\neg c \wedge \neg d)) \wedge ((\neg a \wedge \neg b) \equiv (c \wedge d))$$

Inv(a, b, c, d) states that "what a and b have in common, c and d do not have it and conversely". This expresses a kind of "orthogonality" between the pairs (a, b) and (c, d). Inv(a, b, c, d) is clearly symmetrical and code independent. It can be shown [27] that Inv is the unique logical proportion (among the 120's!) which remains unchanged under any permutation of two terms among the four. Namely Inv$(a, b, c, d) \Leftrightarrow$ Inv$(b, a, c, d) \Leftrightarrow$ Inv$(a, c, b, d) \Leftrightarrow$ Inv(c, b, a, d) (the other permutations of two terms are obtained by symmetry). The patterns that make true Inv are given in the column 4 of Table 4. As can be seen, it is true for the patterns encountered in the truth tables of Ana, Par, Rev, except 0000 and 1111.

Note also in Table 4 that the 6 patterns that make the four proportions true belong to a set of 8 patterns. This set of 8 patterns is characterized by the logical formula $(a \equiv b) \equiv (c \equiv d)$, which corresponds to an analogical-like connective proposed by Klein [17], in relation with anthropological materials. Table 5 exhibits the pair of patterns that each proportion misses among the 8 patterns. It also shows what may be called the "characteristic" patterns of a proportion, namely the ones making one of the two involved indicators true for (a, b) and one true for (c, d). Those latter patterns correspond to the reading of the proportion.

Table 5. Analogy, Reverse analogy, Paralogy, Inverse paralogy: Characteristic/missing patterns

	Characteristic patterns	Missing patterns
Ana	1010 and 0101	1001 and 0110
Rev	1001 and 0110	1010 and 0101
Par	1111 and 0000	1100 and 0011
Inv	1100 and 0011	1111 and 0000

An illustration of the use of the inverse paralogy is provided by Bongard problems [3] that are visual puzzles where we have two sets A and B of relatively simple pictures. All the pictures in set A have a common feature, which is lacking in all the ones in set B. The problem is to find the common feature. This corresponds to one of the characteristic patterns of Inv [30].

4.3 The 4 Heterogeneous Proportions

The 4 heterogeneous proportions are obtained by putting the $\equiv$ connectives between indicators of different kinds for (a, b) and for (c, d). Their logical expressions are given below:

$$\text{-}H_1(a, b, c, d) = (\neg a \wedge b \equiv \neg c \wedge \neg d) \wedge (a \wedge \neg b \equiv c \wedge d)$$
$$\text{-}H_2(a, b, c, d) = (\neg a \wedge b \equiv c \wedge d) \wedge (a \wedge \neg b \equiv \neg c \wedge \neg d)$$
$$\text{-}H_3(a, b, c, d) = (\neg a \wedge \neg b \equiv \neg c \wedge d) \wedge (a \wedge b \equiv c \wedge \neg d)$$
$$\text{-}H_4(a, b, c, d) = (\neg a \wedge \neg b \equiv c \wedge \neg d) \wedge (a \wedge b \equiv \neg c \wedge d)$$

Table 6. H_1, H_2, H_3, H_4 Boolean truth tables

H1				H2				H3				H4			
1	1	1	0	1	1	1	0	1	1	1	0	1	1	0	1
0	0	0	1	0	0	0	1	0	0	0	1	0	0	1	0
1	1	0	1	1	1	0	1	1	0	1	1	1	0	1	1
0	0	1	0	0	0	1	0	0	1	0	0	0	1	0	0
1	0	1	1	0	1	1	1	0	1	1	1	0	1	1	1
0	1	0	0	1	0	0	0	1	0	0	0	1	0	0	0

They are clearly code independent. The six patterns that make them true are given in Table 6. The four heterogeneous logical proportions have a quite different semantics from the ones of homogeneous proportions. They express that there is an intruder among $\{a, b, c, d\}$, which is not a (H_1), which is not b (H_2), which is not c (H_3), and which is not d (H_4) respectively. The reader is referred to [29] for the study of the properties of heterogeneous logical proportions. They have been shown to be appropriate for solving puzzles of the type "Finding the odd one out" [29]. Moreover they are at the basis of an "oddness" measure, which has been shown to be of interest in classification, following the straightforward idea of classifying a new item in the class where it appears to the least at odds [6].

5 Graded Analogical Proportion

Attributes or features are not necessarily Boolean, and graded extensions of logical proportions are of interest. In the following we only focus on analogical proportion. We assume that attributes are now valued in $[0, 1]$ (possibly after renormalization). There are potentially many ways of extending expressions such as Eqs. (1) or (3). Still there are mainly two options that make sense [8].

The first one is obtained by replacing (i) the central $\wedge$ in (1) by min, (ii) the two $\equiv$ symbols by $\min(s \to_{Ł} t, t \to_{Ł} s) = 1 - \mid s - t \mid$, where $s \to_{Ł} t = \min(1, 1 - s + t)$ is Łukasiewicz implication, (iii) the four expressions of the form $s \wedge \neg t$ by the bounded difference $\max(0, s - t) = 1 - (s \to_{Ł} t)$, which is associated to Łukasiewicz implication, using $1 - (\cdot)$ as negation. The resulting expression is then

$$a : b ::_{Ł} c : d = \begin{cases} 1 - \mid (a - b) - (c - d) \mid, \\ \quad \text{if } a \geq b \text{ and } c \geq d, \text{ or } a \leq b \text{ and } c \leq d \\ 1 - \max(\mid a - b \mid, \mid c - d \mid), \\ \quad \text{if } a \leq b \text{ and } c \geq d, \text{ or } a \geq b \text{ and } c \leq d \end{cases} \quad (4)$$

It coincides with $a : b :: c : d$ on $\{0, 1\}$. As can be seen, this expression is equal to 1 if and only if $(a - b) = (c - d)$, while $a : b ::_{Ł} c : d = 0$ if and only if (i)

$a-b=1$ and $c \leq d$, or if (ii) $b-a=1$ and $d \leq c$, or if (iii) $a \leq b$ and $c-d=1$, or if iv) $b \leq a$ and $d-c=1$. Thus, $a:b ::_{\text{Ł}} c:d=0$ when the change inside one of the pairs (a,b) or (c,d) is maximal, while the other pair shows either no change or a change in the opposite direction. It can be also checked that code independency continue to hold under the form $a:b ::_{\text{Ł}} c:d = 1-a:1-b ::_{\text{Ł}} 1-c:1-d$.

We have pointed out that the algebraic difference between a and b equated with the difference between c and d, namely $a-b=c-d$, provides a constraint that is satisfied by the 6 patterns making true the analogical proportion $a:b::c:d$ in the Boolean case, and by none of the 10 others. However, $a-b$ may not belong to $\{0,1\}$ when $a,b \in \{0,1\}$. When considering the graded case, the situation remains the same: $a-b$ is not close either in $[0,1]$, but $a:b ::_{\text{Ł}} c:d=1$ if and only if $a-b=c-d$; moreover, the modeling of the analogical proportion by the constraint $a-b=c-d$ does not provide a graded evaluation of how far we are from satisfying it, as it is the case with the above extension.

There is another meaningful graded extension of the analogical proportion, which is directly obtained from Eq. (3) by also taking 'min' for the internal conjunction and 'max' for the internal disjunction. It yields the so-called "conservative" extension [8]:

$$a:b ::_C c:d = \min(1-|\max(a,d)-\max(b,c)|, 1-|\min(a,d)-\min(b,c)|) \quad (5)$$

Note that $a:b ::_C c:d=1 \Leftrightarrow \min(a,d)=\min(b,c)$ and $\max(a,d)=\max(b,c)$. This means that the patterns (s,s,t,t), and (s,t,s,t) (and (s,s,s,s)) are then the unique way to have the analogical proportion fully true (equal to 1). It can be checked that $a:b ::_{\text{Ł}} c:d=1 \Rightarrow a:b ::_C c:d=1$. For instance, $0:0.5 ::_{\text{Ł}} 0.5:1=1$, while $0:0.5 ::_C 0.5:1=0.5$. Besides, $a:b::c:d=0$ if and only if $|\min(a,d)-\min(b,c)|=1$ or $|\max(a,d)-\max(b,c)|=1$, i.e. the only patterns fully falsifying the analogical proportion are of the form $1:0::x:1$ or $0:1::x:0$ (and the other patterns obtained from these two by symmetry and central permutation).

6 Analogical Inference

The equation $a:b::c:x$ in $\mathbb{B}$ has not always a solution. Indeed neither $0:1::1:x$ nor $1:0::0:x$ have a solution (since 0111, 0110, 1000, 1001 are not valid patterns for an analogical proportion). The solution exists if and only if $(a \equiv b) \vee (a \equiv c)$ holds. When the solution exists, it is unique and given by solution $x = c \equiv (a \equiv b)$. This corresponds to the original view advocated by Klein [17] who applied even to the cases $0:1::1:x$ and $1:0::0:x$, where it yields $x=0$ and $x=1$ respectively; as already said S. Klein was making no differences between Ana, Par and Rev.

This equation solving mechanism directly applies to Boolean vectors in $\mathbb{B}^n$, i.e., looking for $\boldsymbol{x}=(x_1,\cdots,x_n)$ such as $\boldsymbol{a}:\boldsymbol{b}::\boldsymbol{c}:\boldsymbol{x}$ holds, amounts to solving the n equations $a_i:b_i::c_i:x_i$. When the n equations are solvable, we can observe that the analogical proportion solving process may be *creative* (an informal quality usually associated with the idea of analogy) in the sense

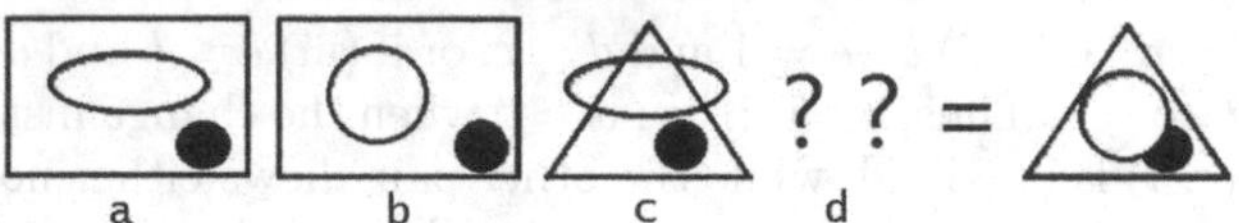

Fig. 2. A simple analogical sequence of pictures

	hS	hBD	hT	hC	hE
$\boldsymbol{a}$	1	1	0	0	1
$\boldsymbol{b}$	1	1	0	1	0
$\boldsymbol{c}$	0	1	1	0	1
$\boldsymbol{x}$	?	?	?	?	?

Fig. 3. A Boolean coding for Fig. 2

that it may be the case that $\boldsymbol{x} \neq \boldsymbol{a}$, $\boldsymbol{x} \neq \boldsymbol{b}$, and $\boldsymbol{x} \neq \boldsymbol{c}$. For instance, we obtain $\boldsymbol{x} = (x_1, x_2) = (0, 0)$, from $\boldsymbol{a} = (a_1, a_2) = (1, 1)$, $\boldsymbol{b} = (b_1, b_2) = (1, 0)$, and $\boldsymbol{c} = (c_1, c_2) = (0, 1)$.

This can be applied to completion tests such as the example of Fig. 2. The problem may be encoded using the 5 Boolean predicates *hasSquare*(hS), *hasBlack Dot*(hBD), *hasTriangle*(hT), *hasCircle*(hC), *hasEllipse*(hE) in that order.

This leads to the code of Fig. 3. Applying componentwise the solving process, we get $\boldsymbol{x} = (0, 1, 1, 1, 0)$ which is the code of the expected solution. The approach is constructive since the missing picture $\boldsymbol{x}$ is obtained by computation from $\boldsymbol{a}, \boldsymbol{b}, \boldsymbol{c}$. This contrasts with the classical approaches to this problem, pioneered by Th. Evans [9] where d is to be chosen among a set of candidate pictures which contains a picture considered as being the right answer, and where the change between $\boldsymbol{a}$ and $\boldsymbol{b}$ is compared with the change between $\boldsymbol{c}$ and $\boldsymbol{x}$ for each $\boldsymbol{x}$, leading to choose $\boldsymbol{x}$ as the one maximizing the similarity between the changes. Clearly, we may imagine some sequence of pictures $\boldsymbol{a}, \boldsymbol{b}, \boldsymbol{c}$ which cannot be completed by a fourth picture $\boldsymbol{x}$ in the sense of analogical proportion since the equation $\boldsymbol{a} : \boldsymbol{b} :: \boldsymbol{c} : \boldsymbol{x} = \mathbf{1}$ is not always solvable. When there is no analogical solution, we may think of using another homogeneous logical proportions. It should be also clear that the approach may not be suitable for solving quizzes obeying to a functional pattern of the form $\boldsymbol{a} : f(\boldsymbol{a}) :: \boldsymbol{b} : f(\boldsymbol{b})$, when the features considered for defining the vectors do not account for the modeling of f; then the function f has to be guessed on the basis of some simplicity principle [1].

The equation solving process may be also restricted to a subpart of $\boldsymbol{x}$. This is the basic inference pattern underlying the analogical proportion-based inference, which can be described as follows: if an analogical proportion holds between p components of four vectors, then this proportion may hold for the last remaining components as well. This inference principle [35] can be formally stated as follows:

$$\frac{\forall i \in \{1, ..., p\}, \quad a_i : b_i :: c_i : d_i \text{ holds}}{\forall j \in \{p+1, ..., n\}, \quad a_j : b_j :: c_j : d_j \text{ holds}}$$

This is a generalized form of analogical reasoning, where we transfer knowledge from some components of our vectors to their remaining components, tacitly assuming that the values of the p first components determine the values of the others. Then analogical reasoning amounts to finding completely informed triples $(\boldsymbol{a}, \boldsymbol{b}, \boldsymbol{c})$ suitable for inferring the missing value(s) of an incompletely informed item. In case of the existence of several possible triples leading to possibly distinct plausible conclusions, a voting procedure may be used, as in case-based reasoning where the inference is based on a collection of single cases (i.e., the nearest neighbors) rather than on a collection of triples. This inference pattern can be generalized when the p attributes include numerical ones, by computing an average score of the qualities of the analogical proportions over the p components (using one of the graded extensions of Sect. 5), and choosing the prediction such that the sum of the average scores to which it is associated, is maximal.

7 Applications

We briefly survey some existing or potential applications of analogical proportions.

Classification. Classification is an immediate application of the above inference principle where one has to predict a class $cl(\boldsymbol{x})$ (viewed as a nominal attribute) for a new item $\boldsymbol{x}$. Then one looks for triples $(\boldsymbol{a}, \boldsymbol{b}, \boldsymbol{c})$ of items with a known class, for which the class equation $cl(\boldsymbol{a}) : cl(\boldsymbol{b}) :: cl(\boldsymbol{c}) : cl(\boldsymbol{x})$ is solvable, and for which analogical proportions hold with $\boldsymbol{x}$ on the attributes describing the items. It has been first successively applied to Boolean attributes [4,21] and then extended to nominal and to numerical ones [5]. Recent formal studies have shown that analogical classifiers always give exact predictions in the special cases where the classification process is governed by an affine Boolean function (which includes x-or functions) and only in this case, which does not prevent to get good results in other cases (as observed in practice), but which is still to be better understood [7,16]. This suggests that analogical proportions enforces a form of linearity, just as numerical proportions fit with linear interpolation.

Raven Tests. They are IQ tests, where one is faced with a 3×3 matrix with 8 cells containing pictures, where one has to guess what is the right contents of the empty ninth cell, among 8 proposed solutions. An example[1] is given with its solution (a simple big square) in Fig. 4. The idea is to postulate that in a line (and maybe in a column), the picture of the 3rd cell is to the pictures of the first two pictures as the picture of the 3rd cell is to the pictures of the first two pictures in the next line (or column). It amounts to dealing with proportions of the form $(\boldsymbol{cell}_1, \boldsymbol{cell}_2) : \boldsymbol{cell}_3 :: (\boldsymbol{cell}'_1, \boldsymbol{cell}'_2) : \boldsymbol{cell}'_3$ (where the $\boldsymbol{cell}_i$'s refer to

[1] For copyright reasons and to protect the security of the test problems, the original Raven test has been replaced by an isomorphic example (in terms of logical encoding).

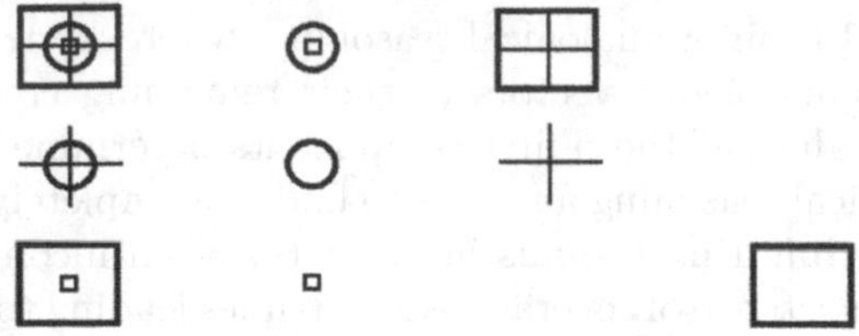

Fig. 4. Modified Raven test 12 and its solution

feature-based vectors describing the feature. Then the application of the analogical inference amounts to copying patterns observed in other lines or columns, feature by feature. Again the solution is built, and not chosen. See [1] for details and discussions.

Analogy-Based Decision. Let us just outline the idea using generic scenarii; see [2] for details. Suppose that a decision δ was experienced in two different situations sit_1 and sit_2 in the presence or not of special circumstances, leading to *good* or *bad* results respectively depending on the absence or on the presence of these special circumstances. Suppose we have in our repository the first three lines of the following table (cases a, b, c), while we wonder if we should consider applying decision δ or not in sit_2 when no special circumstances are present (case d). The analogical inference leads here to the prediction that the result should be *good*.

case	*situation*	*special circumstances*	*decision*	*result*
a	sit_1	*yes*	δ	*bad*
b	sit_1	*no*	δ	*good*
c	sit_2	*yes*	δ	*bad*
d	sit_2	*no*	$\boldsymbol{\delta}$	***good***

Note that if we apply a case-based decision view, case d might be found quite similar to case c, since they are identical on all the features used for describing situation sit_2, and differs only on the maybe unique feature describing the so-called "special circumstances"; this would lead to favor the idea that decision δ in case d would also lead to a *bad* result as in case c. Still, a more careful examination of cases a, b, c may lead to an opposite conclusion. Indeed it is natural to implicitly assume here that the possibly many features gathered here under the labels "situation" and "special circumstances" are enough for describing the cases and for determining the quality of the result of decisions applied to the cases. Thus, the fact that in sit_1, the quality of the result of decision δ is *bad* (resp. *good*) is explained by the presence (resp. absence) of "special circumstances". Then the analogical inference enforces here that we should have the same behavior in sit_2.

Rather than analogically predicting the evaluation of the output of a potential decision in a new situation, one may suppose that we start with a repertory of recommended actions in a variety of circumstances, and then one may also think of trying to take advantage of the creative capabilities of analogy for adapting a decision to the new situation. This may be useful when the final decision has diverse options. Such as *Serve a tea* with or without sugar, with or without milk. Let us consider this example to illustrate the idea. As stored in the table below, in situation sit_1 with contraindication ($c\ i$), it is recommended to serve tea only, in situation sit_1 with no $c\ i$, tea with sugar, while in situation sit_2 with $c\ i$ one serves tea with milk. What to do in situation sit_2 with no $c\ i$? Common sense suggests tea with sugar and milk, maybe. It is what analogical proportion equations says: indeed $\delta : \delta :: \delta : x$, $0 : 1 :: 0 : y$ and $0 : 0 :: 1 : z$ yield $xyz = \delta 11$ as in the table below.

case	*situation*	*contraindication*	*decision*	*option*1	*option*2
a	sit_1	*yes*	δ	0	0
b	sit_1	*no*	δ	1	0
c	sit_2	*yes*	δ	0	1
d	sit_2	*no*	$\boldsymbol{\delta}$	**1**	**1**

Analogical Inequalities. An analogical proportion states that the results of the comparisons of a and b on the one hand, and of c and d on the other hand, in terms of dissimilarity indicators, are the same. Analogical inequalities [31] weaken such statements of identity into statements of the form "a is to b at least as much as c is to d". Starting from the Boolean expression (1) of the analogical proportion, we replace the two symbols $\equiv$ expressing sameness by two $\rightarrow$ for modeling the fact that the result of the comparison of c and d is larger or equal to the result of the comparison of a and b. Namely, we obtain

$$a : b \ll c : d = ((a \wedge \neg b) \rightarrow (c \wedge \neg d)) \wedge ((\neg a \wedge b) \rightarrow (\neg c \wedge d)) \qquad (6)$$

It can be checked from the definition that the following expected properties hold:

- $a : b \ll a : b$
- $a : b :: c : d \Rightarrow a : b \ll c : d$
- $a : b :: c : d \Leftrightarrow ((a : b \ll c : d) \wedge (c : d \ll a : b))$
- $(a : b \ll c : d) \Leftrightarrow (\neg a : \neg b \ll \neg c : \neg d)$

Indeed, $a : b \ll c : d$ is weaker than $a : b :: c : d$. More precisely $a : b \ll c : d$ holds true for the 6 patterns that makes analogical proportion true, plus the 4 patterns 0001, 0010, 1110, 1101. These latter patterns correspond to the 4 situations where $a \equiv b$ and $c \not\equiv d$. In these 4 situations a and b are indeed strictly closer than c and d, and these are the only cases in $\{0, 1\}$. It can be checked that $a : b \ll c : d$ is true if and only if $(a : b :: c : d) \vee (a \equiv b)$ is true.

When extended to the multiple-valued case [31], we obtain graded analogical inequalities that might be of interest in visual multiple-class categorization tasks for the handling of pieces of knowledge about semantic relationships between classes of the form "a is to b at least as much as c is to d" where a, b, c, d refer to the value of a feature of interest in 4 different images [18].

8 Conclusion: Link and Differences with Case-Based Reasoning

As suggested in this overview, and already emphasized in [26], analogical proportion-based inference departs from case-based reasoning since the former takes advantage of triples for extrapolating plausible conclusions, while the latter exploits the similarity of the new case with a collection of stored cases considered one by one. Indeed, although "$< solution_1 >$ is to $< problem_1 >$ as $< solution_2 >$ is to $< problem_2 >$" may be regarded as an analogical proportion, the view presented here assumes that the vectors representing the four items in the analogical proportion "$\boldsymbol{a}$ is to $\boldsymbol{b}$ as $\boldsymbol{c}$ is to $\boldsymbol{d}$" are all defined on *the same* set of features. As suggested when discussing analogy-based decision, we would rather suggest to exploit analogical proportions of the form $(< problem_1 >, < solution_1 >) : (< problem_2 >, < solution_2 >) :: (< problem_3 >, < solution_3 >) : (< problem_0 >, < solution_0 >)$ for extrapolating $< solution_0 >$ from 3 known cases $(\{(< problem_i >, < solution_i >) \mid i = 1, 3\})$ by solving equation $< solution_1 >:< solution_2 >::< solution_3 >:< solution_0 >$ (where $< solution_0 >$ is unknown), provided that $< problem_1 >:< problem_2 >:: < problem_3 >:< problem_0 >$ holds.

The view advocated here is also in line with the use of the creative power of analogical reasoning for, e.g., creating a new recipe from known ones, as suggested by the following example where one knows about lemon pie ($\boldsymbol{a}$), lemon cream ($\boldsymbol{b}$), and apple pie ($\boldsymbol{c}$) (roughly described by obvious features here), leading to the creation of the apple cream, in a spirit not so far of adaption methods in case-based reasoning [10].

	pastry	cream	lemon	apple	dessert	fruit	juice
lemon pie	1	0	1	0	1	1	1
lemon cream	0	1	1	0	1	1	1
apple pie.	1	0	0	1	1	1	1
apple cream	0	1	0	1	1	1	1

Moreover, the computation of the result of the inference may give birth to some explanation: pies and creams are two desserts, both lemon and apple are juicy fruits, lemon pie and lemon cream exist, apple pie exists, why not trying apple cream ?

The handling of analogical proportions "$\boldsymbol{a}$ is to $\boldsymbol{b}$ as $\boldsymbol{c}$ is to $\boldsymbol{d}$" where $\boldsymbol{a}$ and $\boldsymbol{c}$ do no belong to the same conceptual universe as $\boldsymbol{b}$ and $\boldsymbol{d}$, as in the sentence "Oslo is to Norway as Paris is to France" is more tricky [20,23] and still under study.

References

1. Correa Beltran, W., Prade, H., Richard, G.: Constructive solving of Raven's IQ tests with analogical proportions. Int. J. Intell. Syst. **31**(11), 1072–1103 (2016)
2. Billingsley, R., Prade, H., Richard, G., Williams, M.-A.: Towards analogy-based decision - a proposal. In: Jaudoin, H., Christiansen, H. (eds.) Proceedings of the 12th Conference on Flexible Query Answering Systems (FQAS 2017), London, Jun. 21–23, LNAI. Springer (2017, to appear)

3. Bongard, M.M.: Pattern Recognition. Spartan Books, Rochelle Park, Hayden Book (1970)
4. Bounhas, M., Prade, H., Richard, G.: Analogical classification: a new way to deal with examples. In: ECAI 2014–21st European Conference on Artificial Intelligence, 18–22 August 2014, Prague, Czech Republic, vol. 263, Frontiers in Artificial Intelligence and Applications, pp. 135–140. IOS Press (2014)
5. Bounhas, M., Prade, H., Richard, G.: Analogical classification: handling numerical data. In: Straccia, U., Calì, A. (eds.) SUM 2014. LNCS, vol. 8720, pp. 66–79. Springer, Cham (2014). doi:10.1007/978-3-319-11508-5_6
6. Bounhas, M., Prade, H., Richard, G.: Oddness/evenness-based classifiers for Boolean or numerical data. Int. J. Approx. Reasoning **82**, 81–100 (2017)
7. Couceiro, M., Hug, N., Prade, H., Richard, G.: Analogy-preserving functions: a way to extend Boolean samples. In: Proceedings of the 26th International Joint Conference on Artificial Intelligence (IJCAI 2017), Melbourne (2017)
8. Dubois, D., Prade, H., Richard, G.: Multiple-valued extensions of analogical proportions. Fuzzy Sets Syst. **292**, 193–202 (2016)
9. Evans, T.G.: A program for the solution of a class of geometric-analogy intelligence-test questions. In: Minsky, M.L. (ed.) Semantic Information Processing, pp. 271–353. MIT Press, Cambridge (1968)
10. Fuchs, B., Lieber, J., Mille, A., Napoli, A.: Differential adaptation: an operational approach to adaptation for solving numerical problems with CBR. Knowl.-Based Syst. **68**, 103–114 (2014)
11. Gentner, D.: Structure-mapping: a theoretical framework for analogy. Cogn. Sci. **7**(2), 155–170 (1983)
12. Gentner, D., Holyoak, K.J., Kokinov, B.N., Mind, T.A.: The Analogical Mind: Perspectives from Cognitive Science. Cognitive Science, and Philosophy. MIT Press, Cambridge (2001)
13. Hesse, M.: On defining analogy. Proc. Aristotelian Soc. **60**, 79–100 (1959)
14. Hofstadter, D., Mitchell, M.: The Copycat project: a model of mental fluidity and analogy-making. In: Hofstadter, D., The Fluid Analogies Research Group (eds.) Fluid Concepts and Creative Analogies: Computer Models of the Fundamental Mechanisms of Thought, pp. 205–267. Basic Books Inc, New York (1995)
15. Hofstadter, D., Sander, E., Surfaces, E.: Analogy as the Fuel and Fire of Thinking. Basic Books, New York (2013)
16. Hug, N., Prade, H., Richard, G., Serrurier, M.: Analogical classifiers: a theoretical perspective. In: Kaminka, G.A., Fox, M., Bouquet, P., Hüllermeier, E., Dignum, V., Dignum, F., van Harmelen, F. (eds.) Proceedings of the 22nd European Conference on Artificial Intelligence (ECAI 2016), The Hague, 29 August–2 September, pp. 689–697. IOS Press (2016)
17. Klein, S.: Analogy and mysticism and the structure of culture (and Comments & Reply). Curr. Anthropol. **24**(2), 151–180 (1983)
18. Law, M.T., Thome, N., Cord, M.: Quadruplet-wise image similarity learning. In: Proceedings of the IEEE International Conference on Computer Vision (ICCV) (2013)
19. Lepage, Y.: Analogy and formal languages. Electr. Not. Theor. Comp. Sci. 53, 180–191 (2002). Proc. joint meeting of the 6th Conf. on Formal Grammar and the 7th Conf. on Mathematics of Language, (L. S. Moss, R. T. Oehrle, eds.)

20. Miclet, L., Barbot, N., Prade, H.: From analogical proportions in lattices to proportional analogies in formal concepts. In: Schaub, T., Friedrich, G., O'Sullivan, B. (eds.) Proceedings of the 21st European Conference on Artificial Intelligence (ECAI 2014), Prague, August 18–22, vol. 263, Frontiers in Artificial Intelligence and Applications, pp. 627–632. IOS Press (2014)
21. Miclet, L., Bayoudh, S., Delhay, A.: Analogical dissimilarity: definition, algorithms and two experiments in machine learning. J. Artif. Intell. Res. (JAIR) **32**, 793–824 (2008)
22. Miclet, L., Delhay, A.: Relation d'analogie et distance SUR un alphabet défini par des traits. Technical Report 1632, IRISA, July 2004
23. Miclet, L., Nicolas, J.: From formal concepts to analogical complexes. In: Proceedings of the 12th International Joint Conference on Concept Lattices and their Applications (CLA 2015), Clermont-Ferrand (2015)
24. Miclet, L., Prade, H.: Handling analogical proportions in classical logic and fuzzy logics settings. In: Sossai, C., Chemello, G. (eds.) ECSQARU 2009. LNCS, vol. 5590, pp. 638–650. Springer, Heidelberg (2009). doi:10.1007/978-3-642-02906-6_55
25. Piaget, J.: Logic and Psychology. Manchester University Press, New York (1953)
26. Prade, H., Richard, G.: Analogy-making for solving IQ Tests: a logical view. In: Ram, A., Wiratunga, N. (eds.) ICCBR 2011. LNCS, vol. 6880, pp. 241–257. Springer, Heidelberg (2011). doi:10.1007/978-3-642-23291-6_19
27. Prade, H., Richard, G.: From analogical proportion to logical proportions. Log. Univers. **7**(4), 441–505 (2013)
28. Prade, H., Richard, G. (eds.): Computational Approaches to Analogical Reasoning: Current Trends. Studies in Computational Intelligence, vol. 548. Springer, Heidelberg (2014)
29. Prade, H., Richard, G.: Homogenous and heterogeneous logical proportions. IfCoLog J. Logics Appl. **1**(1), 1–51 (2014)
30. Prade, H., Richard, G.: On different ways to be (dis)similar to elements in a set. Boolean analysis and graded extension. In: Carvalho, J.P., Lesot, M.-J., Kaymak, U., Vieira, S., Bouchon-Meunier, B., Yager, R.R. (eds.) IPMU 2016. CCIS, vol. 611, pp. 605–618. Springer, Cham (2016). doi:10.1007/978-3-319-40581-0_49
31. Prade, H., Richard, G.: Analogical inequalities. In: Papini, O., Antonucci, A., Cholvy, L. (eds.) Proceedings of the 14th European Conference on Symbolic and Quantitative Approach to Reasoning with Uncertainty (ECSQARU 2017), Lugano, July 10–14, LNAI. Springer (2017, to appear)
32. Prade, H., Richard, G.: Boolean analogical proportions - Axiomatics and algorithmic complexity issues. In: Papini, O., Antonucci, A., Cholvy, L. (eds.) Proceedings of the 14th European Conference on Symbolic and Quantitative Approach to Reasoning with Uncertainty (ECSQARU 2017), Lugano, July 10–14, LNAI. Springer (2017, to appear)
33. Rumelhart, D.E., Abrahamson, A.A.: A model for analogical reasoning. Cognitive Psychol. **5**, 1–28 (2005)
34. Sowa, J.F., Majumdar, A.K.: Analogical reasoning. In: Ganter, B., Moor, A., Lex, W. (eds.) ICCS-ConceptStruct 2003. LNCS, vol. 2746, pp. 16–36. Springer, Heidelberg (2003). doi:10.1007/978-3-540-45091-7_2
35. Stroppa, N., Yvon, F.: Analogical learning and formal proportions: definitions and methodological issues. Technical Report D004, ENST-Paris (2005)
36. Winston, P.H.: Learning and reasoning by analogy. Commun. ACM **23**(12), 689–703 (1980)

Regular papers

A Hybrid CBR Approach for the Long Tail Problem in Recommender Systems

Gharbi Alshammari[1(✉)], Jose L. Jorro-Aragoneses[2], Stelios Kapetanakis[1], Miltos Petridis[3], Juan A. Recio-García[2], and Belén Díaz-Agudo[2]

[1] School of Computing, Engineering and Mathematics, University of Brighton, Brighton, UK
{g.alshammari,s.kapetanakis}@brighton.ac.uk

[2] Department of Software Engineering and Artificial Intelligence, Universidad Complutense de Madrid, Madrid, Spain
{jljorro,jareciog,belend}@ucm.es

[3] Department of Computing, University of Middlesex, London, UK
m.petridis@mdx.ac.uk

Abstract. Recommender systems is an important tool to help users find relevant items to their interests in a variety of products and services including entertainment, news, research articles, and others. Recommender systems generate lists of recommendations/suggestions based on information from past user interactions, choices, demographic information as well as using machine learning and data mining. The most popular techniques for generating recommendations are through content-based and collaborative filtering with the latter used to provide user to user recommendations. However, collaborative filtering suffers from the long tail problem, i.e., it does not work correctly with items that contain a small number of ratings over large item populations with respectively large numbers of ratings. In this paper, we propose a novel approach towards addressing the long tail recommendation problem by applying Case-based Reasoning on "user history" to predict the rating of newly seen items which seem to belong to the long tail. We present a hybrid approach and a framework implemented with jCOLIBRI to evaluate it using the freely available Movielens dataset [8]. Our results seem promising and they seem to improve the existing prediction outcomes from the available literature.

1 Introduction

Recommender systems seem a key step in tackling the increasing difficulty users have with handling large information volumes, cognitive overload and data sparsity while attempting to find the right information at the right time [2]. Nowadays, recommender systems are used by a continually increasing stream of applications in an attempt to maximise the provision of successful user suggestions. Usually, these suggestions rely heavily on profiling, user demographic information and user behaviour analysis by applying data-mining and machine-learning techniques.

D.W. Aha and J. Lieber (Eds.): ICCBR 2017, LNAI 10339, pp. 35–45, 2017.
DOI: 10.1007/978-3-319-61030-6_3

At present, two recommendation strategies seem to be the most prominent, that of: Content Based Filtering (CBF) or Content-based recommender systems; and the Collaborative Filtering (CLF) strategies. A content-based example may be an online image gallery where users can provide suggestions either explicitly or implicitly. A user profile can then be generated and gradually populated with the preferences of the user and, as a consequence, to steadily be able to provide more accurate recommendations. Such an approach relies not only on ongoing knowledge acquisition, but also on maintenance from the designers of the system. This is due to the fact that the recommender system's knowledge does not exist in the first place. Alternatively, collaborative filtering is dependent on the observation that item knowledge is not required, since the possible availability of other user profiles may be able to "suggest" a number of "right" items to a user whose profile is sufficiently similar to any profile available within a pool of users. The degree of similarity can be calculated based on user ratings of past events, item ratings, user opinions etc. which are related to the target user.

Several recommender systems seem to focus on the highest rating of popular or long-standing items. This method works well when there are sufficient numbers of users to rate such items. However, a challenge emerges when an item has not been rated "enough", i.e., when there are insufficient user-ratings available. This is the long-tail problem in recommender systems which typically refers to less popular, or newly-added items. Such items seem to belong to the distribution long-tail (long tail recommendation problem - LTRP) [14]. They should not, therefore, be ignored since: a) they could contribute to more specific recommender gain(s) and b) they could solve the cold start problem where sparse-rating items exist. [14] has presented a solution to the LTRP by splitting the recommendation into head and tail and by providing a clustering approach for the tail items. In [15] this way, similar users have been clustered with similar clusters in order to mitigate the CLF data-sparsity challenge. [14] Clustering was also employed so as to improve the accurate prediction of CLF in which items are divided into similar groups and where CLF is applied to each group separately.

Our work presents a new approach to the long-tail recommendation problem (LTRP) by relying on past user experience to retrieve items of similar rating for a new user. This is done by combining collaborative and content-based filtering in order to mitigate the increased error rate of a recommender system while dealing with low-ranked items in the tail. We have defined such low-ranked items as unpopular items, and the high-ranked items as popular ones. We propose that presenting a user history as a case base can help to reduce the LTRP as well as increasing the probability of identifying similar items from a user's past ratings of similar items. These cases can be identified using case-based reasoning on top of a hybrid CBF-CLF recommender system. Our proposed model and experiments on a freely-available dataset of more than 100,000 cases, have shown promising results and, in a number of cases, compare better to CBR and CLF baselines, respectively, as well as to the presented approaches within the literature.

This paper is structured as follows: Sect. 2 presents work related to recommender systems, the long tail recommendation problem and recommender systems, as well as Case-based Reasoning. Section 3 details our hybrid recommender system approach and defines our proposed model for mitigating the LTRP problem. Section 4 presents our experiment topology, our promising experiment results and the domain specific challenges using the publicly available Movielens dataset. Finally, Sect. 5 focuses on the limitations of this work as well as several possible paths for future work.

2 Related Work

Recommender systems focus on tackling the problem of finding user relevant and/or desirable products without the user having prior knowledge of, preference for, or interaction with them. The literature presents substantial work in this field and focuses largely on the (two) most popular recommender system methods: collaborative filtering [12] and content-based filtering [19]. In collaborative filtering, recommender systems rely on the "wisdom of the crowds" and is based on the assumption that similar users should have the same preferences and as a result "rate the same" similar items. On the other hand, content-based filtering is based on the item description in relation to users as well as on their preferences as they relate to information retrieval and filtering techniques, e.g., the TF-IDF (vector space) representation.

When a collaborative system uses available ratings from "similar users" it is called User-User collaborative filtering [6]. Examples of User-User collaborative filtering can be seen in [22] where mobile activities were recommended to users based on the other locations. [3] also proposed a User- User approach as an appropriate method of predicting internet user trends based on user opinions.

Contrary to collaborative systems, in content-based systems, recommendations are made using an "Item-User" similarity [16] were attributes of an item (e.g. for a book, author, price, number of pages) may suggest another item of interest to a user. For example, [17, 18] propose content-based approaches in order to recommend news stories based on their users Twitter history.

In addition to the CLF and CBF techniques, the literature presents a number of approaches using Hybrid recommender systems (HRSs) [13]. HRSs seem to solve CLF and/or CBF deficiencies in several recommendation applications. It is for this reason the literature shows an increased number of applications over the years. Furthermore, there seems to be an optimisation standard in matching the shortcomings of the main recommendation techniques [20].

Case-based reasoning appears to be a popular technique in creating both recommender system algorithms applications. For example, [7] detailed how online stores have used past information via case-based reasoning in order to increase sales of books, movies, mobile phones and other devices. [9] proposed feature weighting case based reasoning as recommender systems with neural networks to predict the customer characteristics and thus market behaviour.

A major challenge in recommender system is how to recommend less popular items, a problem also referred to as the long tail problem [1]. The literature presents a number of ways of solving this problem, such as through the use of clusterings [14, 15] techniques in order to boost item rating in the long tail. In addition, authors in [5] use matrix factorization algorithms and neighbourhood method so as to evaluate the performance of the recommendation of items in the long tail. Graph-based algorithms have also been proposed [21] for the long tail recommendation by using user-item information along with undirected edge- weighted graphs for long tail item recommendation. In [4] a CBR system, it is suggested that unknown artist and tracks are recommended. The proposed system in that study could identify whether an item resided in the long tail and if it were attempting to improve its provided meta-data through the addition of tag knowledge.

As we have seen the literature suggests a number of approaches in collaborative, content-based and hybrid recommendations. However, most of the existing recommendations algorithms do not apply CBR techniques to solve the long tail problem in the recommendations. The use of a CBR method has many advantages over the others. Firstly, this method does not need any pre-calculation of the items. For example, in the literature we find a number of methods that need pre-calculation before the recommendation such as occurs in clustering. Another advantage is that this method does not require the saving or creation of further information. If the system does not calculate the rating based on other users, it uses the history of the user and this information is the same information as in the CBF. And finally, in [21] it is explained that many algorithms decrease the accuracy when recommending long tail items, but in our case, the accuracy is increased using CBR system.

In the next section, we will present our suggested approach based on a hybrid recommender system that solves the Long Tail recommendation problem using a CBR framework and combined collaborative filtering and content-based methods.

3 Hybrid Recommender System

This section explains the hybrid system that is designed to resolve the long tail problem using CBR. In Fig. 1 we can see that our proposed architecture has two (2) different modules: a collaborative filtering component that calculates the predicted rating based on other similar users and a content-based one. The content-based component calculates the predicted rating using other similar movies that the user has rated in the past. Both modules have been developed using jCOLIBRI [10], a Java framework that allows for the rapid prototyping of cases, information retrieval as well as continuous integration to the CBR cycle.

The system receives a query (Q) that identifies the target user (u) and movie (m). Ratings are in a scale from 1 to 5 and the goal is to compute the estimated rating for the movie $r(m,u)'$

$$Q = <u, m> \tag{1}$$

The first step in the system is to decide which method is more effective in correctly calculating the rating prediction. This decision is based on the number of ratings received by the target movie. In order to make this decision, the system computes a vector (R_m) that represents the ratings of a concrete movie m.

$$R_m = \langle m, r\rangle_m = \left(\langle m, r(m, u_1)\rangle, \ldots \langle m, r(m, u_j)\rangle\right) \tag{2}$$

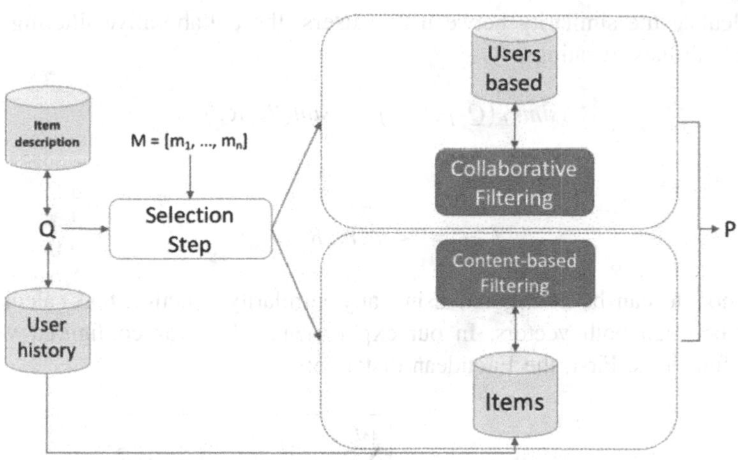

Fig. 1. Recommender system architecture.

In this first step, the system obtains the number $|R_m|$ of ratings that the query movie (m) has. Then, it compares this value with a threshold constant (δ). If the number of ratings of m is higher than δ, then this movie is not in the long tail problem and the collaborative filtering module can be used. On the other hand, if the number of ratings is lower than δ, the system can not search for similar users that rate this movie, due to the fact that the system does not have a sufficient number of ratings and should, therefore, use the content based module.

Next, we will explain each module in detail and set out how the rating prediction of a movie is calculated.

3.1 Collaborative Filtering Module Based on Users

The first module used in the hybrid system is a collaborative filtering CBR system. The main goal of this module is to calculate the rating prediction based on the ratings of similar users u'. This module computes a list with all the movies previously rated by any user (R_u) and compares the list in order to obtain the user similarity.

$$Q_{CF} = <u, m, R_u> \tag{3}$$

where

$$R_u = \langle m, r\rangle_u = (\langle m_1, r(m_1, u)\rangle, \ldots, \langle m_n, r(m_u, u)\rangle) \tag{4}$$

$$r \in [1..5] \tag{5}$$

This module uses the k Nearest-Neighbours (kNN) algorithm to calculate the rating prediction. The users obtained by the kNN must contain a rating for the target movie

m. To calculate the similarity between two users, the collaborative filtering module compares both lists of ratings:

$$sim_{CF}(Q_{CF}, C_{CF}) = sim(R_u, R_{u'}) \tag{6}$$

where

$$C_{CF} = \ <u', m, R'_u> \tag{7}$$

This module can be configured using any similarity function that calculates the similarity between both vectors. In our experiments, this was configured with two similarity functions. First, the Euclidean distance:

$$sim_{Euc}(R_u, R_{u'}) = 1 - \sqrt{\sum_{m=0}^{|M|} (r(m,u) - r(m,u'))^2} \tag{8}$$

where

$$M = R_u \cap R_{u'} \tag{9}$$

The other similarity function explored in the experimental evaluation is the Pearson correlation [11]:

$$sim_{Pea}(R_u, R_{u'}) = \frac{\sum_{m=0}^{|M|} \left(r(m,u) - \overline{R_u}\right)\left(r(m,u') - \overline{R'_u}\right)}{\sigma_u \sigma_{u'}} \tag{10}$$

where

$$M = R_u \cap R_{u'} \tag{11}$$

When the system has retrieved the k most similar users that have rated the target movie m, it calculates the rating prediction using the other rates derived from these users. This prediction is calculated with the weighted average of the rating and the similarity measure.

$$r(m,u)' = \frac{\sum_{i=0}^{k} r_i(m,u') * sim_i(R_u, R_{u'})}{\sum_{i=0}^{k} sim_i(R_u, R_{u'})} \tag{12}$$

Finally, $r(m,u)'$ is the result returned by this module as the rating prediction. We will now explain the second method.

3.2 Content Based Module Based on User Ratings History

The second CBR module is used in order to resolve the long tail problem, i.e., the situation where the system lacks a sufficient number of ratings for the target movie *m* from other users. This module calculates the rating with a content based similarity function based on the description of the movie rated by the user. This system creates a personal case base for the target user. Each case (C_{CB}) contains a list of genres that describes the movie.

$$C_{CB} = < u, m, G_m > \tag{13}$$

$$G_m = \{g_1, \ldots g_i, \ldots, g_n\} \tag{14}$$

Now, given a query, movies are compared according to the number of common genres.

$$sim(m, m') = \frac{G_m \cap G'_m}{G_m \cap G'_m} \tag{15}$$

Using the *k* most similar movies, the CBR module calculates the rating prediction using Eq. 12.

The next section explains the experiments we designed to evaluate this system as well as the corresponding results.

4 Experimental Evaluation

In this section, we explain the experiments realized to evaluate our CBR approach. First, we detail all the steps in the experiment and, finally, we show the results and explain them.

4.1 Experimental Setup

The dataset used in these experiments is the MovieLens dataset [8]. It is a common dataset used to evaluate recommender systems. It contains the result of users interacting with MovieLens recommender system from 19th of September 1997 to 22nd of April 1998. This dataset consists of 100,000 ratings in the range of 1 to 5 from 943 users based on 1682 movies. The users who have fewer than 20 ratings were removed.

In these experiments, we evaluate the accuracy rate of the hybrid system. In addition, we compare these results with an approach that only uses collaborative filtering or content based methods. To evaluate the accuracy rate of each system, we apply a Leave-One-Out evaluation. For each query, we calculate the difference between the rating predicted by the system and the actual rating. This evaluation was executed with the following conditions:

Collaborative Filtering. In this test, we calculate the accuracy rate using only a collaborative filtering method based on users.

Content Based. Next, we repeated the experiment with a content-based method based on users. This method, uses the statistical average of ratings per genre defined in each user description.

Combined Method. In this experiment, we combined both recommender systems with a similarity function. In this case, the similarity between 2 users is the average of the similarity calculated with the collaborative filtering method and the content-based method.

CBR Approach. Finally, we have evaluated the system explained in this paper.

All tests were varied by systematically changing two elements: the similarity function used (less in the content-based module based on items) and the *k* in the kNN algorithm. The similarity functions used were the Euclidean similarity and the Pearson similarity shown in Sect. 3 above. The values of the *k* were: 3, 5, 10, 15 and 25.

4.2 Results

In this section we present and explain the results obtained by these experiments.

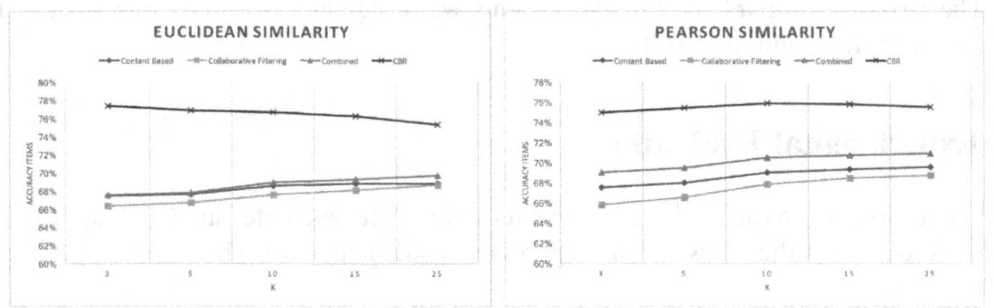

Fig. 2. Accuracy rate for each algorithm based on the similarity function and kNN.

Figure 2 shows the accuracy rate of each method using different similarity functions, and different values of the k in the kNN algorithm. These results show that the CBR method presented in this paper improves the general accuracy of the prediction. The other methods have an accuracy of less than 70% with both similarity functions. The system proposed improves the accuracy by between 6–10% depending on the similarity function and the *k* selected in the kNN algorithm.

In addition, Fig. 3 shows the performance of our method. It also shows the Mean Absolute Error (MAE) of each method. In addition to this, it demonstrates that the error rate of the CBR system is lower than in the other approaches, i.e., the CBR system can predict unknown ratings better in the Movielens dataset.

In Fig. 4, the Root Mean Square Error (RMSE) which demonstrates that the performance of our system improves compared to the baseline.

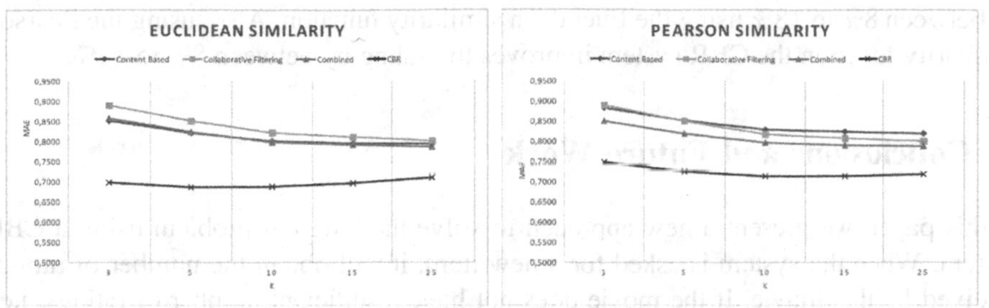

Fig. 3. Mean Absolute Error (MAE) for each algorithm based on the similarity function and kNN.

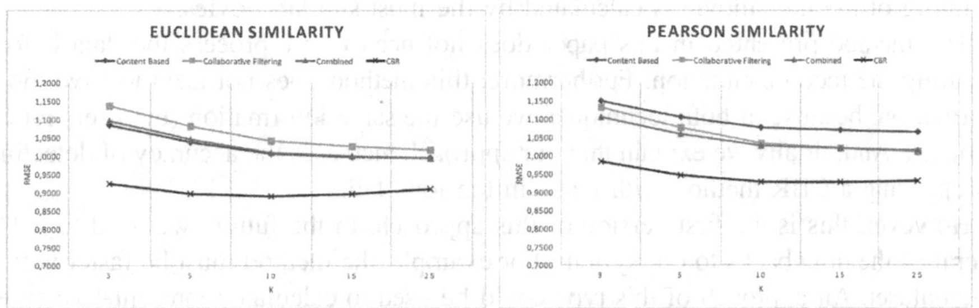

Fig. 4. Root Mean Square Error (RMSE) for each algorithm based on the similarity function and kNN.

To calculate the improvement rate of our system, we used the improvement function described in [14]. This calculates the improvement rate based on comparing the RMSE of our system with the others:

$$Improvement\ rate = \frac{RMSE_{base} - RMSE_{CBR}}{RMSE_{base}} \quad (16)$$

Figure 5 shows the comparison of the improvement rate of the CBR system versus the baseline in the results obtained. This figure shows that our system improves the prediction

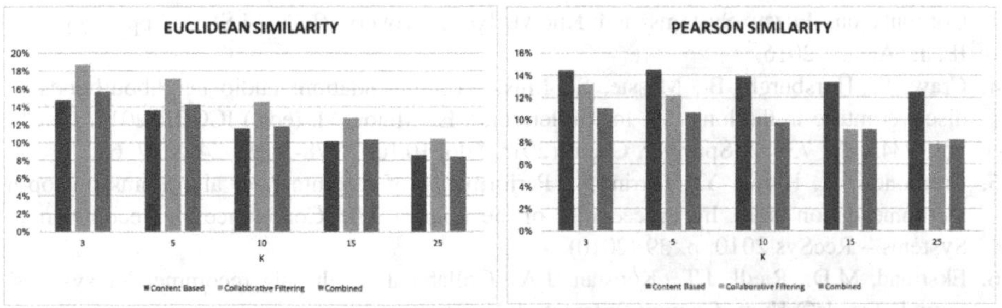

Fig. 5. Improvement Rate of the CBR system with the baseline.

by between 8% to 18% using the Euclidean similarity function. Also, using the Pearson Similarity function the CBR system improves the rating by between 8% to 14%.

5 Conclusions and Future Work

In this paper, we present a new approach to solve the long tail problem using a CBR system. When the system is asked for a new item, it will obtain the number of ratings received by this movie. If the movie does not have a sufficient number of ratings, i.e., the movie has in the long tail problem, then it will use the CBR approach to calculate the rating. In this case, the CBR system uses as cases movies that the user has rated in the past. Then, this method retrieves the most similar movies based on the genres. Next, the rating of the new movie is calculated by the most similar movies.

The method presented in this paper does not need to pre-process the data before executing the recommendation. Furthermore, this method does not need to save more information, because in both techniques we use the same information (the users rating histories). And, finally we explain that our approach increases the accuracy of detecting and applying a CBR method with items in the long tail.

However, this is the first version of this approach. In the future, we need to solve several of the drawbacks to our system. For example, the method must be faster with a large dataset. An approach of this type could be used to calculate representative cases for each item, or divide all cases into groups of users with the same features. In addition, another challenge is to identify" where do we have" a long tail problem. In this paper, we define a constant in order to determine this. However, it may be that the long tail problem is not defined by the number of ratings, although it is possible that there may not be enough different types of users who rate an item.

References

1. Anderson, C.: The long tail: why the future of business is selling less of more by Chris Anderson. J. Prod. Innov. Manag. **24**(3), 1–30 (2007)
2. Bridge, D., Goker, M.H., McGinty, L., Smyth, B.: Case-based recommender systems. Knowl. Eng. Rev. **20**(03), 315 (2005)
3. Chen, X., Xia, M., Cheng, J., Tang, X., Zhang, J.: Trend prediction of internet public opinion based on collaborative filtering. In: 2016 12th International Conference on Natural Computation, Fuzzy Systems and Knowledge Discovery (ICNC-FSKD), pp. 583–588. IEEE, August 2016
4. Craw, S., Horsburgh, B., Massie, S.: Music recommendation: audio neighbourhoods to discover music in the long tail. In: Hüllermeier, E., Minor, M. (eds.) ICCBR 2015. LNCS, vol. 9343, pp. 73–87. Springer, Cham (2015). doi:10.1007/978-3-319-24586-7_6
5. Cremonesi, P., Koren, Y., Turrin, R.: Performance of recommender algorithms on top-n recommendation tasks. In: Proceedings of the Fourth ACM Conference on Recommender Systems – RecSys 2010, p. 39 (2010)
6. Ekstrand, M.D., Riedl, J.T., Konstan, J.A.: Collaborative filtering recommender systems. Found. Trends® Hum.-Comput. Interact. **4**(2), 81–173 (2011)

7. Gedikli, F., Jannach, D., Ge, M.: How should i explain? A comparison of different explanation types for recommender systems (2014)
8. Harper, F.M., Konstan, J.A.: The movielens datasets: history and context. ACM Trans. Interact. Intell. Syst. **5**(4), 19:1–19:19 (2015). http://doi.acm.org/10.1145/2827872
9. Im, K.H., Park, S.C.: Case-based reasoning and neural network based expert system for personalization. Expert Syst. Appl. **32**(1), 77–85 (2007)
10. Recio-García, J.A., González-Calero, P.A., Díaz-Agudo, B.: jcolibri2: A framework for building case-based reasoning systems. Sci. Comput. Program. **79**, 126–145 (2014)
11. Kelleher, J., Bridge, D.: An accurate and scalable collaborative recommender. Artif. Intell. Rev. **21**(3), 193–213 (2004)
12. Linden, G., Smith, B., York, J.: Amazon.com recommendations: Item-to-item col- laborative filtering. IEEE Internet Comput. **7**(1), 76–80 (2003)
13. MelvilleP, M.R., R, N.: Content boosted collaborative filtering for improved recommendations, pp. 187–192 (2002)
14. Park, Y.J.: The adaptive clustering method for the long tail problem of recommender systems. IEEE Trans. Knowl. Data Eng. **25**(8), 1904–1915 (2013)
15. Park, Y.J., Tuzhilin, A.: The long tail of recommender systems and how to leverage it. In: Proceedings of the 2008 ACM Conference on Recommender Systems, pp. 11–18. ACM (2008)
16. Pazzani, M.J., Billsus, D.: Content-based recommendation systems. In: Brusilovsky, P., Kobsa, A., Nejdl, W. (eds.) The Adaptive Web, pp. 325–341. Springer, Heidelberg (2007). doi:10.1007/978-3-540-72079-9_10
17. Phelan, O., McCarthy, K., Bennett, M., Smyth, B.: Terms of a feather: content-based news recommendation and discovery using Twitter. In: Clough, P., Foley, C., Gurrin, C., Jones, G. J.F., Kraaij, W., Lee, H., Mudoch, V. (eds.) ECIR 2011. LNCS, vol. 6611, pp. 448–459. Springer, Heidelberg (2011). doi:10.1007/978-3-642-20161-5_44
18. Phelan, O., McCarthy, K., Smyth, B.: Using Twitter to recommend real-time topical news. In: Proceedings of the Third ACM Conference on Recommender Systems – RecSys 2009, p. 385. ACM Press, New York, USA (2009)
19. Semeraro, G., Lops, P., Basile, P., de Gemmis, M.: Knowledge infusion into content-based recommender systems. In: Proceedings of the Third ACM Conference on Recommender Systems (RecSys 2009), pp. 301–304. ACM, New York (2009)
20. Sun, J., Zhao, Q., Antony, S., Chen, S.: Personalized recommendation systems: an application in case-based reasoning (2015)
21. Yin, H., Cui, B., Li, J., Yao, J., Chen, C.: Challenging the long tail recommendation. Proc. VLDB Endowment **5**(9), 896–907 (2012). http://dl.acm.org/citation.cfm?id=2311916
22. Zheng, V.W., Cao, B., Zheng, Y., Xie, X., Yang, Q.: Collaborative filtering meets mobile recommendation: a user-centered approach. In: Twenty-Fourth Conference on Artificial Intelligence, pp. 236–241 (2010)

Extending the Flexibility of Case-Based Design Support Tools: A Use Case in the Architectural Domain

Viktor Ayzenshtadt[1,3](✉), Christoph Langenhan[4], Saqib Bukhari[3], Klaus-Dieter Althoff[1,3], Frank Petzold[4], and Andreas Dengel[2,3]

[1] Institute of Computer Science, University of Hildesheim, Samelsonplatz 1, 31141 Hildesheim, Germany
[2] Kaiserslautern University, P.O. Box 3049, 67663 Kaiserslautern, Germany
[3] German Research Center for Artificial Intelligence, Trippstadter Strasse 122, 67663 Kaiserslautern, Germany
{viktor.ayzenshtadt,saqib.bukhari,klaus-dieter.althoff, andreas.dengel}@dfki.de
[4] Chair of Architectural Informatics, Faculty of Architecture Technical University of Munich, Arcisstrasse 21, 80333 Munich, Germany
{langenhan,petzold}@ai.ar.tum.de

Abstract. This paper presents results of a user study into extending the functionality of an existing case-based search engine for similar architectural designs to a flexible process-oriented case-based support tool for the architectural conceptualization phase. Based on a research examining the target group's (architects) thinking and working processes during the early conceptualization phase (especially during the search for similar architectural references), we identified common features for defining retrieval strategies for a more flexible case-based search for similar building designs within our system. Furthermore, we were also able to infer a definition for implementing these strategies into the early conceptualization process in architecture, that is, to outline a definition for this process as a wrapping structure for a user model. The study was conducted among the target group representatives (architects, architecture students and teaching personnel) by means of applying the paper prototyping method and Business Processing Model and Notation (BPMN). The results of this work are intended as a foundation for our upcoming research, but we also think it could be of wider interest for the case-based design research area.

Keywords: CBR and creativity · Process-oriented CBR · Knowledge modeling · Business process modeling · Case-based design

1 Introduction

In this paper, we address the early conceptual design phase in architecture, where searching for helpful, inspirational, and similar previous designs and solutions

D.W. Aha and J. Lieber (Eds.): ICCBR 2017, LNAI 10339, pp. 46–60, 2017.
DOI: 10.1007/978-3-319-61030-6_4

can support the design and development process by offering insights and conclusions from existing solutions. The ability to compare and evaluate relevant reference examples of already built or designed buildings helps designers assess their own design explorations and informs the design process. Most computational search methods available today rely on textual rather than graphical approaches to representing information. However, textual descriptions are not sufficient to adequately describe spatial configurations such as floor plans. To address these shortcomings, a novel approach was introduced by Langenhan et al. [13] which facilitates the automatic lookup of reference solutions from a repository using graphical search keys. In the basic research project, Metis[1], these issues were examined further using methods of case-based reasoning (CBR), multi-agent systems (MAS), and computer-aided architectural design (CAAD).

As part of the project activities, a distributed case-based retrieval engine *MetisCBR* [2] was developed that retrieves similar building designs using case-based agents that apply search methods implemented in the CBR framework myCBR. In its general mode of operation, MetisCBR's core functionality seeks the most suitable *strategy* for each query from a set of such strategies. Currently, these strategies *do not have a structural definition according to architectural requirements*, that is, MetisCBR has only basic similarity assessment strategies that were designed using an adapted bottom-up method based on the basic elements of the domain model described in Ayzenshtadt et al. [1]. However, as we are planning to extend MetisCBR to a *process-oriented* case-based design support tool for the architectural conceptualization phase, such structural definitions are needed to provide a common interface for implementating different high- and low-level processes. To address this issue, we conducted a *process modeling* study among the target group representatives to examine their similarity assessment processes (low-level) and the inclusion of the similarity assessment in the whole conceptualization process (high-level). Our main aims for this research were:

- Determine common features in a multitude of architects' own strategies and infer a methodology for defining such strategies in our system.
- Find a common structure for the conceptualization process (that we call a *user model*) with inclusion of similarity assessment strategies for the further design of user models for the system.

This paper is structured as follows: in Sect. 2, we describe our previous work in the Metis project and other related research in this area. In Sect. 3, we show which suggestions from previous research in this field led us to work on the strategic and process-related aspects for the system. In Sect. 4, we describe in detail the modeling study we conducted, presenting the background and short descriptions of the main elements we used (POCBR, BPMN, paper prototyping) and then describing the study's main phases and summarized results. In Sect. 5,

[1] *Metis – Knowledge-based search and query methods for the development of semantic information models for use in early design phases.* Funded by DFG (German Research Foundation).

we provide definitions of the foundations (strategy and process) for user models for our system. Finally, in the last section we conclude with a review of this paper and an outlook on future work.

2 Related Work

The research area of case-based design (CBD) has a long history in the communities of both CBR and knowledge-based design research fields. As an important part of early CBR research, CBD (and its application in architecture especially) has gained much interest from the beginning of the advanced domain-bounded CBR research. Many projects have been initiated since then and several approaches and applications have been developed for both basics as well as advanced methodologies in this research domain. In this section, we review research conducted our Metis project began and the work accomplished since over the course of the project activities.

Research conducted prior to our project includes a number of essential approaches and fundamental work, now well-known in the research community. An example of such fundamental work is [14] and an overview of the prior approaches can be found, for example, in [18]. FABEL [22], CaseBook [8], or DIM [10] are examples of approaches that apply CBR to design problems. One of the most comprehensive work in studies of the application of CBR to the architectural design process is Richter's work [17], and [19] contains a summary of this research including an overview of suggestions for improving such applications.

In the Metis project, which was initiated to enhance architectural design by providing knowledge-based retrieval methods to support the early conceptualization phase, a number of different approaches to searching for similar architectural design solutions were developed. MetisCBR, mentioned above, is one such approach, while others include an adapted VF2 approach (described in [5]) for (sub)graph matching, index-based retrieval with the *Cypher* language queries of the graph database Neo4j, and the enhancement [23] of the original Messmer-Bunke algorithm [15]. Comparative evaluations of MetisCBR and other retrieval methods were undertaken, for example, in [3,20]. Retrieval support tools, such as a web-based floor plan editor *(Metis WebUI)* [4] and a content management system *mediaTUM*, were also developed.

The theoretical foundation underlying our systems is the paradigm of *Semantic Fingerprint* [12] and *AGraphML* [11] (a representation format for graph-based floor plans). For example, a case in MetisCBR is a semantic fingerprint of a floor plan that is imported as an AGraphML from mediaTUM and represented in the myCBR internal case format according to the domain model described in [1].

3 Problem Definition in the CBD Domain

Richter [19] presents the results of research conducted in the field of case-based reasoning in architecture and makes a number of recommendations for further research in this area. One of the main recommendations is that *query strategies should be optimized* in case-based design support systems. In our work, we tried to find an initial solution to this problem. First, we try to obtain knowledge for constructing such strategies from the actual knowledge carriers by means of a modeling study. From an analysis of this study, we then infer and propose a structural definition for query strategies and, more general, superstructural definition of processes (user models) which we can then use for MetisCBR, but this can also potentially serve as a foundation for other approaches.

Another suggestion made in [19] is that an unaddressed issue of quality assessment in case-based design (especially in architecture) should be investigated. We also address this in our study, but in addition to similarity assessment. Although, the question of quality should be investigated in a separate context, our work may provide a starting point or serve as inspiration for more detailed research in this direction.

The last suggestion from [19] we deal with is the variability of the CBD approaches. This is also a question that needs to be fully examined in a separate context, for example to identify which degrees of variability are required in which context of the conceptualization phase. We also address this suggestion in our study, and initially investigate how this feature can be included (e.g., in strategies).

4 The Process Modeling Study

4.1 Background and Main Aspects

As mentioned in Sects. 1 and 3, our process modeling study was conducted to define the foundations for developing user models for enhancing our system from a search engine to a design support tool. While developing ideas for solutions to the questions mentioned in Sect. 3, i.e., how best to strategically improve the system to provide the most suitable design recommendations at the right point in time in the conceptualization process, the most logical and valuable source of potential answers seemed to be expert knowledge from the target group (representatives of the architectural design domain). Likewise, we expected that there would be many commonalities in strategies among different representatives of the target group. To gain this knowledge, to confirm or disprove our assumptions, and to conduct the study we developed a small methodology that consists of a questionnaire, modeling, and cross-evaluation (see more in Sect. 4.2). During the development of this methodology, we tried to consider how components of thinking (such as *categorization, comparison*, and *condition* that are explained, e.g., in [7]) could be combined to extract the most valuable aspects of knowledge for our aims. In the next sections, we describe the main aspects and components of our study and how we use (and/or are going to use) them in our research and development.

Process-Oriented CBR. POCBR is an approach for applying case-based reasoning to process-oriented information systems (POIS). Today, the applications of POCBR extend beyond classical POIS domains, such as workflow management, to include other domains, such as medical healthcare, e-science, or cooking [16] and in recent years, POCBR has been the subject of research for major CBR problem fields: retrieval is covered, for example, in [9], and adaptation in [16].

To enhance MetisCBR from a CBR-based retrieval system to a CBR-based design support tool, we decided to extend it into a POIS with a number of implemented processes (user models), where the most suitable user model is activated when certain user behavior is detected (i.e., the user behavior will be a case with actions as attributes that, when sufficiently matched, activate the model).

BPMN. *Business Process Model and Notation* (BPMN) is a graph-oriented process modeling language for the visualization of business workflows with predefined elements and notations. It has become a very widely-used standard form of notation for the analysis of processes on the corresponding domain's high level system design [6]. The current Version of BPMN is *2.0.* BPMN consists of a number of element groups, the most important of which are *Flow Objects* for denoting tasks and events, and *Connection Objects* for denoting the connections between the elements. A multitude of software tools is available for BPMN-based process modeling.

For our post-study analysis, i.e., the transformation of processes modeled by our participants into a digital form, we used *Camunda Modeler*[2]. In a comparative evaluation of open source tools for building research prototypes, it was determined that Camunda Modeler is a cross-platform application that provides full BPMN 2.0 support [21].

Paper Prototyping. *Paper Prototyping* (or *Rapid Paper Prototyping*) is a method for evaluating user interfaces in early stages of the software development process. It is commonly-used by usability engineers for implementing user-centered design and to test the functions of future software products among the potential user group. The principle relies on the concept of a printed or sketched version of the software's user interface that prototypically represents its functions. The user interacts with these to detect usability problems in early phases of implementation of the software.

For our study, we adapted and modified the paper prototyping method to allow our participants to model their similarity assessment and conceptualization processes with several sketched elements of the BPMN. In contrast to a usability study for a software prototype, our participants did not have a concrete prototype to test, but were instead asked to *model a prototype of their processes using the sketched BPMN elements.*

4.2 Study Process and Results

Five representatives of different areas of the architecture domain agreed to take part in the study, including architects, architecture PhD students, and

[2] https://camunda.org/features/modeler/.

architecture teaching staff. On average, we spent approximately 2 hours per interview for each of the participants. In the next sections, we describe the methodology we developed and used, the questions and tasks we used when working with the participants, as well as the corresponding results.

Methodology. Our methodology for conducting the study consisted of four main phases described below (see also Fig. 1. Detailed descriptions of phases and corresponding results are provided in the following sections):

1. *Criteria Survey:* the participants were asked to name the criteria for rating the quality and similarity of architectural designs.
2. *Similarity Assessment Modeling:* the participants were asked to manually select the most similar design(s) from a collection of designs for a given pre-defined query. After the selection they were asked to model their process, i.e., to reconstruct their cognitive similarity assessment process using the sketched BPMN element prototypes. This phase consisted of three sub-phases that correspond to three complexity levels of a floor plan.
3. *Conceptualization Process Modeling:* the participants were asked to model their entire (early) conceptualization process, including how they incorporate similarity assessment.
4. *Cross-Evaluation:* the current participant was asked to evaluate the similarity assessment process of one of the previous participants.

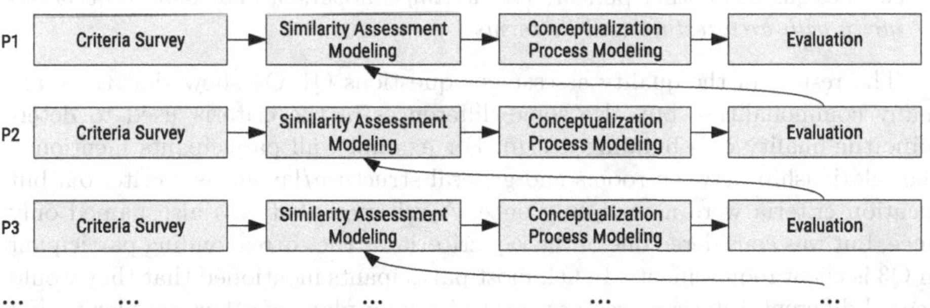

Fig. 1. The main steps of the methodology we applied during the process modeling study. P[n] denotes a participant.

Preliminary Questionnaire. Before the main phases, we conducted a preliminary questionnaire phase to ascertain the participants' familiarity with CBR or at least the term *case-based reasoning*, and also if they have applied or worked with CBR applications during their job-related activities. This was essential for the subsequent interview, especially with respect to the terms used in the interview (e.g., CBR technical terms for participants who are familiar with CBR).

In general, most of our participants were familiar with the main concepts of the CBR paradigm, but only 40% of the participants were aware of using CBR applications in their professional work. The non-academia participants were not aware of CBR at all and told that architectural practices rarely apply CBR-based or similar reasoning and/or retrieval applications in practice, confirming results of Richter's research in [19].

Phase 1 – Building Design (Floor Plan) as a Case: Criteria Survey. The criteria survey was the first phase of the interview with a participant. In this phase, we asked participants about the criteria they use to rate quality and similarity of architectural designs. The theoretical background of this phase is that *a case in our CBR-based retrieval engine is a fingerprint of an architectural building design (floor plan)* as defined in [1]. The following questions were asked in this phase (referred below as Q[n], e.g., Q1):

1. Which criteria do you use to assess the *quality* of an architectural design?
2. Consider this floor plan. *(The participant is given a printed floor plan for analysis.)* How would you rate its quality using your quality criteria?
3. Which of the quality criteria do you consider to be the key criteria?
4. For more complex floor plans, would you change the priority distribution of your criteria? Are there criteria that you would consider important only for abstract floor plans? *(The participant is shown an abstract and a more complex floor plan for comparison. In Sect. 4.2 we show the difference between abstract and complex floor plans.)*
5. Which criteria do you use to assess the *similarity* of an architectural design?
6. Which terms would you use to/how would you describe the similarity between two designs to another person? *(Assuming this person has some basic familiarity with architecture and its terms.)*

The results of the quality assessment questions Q1–Q4 show that there are many commonalities, but also some differences in the criteria used to determine the quality of a building design. For example, all participants mentioned the relationship between rooms and general structure/layout as a criterion, but location criteria were named only once. A criterion that was also named only once, but was considered one of the key criteria by the corresponding participant in Q3 is client requirements. In Q4, most participants mentioned that they would accord different criteria a greater priority for floor plans of other complexity. For example, the form/shape of the rooms on the abstract level could become more important.

In Q5, most participants said they would use virtually the same criteria as in Q1–Q4 to assess the similarity of two floor plans. However, some participants did introduce some new terms for similarity assessment only. For example, the criterion of cost-economy (which surprisingly did not feature in the quality assessment questions). In Q6, participants said they would additionally use examples and situations to explain the difference between two designs to another person (the results of Q6 have been also preserved for our upcoming research into explanation-aware systems in architecture).

For our analysis of the results of Q1–Q6, we categorized the criteria named by the participants, as shown in Fig. 2.

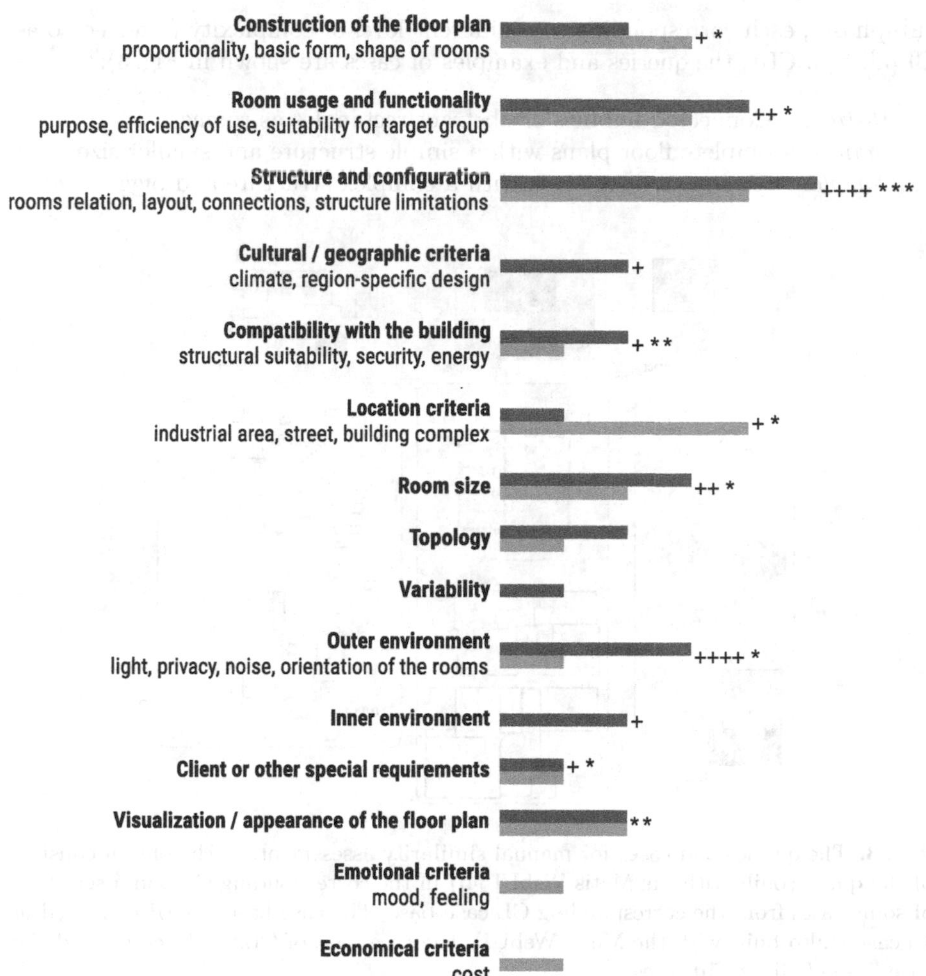

Fig. 2. Results of the categorization of Q1–Q6. The length of the lines indicates how often they were mentioned as quality criteria (red) and as similarity criteria (blue). [+] indicates the frequency of mention as the *key criterion* (Q3). [*] indicates the frequency of mention as criteria that changes its priority when it comes to the change of complexity level of a floor plan (Q4). (Color figure online)

Phase 2 – Similarity Assessment Modeling. In the next step of our study, our intention was to reconstruct and understand how our target user group (architects) would *manually select the most similar design* from a collection of architectural designs for a given predefined query. To accomplish this, we asked the participants to assume the role of our CBR-based retrieval system, that is, to imagine him- or herself as the system assigned with task of selecting the most similar floor plan to the query drawn by a user.We undertook this phase in three sub-phases, each corresponding to a different level of complexity (referred to as CL[n], e.g., CL1, the queries and examples of cases are shown in Fig. 3):[3]

1. *Abstract* – Connected bubbles or abstract rectangles as rooms
2. *Simple* – Complete floor plans with a simple structure and smaller size
3. *Complex* – Complete floor plans with a complex structure and bigger size.

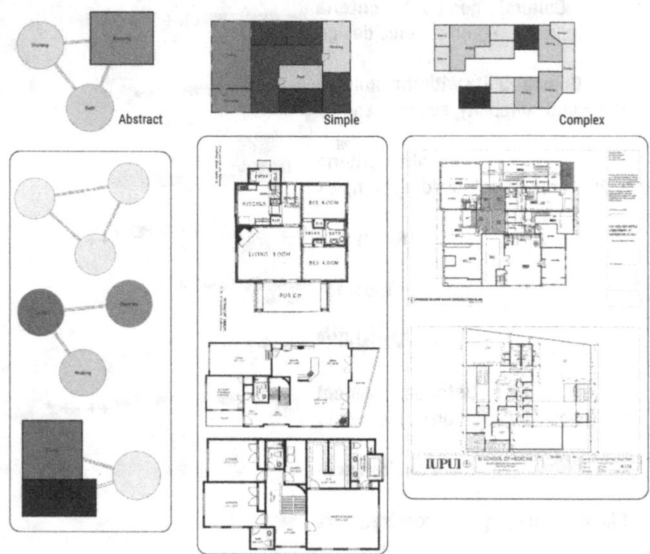

Fig. 3. The queries and cases for manual similarity assessment. Each column consists of the query (built with the Metis WebUI [4]) in the corresponding CL and a selection of some cases from the corresponding CL case base. The case base of CL1 consisted of 10 cases (also built with the Metis WebUI), the case base of CL2 of 11 cases, and the case base of CL3 of 10 cases.

After each manual selection process, we asked the participants to apply paper prototyping and model their cognitive process of selection using the sketched elements of the BPMN. Because the number of BPMN elements is quite large and it can be time-consuming to explain all of them to a participant, we decided

[3] The designs for CL2 and CL3 were taken from Flickr. In Fig. 3: *1391 Second Floor Plan* https://www.flickr.com/photos/philmanker/3516873511/ by Phil Manker, CC-BY 2.0; *Architecture and Building, 1922* https://www.flickr.com/photos/revivaling/5549896664/ by Learn From. Build More., CC-BY-SA 2.0; *A104: Level 2 Dimensioned Floor Plan* https://www.flickr.com/photos/therichardlife/5574176101/ by Stefanie Richard, CC-BY-SA 2.0; *216 Brookwood floor plan - Main Floor* https://www.flickr.com/photos/homesbycharlotte/26899442344/ and *216 Brookwood floor plan - 2nd Floor* https://www.flickr.com/photos/homesbycharlotte/27409880812/ by Charlotte Turner, Public Domain/all scaled from original. All visited on 23.04.17.

not to use all the BPMN elements, but to restrict the selection of elements to the basic ones. Figure 4 shows the BPMN elements used for modeling, and Fig. 5 a result of the process modeling by one of the participants. During the selection and modeling process, we also asked the participants to *think aloud* to give us more insights into their thinking process while selecting and modeling. After each modeling process, we asked the participants if they applied their own criteria (named in Q1–Q6) determining the similarity and quality of results. To analyze this phase after the experiment, we reconstructed the paper-prototyped processes with the Camunda Modeler software, mentioned in Sect. 4.1.

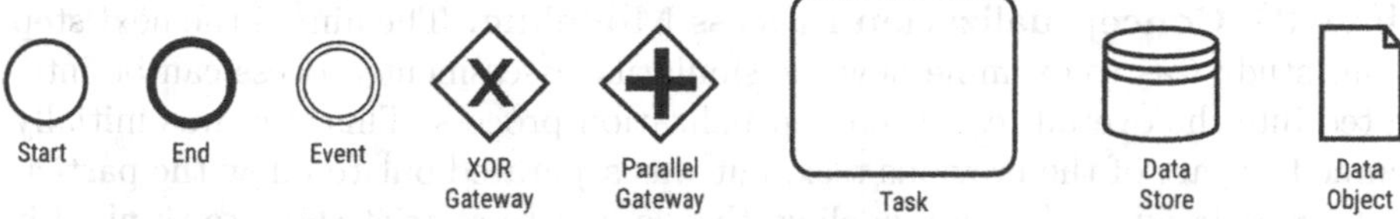

Fig. 4. The BPMN elements used for paper prototyping of the processes.

Fig. 5. Similarity assessment process by one of the participants modeled using the sketched BPMN elements.

During the manual selection of the most similar floor plan, in 14 of the 15 comparison processes the criteria named in Q1–Q6 were applied. An analysis of the results of the modeling phase shows that several tasks in all of the processes have at least similar intentions. For example, a criteria-based comparison takes place in each of the processes, but for some of the participants (minority) the set of criteria is immutable, whereas others tend not to restrict this set of criteria.

The main difference was mostly the method of application: *sequential* as well as *parallel* comparison took place, mostly in a mix where some criteria were used initially for pre-selection (e.g., topology and functionality, or room count and functionality) followed by a parallel process of comparison with other criteria. However, purely sequential processes were also modeled for each complexity level. The flexibility of criteria played a role for participants who did not want to restrict their comparison to a set of pre-defined criteria. It is also notable, that some of the participants excluded some criteria when dealing with a greater complexity level. Also, expert knowledge and meta information about the floor plans were also drawn on for the comparison processes.

Phase 3 – Conceptualization Process Modeling. The aim of the next step of our study was to examine how the similarity assessment process can be integrated into the overall (early) conceptualization process. This step was initially planned as part of the previous task, but was separated out to allow the participants more freedom during modeling, that is, not to restrict them to think of it as an additional question. In the modeling, the only requirement was to reflect on how the similarity assessment process fits into the conceptualization process. Participants were free to choose whether to model this process using paper prototyping or simply drawing on paper with or without the BMPN elements. The majority of the participants chose to draw on paper, but most of them used the BPMN notation to visualize their processes (these were also transferred to digital form for later analysis).

An analysis of the models of (early) conceptualization processes reveals that, generally speaking, the iterative nature of the process is obviously natural to all the participants. We identified two general structural setups of the processes:

- The process is sequential with a number of subsequent sub-processes, where some sub-processes are of iterative type.
- The process is an enclosing iteration that consists of sub-tasks, which can also be iterative.

The similarity assessment was placed at different positions in the overall conceptualization processes. For example, one of the participants positioned it in the beginning of the conceptualization phase, applying it only once for abstract floor plans with bubble-shaped (i.e., undefined shape) rooms. Another bias case is the dynamic positioning of similarity assessment, that is conducted either during the analysis of requirements or during the synthesis of possible solutions. The normal case however, was to place the similarity assessment either in the middle or in the final phases of the process after the determining client requirements and identifying the key issues, and before the evaluation by the client (other phases being, for example, cost calculation or 3D conceptualization).

Phase 4 – Cross-Evaluation. The final step of our study was to cross-evaluate the participants' similarity assessment processes. To accomplish this, we asked the participants to compare their process against a random process from one of the previous participants. The current participant was asked to identify differences, commonalities, pros and cons, advantages and disadvantages, as well as anything else that came to mind during the comparison. Using this evaluation method, we tried to obtain a competent opinion on the similarity process models, to see how to improve if one process becomes an inspiration or a template for a strategy in a user model in our system.

The participants mostly criticized the lack of different types of knowledge that could help in the comparison process: for example, one of the processes lacked the expert knowledge component (i.e., comprehensive professional knowledge in the architectural domain), which the evaluating participant viewed as being an essential part of such processes. Similarly, the lack of control of criteria (e.g., when have enough criteria been compared to achieve a sufficient degree of similarity) and the non-dynamic nature of criteria in some of the processes was criticized. On the positive side, the flexibility of some of the processes was emphasized, for example, a flexible threshold for criteria match evaluation. Another positive aspect was the application of a more systematic approach, than the evaluating participant's own process.

5 Definitions Inferred from the Results

To achieve our actual goal – the *definitions of the foundations for user models* – we generalized the results of the questions and modeling phase to infer structures for the foundations, as defined in the following sections.

5.1 Strategy

Strategy is a basic element of the user model. Strategy will be used as a controlling structure for the actual algorithm for searching for similar floor plans. That is, the algorithm should satisfy all the requirements of the definition to become a strategy in our system. We define strategy as follows:

Definition 1. *Strategy is a quadruple $S = (C, K, \mu, F)$, where C is criteria, K is knowledge, μ is similarity measure, and F is flexibility. $C = C_s \cup C_d$ (criteria can be of dynamic and static type), where $C_s \vee C_d \neq \varnothing$. $K = K_m \cup K_e$ (meta knowledge about the cases in the case base and expert knowledge in the domain, e.g., in architecture), where $K_m \vee K_e \neq \varnothing$. $\mu = \mu_s \cup \mu_p$ (similarity measures can be of parallel or sequential type), where $\mu_s \vee \mu_p \neq \varnothing$. $F = (f_c, f_\mu)$, where f_c is the value of the strategy's flexibility that corresponds to the criteria and f_μ is the value for the conditional variability of μ, i.e., the variability of the similarity value's conditional values (such as weight or degree[4]) under certain constraints (e.g., different complexity levels of the floor plan).*

To explain the application of this definition, we defined an exemplary strategy that satisfies all the requirements of the definition (see Fig. 6).

[4] In our research, we use the following classification of degrees of similarity since [1]: *very similar, similar, sufficiently similar,* and *unsimilar.*

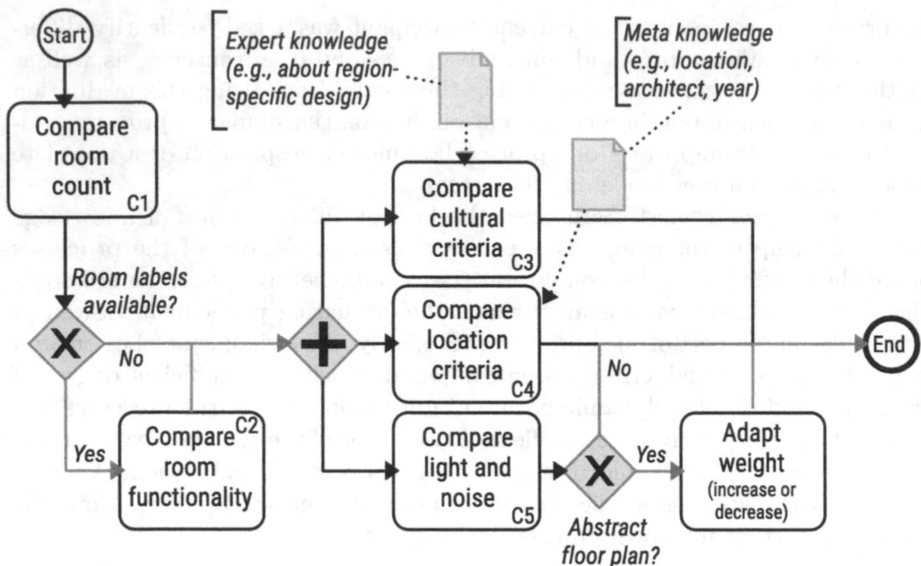

Fig. 6. An exemplary strategy that satisfies all of the requirements named in Definition 1. Here, C1 and C3–C5 are the static criteria that are always applied as comparison criteria. C2, however, is a dynamic criteria that depends on the availability of room labels, i.e., functions. Expert and meta knowledge help to resolve the comparison of C3 and C4. C1 and C2 are resolved with sequential similarity measures, i.e., C2 follows C1. In contrast, C3–C5 are resolved with a parallel type of similarity measure (e.g., with agents that work concurrently and then apply weights and calculate an amalgamated similarity value out of these three). Assuming, we have applied $f_c = 0.6$, we get a flexibility that 3 of 5 criteria should be at least sufficiently similar for a floor plan to be considered for inclusion in retrieval results, where the weight of similarity value of C5 depends on the complexity of the floor plan (alternatively, C5 can be defined as a dynamic criterion with complexity of floor plan as its condition).

5.2 Process

Process is a wrapper for the user model and is intended to represent the (early) conceptualization phase as a template that will be activated when user's actions and behavior indicate a sufficiently similar match in the set of processes implemented in the system (what we will consider an action and a behavior, is one of the subjects of our upcoming research, see also Sect. 6).

Definition 2. *Process is a triple $P = (S, t, A)$, where S is a set of strategies as defined in Definition 1, t is the type of the process (e.g., sequential, semi-sequential, enclosing iteration), and A is the set of actions. $A = A_s \cup A_i \cup A_e$ (actions can be of starting, ending, and intermediate type), where $A_s \wedge A_e \neq \varnothing$. Strategies are linked to actions with a surjective mapping $S \twoheadrightarrow A$, i.e., $\forall a \in A \exists s \in S$ (for each of the strategies at least one action exists that this strategy is mapped to).*

6 Conclusion and Future Work

In this paper, we presented a study that investigates the search for similar architectural references during the early conceptualization process in architecture, and from this inferred definitions for strategic foundations for structures for user models in our system MetisCBR. We conducted this study with various representatives of the architectural domain. The study surveyed quality and similarity criteria, similarity assessment modeling and conceptualization phase modeling (both with BPMN elements), and then undertook a cross-evaluation. The results have shown that it is possible to infer definitions of *strategy* and *process*, and therefore to provide structures for query strategies (which is recommended in [19]). The study and the definitions also address the problems of quality criteria and variability discussed in [19].

Our future work will include the investigation of what an action (e.g., step, intermediate step, or iteration) and behavior can be in the specific context of architectural design (as mentioned in Sect. 5). We will also work on developing an explanation module for the target group-specific explanation of retrieval results with special explanation patterns – for this research we will use some results of the experiment presented in this paper (see Sect. 4.2). Our next step in the context of this paper is the implementation of the strategies according to the results and definitions derived in this paper, and then to undertake a performance comparison with the system's former strategies.

References

1. Ayzenshtadt, V., Langenhan, C., Bukhari, S.S., Althoff, K.D., Petzold, F., Dengel, A.: Distributed domain model for the case-based retrieval of architectural building designs. In: Petridis, M., Roth-Berghofer, T., Wiratunga, N. (eds.) Proceedings of the 20th UK Workshop on Case-Based Reasoning. UK Workshop on Case-Based Reasoning (UKCBR-2015), pp. 15–17. Cambridge, UK, December 2015
2. Ayzenshtadt, V., Langenhan, C., Bukhari, S.S., Althoff, K.D., Petzold, F., Dengel, A.: Thinking with containers: a multi-agent retrieval approach for the case-based semantic search of architectural designs. In: Filipe, J., van den Herik, J. (eds.) 8th International Conference on Agents and Artificial Intelligence (ICAART-2016), pp. 24–26. SCITEPRESS, Rome, February 2016
3. Ayzenshtadt, V., Langenhan, C., Roith, J., Bukhari, S., Althoff, K.-D., Petzold, F., Dengel, A.: Comparative evaluation of rule-based and case-based retrieval coordination for search of architectural building designs. In: Goel, A., Díaz-Agudo, M.B., Roth-Berghofer, T. (eds.) ICCBR 2016. LNCS (LNAI), vol. 9969, pp. 16–31. Springer, Cham (2016). doi:10.1007/978-3-319-47096-2_2
4. Bayer, J., Bukhari, S.S., Langenhan, C., Liwicki, M., Althoff, K.-D., Petzold, F., Dengel, A.: Migrating the classical pen-and-paper based conceptual sketching of architecture plans towards computer tools—prototype design and evaluation. In: Lamiroy, B., Dueire Lins, R. (eds.) GREC 2015. LNCS, vol. 9657, pp. 47–59. Springer, Cham (2017). doi:10.1007/978-3-319-52159-6_4
5. Cordella, L.P., Foggia, P., Sansone, C., Vento, M.: A (sub) graph isomorphism algorithm for matching large graphs. IEEE Trans. Pattern Anal. Mach. Intell. **26**(10), 1367–1372 (2004)
6. Dijkman, R.M., Dumas, M., Ouyang, C.: Semantics and analysis of business process models in BPMN. Inf. Softw. Technol. **50**(12), 1281–1294 (2008)

7. Glatzeder, B., Goel, V., von Müller, A.: Towards a Theory of Thinking: Building Blocks for a Conceptual Framework. Springer Science and Business Media, Berlin (2010)
8. Inanc, B.S.: Casebook. an information retrieval system for housing floor plans. In: The Proceedings of 5th Conference on Computer Aided Architectural Design Research (CAADRIA), pp. 389–398 (2000)
9. Kendall-Morwick, J., Leake, D.: A study of two-phase retrieval for process-oriented case-based reasoning. In: Montani, S., Jain, L.C. (eds.) Successful Case-Based Reasoning Applications-2, pp. 7–27. Springer, Heidelberg (2014). doi:10.1007/978-3-642-38736-4_2
10. Lai, I.C.: Dynamic idea maps: a framework for linking ideas with cases during brainstorming. Int. J. Archit. Comput. **3**(4), 429–447 (2005)
11. Langenhan, C.: A federated information system for the support of topological bim-based approaches. In: Forum Bauinformatik, Aachen (2015)
12. Langenhan, C., Petzold, F.: The fingerprint of architecture-sketch-based design methods for researching building layouts through the semantic fingerprinting of floor plans. Int. Electron. Sci.-Educ. J.: Archit. Mod. Inf. Technol. **4**, 13 (2010)
13. Langenhan, C., Weber, M., Liwicki, M., Petzold, F., Dengel, A.: Graph-based retrieval of building information models for supporting the early design stages. Adv. Eng. Inform. **27**(4), 413–426 (2013)
14. Maher, M., Balachandran, M., Zhang, D.: Case-Based Reasoning in Design. Lawrence Erlbaum Associates, Mahwah (1995)
15. Messmer, B.T., Bunke, H.: A decision tree approach to graph and subgraph isomorphism detection. Pattern Recognit. **32**(12), 1979–1998 (1999)
16. Müller, G., Bergmann, R.: Learning and applying adaptation operators in process-oriented case-based reasoning. In: Hüllermeier, E., Minor, M. (eds.) Case-Based Reasoning Research and Development. LNCS, pp. 259–274. Springer, Cham (2015). doi:10.1007/978-3-319-24586-7_18
17. Richter, K.: Augmenting Designers' Memory: Case-Based Reasoning in Architecture. Logos-Verlag, Berlin (2011)
18. Richter, K., Heylighen, A., Donath, D.: Looking back to the future – an updated case base of case-based design tools for architecture. Knowl. Model.-eCAADe. **25**, 285–292 (2007)
19. Richter, K.: What a shame-why good ideas can't make it in architecture: a contemporary approach towards the case-based reasoning paradigm in architecture. In: FLAIRS Conference (2013)
20. Sabri, Q.U., Bayer, J., Ayzenshtadt, V., Bukhari, S.S., Althoff, K.D., Dengel, A.: Semantic pattern-based retrieval of architectural floor plans with case-based and graph-based searching techniques and their evaluation and visualization. In: 6th International Conference on Pattern Recognition Applications and Methods (ICPRAM 2017), Porto, Portugal, 24–26 February 2017
21. Seel, C., Dörndorfer, J., Schmidtner, M., Schubel, A.: Vergleichende Analyse von Open-Source-Modellierungswerkzeugen als Basis für Forschungsprototypen. In: Barton, T., Herrmann, F., Meister, V.G., Müller, C., Seel, C. (eds.) Prozesse, Technologie, Anwendungen, Systeme und Management, p. 35 (2016)
22. Voss, A.: Case design specialists in FABEL. In: Issues and Applications of Case-Based Reasoning in Design, pp. 301–335 (1997)
23. Weber, M., Langenhan, C., Roth-Berghofer, T., Liwicki, M., Dengel, A., Petzold, F.: Fast subgraph isomorphism detection for graph-based retrieval. In: Ram, A., Wiratunga, N. (eds.) Case-Based Reasoning Research and Development. LNCS, pp. 319–333. Springer, Heidelberg (2011). doi:10.1007/978-3-642-23291-6_24

A Reasoning Model Based on Perennial Crop Allocation Cases and Rules

Florence Le Ber[1](✉), Xavier Dolques[1], Laura Martin[2], Alain Mille[3], and Marc Benoît[2]

[1] ICUBE, Université de Strasbourg, ENGEES, CNRS, Illkirch-Graffenstaden, France
{florence.leber,xavier.dolques}@engees.unistra.fr
[2] SAD ASTER, UR 055, INRA Mirecourt, Mirecourt, France
Marc.Benoit@mirecourt.inra.fr
[3] LIRIS, Université Lyon1, UMR CNRS 5205, Lyon, France
alain.mille@univ-lyon1.fr

Abstract. This paper presents a prototype of case-based reasoning, built for the agricultural domain. Its aim is to forecast the allocation of a new energy crop, the miscanthus. Interviews were conducted with french farmers in order to know how they make their decisions. Based on interview analysis, a case base and a rule base have been formalized, together with similarity and adaptation knowledge. Furthermore we have introduced variations in the reasoning modules, for allowing different uses. Tests have been conducted. Results showed that the model can be used in different ways, according to the aim of the user, and e.g. the economic conditions for miscanthus allocation.

Keywords: Case based reasoning · Adaptation · Decision rules · Agriculture

1 Introduction

To face the decrease of fossil energy supplies and to reduce the greenhouse gas emissions, new biomass energy[1] resources become of a great interest in Europe, Their spatial extension seems then unavoidable. For instance, the miscanthus (*Miscanthus ex giganteus*), has interesting caloric values and constitutes a great potential for biofuel and heating plant. As the production of such biomass crops is perennial (15 to 20 years of production by land) and exclusively dedicated to the energy use, it is necessary to anticipate their allocation to prevent a forecasted perennial food/non-food competition.

Modeling is useful to anticipate the extension of biomass crops and bring decision-making support for politics. Several land-use change models deal with this specific problem [1]. Most of these models simulate large-scale allocation

[1] Biomass energy corresponds to organic matter, essentially from agricultural and forest products (e.g. sugar beet, wood), co-product (e.g. wheat straw) and wastes (e.g. liquid manure).

D.W. Aha and J. Lieber (Eds.): ICCBR 2017, LNAI 10339, pp. 61–75, 2017.
DOI: 10.1007/978-3-319-61030-6_5

processes, taking into account numerous biophysical variables but only few human decision-making processes linked to the land system management of farmers, whereas it is a major driving factor of miscanthus allocation process [2]. Indeed, representing and modelling human behaviour and decisional processes regarding land use change is difficult and constitutes a main research challenge [3].

The goal of our research is thus to model the miscanthus allocation according to farmers practices and decision-making process. Because the allocation of miscanthus is too recent to use national statistics, we decided to build our model based on a case study in Burgundy (East of France). Furthermore, because the allocation of miscanthus is too recent to be fully understood, we decided to rely on a case-based reasoning approach [4,5] as a pathway to use current practices for predicting land use change.

Case-based reasoning (CBR), introduced in [6], indeed allows to model application domains where general knowledge is incompletely formalized and where expertise mainly relies on experiences. Examples of such domains are medicine, chemistry, engineering, risks prevention or cooking, domains where case based reasoning systems have been successfully implemented.

Building a CBR system requires primarily to work on knowledge modeling. This step can rely on documents, that are more or less normalized, like a patient file, or a cooking recipe (examples are given in [7]). Data mining or text analysis techniques can also be used [8]. However, in several application domains, a long collaborative analysis work is required between computer scientist and domain experts. In our application, knowledge modeling has first required to survey farmers who had -or not- chosen to plant Micanthus in their farm fields. These surveys have resulted in two types of knowledge: on one hand, knowledge on farm field characteristics, as described by farmers, on the other hand, explanations given by farmers on the reasons why they choose or not to plant miscanthus in a specific field, and on conditions that could change their decision [2].

Relying on these pieces of knowledge, we have built a CBR proptotype, named SAMM, i.e. Spatial Allocation Modelling of Miscanthus, which aims at forecasting miscanthus allocation in farm fields. This prototype is composed with a case base, a rule base, a knowledge base (containing both similarity and adaptation knowledge) and a reasoning module, including various strategies.

The paper is structured as follows. Section 2 presents an overview of related work. The SAMM prototype is detailed in Sects. 3 and 4. Experiments are conducted in Sect. 5. The paper ends with a conclusion and some perspectives.

2 Related Work

Our work belongs to a research domain at the intersection of artificial intelligence and agronomy. More precisely, it relies on a previous work [9] which focused on the comparison of farm surveys and proved the interest of CBR as a modeling tool for landscape agronomy [10]. In this case model, the problem was a farm spatial organization (e.g., location of farm plots, roads, farm buildings) and the solution was a farm functional organization. The assumption was that similar spatial organizations corresponded to similar functional organizations.

More largely, there exists numerous systems linking CBR and environmental sciences, most of them based on numerical approaches, close to machine learning methods. In [11] for instance, spatial relations between neighboring areas are used to compute a similarity measure between them and forecast their land use (buildings, forest or crops). Older work already used CBR to analyse geographic data, e.g. for soil classification [12].

When CBR systems are based on numerical approaches, knowledge of stakeholders is slightly included, whereas our objective is to mainly rely on this knowledge. The system described in [9] is based on stakeholder knowledge that was already synthesized by researchers. The CARMA system is used for diagnosis and treatment of crop destructive animals [13], by adapting models built on expert knowledge, and was generalised over several american states. [14] handled stakeholder knowledge within a modeling approach close to CBR, but no system was implemented, due to the complexity of forms of knowledge to be modelled (management of sheep herds). There are actually very few systems that explicitly includes stakeholder experiences to be shared, as done in the system described by [15], which gathers community knowledge about rangeland management in New-Zealand.

Our work also deals with explanations and thus can be linked to *Explanation Based Reasoning* [16,17]. Indeed, in our work, farmer explanations about their choice are modelled by rules, that can be used to propose –what is done by the current prototype– and explain a solution –what could be done in the future. Such a reasoning approach has been initially developed for argumentation building based on american case law [6]. To our knowledge no application to the agricultural domain has been developed sofar.

A machine learning approach has been used to extract statistical rules explaining the spatial location of miscanthus, based on the characteristics of farm fields [18]. On the contrary, we have chosen to develop a CBR approach that rely on decision rules stated by farmers.

3 Case and Knowledge Bases

The originality in SAMM is to use both a case and a rule base. Rules are linked to cases and used for the adaptation step.

3.1 The Case Base

In our application, a case is defined as a specific experience of miscanthus allocation (or non allocation) in a farm field. The problem-solution pair is a farm field and its allocation potential for miscanthus. Each case is represented with a vector of qualitative values, divided in two parts:

1. the problem part gives the farm field characteristics, as described by the farmer; there are 32 possible attributes with 159 values influencing the allocation potential of miscanthus, classified into the six following categories:

   - agronomy, 14 attributes, e.g. last land-use, soil characteristics, agronomic potential, slope, soil water regime of the field;
   - geometry, 2 attributes, shape and size of the field;
   - access, 7 attributes, e.g. distance to farmstead, crossing zones;
   - neighbourhood, 6 attributes, e.g., tree, village, crop neighbourhood;
   - ownership, 2 attributes, land status and perennial use of the field;
   - environmental measures, 1 attribute, protected site.

   The subset of attribute-values describing the problem is denoted DP (for *problem descriptors*) and can be formalized as a set of pairs $(a, v) \in \mathcal{A} \times \mathcal{V}$, where $\mathcal{A}$ and $\mathcal{V}$ are respectively the attribute set and the value set.
2. the solution part describes the miscanthus allocation potential of the farm field with a unique variable and three values: the field cannot be allocated (value 0), can be allocated (value 1) or can be allocated under conditions (value 2); then the solution is formalized as a pair (miscanthus, i) where $i = 0, 1$ or 2.

To each case is also associated a label refering to the farmer who manages the farm field.

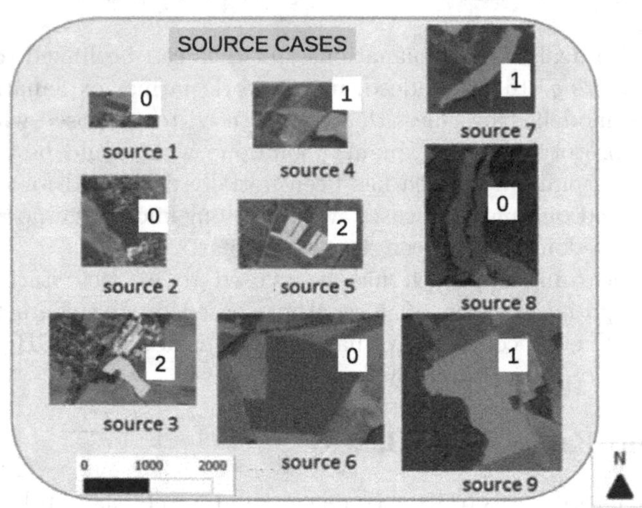

Fig. 1. Some farm fields that are modeled as source cases; the allocation potential is represented with a color (0 = red, 1 = green, 2 = pale green) (Color figure online)

The case base in SAMM includes 82 farm fields of which micanthus allocation potential has been stated by farmers during past interviews. For these problems, the solution is known. They are called *source cases* in the following (see examples on Fig. 1). The farmer associated to a source case is called *source farmer*. The case base organisation is flat, that is all cases belong to a same information level. Nevertheless, to help retrieval and adaptation steps, indexes are used. Each case is thus described with a subset of descriptors, denoted by DI (for *index descriptors*), collecting the elements that were explicitely involved in the farmer decision process. For instance, when a farmer says: *Une parcelle pas drainée: très mauvaise, enfin humide, humide, très humide (...) j'ai tout le temps vu en jachère (...) ce n'est pas le même prix: je dis "hop, je fais le miscanthus dedans"*[2], attributes about the soil water regime of the farm field and its last land-use are included in the set of index descriptors.

Table 1. Two source cases – DP: problem descriptors; DI: index descriptors (marked with x); attributes describe different aspects of the fields: agronomy, geometry, neighbourhood, and ownership

Attributes	Source 1		Source 2	
	DP	DI	DP	DI
Last land-use	Crop rotation		Crop rotation	
Agronomic potential	Good		Middle	
Soil depth			Shallow	
Stony soil			High	
Soil mechanics	Partial crusting			
Area morphology			Steep-sided	x
Field shape	Irregular	x	Irregular	
Tree neighbourhood	Wood	x	Wood	x
Water neighbourhood	Watercourse			
Land status	Property		Renting	
Allocation potential	1		0	

Two case examples are described in Table 1. Only attributes with values for each case are represented, i.e. attributes which were mentioned by the farmer to describe his/her field. The value of other attributes are not known for these farm fields. Note that there are few problem descriptors (DP) (6 or 7 among 32 possible) and still less index descriptors (DI), that are used to explain the farmer decision (here 2 or 3, marked by an x). This involves that most attributes are sparsely represented in the case base: 14 attributes (among 32) are used in less than 9 source cases, 10 are used in 10 to 19 cases and 8 are used in almost or more than half of the source cases. Finally cases have generally few attributes in common. All attributes are discrete (some are binary) since the case descriptions come from the analysis of farmer interviews, and thus are based on verbal nominal data. To elaborate the problems to be solved (called *target problems*), it is necessary to link these nominal attributes with numerical data, like geographical

[2] A not drained field: very bad, that is wet, wet, very wet (...) I always saw fallows there (...) it is not the same price: I said "hop, I make miscanthus inside".

data. Furthermore, it can be difficult to obtain some informations (e.g. the location of the farmstead) but others are frequently used in agronomical applications (last land-use, farm field geometry, soil characteristics and water regime, etc.).

3.2 The Rule Base

The SAMM reasoning system also relies on a set of rules that have been collected after transcription of farmer interviews [2]. They formalize the elements given by farmers when explaining their decision to plant (or not) miscanthus in a farm field. These rules are called decision rules. They are of two types: generic rules are independant from space (they are non-spatial) while spatial rules are linked to a farm field. Non-spatial rules are for example about economical context or environmental regulation; spatial rules are concerned with a farm field own characteristics. Rules are represented as pairs $< \bigwedge$ `conditions` (descriptors), `conclusion` (allocation potential) $>$. Conditions depends on various attributes and values as examplified in the two following rules: $<$ (`distance to farmstead` $=$ `close`) $\wedge$ (`access suitability` $=$ `low`) $\wedge$ (`protected site` $=$ `Natura 2000`), (1) $>$ and $<$ (`soil water regime` $=$ `flood area`), (0) $>$.

Each rule is labelled with an identifier refering to the farmer who expressed it, and thus linked to source cases labelled with the same identifier. It is worth noting that the rule set can be inconsistent since farmers can consider the same elements in different ways, i.e. a same field characteristic can have a positive or negative influence on their decision to implant miscanthus. For instance, the rule $<$ (`distance to farmstead` $=$ `far`), (2) $>$ holds for a farmer A_1, whereas another rule $<$ (`distance to farmstead` $=$ `far`), (0) $>$ holds for a farmer A_2. Furthermore, these rules are essential for the agronomists, since they express how and why farmer do choose or not to plant miscanthus.

The rule base covering all surveyed farmers includes 96 rules: 61 rules with conclusion 0 (the farm field cannot be allocated with miscanthus), 8 rules with conclusion 1 (the farm field can be allocated with miscanthus) and 27 rules with conclusion 2 (the farm field can be allocated with respect to further conditions). Regarding the size of rules, 65% of them have only one condition, 25% have 2 conditions and 10% have 3 or more (until 7) conditions.

The low number of rules with conclusion 1 highlights the novelty and scarcity of miscanthus plots. We therefore focus below mainly on the situations where the farm field can be allocated under conditions or cannot be allocated.

3.3 Similarity and Adaptation Knowledge

Similarity knowledge has been elaborated based on the content analysis of interviews made with farmers. Various levels and types of similarity are considered (in the following s refers to the *source* case and t refers to the *target* case):

– a global numeric level: the number of common descriptors (whatever value) of both *source* and *target* problems, denoted n; the sets DP_s of *source* and DP_t of *target* can be considered, then:

$$n_{DP}(s,t) = |\{a \in \mathcal{A} | \exists (a, v_1) \in DP_s \wedge \exists (a, v_2) \in DP_t\}|$$

or the sets DI_s and DP_t (the set DI is unknown for *target*) are considered, then:

$$n_{DI}(s,t) = |\{a \in \mathcal{A} | \exists (a, v_1) \in DI_s \wedge \exists (a, v_2) \in DP_t\}|$$

- a global semantic level: the global distance between *source* and *target* is the average of local measures (denoted d_l) on common attributes; it can be computed for $n = n_{DP}$ or $n = n_{DI}$:

$$d_g(s,t) = \Sigma_{i=1}^{n} d_l((a_i, v_{is}), (a_i, v_{it})))/n$$

The local measures on attributes are given in a distance matrix (see an excerpt in Table 2). Distances rely on the influences of attribute values towards the miscanthus allocation potential of farm fields (same influence or opposed influence), based on the analysis of farmer interviews. For instance, considering the last land-use, the value "forage crop" is close to the value "crop rotation", but distant from the value "fallows"; finally, distance values have been chosen heuristically as follows:

- $d_l((a, v_s), (a, v_t)) = 5$ if both values have similar influences, positive or negative with respect to miscanthus allocation;
- $d_l = 20$ if both values have different influences, one is positive and the other negative;
- $d_l = 10$ if at least one of the values has a neutral influence.

Table 2. Semantic distance between values of attribute "last land-use"

	Forage crop	Production	Fallows	Grassland	Crop rotation
Forage crop	0	5	20	5	5
Production	5	0	20	5	5
Fallows	20	20	0	20	20
Grassland	5	5	20	0	5
Crop rotation	5	5	20	5	0

In CBR systems, the solution adaptation is achieved when at least one descriptor of the source problem is different from the one of the target problem. Transformational adaptation [19] starts from the solution of the selected source case, and modifies it with respect to the differences between the source and the target problems.

To build such a solution, we here rely on the farmer decision rules. The underlying idea is to copy the source case solution, or to use the rules from the farmer (or a set of farmers) associated to the source case to build the solution.

Adaptation knowledge allows to choose the rule to apply among the relevant ones, according to an adaptation context: the user can favorise the rules with conclusion 0 (when the context is not favourable for miscanthus, e.g. because its price is low with regards to traditional crops) or those with conclusion 2 or 1, if the economical context is favourable for miscanthus.

4 SAMM Reasoning Module

In SAMM proptoype, reasoning is based on the two main steps of CBR, retrieval of source cases similar to the target problem, and adaptation of a source case solution to the target problem.

4.1 Retrieving Source Cases

The retrieval step consists in identifying one or several source cases which can help resolving the target problem. This step is made of three sub-steps: matching of the target problem to source problems, similarity assessment between problems, and selection of a source case. In SAMM prototype, matching the target problem to source cases is a simple vector matching. Similarity assessment between each source problem and the target problem relies on the measures above-defined. Finally various measure combinations can be used for selecting a source case (see Fig. 2).

Four retrieval algorithms have been defined. They are specified by the descriptors used for matching problems and the way similarity measures are combined:

- There are two sets of descriptors: the first set, DP, contains the descriptors of the source problem; the second, DI, contains descriptors that index the source case (see Sect. 3.3);
- There are two ways of combining measures: *(i)* the first one first minimizes global distance and then maximizes the number of common descriptors while *(ii)* the second one first maximizes the number of common descriptors and then minimizes the global distance between source cases and target problem.

In the current version of SAMM, the retrieval step selects the source case that minimizes the distance *(i)* or maximizes the number of descriptors *(ii)*. If several cases are returned, then the second measure is used.

In the following, we denote the four retrieval algorithms as RM1 (DP and *(i)*), RM2 (DP and *(ii)*), RM3 (DI and *(i)*), RM4 (DI and *(ii)*).

4.2 Adapting a Source Case Solution

The adaptation step in SAMM is achieved by copying the source solution to the target solution or by transforming the source solution based on the decision rules. Indeed the retrieval step results in a source case which solution (0, 1 or 2) can be directly copied to the target case. If several source cases are retrieved,

```
Matching source case 10385231 and target case 10499801
distance pb between fallows and fallows : 0
distance pb between partially_hydromorphic and drying : 20
distance pb between sharp_shape_AND_convex and sharp_shape : 10
distance pb between foncS_1 and foncS_0 : 20
distance index between fallows and fallows : 0
distance index between sharp_shape_AND_convex and sharp_shape : 10
number of common attributes (DP) : 4
global distance on problem : 12.5
number of common attributes (DI) : 2
global distance on index : 5.0
```

Fig. 2. A example of similarity computing between a source case and a target problem: similarity is computed first on DP then on DI

the solution is chosen by a majority vote. If none majority can be established a combined solution is proposed, among the following values: 0-1, 1-2, 0-2, 0-1-2.

Various rule sets can be considered for transformation: the rule set of the farmer associated to the retrieved source case, the rule set of a group of farmers (e.g. farmers from the same small agricultural region as the source farmer), the set of all rules. Rules can be applied as soon as each attribute-value (a_i, v_{ir}) in their conditions is similar to one of the target problem descriptors, i.e., for each attribute, the local distance $d_l((a_i, v_{ir}), (a_i, v_{it}))$ is smaller or equal to a given threshold denoted d_{rule}.

When several rules match a target problem it is necessary to select the right rule to be applied. This task is done according to the adaptation context, as said before. Three adaptation algorithms have been therefore implemented. A pessimistic algorithm (ADAPT0) first selects rules with conclusion 0 (see Algorithm 1); an optimistic algorithm (ADAPT12) first selects rules with conclusion 2 if it exists rules with conclusion 0, then it selects rules with conclusion 0, and at least rules with conclusion 1 (see Algorithm 2); a weighted algorithm (ADAPT3) selects rules with the greatest number of conditions (see Algorithm 3). If no rule can be matched with a target problem, then transformation cannot be done and the solution of the retrieved source case is copied into the target solution. If there are several retrieved source cases, the adaptation algorithm is runned for each case and the final result is chosen by a majority vote.

For example, suppose that the retrieved rule set contains the following rules:

R1: < (`soil water regime` = `flood area`), (0) >
R2: < (`field size` = `middle`), (2) >
R3: < (`drainage` = `none`) $\wedge$ (`soil water regime` = `resurgences and sources`), (0) >
R4: < (`last land-use` = `fallows`) $\wedge$ (`soil water regime` = `wet`), (1) >

and is used to solve a target problem t with the following descriptors:

{(`size: middle`); (`drainage: none`); (`soil water regime: wet`); (`excess of water: flood area`); (`last land-use: fallows`)}

Algorithm 1. ADAPT0

```
input  : target problem t, set of matching rules S_x with conclusion x = 0,1,2
output: problem solution, sol_t
sol_t ← −1
if S_0 ≠ ∅ then
  | sol_t ← 0
else if S_1 ≠ ∅ then
  | sol_t ← 1
else if S_2 ≠ ∅ then
  | sol_t ← 2
end
```

Algorithm 2. ADAPT12

```
input  : target problem c, set of matching rules S_x with conclusion x = 0,1,2
output: problem solution, sol_t
sol_t ← −1
if S_0 ≠ ∅ then
  | if S_2 ≠ ∅ then
  |   | sol_t ← 2
  | if S_2 = ∅ then
  |   | sol_t ← 0
else if S_1 ≠ ∅ then
  | sol_t ← 1
else if S_2 ≠ ∅ then
  | sol_t ← 2
end
```

The set of matching rules for t, with respect to the distance threshold $d_{rule} = 0$, is {**R1**, **R2**, **R4**}. Algorithm ADAPT0 returns solution 0 (rule **R1** is applied). Algorithm ADAPT12 returns solution 2 (rule **R2**), and the weighted algorithm ADAPT3 selects rule **R4**, with two conditions, and returns solution 1. With the distance threshold $d_{rule} = 5$, the rule **R3** with conclusion 0 can be used, since the distance between values `flood area` and `resurgences and sources` is 5 (attribute `excess of water`). Rule **R3** has two conditions, as rule **R4**; nevertheless, algorithm ADAPT3 first selects rule **R4** which conclusion is 1.

5 Assessing SAMM Performance

SAMM prototype has been implemented in java within the Eclipse development environment[3]. Currently, the user can load various rule bases or case bases, and thus can use the model on various territories. The user also can test the algorithms and parameters for the retrieval and adaptation steps, thanks to a configuration panel. He/she can evaluate the proposed solutions *via* the system

[3] www.eclipse.org.

Algorithm 3. ADAPT3

input : target problem c, set of matching rules S_x with conclusion $x = 0, 1, 2$
output: problem solution, sol_t
$sol_t \leftarrow -1$, $n_0 = n_1 = n_2 = 0$
for $x = 0$ *to* $2 \wedge S_x \neq \emptyset$ **do**
  $n_x \leftarrow$ average number of conditions for rules in S_x
end
if $\exists i, n_i > 0 \wedge n_i > n_j\ \forall j \neq i$ **then**
  $sol_t \leftarrow i$
else if $n_2 > 0 \wedge n_2 \geq n_j \forall j \neq 2$ **then**
  $sol_t \leftarrow 2$
else if $n_1 = n_0 \neq 0$ **then**
  $sol_t \leftarrow 1$
end

interface or with output files. Figure 3 represents the various algorithms and bases in SAMM with respect to the CBR general cycle [4].

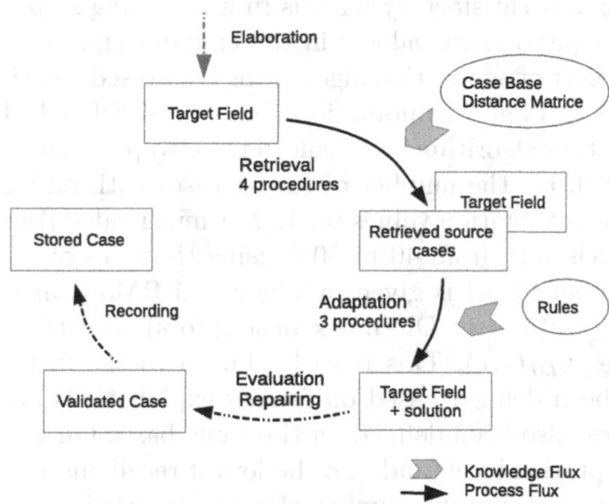

Fig. 3. The CBR cycle in SAMM: dotted lines correspond to currently manual steps

Tests have been achieved based on a subset of 72 cases (18 with solution 0 – no miscanthus allocation, 28 with solution 1 – miscanthus allocation, and 26 with solution 2 – miscanthus allocation under conditions). Tests were carried out by extracting from the case base an individual, which is the target problem, while the remaining cases are the source cases. This procedure was repeated for each case in the case base. Various procedures have been tested, based on the combination of various experimental parameters:

- choice of the retrieval algorithm, four modalities, one for each combination of similarity measures, RM1, RM2, RM3, RM4 (see above);

- choice of rule set, in this experiment, we use rules of the farmer associated to the retrieved source case (denoted RSF for *rules of source farmer*)
- choice of the distance threshold between rule conditions and descriptors of the target problem for selecting the rule to be applied ($d_{rule} = 0$ in this experiment);
- choice of the adaptation algorithm, three modalities, one for each algorithm ADAPT0, ADAPT12, ADAPT3 (see above).

Firstly, results obtained at the end of the retrieval step are examined: at this step, the target solution is a copy of the solution (chosen by a majority vote) of the retrieved source cases. The proposed solution is compared to the original solution of the target case. Precision and recall scores are computed for each combination of parameters and each solution (0, 1, or 2). Precision scores are higher than recall ones; this is (partially) due to the fact that some problems got a multi-valued solution when no majority could be etablished (for 1.4 to 9.7% of cases, depending on the algorithm, see Table 3). The RM4 algorithm is better both for recall and precision. RM1, RM2, and RM3 obtain similar results, but RM3 has the higher level of cases with a multi-valued solution. RM2 has the lower recall and the lower level of cases with multi-valued solutions. Precision and recall value are globally satisfactory for this first reasoning step, and considering the variety of pairs (attributes, values) in the case descriptions.

Secondly, the effect of adaptation algorithms is assessed based on the parameters described above. Twelve combinations have been defined (4 retrieval algorithms × 3 adaptation algorithms × 1 rule set). Two parameters are measured, the adaptation level, i.e. the number of target cases with an adapted solution, and the distribution of solution values (0, 1, 2 or multi-valued) in adapted cases.

Adaptation levels vary from 40 to 50% using the rules of the source farmer (RSF). The best level (50%) is given by algorithm RM3, which first optimises global distance $d_g(s, c)$ on set DI (index descriptors), and then the number of common attributes $n_{DI}(s, c)$. This is to be linked to the fact that case index parameters have been defined based on farmer explanations about farm fields; rule conditions have also been defined on the same basis. On the contrary, RM2 has the lowest adaptation level (and also the lowest recall and precision levels): it first optimises $n_{DP}(s, c)$, i.e. the number of common attributes in DP (problem descriptors).

Table 3. Results for the four retrieval algorithms

	RM1	RM2	RM3	RM4
Recall	55.65	52.48	54.32	56.72
Precision	59.70	59.11	60.37	70.39
% multi-valued solution	4.17	1.39	9.72	5.56

Table 4 shows the distributions of computed solutions for the 12 algorithm-parameter combinations. Results show that pessimistic algorithm ADAPT0 allows to favour solution 0 especially for RM1 and RM3 (the level of solution 0 is 55–56%). Optimistic algorithm ADAPT12 favours solution 2 for RM1, RM2, and RM3 (the level of solution 2 is between 47 and 52%; it is lower for RM4). Weighted algorithm ADAPT3, which also favours solution 2, gives results similar to these of ADAPT12, but the level of solution 1 can be higher (from 19 to 26% for RM4). Solution 1 is seldom chosen for all algorithm combinations (between 8 and 26%): this is partially due to the algorithm structures (especially ADAPT0 and ADAPT12) but also to the low number of rules with conclusion 1 in the rule base (see Sect. 3.2). Besides, multi-valued solutions are not many: this shows that even when several source cases are retrieved, and then several rules can be used, the adaptation algorithms allow to compute a dominant solution.

Table 4. Adaptation: distribution (%) of computed solutions by adaptation algorithms, for each retrieval algorithm, using rules of the source farmer

	ADAPT0	ADAPT12	ADAPT 3	ADAPT0	ADAPT12	ADAPT 3
	RM1			RM2		
Solution 0	0.55	0.39	0.39	0.41	0.34	0.31
Solution 1	0.09	0.09	0.09	0.14	0.14	0.17
Solution 2	0.36	0.52	0.52	0.41	0.48	0.48
Multi. sol.	0	0	0	0.03	0.03	0.03
	RM3			RM4		
Solution 0	0.56	0.42	0.42	0.41	0.42	0.35
Solution 1	0.08	0.08	0.08	0.19	0.19	0.26
Solution 2	0.33	0.47	0.47	0.34	0.35	0.35
Multi. sol.	0.03	0.03	0.03	0.06	0.03	0.03

These results highlight that various algorithm/parameter combinations can be used, according to the user objectives: if he or she wants to forecast the allocation of miscanthus in a negative context, he or she will use the RM1-ADAPT0-RSF or RM3-ADAPT0-RSF combinations; for a positive context (e.g. good economic conditions), he or she will use the ADAPT12-RSF combination with either RM1, or RM3. Finally, to be close to current farmer choices, it will be better to use the RM4-ADAPT3-RSF combination.

This experiment also shows the central role of the user, who has to choose algorithms and parameters, and to examine results step by step: retrieved cases, available rules, proposed solutions. Rules in particular can be analysed to highlight the field characteristics that are important wrt the farmer decision.

6 Conclusions and Perspectives

We have presented a CBR system, SAMM, which aim is to forecast the allocation in french farms of a new biomass energy perenial crop, the miscanthus.

The system includes a case and a rule base, that have been modeled on the basis of farmer interviews. Similarity measures have been defined and various algorithms for the retrieval and adaptation steps have been implemented. The system has been tested on a subset of 72 cases, each case representing a farm field and its allocation potential for miscanthus. Results have shown the various behaviours of the system, and thus have highlighted the predominant role of the user, who chooses how to combine algorithms and parameters according to his/her objectives.

According to agronomist point of vue, SAMM modeling and implementation has allowed two advances: *(i)* the formalisation of farmer decision rules, based on complex farm surveys; *(ii)* the construction of various scenarios for forecasting miscanthus allocation. However, there is a difficulty: since cases are described with few attributes, case source retrieval relies on a weak matching (target and source problems have only 2 or 3 common descriptors). This underlines the importance of the rules used in the adaptation step, which should be carefully chosen. Furthermore, farmer reasoning includes various spatial scales (the farm fields, the field clusters, the farm). Description of rules and cases could be completed to include these various information scales and thus obtain more reliable results. Besides, the problem elaboration step should be implemented, based on existing work [18], to make the prototype usable on a larger territory.

Rules can also be used to explain the behaviour of the system and the proposed solution of a target problem. Actually, our work deals with explanation based reasoning. Explanation is a most interesting notion for exploring decisional knowledge, especially when the context is evolving, as it is when a new crop is introduced in farms, with new characteristics and specific issues. Furthermore, the complexity of farmer reasoning processes pleads for a CBR system with capabilities for interacting with the user rather than only capabilities for forecasting. The current prototype is useful to build new loops of survey, modelling and implementation steps, which will lead to further systems with actual capabilities in interacting with the user, and in forecasting or decision-making.

To help the user, the system could also include text elements from the farmer interviews, or drawings they made of their fields, that would complete the case descriptions and rules. Such improvements are proposed in knowledge approaches dealing with knowledge provenance [20]. The underlying idea is both to help the user and to make the system maintenance easier. Finally, an interesting prospect for our work is to directly associate resolution and training on the case base, allowing the concerned stakeholders to appropriate the modeling process, and to share knowledge between them.

References

1. Hellmann, F., Verburg, P.: Spatially explicit modelling of biofuels crops in Europe. Biomass Bioenergy **35**, 2411–2424 (2008)
2. Martin, L., Wohlfahrt, J., Le Ber, F., Benoît, M.: Perennial biomass crop cultivation and its territorial patterns a case-study of miscanthus in Côte-d'Or (Burgundy, France). Espace Geogr. **2**(41), 133–147 (2012)

3. Rounsevell, M.D., Arneth, A.: Representing human behaviour and decisional processes in land system models as an integral component of the earth system. Global Environ. Change **21**(3), 840–843 (2011)
4. Aamodt, A., Plaza, E.: Case-based reasoning: foundational issues, methodological variations, and system approachs. AICOM **7**(1), 39–59 (1994)
5. Leake, D.B. (ed.): Case-Based Reasoning. Experiences, Lessons, & Future Directions. AAAI Press/The MIT Press, Cambridge (1996)
6. Riesbeck, C.K., Schank, R.C.: Inside Case-Based Reasoning. Lawrence Erlbaum Associates, Hillsdale (1989)
7. Brüninghaus, S., Ashley, K.D.: The role of information extraction for textual CBR. In: Aha, D.W., Watson, I. (eds.) ICCBR 2001. LNCS, vol. 2080, pp. 74–89. Springer, Heidelberg (2001). doi:10.1007/3-540-44593-5_6
8. Adeyanju, I., Wiratunga, N., Lothian, R., Sripada, S., Lamontagne, L.: Case Retrieval Reuse Net (CR2N): an architecture for reuse of textual solutions. In: McGinty, L., Wilson, D.C. (eds.) ICCBR 2009. LNCS, vol. 5650, pp. 14–28. Springer, Heidelbwerg (2009). doi:10.1007/978-3-642-02998-1_3
9. Le Ber, F., Napoli, A., Metzger, J.L., Lardon, S.: Modeling and comparing farm maps using graphs and case-based reasoning. J. Univers. Comput. Sci. **9**(9), 1073–1095 (2003)
10. Benoît, M., Rizzo, D., Marraccini, E., Moonen, A.C., Galli, M., Lardon, S., Rapey, H., Thenail, C., Bonari, E.: Landscape agronomy: a new field for addressing agricultural landscape dynamics. Landscape Ecol. **27**(10), 1385–1394 (2012)
11. Du, Y., Liang, F., Sun, Y.: Integrating spatial relations into case-based reasoning to solve geographic problems. Knowl.-Based Syst. **33**, 111–123 (2012)
12. Holt, A., Benwell, G.: Case-based reasoning and spatial analysis. J. Urban Reg. Inf. Syst. Assoc. **8**, 27–36 (1996)
13. Branting, L.K., Hastings, J., Lockwood, J.: CARMA: a case-based range management advisor. In: Proceedings of the Thirteenth Innovative Applications of Artificial Intelligence Conference (IAAI-2001), Seattle, Washington (2001)
14. Girard, N., Bellon, S., Hubert, B., Lardon, S., Moulin, C.H., Osty, P.L.: Categorising combination of farmers' land use practices: an approach based on examples of sheep farms in the South of France. Agronomie **21**, 435–459 (2001)
15. Bosch, O.J.H., Gibson, R.S., Kellner, K., Allen, W.J.: Using case-based reasoning methodology to maximise the use of knowledge to solve specific rangeland problems. J. Arid Environ. **35**, 549–557 (1997)
16. Schank, R.C., Kass, A., Riesbeck, C.K. (eds.): Inside Case-Based Explanation. LEA, Hillsdale (1994)
17. Leake, D., McSherry, D.: Introduction to the special issue on explanation in case-based reasoning. Artif. Intell. Rev. **24**(2), 103–108 (2005)
18. Rizzo, D., Martin, L., Wohlfahrt, J.: Miscanthus spatial location as seen by farmers: a machine learning approach to model real criteria. Biomass Bioenergy **66**, 348–363 (2014)
19. Carbonell, J.G.: Learning by analogy: formulating and generalizing plans from past experience. In: Michalski, R.S., Carbonell, J.G., Mitchell, T.M. (eds.) Machine Learning. Symbolic Computation, vol. 1, pp. 137–161. Springer, Heidelberg (1983). doi:10.1007/978-3-662-12405-5_5
20. da Silva, P., McGuinnes, D., McCool, R.: Knowledge provenance infrastructure. IEEE Data Eng. Bull. **25**(2), 179–227 (2003)

A SPARQL Query Transformation Rule Language — Application to Retrieval and Adaptation in Case-Based Reasoning

Olivier Bruneau[1], Emmanuelle Gaillard[2], Nicolas Lasolle[1], Jean Lieber[2](✉), Emmanuel Nauer[2], and Justine Reynaud[2]

[1] University of Lorraine, LHSP-AHP, 91 avenue de la Libération, BP 454, 54001 Nancy Cedex, France
olivier.bruneau@univ-lorraine.fr, nicolas.lasolle@telecomnancy.eu
[2] UL, CNRS, Inria, Loria, 54000 Nancy, France
{emmanuelle.gaillard,jean.lieber,emmanuel.nauer, justine.reynaud}@loria.fr

Abstract. This paper presents SQTRL, a language for transformation rules for SPARQL queries, a tool associated with it, and how it can be applied to retrieval and adaptation in case-based reasoning (CBR). Three applications of SQTRL are presented in the domains of cooking and digital humanities. For a CBR system using RDFS for representing cases and domain knowledge, and SPARQL for its query language, case retrieval with SQTRL consists in a minimal modification of the query so that it matches at least a source case. Adaptation based on the modification of an RDFS base can also be handled with the help of this tool. SQTRL and its tool can therefore be used for several goals related to CBR systems based on the semantic web standards RDFS and SPARQL.

Keywords: RDFS · SPARQL · Query transformation · Retrieval · Adaptation · Application

1 Introduction

This paper presents a language and a tool that have proven to be useful for addressing three application problems related to the issues of retrieval and adaptation in case-based reasoning (CBR [19]), when the underlying representation language is the semantic web standard RDFS [3].

CBR aims at solving problems by reusing previously solved problems. It is often considered to be a methodology [21]. Indeed, its principles are independent from a knowledge representation language. The upside of this is that it covers a huge family of problem-solving issues, in many application domains. The downside of it is that the application of CBR to a particular domain, with a given representation language, often requires to reimplement most of the CBR steps. However, many studies have been carried out to fill the gap between general principles and particular applications. Some general CBR shells, like

D.W. Aha and J. Lieber (Eds.): ICCBR 2017, LNAI 10339, pp. 76–91, 2017.
DOI: 10.1007/978-3-319-61030-6_6

JColibri [18] have been implemented and distributed. Furthermore, some tools have been implemented for particular types of problems (e.g., case-based planning [9]) and/or particular types of formalisms (e.g., workflows [15]).

This paper presents a formalism and a tool of general use for the development of CBR systems based on the representation language RDFS, thus it contributes to filling the theory-application gap in CBR. RDFS can be seen as a language combining attribute-value pairs and the use of hierarchies, two features commonly used in CBR systems [13]. The standard query language of RDFS is SPARQL. The language presented in this work is named SQTRL and is a language of rules to transform queries written in SPARQL.

This paper is organized as follows. Section 2 presents preliminaries on RDFS and SPARQL. Then, three application problems that have motivated this work are presented in Sect. 3. Section 4 presents SQTRL and a tool to manage rules of this language. A discussion pointing out some related work is given in Sect. 5. Section 6 concludes and highlights some future work.

2 Preliminaries: RDFS and SPARQL

RDFS (for RDF Schema [3]) is a knowledge representation formalism based on RDF, a resource description framework defined on several syntaxes. In RDF, a *resource* is either an *identified resource* or a *variable* (also called blank node or anonymous resource) that is represented in this paper by an identifier starting with the '?' character (e.g. `?x`) and can be interpreted as an existentially quantified variable. Some resources are *properties*; they are intended to represent binary relations. A literal is a value of a simple datatype (e.g., an integer, a string, etc.). An RDF base is a set of *triples* $\langle s\ p\ o \rangle$, where s—the subject of the triple—is a resource, p—the predicate—is a property, and o—the object—is either a resource or a literal. For example $\mathcal{B} = \{\langle$`tarteTatin ing ?x`$\rangle, \quad \langle$`?x type Apple`$\rangle\}$ means that the tarte Tatin has an `ingredient` of type apple.

Some RDF resources constitute the so-called RDFS vocabulary. For the sake of simplicity, only three such resources are considered in this paper: `rdf:type` (abbreviated in `type`), `rdfs:subClassOf` (abbreviated in `subc`), and `rdfs:subPropertyOf` (abbreviated in `subp`). These resources are properties and have the following meaning

$$\langle a\ \texttt{type}\ C \rangle \text{ means that } a^{\mathcal{I}} \in C^{\mathcal{I}} \qquad \langle C\ \texttt{subc}\ D \rangle \text{ means that } C^{\mathcal{I}} \subseteq D^{\mathcal{I}}$$

$$\langle p\ \texttt{subp}\ q \rangle \text{ means that } p^{\mathcal{I}} \subseteq q^{\mathcal{I}} \text{ i.e., if } (x,y) \in p^{\mathcal{I}} \text{ then } (x,y) \in q^{\mathcal{I}}$$

where $a^{\mathcal{I}}$ is an object represented by a, $C^{\mathcal{I}}$ and $D^{\mathcal{I}}$ are sets represented by the classes C and D, and $p^{\mathcal{I}}$ and $q^{\mathcal{I}}$ are relations represented by properties p and q. For example, $\langle$`tarteTatin type DessertRecipe`$\rangle$ means that the tarte Tatin is a dessert, $\langle$`Apple subc Fruit`$\rangle$ means that an apple is a fruit, and $\langle$`mainIng subp ing`$\rangle$ means that the main ingredient of a recipe is an ingredient of this recipe. The inference rules

$$\frac{\langle a \text{ type } C\rangle \quad \langle C \text{ subc } D\rangle}{\langle a \text{ type } D\rangle}, \frac{\langle C \text{ subc } D\rangle \quad \langle D \text{ subc } E\rangle}{\langle C \text{ subc } E\rangle},$$

$$\frac{\langle p \text{ subp } q\rangle \quad \langle q \text{ subp } r\rangle}{\langle p \text{ subp } r\rangle} \text{ and } \frac{\langle a\ p\ b\rangle \quad \langle p \text{ subp } q\rangle}{\langle a\ q\ b\rangle}$$

are used to define the inference relation $\vdash$. In this paper, every RDFS base $\mathcal{B}$ is considered up to entailment, meaning that if $\mathcal{B} \vdash \tau$, then the triple τ is considered as an element of $\mathcal{B}$.

```
SELECT ?r                                              select
WHERE {?r  type  DessertRecipe   .       the recipes of desserts
       ?r  ing  ?x  .                       having an ingredient
       ?x  type  Fruit   .                           of type fruit
       ?r  prepTimeInMinutes  ?t   .   and a preparation time
       FILTER (?t  <=  90)}                   of at most 90 minutes
```

Fig. 1. Example of SPARQL query.

To access an RDFS base, a SPARQL query is used. Figure 1 shows such a query. More generally, a SPARQL query Q is constituted by a "SELECT line" stating the variable(s) to be unified and a SPARQL body following the keyword WHERE, denoted by body(Q) in this paper. The SPARQL body is a sequence of triples and FILTER assertions separated by dots which mean "and". Given an RDFS base $\mathcal{B}$ and a SPARQL query Q, the execution of Q on $\mathcal{B}$ gives a set of bindings of the variables of Q's SELECT line corresponding to matchings of the Q's body with $\mathcal{B}$. This set is denoted by $\texttt{exec}_\vdash(\texttt{Q}, \mathcal{B})$. For example, if Q is the query of Fig. 1, if $\mathcal{B}$ contains the recipe description of the tarte Tatin (containing apples and with a preparation time of 65 min) and a cooking ontology stating that $\mathcal{B} \vdash \langle \texttt{Apple subc Fruit}\rangle$, then $\texttt{exec}_\vdash(\texttt{Q}, \mathcal{B})$ contains the binding pair (?r, tarteTatin).[1]

Remark About the Notation. An RDFS triple, such as ⟨?x type Fruit⟩ appears with a slightly different syntax as an assertion of a SPARQL query body. However, in this paper, we will use the two notations interchangeably. Of course, this involves the use of translation procedures in the code.

3 Three Application Problems

3.1 Recipe Retrieval in the CBR System TAAABLE

TAAABLE [7] is a CBR system that was originally developed as a contestant of the CCC (the computer cooking contest, organized within most ICCBR conferences since 2008). A contestant of the CCC aims at answering cooking queries such as

[1] This presentation of RDFS and SPARQL is simplified to fit the needs of this paper.

$$\mathcal{Q} = \text{"I want a dessert recipe with pears and butter but without cinnamon."} \tag{1}$$

For this purpose, it reuses a recipe base provided by the contest. To this end, TAAABLE first searches for a recipe from the base that exactly matches $\mathcal{Q}$. If no such recipe exists, TAAABLE minimally modifies $\mathcal{Q}$ into $\mathcal{Q}'$ so that there exists at least one recipe exactly matching $\mathcal{Q}'$. For example, if $\mathcal{Q}' = \sigma(\mathcal{Q})$ with $\sigma = \texttt{pear} \rightsquigarrow \texttt{fruit} \circ \texttt{butter} \rightsquigarrow \texttt{margarine}$ (i.e., σ is the substitution of `pear` by `fruit` and of `butter` by `margarine`), an apple crumble with margarine and without cinnamon can be selected in the recipe base (because the piece of knowledge "apples are fruit" is in the TAAABLE domain ontology), and then adapted to answer $\mathcal{Q}$ by substituting apples with pears and margarine with butter. Other adaptation issues (of ingredient quantity and of the preparation) are also studied in TAAABLE but not considered here.

A semantic wiki called WIKITAAABLE [8] (wikitaaable.loria.fr) has been developed in order to manage the TAAABLE knowledge base which contains its domain ontology (hierarchies of ingredients, etc.) and the recipe base. WIKITAAABLE is implemented thanks to the semantic wiki engine SMW (www.semantic-mediawiki.org) that comes with an RDFS export, hence the idea of using RDFS techniques and tools for TAAABLE. This idea has led to TUUURBINE, a CBR retrieval engine using SPARQL [11]. More precisely, a TUUURBINE query combines SPARQL queries. For example, the query $\mathcal{Q}$ of Eq. (1) is based on the following SPARQL queries:

$\mathtt{Q}_+ =$

```
SELECT ?r
WHERE { ?r  type  DessertRecipe   .
        ?r  ing   ?x  .  ?x  type  Pear    .
        ?r  ing   ?y  .  ?y  type  Butter }
```

$\mathtt{Q}_- =$

```
SELECT ?r
WHERE { ?r  ing  ?x  .  ?x  type  Cinnamon }
```

The recipes matching $\mathcal{Q}$ are in the set $\mathtt{exec}_{\vdash}(\mathtt{Q}_+, \mathcal{B}) \setminus \mathtt{exec}_{\vdash}(\mathtt{Q}_-, \mathcal{B})$ where $\mathcal{B}$ is the base exported from WIKITAAABLE: this gives the recipes matching $\mathtt{Q}_+$ and not $\mathtt{Q}_-$. $\mathcal{B}$ contains the recipe and the ontology. This latter contains the fact that apples and pears are fruits: $\mathcal{B} \vdash \{\langle\texttt{Apple subc Fruit}\rangle, \langle\texttt{Pear subc Fruit}\rangle\}$. The modification of $\mathcal{Q}$ into $\mathcal{Q}'$ consists in a modification of $\mathtt{Q}_+$ into $\mathtt{Q}'_+$ (keeping $\mathtt{Q}_-$ unchanged, i.e., $\mathtt{Q}'_- = \mathtt{Q}_-$):

$\mathtt{Q}'_+ =$

```
SELECT ?r
WHERE { ?r  type  DessertRecipe   .
        ?r  ing   ?x  .  ?x  type  Fruit  .
        ?r  ing   ?y  .  ?y  type  Margarine }
```

Therefore, a first module for transforming SPARQL queries has been developed for TUUURBINE. This module, though it has a general purpose, remained limited and proved to be insufficient for the application described hereafter.

3.2 Approximate Search in the Corpus of Henri Poincaré Letters

The famous mathematician Henri Poincaré has had a long correspondence with many people, including scientists of his time. The letters he has written and received are gathered in the "HP papers corpus" (`henripoincarepapers.univ-lorraine.fr`), which has been scanned and indexed. Currently, this index is being migrated into RDFS annotations. For example, the following triples concern the letter number 12 that has been sent by H. Poincaré to David Hilbert and that is about hyperbolic geometry:

$$\langle \texttt{letter12 isSentBy hPoincaré}\rangle\langle \texttt{letter12 isSentTo dHilbert}\rangle$$

$$\langle \texttt{letter12 hasForTopic ?t}\rangle\langle \texttt{?t type HyperbolicGeometry}\rangle$$

The RDFS base $\mathcal{B}$ of the H. Poincaré corpus contains such annotations about letters as well as information about some persons and organizations (e.g., $\mathcal{B} \vdash \langle \texttt{dHilbert type Mathematician}\rangle$), and an ontology related to the domain (e.g., $\mathcal{B} \vdash \langle \texttt{Mathematician subc Scientist}\rangle$).

Therefore, the letters of this corpus sent to a geometer before 1895 are in $\texttt{exec}_\vdash(\texttt{Q}, \mathcal{B})$

with Q=

```
SELECT ?ℓ
WHERE { ?ℓ  isSentTo  ?x  .  ?x  type  Geometer  .
        ?ℓ  dateOfExpedition  ?d  .
        FILTER (?d< '01/01/1895') }
```

.

Now, it happens that an exact search is not always sufficient. Such situations are described in [4]. Two examples are given below.

First, consider Q =

```
SELECT ?ℓ
WHERE { ?ℓ  isSentBy  hPoincaré  .
        ?ℓ  isSentTo  gMittagLeffler }
```

. If $\texttt{exec}_\vdash(\texttt{Q}, \mathcal{B}) = \emptyset$, that does not mean that no letters has been written by H. Poincaré to Gösta Mittag-Leffler. It could mean that such a letter has been written but was lost.

Now, consider the query Q′ =

```
SELECT ?ℓ
WHERE { ?ℓ  isSentTo  hPoincaré  .
        ?ℓ  isSentBy  gMittagLeffler }
```

.

Q′ is obtained by exchanging sender and recipient in Q. The query transformation $\texttt{Q} \mapsto \texttt{Q}'$ is relevant for historians since, when searching a letter from H. Poincaré to G. Mittag-Leffler, accessing the letters from the latter to the former can be useful, because such a letter can be a response to a letter that has disappeared.

Now, consider the query for letters sent by D. Hilbert at "the end of the XIXth century". This period of time is imprecisely specified, so an interval of time is chosen to model it, e.g., [1890, 1900], hence the query

Q=

```
SELECT ?ℓ
WHERE { ?ℓ  isSentBy  dHilbert  .  ?ℓ  dateOfExpedition  ?d  .
        FILTER (?d>= '01/01/1890')  .  FILTER (?d<= '31/12/1900') }
```

.

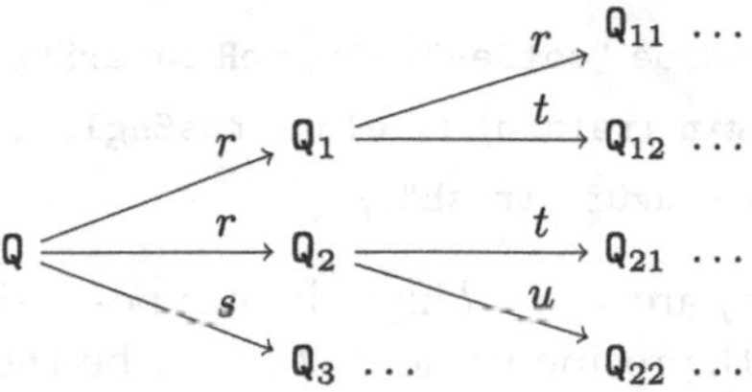

Fig. 2. A SPARQL query state space. Q is the initial query. r, s, t and u are SQTRL rules.

Now, a letter of David Hilbert of 1887 or even 1902 would be an acceptable answer to the informal query, whereas it does not answer the formal query Q. Hence the usefulness of transformations $\texttt{Q} \mapsto \texttt{Q}'$ and $\texttt{Q} \mapsto \texttt{Q}''$ corresponding to the enlargement of the interval of time to [1885, 1900] and [1890, 1905], respectively. $\texttt{Q}'$ (resp., $\texttt{Q}''$) is obtained by substituting `'01/01/1890'` (resp., `'31/12/1900'`) with `'01/01/1885'` (resp., `'31/12/1905'`) in Q.

Other examples have been considered. Some of them consist in generalizing classes in the query (as for the `Pear` to `Fruit` example in Sect. 3.1). Another one consists in replacing a person by another one that is close in a relationship network.

From this study and the previous one has emerged the need to develop a generic tool for managing SPARQL query transformations. A rule language for this purpose—SQTRL—has been developed as well as a system for managing such rules (this is detailed in Sect. 4). The principle of approximate search in the H. Poincaré letters using SQTRL rules is the one of a search in a state space where a state is a SPARQL query, the initial state is the initial query, and transitions correspond to rule applications, as illustrated in Fig. 2. Such a space is searched by increasing "transformation costs" using e.g., dynamic programming. These costs are associated to rules and are assumed to be additive. For example, the transformation from Q to $\texttt{Q}_{21}$ in Fig. 2 is $\texttt{cost}(r) + \texttt{cost}(t)$. This principle is also applied for retrieval in TAAABLE.

3.3 Cocktail Name Adaptation in the CBR System TAAABLE

TAAABLE has been applied for cocktail recipes for the CCCs of 2014 and 2015. For example, given the query $\mathcal{Q}$ = "I want a cocktail with schnapps and hot chocolate", TUUURBINE, the retrieval engine of TAAABLE, finds a recipe R named `"Irish coffee"` and the adaptation consists in applying the substitution $\sigma = \texttt{whisky} \rightsquigarrow \texttt{schnapps} \circ \texttt{coffee} \rightsquigarrow \texttt{hotChocolate}$ on the ingredients of R. The adapted recipe is denoted by $\sigma(R)$. The cocktail name adaptation problem is how to name the adapted cocktail recipe $\sigma(R)$, given the name of R (the string `"Irish coffee"`) and the substitution σ. This issue has been addressed in [12]. It is based on an RDFS base β_R associated with the recipe R and giving a (partial or complete) explanation of its name. In the example:

$$\beta_R = \{\langle \texttt{coffee englishName "coffee"}\rangle, \langle \texttt{?nameR superStringOf "coffee"}\rangle,\\ \langle \texttt{whisky hasOrigin ireland}\rangle, \langle \texttt{ireland hasEnglishAdjective "Irish"}\rangle,\\ \langle \texttt{?nameR superStringOf "Irish"}\rangle\}$$

where `coffee` and `whisky` are some R ingredient types and `?nameR` is the variable that is associated with the recipe name. Let $\beta_{\sigma(R)}$ be the RDFS base obtained by applying the substitution σ on β_R:

$$\beta_{\sigma(R)} = \{\langle \texttt{hotChocolate englishName "coffee"}\rangle, \langle \texttt{?nameR superStringOf "coffee"}\rangle,\\ \langle \texttt{schnapps hasOrigin ireland}\rangle, \langle \texttt{ireland hasEnglishAdjective "Irish"}\rangle,\\ \langle \texttt{?nameR superStringOf "Irish"}\rangle\}$$

$\beta_{\sigma(R)}$ does not match the domain knowledge in the sense that $\texttt{exec}_\vdash(\texttt{Q}_{\sigma(R)}, \mathcal{B}) = \emptyset$, with

$\texttt{Q}_{\sigma(R)} =$

```
SELECT ?anyVariable
WHERE { hotChocolate  englishName  "coffee"  .
        schnapps  hasOrigin  ireland  .
        ireland  hasEnglishAdjective  "Irish" }
```

(`?anyVariable` being a variable that is used only for syntax purpose). The transformation here consists in substituting identified resources or literals with variables, knowing that the resources occurring in $\sigma(R)$ (i.e., `hotChocolate` and `schnapps`). In the example, the substitutions `"coffee"` $\rightsquigarrow$ `?x`, `ireland` $\rightsquigarrow$ `?y` and `"Irish"` $\rightsquigarrow$ `?z` can be done (among others: one could have also substituted a property by a variable). A final state of this search is a query $\texttt{Q}_{\texttt{gen}}$ such that $\texttt{exec}_\vdash(\texttt{Q}_{\texttt{gen}}, \mathcal{B}) \neq \emptyset$, for example:

$\texttt{Q}_{\texttt{gen}} =$

```
SELECT ?anyVariable
WHERE { hotChocolate  englishName  ?x  .
        schnapps  hasOrigin  ?y  .
        ?y  hasEnglishAdjective  ?z }
```

Here, it is assumed that the execution of this query gives exactly one binding:

$$\texttt{exec}_\vdash(\texttt{Q}_{\texttt{gen}}, \mathcal{B}) = \{\{\texttt{?x}, \texttt{"hot chocolate"}\}, \{(\texttt{?y}, \texttt{germany}\}, \{(\texttt{?z}, \texttt{"German"})\}\}$$

Composing the generalizations done from $\beta_{\sigma(R)}$ to $\beta_{\texttt{gen}}$ and this binding, it comes the substitutions $\sigma_1 =$ `"coffee"` $\rightsquigarrow$ `"hot chocolate"` and $\sigma_2 =$ `"Irish"` $\rightsquigarrow$ `"German"`. Since the literals `"coffee"` and `"hot chocolate"` are linked in β_R with `?nameR`, the adaptation consists in applying these substitutions, hence the proposed name of the adapted recipe:

$$\sigma_1(\sigma_2(\texttt{"Irish coffee"})) = \texttt{"German hot chocolate"}$$

4 SQTRL: A Language for SPARQL Query Transformation Rules

The language SQTRL presented below has emerged from the need to transform SPARQL queries as presented above. After the presentation of this language, examples of SQTRL rules that cover the examples of Sect. 3 are given. Finally, the system that manages these rules is briefly described.

4.1 SQTRL: Syntax and Application of the Rules

A SQTRL rule is defined in an XML syntax as follows (the texts in italics have to be substituted by the appropriate strings):

```
<rule name=name of the rule>
   <context>RDFS triples under the SPARQL syntax</context>
   <left>RDFS triples under the SPARQL syntax</left>
   <right>RDFS triples under the SPARQL syntax</right>
   <cost>a float</cost>
   <explanation>a text possibly using variables</explanation>
</rule>
```

If r is such a rule, the contents in the fields with tags `<context>`, `<left>`, `<right>` and `<cost>` are denoted by context(r), left(r), right(r) and cost(r).

For example, the following rule substitutes a class C by a class D provided that C occurs as an object in a triple of the query Q body and that C is a subclass of D:

$r =$ (2)

```
<rule name=Generalize an object class>
   <context>?C  subc  ?D</context>
   <left>?x  ?p  ?C</left>
   <right>?x  ?p  ?D</right>
   <cost>1.0</cost>
   <explanation>Generalize ?C in ?D</explanation>
</rule>
```

Thus, context(r) = `?C subc ?D`, left(r) = `?x ?p ?C`, right(r) = `?x ?p ?D`, and cost(r) = 1.

Let Q be a SPARQL query and $\mathcal{B}$ be an RDFS base representing relevant domain knowledge. For the example,

Q =

```
SELECT ?r
WHERE { ?r  type  TartDishRecipe  .
        ?r  ing  ?i  .  ?i  type  Pear }
```

and $\mathcal{B}$ such that $\mathcal{B} \vdash$ {⟨Pear subc Fruit⟩, ⟨TartDish subc DishWithPastry⟩}. Let r be a SQTRL rule, for the example, it is the rule defined in Eq. (2). The application of r to `Q` given $\mathcal{B}$ is the set $\texttt{apply}(r, \texttt{Q}, \mathcal{B})$ of the queries $\texttt{Q}'$ obtained by transforming `Q` using rule r, knowing $\mathcal{B}$. If $\texttt{apply}(r, \texttt{Q}, \mathcal{B}) = \emptyset$, the rule r is said to be non applicable on `Q` given $\mathcal{B}$.

Figure 3 presents the algorithm for computing $\texttt{apply}(r, \texttt{Q}, \mathcal{B})$ and illustrates the algorithm with the example values r, `Q` and $\mathcal{B}$ given above.

Note About the Computing Time. Applying an SQTRL rule amounts mainly to execute a few SPARQL queries. Now, executing a SPARQL query amounts mainly is finding subgraph isomorphisms (between the body of the query to the graph representing the RDFS base), which is known to be an NP-complete problem. However, one must keep in mind that the size of the parameter to be taken into account is mainly the size of the query, which, for rule application, corresponds to the size of the context and of the left part of the rule. Thus, unless the rule is huge—a situation that has not occurred in our applications—the application of a rule is very quick, especially thanks to the use of an efficient SPARQL execution tool. So, our intuition is that the application of SQTRL rules is fast in practice though an accurate complexity estimation as well as some experiments remain to be done.

The same argument can be given for case retrieval and case adaptation: though they are based on search in a state space, if the size of the initial query is reasonable, then the search in this space does not take too much time in practice, and that is what we have experimented with TAAABLE. If the closest case is very dissimilar to the query, this involves a result that is likely to be of poor quality: for example, if the user queries TAAABLE for a vegetarian recipe with pastry and pineapple and the only recipe in the case base is for beef Stroganov, then it is unlikely that the users will be satisfied with the adaptation query (despite the adaptation capabilities of TAAABLE). For this reason, a timeout interruption is used: after a too long computing time for retrieval, it is considered that the adaptation procedure will not be able to make enough modifications to achieve a satisfying result.

4.2 Examples of SQTRL Rules

This section presents some SQTRL rules that cover the examples presented in Sect. 3.

Generalization Rules. Let $\texttt{Q}_1$ and $\texttt{Q}_2$ be two SPARQL queries with the same SELECT line. $\texttt{Q}_1$ is said to be less general than $\texttt{Q}_2$—denoted by $\texttt{Q}_1 \sqsubseteq \texttt{Q}_2$—if for every RDFS base $\mathcal{B}$, $\texttt{exec}_{\vdash}(\texttt{Q}_1, \mathcal{B}) \subseteq \texttt{exec}_{\vdash}(\texttt{Q}_2, \mathcal{B})$. A generalization rule is a rule r such that for every SPARQL query `Q`, every RDFS base $\mathcal{B}$ and every $\texttt{Q}' \in \texttt{apply}(r, \texttt{Q}, \mathcal{B})$, $\texttt{Q} \sqsubseteq \texttt{Q}'$. The rule r of Eq. (2) is such a rule, as way as the rule `"Generalize a subject class"` obtained by replacing $\texttt{left}(r)$ by `?C ?p ?x` and $\texttt{right}(r)$ by `?D ?p ?x`. Similarly, the following rule is a

```
function apply(r : SQTRL rule, Q : SPARQL query, B : RDFS base)
begin
    ▷ Initialize producedQueries, the output of the algorithm
    producedQueries ← ∅
    ▷ Unification of the context of r with B
①   ctxtToBase ← exec⊢ ( SELECT V(context(r))
                          WHERE {context(r)} , B )
    for binding ∈ ctxtToBase do
②       boundLeft ← application of binding to left(r)
③       boundRight ← application of binding to right(r)
        ▷ Unification of the bound left part of r to the body of Q
④       leftToQuery ← exec⊢ ( SELECT V(boundLeft)
                               WHERE {boundLeft} , body(Q) )
        for binding′ ∈ leftToQuery do
⑤           boundLeft′ ← application of binding′ to boundLeft
⑥           boundRight′ ← application of binding′ to boundRight
            Q′ ← copy of Q
            Remove the triples of boundLeft′ from the body of Q′
⑦           Add the triples of boundRight′ to the body of Q′
            producedQueries ← producedQueries ∪ {Q′}
        end
    end
    return producedQueries
end
```

① $\texttt{ctxtToBase} = \texttt{exec}_{\vdash}\left(\begin{array}{|l|}\hline \texttt{SELECT ?C ?D} \\ \texttt{WHERE \{?C subc ?D\}} \\ \hline\end{array}, \mathcal{B}\right) = \{\texttt{binding}_1, \texttt{binding}_2\}$
with $\texttt{binding}_1 = \{(\texttt{?C}, \texttt{Pear}), (\texttt{?D}, \texttt{Fruit})\}$
and $\texttt{binding}_2 = \{(\texttt{?C}, \texttt{TartDishRecipe}), (\texttt{?D}, \texttt{DishWithPastryRecipe})\}$.
The following of the explanation is with $\texttt{binding} = \texttt{binding}_1$.

② Since $\texttt{left}(r) =$ `?x ?p ?C`, boundLeft = `?x ?p Pear`.

③ Since $\texttt{right}(r) =$ `?x ?p ?D`, boundRight = `?x ?p Fruit`.

④ $\texttt{leftToQuery} = \texttt{exec}_{\vdash}\left(\begin{array}{|l|}\hline \texttt{SELECT ?x ?p} \\ \texttt{WHERE \{?x ?p Pear\}} \\ \hline\end{array}, \texttt{body(Q)}\right)$ which gives the set {binding′} (only one binding in this example) with binding′ = {(?x, ?i), (?p, type)}.

⑤ boundLeft′ = `?i type Pear`.

⑥ boundRight′ = `?i type Fruit`.

⑦ Q′ =

```
SELECT ?r
WHERE {?r  type  TartDishRecipe  .
          ?r  ing  ?i  .  ?i  type  Fruit}
```

Fig. 3. The SQTRL rule application: the algorithm and an example ($\mathcal{V}(s)$ denotes the set of variables occurring in a sequence s of RDFS triples).

rule that generalizes predicates:

```
<rule name=Generalize a property in predicate position>
   <context>?p  subp  ?q</context>
   <left>?x  ?p  ?y</left>
   <right>?x  ?q  ?y</right>
   <cost>1.0</cost>
   <explanation>Generalize ?p in ?q</explanation>
</rule>
```

A second way to generalize a query is by removing a triple:

```
<rule name=Remove a triple from the body of the query>
   <context></context>
   <left>?x  ?p  ?y</left>
   <right></right>
   <cost>1.0</cost>
   <explanation>Remove the triple ?x  ?p  ?y</explanation>
</rule>
```

It can be noted that removing a triple can "disconnect" variables that were previously connected by a path. For example, if body(Q) = `s p ?x . ?x ?q ?y . ?y ?r ?z` then the removal of `?x ?q ?y` disconnects ?z from s. A variant of the above rule exists that prevent such situation.

A third type of generalization rules works with filters, when a filter has the form `FILTER (?x ⋈ v)` where $\bowtie \in \{$<, <=, >=, >$\}$ and v is a value of a numerical type (or a date). It consists in replacing v by $v + c$ where c is some numerical constant such that $c > 0$ if $\bowtie \in \{$<, <=$\}$ and $c < 0$ else. Such a rule does not follow the syntax of other rules (it has a specific syntax), and has been motivated by the modeling of "the end of the XIXth century" issue (cf. Sect. 3.2).

A final type of generalization rules, that has been used in Sect. 3.3, consists in substituting a constant c (i.e., a resource or a literal) by a variable ?c:

```
<rule name=Generalize subject c by ?c>
   <context></context>
   <left>c  ?p  ?x</left>
   <right>?c  ?p  ?x</right>
   <cost>1.0</cost>
   <explanation>Generalization of c in ?c</explanation>
</rule>
```

Similar rules are defined for substituting a constant by a variable in predicate and object positions. Using such rules can be used to transform $\mathtt{Q}_{\sigma(R)}$ into $\mathtt{Q}_{\mathtt{gen}}$ (cf. Sect. 3.3).

These generalization rules can be qualified as application-independent as they can be used in various applications. By contrast, some other SQTRL rules are strongly related to applications.

Application-Dependent Rules. The generalized rules presented above cover some of the examples presented in Sect. 3 but not all of them. For example, in Sect. 3.1, it is said that, for a dessert recipe, butter can be replaced by margarine and vice-versa, which can be formalized thanks to two rules, the first one being

```
<rule name=Replace butter by margarine in a dessert>
  <context>?r  type  DessertRecipe</context>
  <left>?r  ing  ?x  .  ?x  type  Butter</left>
  <right>?r  ing  ?x  .  ?x  type  Margarine</right>
  <cost>0.1</cost>
  <explanation>Replace butter with margarine</explanation>
</rule>
```

and the other one being a similar rule obtained by exchanging the terms `Butter` and `Margarine` in the first rule. This kind of rules can also be used as adaptation rules: given a dessert recipe R with butter and a query of a dessert recipe with margarine, the above rule can be applied to adapt R to answer the query. Thus, SQTRL can also be used as a language for adaptation rules. In a way, it could be said that retrieval *adapts* the query to fit at least one case from the case base that is then adapted to fit the query.

Another domain-dependent rule is the one exchanging sender and recipient in H. Poincaré letters, which can be formalized by:

```
<rule name=Exchange sender and recipient>
  <context></context>
  <left>?x  isSentTo  ?y  .  ?x  isSentBy  ?z</left>
  <right>?x  isSentBy  ?y  .  ?x  isSentTo  ?z</right>
  <cost>1.0</cost>
  <explanation>Exchange sender/recipient: ?y/?z</explanation>
</rule>
```

4.3 A Tool for Managing SQTRL Rules

A tool has been developed to manage SQTRL rules with the following functionalities: creation, serialization and application of a rule. It uses the rule syntax presented above and the turtle syntax for RDFS base. It uses the RDFS and SPARQL management tool KGRAM [6] and is written in Java. It is freely accessible at http://tuuurbine.loria.fr/sqtrl/ (a site with the code and a user manual). A demo has been developed as illustrated by the screenshot of Fig. 4.

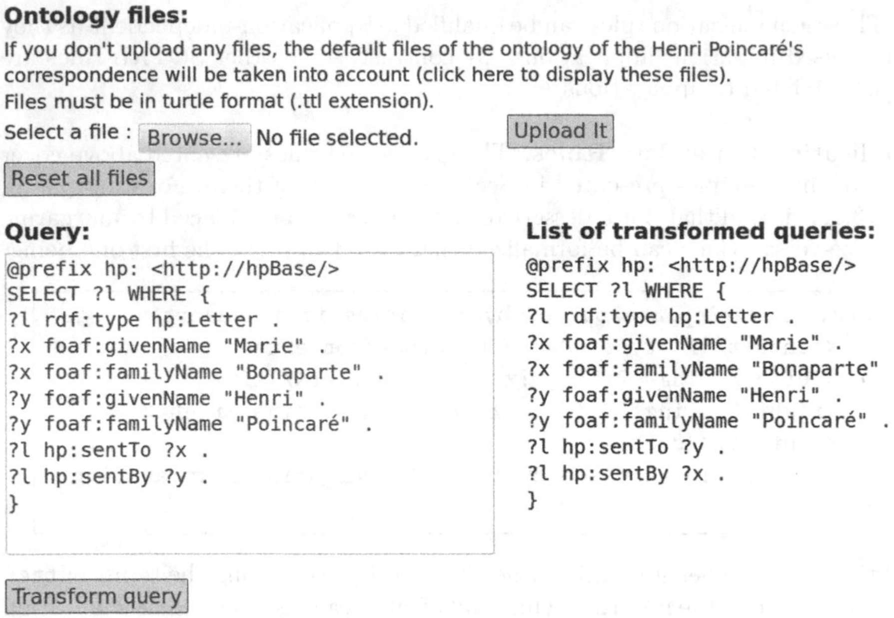

Ontology files:

If you don't upload any files, the default files of the ontology of the Henri Poincaré's correspondence will be taken into account (click here to display these files).
Files must be in turtle format (.ttl extension).

Select a file : Browse... No file selected. Upload It

Reset all files

Query:

```
@prefix hp: <http://hpBase/>
SELECT ?l WHERE {
?l rdf:type hp:Letter .
?x foaf:givenName "Marie" .
?x foaf:familyName "Bonaparte" .
?y foaf:givenName "Henri" .
?y foaf:familyName "Poincaré" .
?l hp:sentTo ?x .
?l hp:sentBy ?y .
}
```

List of transformed queries:

```
@prefix hp: <http://hpBase/>
SELECT ?l WHERE {
?l rdf:type hp:Letter .
?x foaf:givenName "Marie" .
?x foaf:familyName "Bonaparte" .
?y foaf:givenName "Henri" .
?y foaf:familyName "Poincaré" .
?l hp:sentTo ?y .
?l hp:sentBy ?x .
}
```

Transform query

Fig. 4. A screenshot of the SQTRL tool demo. The query on the left has been transformed in the list with only one query on the right, the applied rule being the one named `"Exchange sender and recipient"`.

5 Discussion and Related Work

Case retrieval based on minimal generalization of the query is not a new idea in CBR. For example, it was applied with SQL queries for a translation system based on CBR in the early 1990s [20]. It was also applied in a graph formalism for representing molecular structures a few years later [14]. TAABLE has been using this principle since its first version, though the use of RDFS and SPARQL has only been developed since 2014. The originality of this work is that it presents a well-defined rule language for transforming queries and that it is based on RDFS and SPARQL that are semantic web standards that are getting more and more used and with which data and knowledge are represented and accessed freely within the Linked Open Data [2].

Case adaptation based on minimal modification of the source case is not a new idea either. Actually, when the modification is based on generalizations, it is related to generalization methods found in the early years of machine learning [10] and applied to CBR [19]. Here SQTRL rules are used to implement this idea (for both generalization-based and non generalization-based modifications) when (a part of) the source case can be represented as an RDFS base or a SPARQL query. Another work based on this principle is revision-based adaptation [5], i.e., adaptation based on the use of a belief revision operator [1]. Such an operator $\dotplus$ associates to two belief bases ψ and μ a belief base $\psi \dotplus \mu$ equivalent to $\psi' \wedge \mu$ where ψ' is the minimal modification of ψ to make it consistent with μ (given some modification metric). Therefore, the idea of $\dotplus$-adaptation is to make the revision of the source case by the query (knowing that both have to be consistent with the domain knowledge). The difference here is that RDFS bases are hardly inconsistent.[2] Therefore, belief revision of RDFS bases in the classical sense has little interest. Currently, alternative ways of defining revision in RDFS are investigated which could lead to a unification of the approach presented here with revision-based adaptation.

This work has strong connections with the theory developed last years that is based on amalgams and on refinement operators [16,17]. Indeed, refinement operators can be likened to SQTRL rules and are used both for case retrieval [16] and for (single and multiple) case adaptation [17]. Two differences between this previous work and the current one can be pointed out. First, the amalgalms and refinement theory is defined independently from a knowledge representation formalism, whereas we present an approach more concrete, with the advantage of being associated with an operational tool. Second, the refinement operators are generalization and specialization operators, whereas SQTRL allows to define non generalization rules. One could argue that such a rule can be "simulated" by the application of two rules, one for generalization and one for specialization. For example, the rule making the substitution `Butter` $\rightsquigarrow$ `Margarine` can be seen as the composition of the generalization $g =$ `Butter` $\rightsquigarrow$ `Fat` and the specialization $s =$ `Fat` $\rightsquigarrow$ `Margarine`. However, applying these two rules has a cost $\texttt{cost}(g) + \texttt{cost}(s)$ that may be too high, if the `Fat` class contains subclasses less close to `Butter` and `Margarine` from a cooking viewpoint, such as `DuckFat`.[3] Therefore a rule substituting "directly" butter by margarine with a lower cost is useful and so are other non generalization rules.

6 Conclusion and Future Work

This paper has presented SQTRL, a query transformation rule language and tool adapted to the SPARQL formalism, and three of its applications. Our claim is that this language and this tool can be applied in many application domains of CBR, provided that the case language and the domain knowledge can be translated into RDFS, which embeds the feature-value formalisms and taxonomy languages frequently used in CBR [13]. In particular, it is planned to use SQTRL for an ongoing work in a medical domain.

The SQTRL tool is in the TUUURBINE web site, but the current version of TUUURBINE does not use it: the integration of these tools is a future work.

---

[2] The only case of inconsistency of an RDFS base is related to a type error property for datatype properties. For example, if the age of an individual stated with property `age` is an integer, then the triple ⟨`juliet age true`⟩ is inconsistent. Such situations of inconsistencies are not relevant here.

[3] These classes are taken from WIKITAAABLE, the semantic wiki that contains TAAABLE ontology: http://wikitaaable.loria.fr.

SQTRL is in its first stable version but, surely, this language will require evolutions. Our policy is to make it evolve when new needs emerge. The particular treatment associated with the filters show that there is room for improvement here. The difficulty is that, in general, the filter term can use the Boolean operators (`and`, `or`, `not`), which raises specific issues if the goal is to make transformations up to equivalence (and not only syntactical ones): if $\mathtt{Q}_1$ and $\mathtt{Q}_2$ are equivalent queries (that is $\mathtt{Q}_1 \sqsubseteq \mathtt{Q}_2$ and $\mathtt{Q}_2 \sqsubseteq \mathtt{Q}_1$) and r is a rule that can be applied to $\mathtt{Q}_1$ to give birth to $\mathtt{Q}_1'$, then r should be applicable to $\mathtt{Q}_2$ to give birth to a query $\mathtt{Q}_2'$ equivalent to $\mathtt{Q}_1'$.

In this paper, the cost fields of the rules are given arbitrarily. One question is how to fix them, which is a complex knowledge acquisition issue often met in CBR (similar to the choice of weights in a similarity measure). Another point is that, in this first version of SQTRL, costs are constant, whereas one can consider that they should be parameterized by the bindings. For example, using the generalization rule of Eq. (2), one can argue that it is less costly to make the generalization `WilliamsPear` $\rightsquigarrow$ `Pear` than the generalization `Apple` $\rightsquigarrow$ `Fruit` in the cooking domain. Therefore, having a cost field that is linked with a function will probably be useful.

Acknowledgments. The authors would like to thank the anonymous reviewers for their comments that have helped to improve this paper.

References

1. Alchourrón, C.E., Gärdenfors, P., Makinson, D.: On the logic of theory change: partial meet functions for contraction and revision. J. Symb. Logic **50**, 510–530 (1985)
2. Bizer, C., Heath, T., Berners-Lee, T.: Linked data - the story so far. Semant. Serv. Interoperability Web Appl.: Emerg. Concepts, 205–227 (2009)
3. Brickley, D., Guha, R.V.: RDF Schema 1.1, W3C recommendation (2014). https://www.w3.org/TR/rdf-schema/. Last consultation March 2017
4. Bruneau, O., Garlatti, S., Guedj, M., Laubé, S., Lieber, J.: SemanticHPST: applying semantic web principles and technologies to the history and philosophy of science and technology. In: Gandon, F., Guéret, C., Villata, S., Breslin, J., Faron-Zucker, C., Zimmermann, A. (eds.) ESWC 2015. LNCS, vol. 9341, pp. 416–427. Springer, Cham (2015). doi:10.1007/978-3-319-25639-9_53
5. Cojan, J., Lieber, J.: Applying belief revision to case-based reasoning. In: Prade, H., Richard, G. (eds.) Computational Approaches to Analogical Reasoning: Current Trends. SCI, vol. 548, pp. 133–161. Springer, Heidelberg (2014). doi:10.1007/978-3-642-54516-0_6
6. Corby, O., Gaignard, A., Faron-Zucker, C., Montagnat, J.: KGRAM versatile data graphs querying and inference engine. In: Proceedings of the IEEE/WIC/ACM International Conference on Web Intelligence, Macau, December 2012
7. Cordier, A., et al.: Taaable: a case-based system for personalized cooking. In: Montani, S., Jain, L.C. (eds.) Successful Case-based Reasoning Applications-2. SCI, vol. 494, pp. 121–162. Springer, Heidelberg (2014). doi:10.1007/978-3-642-38736-4_7

8. Cordier, A., Lieber, J., Molli, P., Nauer, E., Skaf-Molli, H., Toussaint, Y.: Wiki-Taaable: a semantic wiki as a blackboard for a textual case-based reasoning system. In: SemWiki 2009–4th Semantic Wiki Workshop, Heraklion, Greece, May 2009
9. Cox, M.T., Muñoz-Avila, H., Bergmann, R.: Case-based planning. Knowl. Eng. Rev. **20**(3), 283–287 (2005)
10. Diettcrich, T.G., Michalski, R.S.: A comparative review of selected methods for learning from examples. In: Michalski, R.S., Carbonell, J.G., Mitchell, T.M. (eds.) Machine Learning, pp. 41–81. Springer, Heidelberg (1983). doi:10.1007/978-3-662-12405-5_3
11. Gaillard, E., Infante-Blanco, L., Lieber, J., Nauer, E.: Tuuurbine: a generic CBR engine over RDFS. In: Lamontagne, L., Plaza, E. (eds.) ICCBR 2014. LNCS, vol. 8765, pp. 140–154. Springer, Cham (2014). doi:10.1007/978-3-319-11209-1_11
12. Kiani, N., Lieber, J., Nauer, E., Schneider, J.: Analogical transfer in RDFS, application to cocktail name adaptation. In: Goel, A., Díaz-Agudo, M.B., Roth-Berghofer, T. (eds.) ICCBR 2016. LNCS, vol. 9969, pp. 218–233. Springer, Cham (2016). doi:10.1007/978-3-319-47096-2_15
13. Kolodner, J.: Case-Based Reasoning. Morgan Kaufmann, Inc., Burlington (1993)
14. Lieber, J., Napoli, A.: Using classification in case-based planning. In: Wahlster, W., (ed.) Proceedings of the 12th European Conference on Artificial Intelligence (ECAI 1996), pp. 132–136. Wiley, Budapest (1996)
15. Minor, M., Bergmann, R., Görg, S.: Case-based adaptation of workflows. Inf. Syst. **40**, 142–152 (2014)
16. Ontañón, S., Shokoufandeh, A.: Refinement-based similarity measures for directed labeled graphs. In: Goel, A., Díaz-Agudo, M.B., Roth-Berghofer, T. (eds.) ICCBR 2016. LNCS, vol. 9969, pp. 311–326. Springer, Cham (2016). doi:10.1007/978-3-319-47096-2_21
17. Ontañón, S., Plaza, E.: Amalgams: a formal approach for combining multiple case solutions. In: Bichindaritz, I., Montani, S. (eds.) ICCBR 2010. LNCS, vol. 6176, pp. 257–271. Springer, Heidelberg (2010). doi:10.1007/978-3-642-14274-1_20
18. Recio-Garía, J.A., Sánchez, A., Díaz-Agudo, B., González-Calero, P.A.: JColibri 1.0 in a nutshell. A software tool for designing CBR systems. In: Petridis, M., (ed.) Proceedings of the 10th UK Workshop on Case Based Reasoning, pp. 20–28. CMS Press, University of Greenwich (2005)
19. Riesbeck, C.K., Schank, R.C.: Inside Case-Based Reasoning. Lawrence Erlbaum Associates Inc., Hillsdale (1989)
20. Shimazu, H., Kitano, H., Shibata, A.: Retrieving cases from relational data-bases: another stride towards corporate-wide case-based systems. In: Proceedings of the 13th International Joint Conference on Artificial Intelligence (IJCAI 1993), Chambéry, pp. 909–914 (1993)
21. Watson, I.: Is CBR a technology or a methodology? In: Pasqual del Pobil, A., Mira, J., Ali, M. (eds.) IEA/AIE 1998. LNCS, vol. 1416, pp. 525–534. Springer, Heidelberg (1998). doi:10.1007/3-540-64574-8_438

Similar Users or Similar Items? Comparing Similarity-Based Approaches for Recommender Systems in Online Judges

Marta Caro-Martinez and Guillermo Jimenez-Diaz(✉)

Department of Software Engineering and Artificial Intelligence,
Universidad Complutense de Madrid, Madrid, Spain
{martcaro,gjimenez}@ucm.es

Abstract. Online judges store hundreds of programming problems but they lack recommendation tools to help users to find relevant problems to solve. In this paper, we extend the exploration of the use of the implicit knowledge derived from the relationships created between users and problems when the users submit their solutions to the online judge. Inspired by collaborative filtering techniques, in this work we compare a user-based and a problem-based approach, both supported by node similarity metrics coming from social network analysis, and we study the inclusion of voting systems in order to rank the problems that best fit for a user in the online judge. Our experiments reveal that the selection of the highest-performing similarity metric is determined by the recommendation method. We also show that the user-based approach outperforms the problem-based approach only when the proposed voting systems are used.

1 Introduction

Online judges are repositories with hundreds of programming exercises and problems used in programming contests and learning sessions [5]. A programming exercise has a statement about the problem to solve and a private set of test cases. When a user submits a solution –a source code– for a problem, the online judge compiles and executes the solution against the test cases, providing a verdict about its correctness.

Online judges are commonly used for training on-site programming contests, becoming a valuable system for expert users who want to increase their performance on competitive programming. Unfortunately, these systems pay little attention to novice users, who just want to practice algorithms or data structures. Usually, these users are overwhelmed by the amount of problems in the repository and they have no idea about which problem they should try to solve.

Despite this problem, online judges rarely provide recommendation mechanisms to help these newbies. Some online judges suggest problems using the

Supported by UCM (Group 910494) and Spanish Committee of Economy and Competitiveness (TIN2014-55006-R).

D.W. Aha and J. Lieber (Eds.): ICCBR 2017, LNAI 10339, pp. 92–107, 2017.
DOI: 10.1007/978-3-319-61030-6_7

Global Ranking Method, which just recommends the problem with the most correct solutions that the user has not resolved yet. This method lacks personalization because it will recommend almost the same problems to the all the users, no matter which problems they have already solved.

A brief review about recommender systems shows that recommendation methods usually exploit user opinions in terms of ratings –collaborative filtering– or item descriptions –content-based– in order to provide personalized recommendations. Online judges hardly ever provide rating tools, which allow users to express their preferences. Moreover, the problems in the repository, apart from the statement description, lack additional information, besides a few tags about the programming concepts required in the solution.

In our previous work [4], we described a recommendation approach that exploited the knowledge implicit in the submissions and its verdicts. The submissions represent the interactions between users and problems in the online judge and they can derive an interaction graph. In that work, we proposed the creation of a problem-problem network in order to find similar problems to the ones solved by a user, and the use of similarity metrics extracted from link prediction techniques in order to recommend new problems. However, as collaborative filtering proposes, the recommendation process can be reformulated as finding similar users to the current one, in order to recommended the problems solved by these users but not already attempted by the current user. Our current work defines this user-based approach, which relies on the construction of a user-user network, and compares it with the problem-based approach. The experimental evaluation of both recommendation methods will use a dataset with the submissions in *Acepta el Reto* (Spanish translation of *Take on the challenge*), an online judge developed by our research group.

The remaining of this paper is organized in the following sections. First, we describe the online judge systems (Sect. 2) and, right after, we detail the recommendation approaches proposed for these systems (Sect. 3) and the similarity metrics employed by these approaches (Sect. 4). The comparative evaluation of the user-based and problem-based approaches is related in Sect. 5 and the paper concludes with some related work (Sect. 6) and a summary of the main conclusions and the derived future work (Sect. 7).

2 Online Judges

Online judges are online repositories with hundreds or even thousands of programming exercises. Each programming exercise includes a public statement, which describes the problem to solve, and a private set of test cases that will be used to automatically validate the solutions submitted to the system. Users choose a problem and then try to solve it by submitting code solutions in one of the programming languages accepted by the judge. The system compiles the source code and runs it against many *test cases*, whose solutions are known by the judge. The output generated by the submitted solution is compared with the official solution and a verdict is provided. Examples of such systems are the UVa Online Judge[1] or *Acepta el reto*[2].

Depending on the comparison result, the online judge provides one of the following verdicts:

AC (*Accepted*): The solution submitted was correct because it produced the right answer and it did not exceed the time and memory usage constraints.
PE (*Presentation Error*): The solution was almost correct, though it failed to write the output in the exact required format (having an excess of blanks or line endings, for example).
CE (*Compile Error*): The solution did not compile.
WA (*Wrong Answer*): The program failed to write the correct answer for one or more test cases.
RTE (*Runtime Error*): The program crashed during the execution (because of segmentation fault, floating point exception...).
TLE (*Time Limit Exceeded*): The execution took too much time.
MLE (*Memory Limit Exceeded*): The solution consumed too much memory.
OLE (*Output Limit Exceeded*): The program tried to write too much information. This usually occurs if it goes into an infinite loop.

Generally, users suffering a negative verdict try to fix their solution and they resubmit it. Moreover, it is not unusual to receive resubmissions of users making changes to their *accepted* code, in order to improve their ranking position creating optimized code. It could happen that those assumed improvements lead into a negative verdict. However, from the system point of view, the user will have the problem still accepted, despite the non-AC verdict in her last submission.

Taking the last statement into account, the relationship between the users and the problems stored in the repository can be simplified and modelled according to one of the following states:

- *Solved*: the user submitted several solutions for a problem and at least one is correct. In this category we can consider both AC and PE verdicts, since PE are close to being correct.
- *Attempted*: the user submitted one or more solutions for the problem, but none of them are correct.
- *Unattempted*: the user did not submit any solution for a problem.

The next section will describe how the submission dataset is handled to create the interaction graphs employed by the proposed recommendation approaches.

3 Recommendation Methods Based on Interaction Graphs

In an online judge system, submissions contain the information about the user-problem interactions. From the system point of view, submissions can be represented as a set of tuples $R = (t, u, p, v)$ where t is the timestamp when the

[1] https://uva.onlinejudge.org.
[2] https://www.aceptaelreto.com (in Spanish).

submission was sent, p and u are the problem and the user respectively, and v is the verdict emitted by the online judge for the solution that u provided for p.

User-problem interactions can be abstracted and represented into a user-problem non-weighted bipartite graph G, where nodes belong exclusively to one of two disjoint sets, the problem-set $P = \{p_1, \ldots, p_m\}$ or the user-set $U = \{u_1, \ldots, u_n\}$. Therefore, we define an adjacency matrix $A = \{a_{ij}\}$, where $a_{ij} = 1$, if the user u_i attempted to solve (or correctly solved, depending on the use of the matrix) the problem p_j.

Instead of using a bipartite graph, we can define a non-bipartite graph where the user-problem interactions will be transformed into implicit relationships among problems, as proposed in our previous work [4]. Analogously, we can define the non-bipartite graph that represents the implicit relationships among users. The *network projection* is the process that aims to transform a bipartite graph into a non-bipartite one.

According to the way that the interaction graph is projected, we will create a user-user graph or a problem-problem graph. Therefore, we propose two different recommendation approaches using the corresponding projected graph.

3.1 Problem-Based Approach

The problem-based recommendation approach was proposed in our previous work [4]. It uses the *Problem-projection* graph $G_p = \{N, E\}$, where $N = \{p \in P\}$ are *Problem nodes*, and $E = \{(p_i, p_j) | p_i, p_j \in P, i \neq j\}$ are the edges. In this case, two nodes are connected if they have at least one common user who solved both problems. To avoid losing information from the original network about user interactions we use a simple weighting method, where an edge (p_i, p_j) is weighted with the number of different users who solved both p_i and p_j. The generated graph using this method can be very dense so we filter edges in order to reduce this density and to make the graph easier to handle. We employ a global threshold on weights [11]: we delete all the edges whose weight is below a threshold value tv.

With this graph and the node similarity metrics described later, we compute a problem-problem similarity matrix Sim_m, which contains information about similarity between pairs of problems (a problem-problem matrix). $Sim_m(i, j)$ represents the similarity score between the problems p_i and p_j using the similarity metric m.

In short, the problem-based approach recommends the most similar problems to the ones that a target user u_t solved before. The list of recommended problems is created following these steps:

1. We create the set of problems P_{u_t} with the problems solved by u_t.
2. For each $p_i \in P_{u_t}$ we create a set of problems $P_{sim_{p_i}} = \{(p_j, s_j), \ldots (p_n, s_n)\}$, removing the problems already attempted by u_t. $s_j \equiv Sim_m(i, j)$ represents the similarity between p_i and p_j using the similarity matrix Sim_m. Every set contains the most similar problems to p_i, removing the ones whose similarity with p_i is below a threshold value.

3. We create a list of problems $L_r = [(p_j, s_j), \ldots (p_x, s_x)]$ aggregating all the problems in the previous sets. When a problem p_i exists in different sets, s_i is the highest similarity value among the sets where p_i existed.
4. Finally, L_r is ranked using the similarity score and we will recommend the first k problems.

3.2 User-Based Approach

The user-based approach uses the *User-projection* graph $G_u = \{N, E\}$, where $N = \{u \in U\}$ are *User nodes*, and $E = \{(u_i, u_j)|u_i, u_j \in U, i \neq j\}$ are the edges. In this graph, two nodes are connected if both users solved at least one problem in common. As stated in the previous approach, we use a simple weighting method: an edge (u_i, u_j) is weighted with the number of different common problems solved by both u_i and u_j users. Additionally, we reduce the graph density removing the edges whose weight is below a threshold value tv.

With this graph and the same node similarity metrics, we compute a user-user similarity matrix Sim_m, where $Sim_m(i, j)$ represents the similarity score between users u_i and u_j using the similarity metric m.

In short, the user-based recommendation looks for similar users to the target user u_t and recommends the problems they have already solved by the similar users but not attempted by u_t. In detail, the user-based recommendation process runs as follows:

1. We create a set of similar users U_{sim} to u_t. U_{sim} is the row t of the similarity matrix Sim_m, removing the users whose similarity with u_t is below a threshold value.
2. For each user $u_j \in U_{sim}$, we create a set of problems P_{u_j} selecting the problems solved by u_j, but not attempted by u_t.
3. Finally, we create a list of problems $L_r = [(p_1, s_1) \ldots (p_n, s_n)]$ merging the problems contained in every P_{u_j} set created in the previous step. s_i is the highest similarity value $Sim_m(t, j)$ among the users who proposed p_i.
4. Finally, L_r is ranked using the similarity score and we will recommend the first k problems.

Due to the social nature of this user-user network, we propose the inclusion of a voting system in step 3. This way, instead of using the best similarity value s_i, we propose the use of a voting score v_i for each problem p_i in L_r, which takes into account not only the similarity of an individual but the aggregation of scores provided by several users.

The voting score is computed using a voting system. In this work, we propose and analyse the following voting systems:

1. *Simple voting.* The voting score counts the number of times that the problem p_i appears in the sets of problems P_{u_j} created in step 2.
2. *Weighted voting.* The voting score is the weighted sum of the times that the problem p_i appears in the sets of problems P_{u_j}, where the weight value w for

a problem in P_{u_j} is computed with the similarity of the users who proposed the problem with u_t:

$$w = \frac{U_{sim}(u_j)}{\sum_{u_k \in U_{sim}} U_{sim}(u_k)}$$

3. *Positional voting.* The voting score is again the weighted sum of the times that the problem p_i appears in the sets of problems P_{u_j}. However, the weight value w for a problem in P_{u_j} is computed using the position pos_{u_j} where the user u_j appears in U_{sim}, sorted by similarity in descending order:

$$w = \frac{1}{pos_{u_j} + 1}$$

In order to illustrate how the voting systems work, we provide an example (see Fig. 1). We have 4 users similar to our target user u_t and $P_r = \{p_1, p_2, p_3, p_4, p_5\}$ contains the problem candidates solved by those users but not attempted by u_t.

User	Problems	Similarity Score
u_3	$\{p_1, p_2, p_5\}$	5
u_1	$\{p_1, p_2, p_4\}$	4
u_2	$\{p_2, p_3, p_5\}$	2
u_4	$\{p_1, p_2, p_3\}$	1

(a) Problems and user similarity score.

	Simple Voting
p_1	$1+1+1=3$
p_2	$1+1+1+1=4$
p_3	$1+1=2$
p_4	1
p_5	$1+1=2$

Ranked $L_r = [p_2, p_1, p_3, p_5, p_4]$

(b) Simple Voting

	Weighted Voting
p_1	$5/12 + 4/12 + 1/12 = 0.83$
p_2	$5/12 + 4/12 + 2/12 + 1/12 = 1$
p_3	$2/12 + 1/12 = 0.25$
p_4	$4/12 = 0.33$
p_5	$5/12 + 2/12 = 0.58$

Ranked $L_r = [p_2, p_1, p_5, p_4, p_3]$

(c) Weighted Voting

	Positional Voting
p_1	$1/1 + 1/2 + 1/4 = 1.75$
p_2	$1/1 + 1/2 + 1/3 + 1/4 = 2.08$
p_3	$1/3 + 1/4 = 0.58$
p_4	$1/2 = 0.5$
p_5	$1/1 + 1/3 = 1.33$

Ranked $L_r = [p_2, p_1, p_5, p_3, p_4]$

(d) Positional Voting

Fig. 1. Voting system example

We can see that each voting system generates different ranked lists of problems. For example, if we choose $k = 2$, the list of problems recommended to u_t will be $[p_2, p_1]$ with all of the voting systems, but if we choose $k = 3$, we obtain two different lists: $[p_2, p_1, p_3]$ with the simple voting system and $[p_2, p_1, p_5]$ with the weighted and positional voting systems. This way, we illustrate that the choice of the voting system to use in our recommender system is relevant.

Both approaches work with a similarity function that compares nodes in a graph, no matter if the nodes represent users or problems. The next section will show that the Social Network Analysis can provide us with a set of different node similarity metrics.

4 Node-Based Similarity Metrics

The network representation allows us to analyse the user-problem interactions that occurred in the online judge system using the methods and metrics defined by the Social Network Analysis field [3]. Link prediction is a technique used in social network analysis that aims to predict new links that might be formed between nodes in a future time or to predict missed links in the current network [12]. There are different approaches to predict these links. However, we will focus on similarity-based methods, which compute the proximity or *similarity* between pairs of nodes in order to predict new links. We will define a variation of these metrics in order to calculate a similarity score that expresses how similar two problems or users are in the corresponding graph according to the users' interactions with these problems.

These metrics must be considered as a *score* for a pair of nodes (x, y) instead of a classic similarity metric because, in general, the value of these metrics does not lie in [0,1] range.

For clarity of the descriptions of the proposed similarity metrics, we give some notation:

- $N(x)$ represents the neighbours of node x.
- $|N(x)|$ represents the number of neighbours (or *node degree*) of node x.
- $WD(x)$ represents the *weighted node degree* of node x, which means the sum of the weights in the edges directly connected with node x.
- A_{xy} represents the weight of the edge that links node x and node y.

Most of these metrics are detailed in [6,12] and some of them are defined in two different flavours: unweighted and weighted metrics [7]. Using this notation, now we can describe the similarity metrics used in our study (Table 1).

Edge Weight (EW). This simple metric measures the similarity between two nodes as the weight of the edge that links them. Two nodes are unconnected if $A_{xy} = 0$. Although an unweighted version of this metric exists ($A_{xy} = 1$ if the edge exists; 0 otherwise), we have not used it because it cannot be employed as a similarity metric.

Common Neighbours (CN). This metric measures the similarity between two nodes as the number of neighbours they have in common. We have defined a weighted version (WCN) of this metric.

Jaccard Neighbours (JN). This is an improvement of $CN(x, y)$ as it measures the number of common neighbours of x and y compared with the number of total neighbours of x and y. This metric does not have an equivalent weighted metric.

Adar/Adamic (AA). This metric also measures the intersection of neighbour-sets of two nodes in the graph, but emphasizing in the smaller overlap. We have defined a weighted version (WAA) of this metric.

Table 1. Similarity metrics

	Unweighted	Weighted
EW		$EW(x,y) = A_{xy}$
CN	$CN(x,y) = \|N(x) \cap N(y)\|$	$WCN(x,y) = \sum_{z \in N(x) \cap N(y)} A_{xz} + A_{yz}$
JN	$JN(x,y) = \frac{\|N(x) \cap N(y)\|}{\|N(x) \cup N(y)\|}$	
AA	$AA(x,y) = \sum_{z \in N(x) \cap N(y)} \frac{1}{log\|N(z)\|}$	$WAA(x,y) = \sum_{z \in N(x) \cap N(y)} \frac{A_{xz}+A_{yz}}{log(1+WD(z))}$
PA	$PA(x,y) = \|N(x)\| \cdot \|N(y)\|$	$WPA(x,y) = WD(x) \cdot WD(y)$

Preferential Attachment (PA). This metric is based on the consideration that nodes create links, with higher probability, with those nodes that already have a larger number of links. The similarity between nodes x and y is calculated as the product of the degree of the nodes x and y, so the higher the degree of both nodes, the higher is the similarity between them. The weighted version (WPA) is an improvement of the previous one, where the edge weights are taken into account when computing the degree of nodes x and y.

5 Comparative Evaluation

In this section we present a comparative evaluation of the different design decisions proposed by our approaches. The decisions are made along the following axes:

- The recommendation method used according to the graph employed –user-user graph and problem-problem graph.
- The similarity metric employed to find similar users or problems.
- The ranking method employed to select the best recommendations: based on similarity or based on a voting system (for the user-based approach only).

Finally, we think that the number of recommendations provided (k) is important so we will enrich the evaluation with the analysis of the influence of k in the performance of the recommendations.

5.1 Data and Experimental Setup

Acepta el reto (ACR) is an online judge created in 2014 at the Complutense University of Madrid (UCM). ACR was initially focused on the students of Computer Science at UCM, who had to resolve their programming assignments using this online judge. Nowadays, ACR is used by a large Spanish community and it is employed in several programming contests.

We carried out an exploratory analysis of the ACR submissions dataset in order to familiarize with the data contained in it and to find relevant information for our recommendation purposes. ACR stores 3,678 registered users, 289 problems and around 110,000 submissions (including resubmissions) at the time of this writing (March 2017).

Table 2. Descriptive analysis of the ACR submissions: the original dataset (*Raw*) and the filtered dataset (*Curated*).

Metric	*Raw*	*Curated*
# Submissions	110.364	25,151
# Problems	289	289
# Users	3,678	3,678
Density	0.10	0.02
Earliest submission	2014/02/17	2014/02/17
Latest submission	2017/02/13	2017/02/13
Time span	1092 days	1092 days
Problems		
Maximum # submissions per problem	5,613	1,290
Median # submissions per problem	232	52
Average # submissions per problem	381.88	87.03
Minimum # submissions per problem	8	3
# Problems with at least 10 submissions	276	253
# Problems with at least 50 submissions	216	138
# Problems with at least 100 submissions	146	56
Users		
Maximum # submissions per user	2,576	250
Median # submissions per user	10	3
Average # submissions per user	30.01	6.84
Minimum # submissions per user	1	1
# Users with at least 5 submissions	2,415	1,160
# Users with at least 10 submissions	1,790	606
# Users with at least 20 submissions	1,198	220
Verdicts		
# Submissions with AC-PE	36,824	18,067
# Submissions with CE	7,061	465
# Submissions with Runtime-Limit Error	31,924	1,803
# Submissions with Wrong Answer	33,443	3,146

According to the attempted-solved states described in Sect. 2, we redefine the submission dataset, removing all submissions for a user-problem tuple except the last solved submission for this user-problem tuple, or the last attempt for this user-problem tuple, if a solved submission does not exist. The final number of submissions considered in the dataset drops to 25,151. Table 2 depicts a catalogue of descriptive statistics, as proposed in [2], about both datasets: the original (*Raw*) submission dataset and the filtered (*Curated*) dataset.

For doing our experimental evaluation, we have implemented 40 recommender systems following the combination of all design decisions described above and using the ACR curated dataset. All the experiments described later were repeated for different recommendation lists of size $k \in [1, 10]$.

The evaluation process starts splitting the dataset into two sets using a particular timestamp t:

- A training set, for building the interaction graph used by our recommendation system. It contains the accepted submissions made *before* time t.
- An evaluation set, for validating the recommendations. It contains the attempted submissions made *after* time t.

We select the target users involved in the evaluation considering only *regular users* –those users who used the system regularly. To do that, we will only consider those users who had attempted a minimum number of problems before and after time t. In this experiment, we choose the users who attempted or solved at least 5 problems. The date selected to split the itineraries was 2016-10-20 because that timestamp allowed us to build the largest test set with 65 users with at least 5 problems before and after the timestamp.

The problem-problem and user-user graphs were created using the training set. The former graph has 169 nodes and 14,149 edges, with a density of 49.5%. The user-user graph is bigger, with 2007 nodes and 624,206 edges, but with a lower density, 15.5%. Both graphs were filtered removing the edges whose weight was less than 5, leaving 10,837 edges (37.9%) in the problem-problem graph, and 16,805 edges (0.4%) in the user-user graph.

For each target user and for each recommender, we create a list with the k recommended problems. To test the recommendations, we compare the recommended problems with the problems attempted by the target users, contained in the evaluation set. The recommendations are evaluated using the following standard evaluation metrics for recommender systems [9]:

- *Precision*, *Recall* and *F-Score* in top k recommendations.
- *At least one hit* (1-hit): ratio of recommendations in which at least one recommended problem was attempted by the user. It corresponds to the metric *Success@k* with a success condition of guessing right at least one problem.
- *Mean Reciprocal Rank* (MRR): it evaluates the quality of a ranked list of recommendations based on the position of the first correct item. Since we only provide one list of recommendations per user, the MRR can be computed as $MRR = 1/_{rank_i}$, where $rank_i$ is the position of the first attempted problem in the recommendation list.

5.2 Results

Table 3 summarizes the results of the evaluation using $k = 5$ and computing the average values for all the users involved in the evaluation. If we compare both the user-based and the problem-based approaches using the same method to rank

Table 3. Evaluation results by similarity metrics and recommendation methods, using the similarity for ranking the recommendations of size $k = 5$. Best results between user and problem based-approaches are in bold. The best evaluation metric values are marked with *.

Sim	Method	Precision	Recall	F1	MRR	1-hit
CN	Prob-Prob	0.111	0.055	0.070	0.194	0.400
	User-User	**0.182**	**0.100**	**0.122**	**0.398**	**0.523**
AA	Prob-Prob	0.129	0.063	0.081	0.241	0.446
	User-User	**0.148**	**0.076**	**0.094**	**0.388**	**0.462**
JN	Prob-Prob	0.098	0.044	0.057	0.185	0.323
	User-User	**0.123**	**0.057**	**0.073**	**0.318**	**0.400**
PA	Prob-Prob	**0.126**	**0.061**	**0.078**	**0.329**	**0.462**
	User-User	0.123	0.057	0.073	0.318	0.400
EW	Prob-Prob	**0.268***	**0.136**	**0.172***	**0.486***	**0.677**
	User-User	0.175	0.092	0.114	0.344	0.462
WCN	Prob-Prob	**0.252**	**0.132**	**0.165**	**0.421**	**0.785***
	User-User	0.114	0.051	0.065	0.309	0.369
WAA	Prob-Prob	**0.249**	**0.131**	**0.164**	**0.423**	**0.785***
	User-User	0.114	0.051	0.065	0.309	0.369
WPA	Prob-Prob	**0.265**	**0.137***	**0.172***	**0.428**	**0.785***
	User-User	0.123	0.057	0.073	0.318	0.400

the recommendations (the similarity), we can see that the performance depends on the similarity metric. Problem-based approaches using weighted metrics (EW, WAA, WCN and WPA) always obtain better results in the evaluation metrics than the user-based approaches. However, the unweighted metrics perform best with the user-based approaches, except the Preferential Attachment (PA), which shows slightly better results in combination with the problem-based approach.

In our previous work [4] related with the problem-based approach, Edge Weights (EW) obtained the best results in all the evaluation metrics proposed. With the new dataset, this metric continues obtaining the best results in Precision, Recall, F1 and MRR. However, WCN, WPA and WAA outperform EW with the 1-hit metric.

When selecting the user-based approach and using the similarity to rank the recommended problems, CN similarity metric yields the best results. Unweighted metrics achieve slightly better results than the weighted ones. The exception is EW, which achieves better or similar precision, recall, F1 and 1-hit values in comparison with AA, JN and PA, and better MRR values in comparison with JN and PA.

It is worth highlighting that using the voting systems to rank the recommendations we achieve better recommendation results (see Table 4). In comparison with the results in Table 3, user-based recommendation approaches using a voting

Table 4. Evaluation results of the user-based approach with variations of similarity metrics and ranking methods for recommendations of size $k = 5$. Best results between user and problem based-approaches are in bold. The best evaluation metric values are marked with *.

Sim	Ranking	Precision	Recall	F1	MRR	1-hit
CN	Similarity	0.182	0.100	0.122	0.398	0.523
	Simple	0.274	0.142	0.179	0.472	0.754
	Weighted	0.314	0.166	0.208	0.474	**0.831**
	Positional	**0.357***	**0.192***	**0.239***	**0.588**	**0.831**
AA	Similarity	0.148	0.076	0.094	0.388	0.462
	Simple	0.274	0.142	0.179	0.472	0.754
	Weighted	0.323	0.170	0.214	0.482	0.831
	Positional	**0.351**	**0.189**	**0.235**	**0.567**	**0.877***
JN	Similarity	0.123	0.057	0.073	0.318	0.400
	Simple	0.274	0.142	0.179	0.472	0.754
	Weighted	0.274	0.143	0.181	0.442	0.754
	Positional	**0.348**	**0.186**	**0.232**	**0.589**	**0.800**
PA	Similarity	0.123	0.057	0.073	0.318	0.400
	Simple	**0.274**	0.142	0.179	**0.472**	**0.754**
	Weighted	**0.274**	**0.143**	**0.181**	0.442	**0.754**
	Positional	0.252	0.130	0.164	0.409	0.754
EW	Similarity	0.175	0.092	0.114	0.344	0.462
	Simple	0.345	**0.187**	0.233	0.587	0.815
	Weighted	**0.348**	**0.187**	**0.234**	0.598	**0.846**
	Positional	0.342	0.185	0.230	**0.602***	0.800
WCN	Similarity	0.114	0.051	0.065	0.309	0.369
	Simple	0.274	0.142	0.179	0.472	0.754
	Weighted	**0.289**	**0.152**	**0.191**	**0.488**	**0.800**
	Positional	0.271	0.144	0.180	0.439	0.785
WAA	Similarity	0.114	0.051	0.065	0.309	0.369
	Simple	0.274	0.142	0.179	0.472	0.754
	Weighted	**0.317**	**0.168**	**0.211**	**0.494**	**0.831**
	Positional	0.283	0.147	0.186	0.436	0.800
WPA	Similarity	0.123	0.057	0.073	0.318	0.400
	Simple	0.274	0.142	0.179	0.472	0.754
	Weighted	**0.280**	0.146	0.184	**0.450**	0.785
	Positional	**0.280**	**0.151**	**0.187**	0.422	**0.815**

system yield better results than the problem-based approaches, no matter the similarity metric employed. It seems that the positional voting system works better in combination with the unweighted metrics, while the weighted voting system enhances the recommendation results with the weighted voting systems. This fact makes sense because the positional voting removes the real differences in similarity among the problems in a list, keeping only its position in that list.

As occurred when comparing user and problem-based methods, PA and WPA break the rule. We suppose that the reason for this behaviour is due to their nature because Preferential Attachment takes into account the importance of the nodes in the ends of the edge, instead of the relationship between both nodes in terms of edge weight or their neighbours. However, this fact needs a deeper analysis.

In our previous work [4], we analysed the evolution of the precision, 1-hit and MRR metrics when we increase the number of recommendations (parameter k) from 1 to 10. We repeated the analysis with the approaches proposed in our current work. For the sake of simplicity, Fig. 2a shows the evolution for the best similarity metrics (CN, AA, JN y EW) for the user-based recommendation method when the problems are ranked using the similarity. We obtain a similar behaviour using the problem-based approach. As expected, precision value decreases slowly and 1-hit increases with k because the probability of guessing right at least one problem increases as we make more recommendations. The trade off between the quality of recommendations and the number of choices available to the users is not easy to decide, and we will have to perform some tests with real users to adjust it. Finally, MRR values slightly increase as long as the k parameter grows asymptotically to a constant value.

The evolution of the precision, 1-hit and MRR metrics when we increase the number of recommendations follows the same tendency when the problems provided by a user-based recommender are ranked using a voting system, instead of

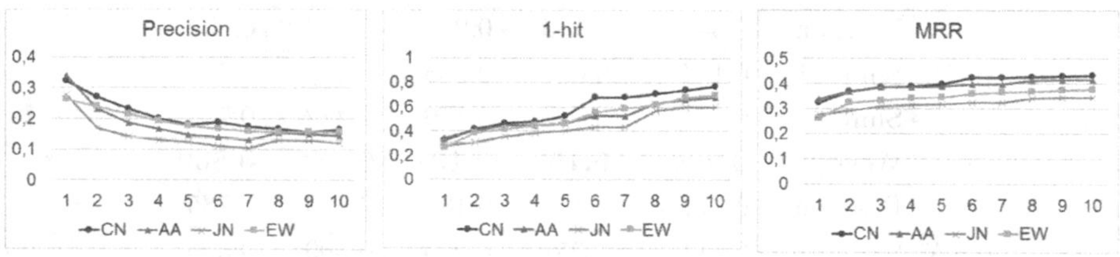

(a) User-based method and ranking recommended problems by similarity.

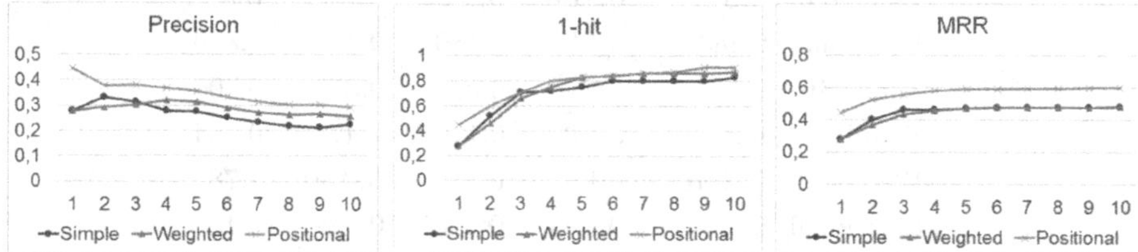

(b) User-based method with CN similarity, modifying the voting system.

Fig. 2. Precision, 1-HIT and MRR evolution when increasing the number of recommended problems k.

the similarity. As we can see in Fig. 2b, 1-hit and MRR values grow asymptotically to a constant value, while the precision decreases when we increase the number of recommended problems. Although the Figure only shows the behaviour for the recommender that uses the similarity metric (CN), which performed the best results, this behaviour is replicated when using the other similarity metrics.

6 Related Work

An online judge like ACR can be seen as an online repository that stores a large amount of resources. These repositories traditionally suffer the problem of how to find resources that best match the user's knowledge or his/her preferences. Recommendation systems help users in this task recommending items *similar* to those a user has liked in the past [8]. *Similarity* is therefore one of the most important metrics in these systems.

Some recommendation approaches that aim to suggest resources apply collaborative filtering, a technique that imposes that the user must rate the resources in order to find similar preferences. Commonly, online judges like ACR hardly ever provide mechanisms to rate problems, so these techniques cannot be applied. However, the review of these techniques and the alternative user-based and item-based approaches described by Sarwar *et al.* [10] inspired us to explore both solutions in our recommendation problem.

On the other hand, content-based techniques require descriptions of the item characteristics and user profiles that describe the interests of that user. ACR problems are tagged with metadata about the programming concepts needed to solve them. However, our previous work [4] highlighted that using the implicit similarity between problems in the problem-based approach can yield better results than using a content-based approach based on this problem metadata.

Other authors consider that the process of recommending items to users can be considered as a link prediction problem in the user-item bipartite networks [1]. For this reason, we have reviewed some literature in the use of link prediction applied in recommender systems, concentrating our work on the similarity-based methods, which employ different similarity metrics in order to predict new links. Liben-Nowell and Kleinberg [6] systematically compared some neighbour, path and random walk based node similarity indices for link prediction problem in co-authorship networks. These algorithms have been also applied in the user-product bipartite graphs in recommender systems and its performance has been evaluated with a Flickr dataset, outperforming collaborative filtering methods in some cases [1].

Finally, most of the studies on link prediction focused on unweighted networks but ignored the link weights. Proximity between nodes can be estimated better by using both graph proximity measures and the weights of existing links. The work in [7] proposed a simple way to extend similarity metrics for binary networks to weighted metrics. However, the latter performed even worse in several real networks. In contrast, our experimental results stress that weighted and unweighted metrics perform different depending on the recommendation approach and, therefore, the graph employed in the recommendation process.

7 Conclusions and Future Work

Online judges lack tools that help a user to find relevant programming problems to solve. Some online judges, like ACR, have the problems categorized with labels, but this information is limited in order to help users to find which problem should resolve next.

In this paper, we have extended our previous work [4] that exploits the implicit knowledge included in the user-problem relationships created when users submit solutions to the problems stored in the online judge. In that work we proposed a problem-based approach that looks for similar problems to the ones that a user has previously solved. In our current work, we have proposed a user-based approach, which relies on selecting problems they have already solved for similar users to the target user of the recommendation. As in our previous work, we have employed different similarity metrics, inspired in the similarity-based link prediction techniques coming from social network analysis.

We have compared the experimental results generated by these approaches in combination with several similarity metrics and the analysis shows that the selection of the highest-performing similarity metric is crucial in order to achieve the best results. Problem-based approaches yield better using weighted metrics, while user-based approaches obtain better results with the unweighted metrics. We expect that our future works on the analysis of these metrics will reveal more insights about this behaviour.

Additionally, we have proposed three alternative voting systems that significantly improve the results of the user-based approach in comparison with the problem-based recommendation, no matter the similarity metric employed. Future work will continue exploring these voting systems and analysing the impact of including them in the problem-based approach.

Finally, the edge weights in the graphs employed in our approaches have a great impact in the similarity measures and, therefore, in the recommendation. A preliminary analysis reveals that the methods employed to compute these weights and to reduce the graph density using edge weights affects the recommendation results. For this reason, we plan to review and apply new techniques employed in social network analysis for weighting and filtering social graphs.

References

1. Chiluka, N., Andrade, N., Pouwelse, J.: A link prediction approach to recommendations in large-scale user-generated content systems. In: Clough, P., Foley, C., Gurrin, C., Jones, G.J.F., Kraaij, W., Lee, H., Mudoch, V. (eds.) ECIR 2011. LNCS, vol. 6611, pp. 189–200. Springer, Heidelberg (2011). doi:10.1007/978-3-642-20161-5_19
2. Dooms, S., Bellogín, A., De Pessemier, T., Martens, L.: A framework for dataset benchmarking and its application to a new movie rating dataset. ACM Trans. Intell. Syst. Technol. **7**(3), 1–28 (2016)
3. Furht, B.: Handbook of Social Network Technologies and applications. Springer Science & Business Media, New York (2010)

4. Jimenez-Diaz, G., Gómez Martín, P.P., Gómez Martín, M.A., Sánchez-Ruiz, A.A.: Similarity metrics from social network analysis for content recommender systems. In: Goel, A., Díaz-Agudo, M.B., Roth-Berghofer, T. (eds.) ICCBR 2016. LNCS (LNAI), vol. 9969, pp. 203–217. Springer, Cham (2016). doi:10.1007/978-3-319-47096-2_14
5. Kurnia, A., Lim, A., Cheang, B.: Online judge. Comput. Educ. **36**(4), 299–315 (2001)
6. Liben-Nowell, D., Kleinberg, J.: The link-prediction problem for social networks. J. Am. Soc. Inf. Sci. Technol. **58**(7), 1019–1031 (2007)
7. Lü, L., Zhou, T.: Link prediction in weighted networks: the role of weak ties. Europhys. Lett. **89**(1), 18001 (2010)
8. Ricci, F., Rokach, L., Shapira, B. (eds.): Recommender Systems Handbook. Springer US, Boston (2015)
9. Said, A., Bellogín, A.: Comparative recommender system evaluation. In: Proceedings of the 8th ACM Conference on Recommender Systems - RecSys 2014, pp. 129–136 (2014)
10. Sarwar, B., Karypis, G., Konstan, J., Riedl, J.: Item-based collaborative filtering recommendation algorithms. In: Proceedings of the 10th International Conference on World Wide Web, pp. 285–295. ACM (2001)
11. Ángeles-Serrano, M., Boguná, M., Vespignani, A.: Extracting the multiscale backbone of complex weighted networks. Proc. Nat. Acad. Sci. **106**(16), 6483–6488 (2009)
12. Wang, P., Xu, B., Wu, Y., Zhou, X.: Link prediction in social networks: the state-of-the-art. Sci. Chin. Inf. Sci. **58**(1), 1–38 (2014)

Tetra: A Case-Based Decision Support System for Assisting Nuclear Physicians with Image Interpretation

Mohammad B. Chawki[1], Emmanuel Nauer[2(✉)], Nicolas Jay[1,2], and Jean Lieber[2]

[1] Service d'évaluation et d'information médicales, Centre Hospitalier Régional Universitaire de Nancy, Nancy, France
{m.chawki,n.jay}@chru-nancy.fr
[2] UL, CNRS, Inria, 54000 Nancy, France
{emmanuel.nauer,jean.lieber}@loria.fr

Abstract. This paper shows how nuclear image interpretation is improved by Tetra, a case-based decision support system. Tetra exploits two kinds of knowledge sources: ontologies and knowledge embedded in past nuclear imaging reports, each imaging report being associated with a case, described by some features and its associated diagnoses. Ontologies are used, in addition with vocabulary resources, to semantically annotate the imaging reports. Links between case features and diagnoses in the training case base have been computed. In practice, when a new image test is run, Tetra exploits this *features/diagnosis* links, as well as the generalization/specialization relation of the ontologies to retrieve the cases that are the most similar to the new image test and to compute the most probable diagnoses. 8000 nuclear imaging reports have been collected to create a case base and almost 1000 other imaging reports have been used for the system evaluation, which shows that Tetra gives good results for the two diagnoses (necrosis and ischemia) which have been considered in this work. The first results shows that an ontology-based similarity computation between cases in order to display the most similar cases as well as the diagnosis probability computation helps the nuclear physician in her image interpretation task.

Keywords: Case-based reasoning · Case similarity · Knowledge · Medical diagnosis · Decision support system · Nuclear medicine · Imaging report

1 Introduction

This paper shows how the nuclear image interpretation is improved by Tetra, a case-based decision support system. Indeed, imaging, in particular in nuclear medicine, is getting more and more complex over the years. Each year, new

D.W. Aha and J. Lieber (Eds.): ICCBR 2017, LNAI 10339, pp. 108–122, 2017.
DOI: 10.1007/978-3-319-61030-6_8

> Mr. Ibrahim, 55 years old, has diabetes and arterial hypertension for 15 years. He has come today to have a myocardial scintigraphy, in the context of screening. He is currently asymptomatic. Seeing the images, the resident of nuclear medicine hesitates and seeks the advice of a senior physician, specialist of myocardial scintigraphy, who is busy. Once free, the professor remembered two similar cases: Mr. Pierre who had a myocardial ischemia without necrosis, but also Mr. Lee who had sequelae of myocardial necrosis, hardly visible on the scintigraphy. Once Mr. Ibrahim's images have been compared to images of these similar cases, the resident concludes easily.

Fig. 1. A typical example of a use case (which has been anonymized) requiring expert knowledge for nuclear image interpretation.

radiotracers[1] and machines are developed and tested. Despite this rapid evolution, few studies address the issue of image interpretation and imaging report. Even if some works propose to improve protocol appropriateness using decision support systems, no work addresses, to the best of our knowledge, this issue from a knowledge point of view, for example, by exploiting knowledge stored in ontologies or knowledge extracted from past experiences of image interpretations. The TETRA system, presented in this paper, aims at exploiting two kinds of knowledge sources: ontologies and past nuclear imaging reports, each imaging report being considered as a case, described by some features and its associated diagnosis. The idea is to exploit these two kinds of knowledge in order to display the most similar cases as well as the diagnosis probability to help a nuclear physician in her image interpretation task. For this, we propose first a way to compute links between case features and diagnosis. These links are used, as well as the ontological generalization/specialization relation, to retrieve the cases the most similar to the new image test and to compute the most probable diagnoses. A first evaluation shows that this approach gives very good results for the two diagnoses (necrosis and ischemia) which have been considered in this work.

This paper is organized as follows. Section 2 introduces a use case and gives the motivation of this work. Section 3 describes the TETRA system. Section 4 presents and discusses first results about the evaluation of TETRA. Section 5 concludes the paper.

2 Context and Objectives

Use-case. Figure 1 introduces a typical example of use case about the activity of a resident of nuclear medicine. When an image has to be interpreted, expert knowledge is required. Past interpretation experiences (i.e. cases) are important in such an activity because they contain knowledge that will be helpful for the resident. This latter can compare the images and the clinical context of her patient with images of these past interpretation episodes, which facilitates the interpretation and the writing of the imaging report. All this activity could be

[1] A radiotracer is a molecule in which one atom has been replaced by a radioisotope, in order to trace the path of this molecule and to explore some biological pathways.

done automatically and quickly by an adapted computer system, and especially by a case-based reasoning (CBR) system [1].

Motivation. Developing a new CBR system taking into account the imaging protocol specific to each radiotracer, the imaging reports, the Electronic Health Records (EHR) and clinical chemistry tests is a way to facilitate practicing by showing to the practitioner the most similar profiles already treated, either locally or in another medical center.

The knowledge which is required to be a competent practitioner has increased exponentially since the beginning of nuclear medicine. Many new radiotracers have been developed lately, which gradually have transformed nuclear medicine into molecular imaging, even though we sorely lack information for image interpretation. Indeed, each radiotracer has its own false positives and its own false negatives, depending on medical background, clinical signs and results of laboratory tests, and a unique way of interpreting these data. In the meantime, the training time for residents did not increase and training becomes more and more difficult and complex.

Combining the clinical data available in the refering clinician order entry and the imaging report in order to improve the selection of diagnoses has already been proposed [2]. In this work, the expertise of the clinicians provides information about the patient and the context in which the image examination took place. This information is used by the radiologists to improve the image interpretation. From a pratical point of view, this kind of work requires a lot of time from the physicians to establish the diagnosis. For this purpose, a meta-analysis has shown that EHR based interventions improve appropriateness and reduces the use of radiological tests [3]. These two works show that exploiting clinical information improve image interpretation. However, this kind of approach does not take benefits from existing domain knowledge, such as the one that could be found in ontologies, nor on semantic similarity approaches that could be used to automatize some correlation computations.

In computer science, many studies use ontologies to structure knowledge in a computer usable form (e.g. a class hierarchy) to build smarter systems based on knowledge exploitation. For example, in the domain of cooking, the TAAABLE system exploits an ontology of the cooking domain to adapt cooking recipies [4]. In medicine, [5] proposes to use SNOMED-CT (Systematic NOMencalture MEDical - Clinical Terms, http://www.snomed.org/) which is a medical ontology known for its accuracy, to annotate EHR using Natural Language Processing (NLP), and to compute EHR similarities.

Moreover, all medicine is currently based on evidence-based medicine (EBM) [6]. EBM is a paradigm stipulating that decision in medicine must be based on evidence contained in reliable studies, that is to say studies with significant statistical power and therefore many patients. CBR is thus a very interesting way to solve medical problems, because of the number of cases and pieces of information within these cases that can be used. But CBR alone is not sufficient to compute reliable solutions [7], especially with the modern EHR, that contain more and more information for each patient [8]. Many solutions have been found to complete CBR paradigm in the medical domain: [9] proposes the use of prototypes which are the generalization of cases and [10] has shown applications for patient suffering from diabetes, whereas [11] proposed to exploit rules for reasoning. In these works, ontologies and statistics complete CBR.

The idea of using clinical particularities to search and expose to the physician the most similar cases compared to the physician's new case has already been tested. It has been shown that the best model between some techniques of machine learning and the simple use of minimum Euclidean distance was this latter [12]. Despite the fact that no ontology was used in this study, good results have been shown. With a system exploiting ontology knowledge and specialization/generalization relationship in particular, results may be even better because some more general features will arise from specialised ones. The KnowBaSICS-M system searches for UMLS ontology concepts in a physician text describing a *Medical Computational Problem* using NLP and it searches for the solutions of the most similar problems compared to the physician one. Precision and accuracy of the solutions are good, especially because of explicit filtering of concepts by physicians about their requests [13]. This work shows also that extracting UMLS or, more generally, ontology concepts by simple NLP techniques in plain text is not a difficult task. A team of the University of Missouri even created a physician decision support system called OntoQuest which purpose is to display a list of similar historical patients with decisions made by physicians about them. Similarity was calculated using sets of ICD-9 diagnoses with several algorithms. Three physicians (a pathologist, a pediatrician and an ENT[2] physician) were asked to score the similarity for 100 patient pairs randomly chosen. The overall coefficient of correlation between the median physician similarity and the ontological similarity is 0.88 which is a good correlation [14]. All these studies show good results, but it is difficult to judge their benefit because the evaluation is often qualitative or subjective. No study addresses how the image diagnosis could be improved using EHR, and no study addresses how nuclear medicine diagnosis could be improved using a decision support system. In fact, very few studies address nuclear medicine problem from a computer science point of view.

3 The Tetra system

The Tetra system presented in this work exploits ontologies. The ontologies are used to characterize cases, either directly or by semantically annotating nuclear imaging reports in order to create a case base. Each case represents an image interpretation episode. Figure 2 gives an example of a case, described by a set of features (previous diseases, radiological signs and biological information), and its associated diagnosis. When a new nuclear imaging test is run, it is compared to previous cases according to a case similarity based on their features and on the ontological knowledge. The most similar cases, as well as the more probable diagnosis are presented to the nuclear physician in order to help her to interpret it more efficiently and more quickly.

[2] ENT: ear, nose, and throat.

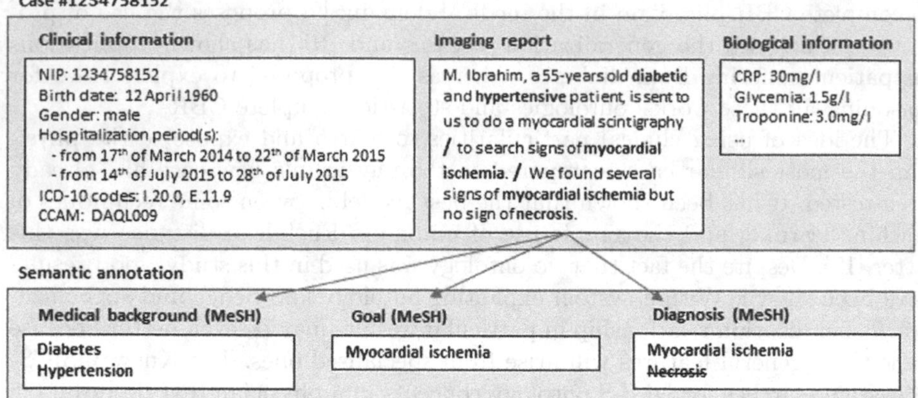

Case #1234758152
Clinical information
NIP: 1234758152
Birth date: 12 April 1960
Gender: male
Hospitalization period(s):
- from 17th of March 2014 to 22th of March 2015
- from 14th of July 2015 to 28th of July 2015
ICD-10 codes: I.20.0, E.11.9
CCAM: DAQL009
Imaging report
M. Ibrahim, a 55-years old diabetic and hypertensive patient, is sent to us to do a myocardial scintigraphy / to search signs of myocardial ischemia. / We found several signs of myocardial ischemia but no sign of necrosis.
Biological information
CRP: 30mg/l
Glycemia: 1.5g/l
Troponine: 3.0mg/l
...
Semantic annotation
Medical background (MeSH)
Diabetes
Hypertension
Goal (MeSH)
Myocardial ischemia
Diagnosis (MeSH)
Myocardial ischemia
Necrosis

Fig. 2. Example of a case, with its semantic annotation, automatically extracted using NLP techniques.

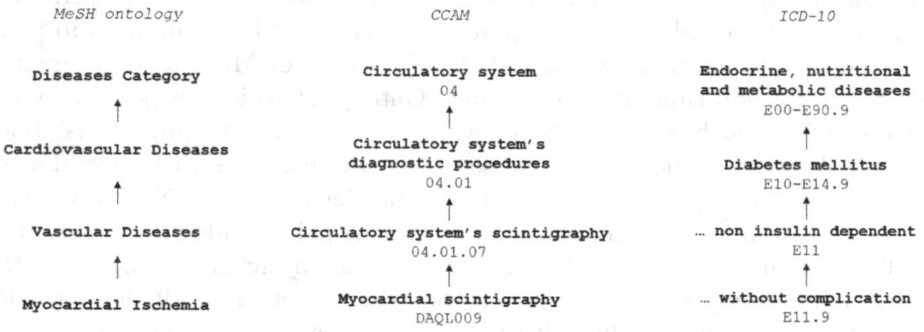

Fig. 3. Examples of MeSH ontology, CCAM and ICD-10 hierarchical organization of diseases.

TETRA knowledge. Three knowledge sources are used:

- ICD-10 (http://www.who.int/classifications/icd/en/): ICD-10 is a classification of medical diagnoses, currently used by most of the health systems in the world, including in France and USA, to encode medical stays. These codes can be used for epidemiology and research purposes. Figure 3 gives an illustration of some hierarchies coming from the ICD-10 disease organization.
- CCAM (http://www.ameli.fr/accueil-de-la-ccam/index.php): an classification of clinical procedures (in French).
- MeSH (https://www.ncbi.nlm.nih.gov/mesh): a thesaurus about diagnoses, signs and symptoms, including terminological data about vocabulary, eg. symptoms.

The two first resources are required because they are directly used in EHR to encode stays and clinical procedures. The last one is used as reference to build the semantic annotation of imaging reports.

TETRA data. All the imaging reports of myocardial scintigraphy from 2014 to early 2016 done in the Centre Hospitalier Régional Universitaire (CHRU) of Nancy have been extracted. That represents 8905 imaging reports. Amoung these 8905 imaging reports, 8000 have been choosen randomly to constitute the TETRA case base, and the remaining 905 have been used for the evaluation. All available EHR for the 8905 cases have been collected. Thus, for each hospitalization of each patient, the full name, the birth date, the list of ICD-10 codes (like: I63.0, G46.8, etc.) and the hospitalization dates are available. Clinical chemistry tests have also been extracted. The full name, the birth date, the complete list of results of biological tests with their dates are obtained for each patient through lines of the form: "`Dupont;Pierre;12/04/1965;21/04/2015;CRP;15.8;mg/L`" which meaning is *"Mr. Dupont Pierre is born on the* 12th *of April 1965 and his CRP was measured at 15.8 mg/L on the* 21st *of April 2015"*. For each line, the full name, the birth date and the report itself are obtained.

TETRA case base. The case base CB is a set of past nuclear image interpretations, according to a given protocol (e.g. PET-scan). A case $C = (Pb(C), Sol(C))$ associates to a problem description $Pb(C)$ its solution $Sol(C)$. Concretely, $Pb(C)$ represents features of the imaging context and $Sol(C)$ is the set of diagnosis found after the image interpretation. More formally, Pb is a triple (CI, G, B), where CI represents clinical information, G is the goal of the imaging test (i.e. the image indication), and B is a set of biological tests:

- CI, the clinical information, is composed of A, the age of the patient (in years) at the time of the imaging test, Ge, her gender, H, the presence of a precedent hospitalization, R, the information about the examination is a revaluation or not, and D, the set of diseases that had been coded during the patient previous hospitalization stays. These diseases are represented using ICD-10 codes.
- G, the goal of the imaging test, is the set of diagnoses found in the test indication. Diagnoses are represented using MeSH concepts. $Sol(C)$ is also a set of diagnoses represented using MeSH concepts, but this time, found in the image interpretation.
- B, the biological tests, is a set of numerical variables associated to their value and unit (e.g. "glycemia = 1.5g/L").

D is extended by adding all the codes of ICD-10 which are more generic than the ones belonging to D. G (resp. $Sol(C)$) is extended by adding all the MeSH concepts which are more generic than the ones belonging to G (resp. $Sol(C)$). $Pb(C)$ is encoded as a vector $(A, Ge, H, R, D_1 \ldots D_m, G_1 \ldots G_n, B_1 \ldots B_p)$ where $D_1 \ldots D_m$, (resp. $I_1 \ldots I_n$) are all the possible ICD-10 codes (resp. MeSH concepts) dimensions used in all the cases of CB, $B_1 \ldots B_p$ are the biological test dimensions. The value of D_i (resp. G_j) is 1 if the case is described by the D_i disease (resp. G_j indication) and 0 otherwise. A, $B_1 \ldots B_p$, which are numerical values are normalized between 0 and 1 according to their respective minimum min_A and maximum max_Ain the case base:

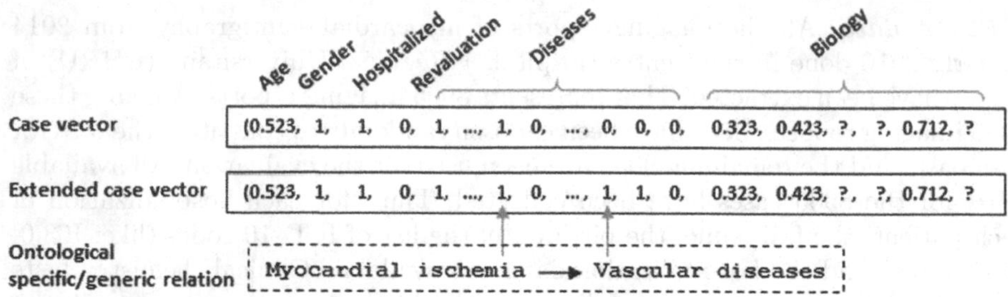

Fig. 4. Example of a vectorial case representation and its extension using hierarchical knowledge.

$$A = \frac{PatientAgeValue - min_A}{max_A - min_A} \qquad B_i = \frac{PatientB_iValue - min_{B_i}}{max_{B_i} - min_{B_i}}$$

The revaluation R is equal to 1 when the examination is a revaluation and 0 when it is a screening. The gender Ge is simply equal to 0 when the patient is a female and 1 when the patient is a male. The "already hospitalized" H is equal to 1 when the patient has already been hospitalized at Nancy's hospital and 0 when she has not. Figure 4 illustrates how a case is represented through a vector and how a patient specific disease (e.g. Myocardial ischemia) produces a vector extension (e.g. on the ischemia dimension), using ontological knowledge. $Sol(C)$ is also encoded as a vector $(D_1, \ldots, D_q)$ where $D_1, \ldots, D_q$ are all the possible diagnosis of the image interpretation. $D_i = 1$ if the image interpretation shows the presence of the diagnosis D_i, otherwise $D_i = 0$. Let d_{pb} (resp. d_{sol}) be the dimension of the vector space used for the problem (resp. the solution) representation and $Pb(c).f$ (resp. $Sol(c).f$) be the value of the feature f of the problem description (resp. solution) of the case c. A same patient can be represented by several cases. This representation allows to consider the case and data surrounding the imaging report like a patient at a given time, so the considered clinical and biological data are always an available part of the patient past at the time of the imaging test.

Semantic annotation process. Nuclear imaging reports, which are full text, are semantically annotated by searching in the text the textual representations associated to the classification concepts, in order to produce the G_i and D_j parts of the cases. Figure 2 gives an example of a case and illustrates the result provided by the annotation process. The imaging report full text is first split in three parts: the patient medical background, the goal of the imaging test, and the image interpretation. MeSH concepts are then extracted from full text. As the MeSH contains a large list of terms associated to each concept, including singular/plurial, lexical, and synonymous variations in several languages, the NLP process consists simply in searching in the text the occurrence of the possible forms of each concept. However, the MeSH vocabulary has required to be completed by some missing variations. For example, the French *"HTA"* abbreviation was missing for the *"arterial hypertension"* concept, and the English *"coronary insufficiency"* synonym was missing for *"coronaropathy"*.

The context of apparition of the concept in the text is also analyzed using single pattern rules to determine its presence (e.g. *"We found several signs of myocardial ischemia"*) or its absence (e.g. *"no sign of necrosis"*). Finally, after a context analysis, only concepts which are present are kept.

Similarity computation. The objective of the system is to provide to the physician, for a new target problem T: information about potential diagnoses ranked by decreasing probabilities, and the most similar cases ranked by decreasing similarity. Two preliminary steps are required to compute these two pieces of information. The first step consists in computing on CB, the relevance of a feature w.r.t. a given diagnosis. The second step consists in computing the similarity between two cases w.r.t a given diagnosis.

Relevance of a feature for a given diagnosis. Computing the relevance of each case feature w.r.t. each diagnosis is required because most of the imaging protocols diagnose several illnesses, generally not contradictory but which causes, signs, symptoms and complications are different. For example, whether the patient has or has not a cancer is not relevant for finding an ischemia during a cardiac tomoscintigraphy, whereas the age or the smoking status are important.

Using a relevance matrix of features to compute the most similar cases according to the diagnosis has already been used several times. Richter et al. used this approach in the PATDEX/2 diagnosis system with only binaries features [15]. Stram et al. proposed to adapt the relevance matrix generation for any type of feature, not only the binaries ones [16]. In these two sudies initial relevances are computed from a given set of cases using statiscal techniques.

The relevance of each case feature for each diagnosis is calculated as follows. $F(d)$, the frequency of the diagnosis d present in the image interpretation is given by the ratio between the number of case containing d in its solution, and the total number of cases of the case base:

$$F(d) = \frac{|\{c \mid d \in Sol(c)\}|}{|CB|}$$

For each diagnosis d, CB is divided into two subsets:

- $C(d)^+$, the set of the positive cases c_i, in which the patient suffers from d (i.e. d appears in $Sol(c_i)$):

$$C(d)^+ = \{c_i \mid d \in Sol(c_i)\}$$

- $C(d)^-$, the set of the negative cases, in which the patient does not suffer from d (i.e. d does not appear in $Sol(c_i)$):

$$C(d)^- = \{c_i \mid d \notin Sol(c_i)\} = CB \setminus C(d)^+$$

Then, the proportion of a feature f, w.r.t. a diagnosis d, in the positive and the negative case subsets are computed as:

$$p^{+}(f|d) = \frac{\sum_{i=1}^{|C(d)^{+}|} Pb(c_i).f}{|C(d)^{+}|} \qquad p^{-}(f|d) = \frac{\sum_{i=1}^{|C(d)^{-}|} Pb(c_i).f}{|C(d)^{-}|}$$

where $|S|$ is the number of elements in a set S.

Finally, a relevance $r(d, f)$ of a case feature f w.r.t. a diagnosis d is computed as the difference of these two proportions:

$$r(d, f) = p^{+}(f|d) - p^{-}(f|d) \in [-1; 1]$$

The relevance of a feature appearing in the same proportion in the set of the positive cases and in the set of the negative cases will be closed to 0, whereas the relevance of a feature more frequent in one of the positive or negative case subset while quite absent in the other will be close to -1 or 1 More the feature relevance is closed to -1 or 1, more the feature will be considered as relevant for d, and thus will influence much more the probability of the diagnosis d.

Case similarity w.r.t a given diagnosis. The similarity $sim_d(c_1, c_2)$ between two cases c_1 and c_2 w.r.t a diagnosis d is computed as:

$$sim_d(c_1, c_2) = 1 - \frac{\sum_{i=1}^{d_{pb}} |Pb(c_1).f_i - Pb(c_2).f_i| \times |r(d, f_i)|}{\sum_{i=1}^{d_{pb}} |r(d, f_i)|} \in [0; 1]$$

Indeed, the similarity between two cases according to a given diagnosis depends on the difference and the relevance of their features. For example, if the supposed diagnosis is ischemia, the gender and the smoking status play a more important role than the presence of cancer or psoriasis.

Diagnosis probability computation. The probability to find a diagnosis d in the image interpretation of the target case T is:

$$p(d, T) = \frac{\sum_{i=1}^{|\mathrm{CB}|} Sol(c_i).d \times sim_d(c_i, T)}{\sum_{i=1}^{|\mathrm{CB}|} sim_d(c_i, T)} \in [0; 1]$$

where $Sol(c_i).d$ is the value on the dimension d of $Sol(c_i)$, and represents a diagnosis which is present (when $Sol(c_i).d = 1$) or is absent (when $Sol(c_i).d = 0$).

Case similarity computation. In order to retrieve the cases which are the most similar to a target case T, similarity between cases must be computed.

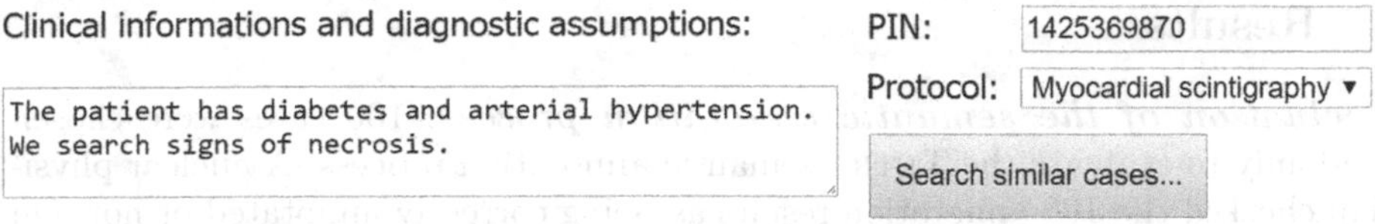

Clinical informations and diagnostic assumptions: PIN: 1425369870

Protocol: Myocardial scintigraphy ▾

The patient has diabetes and arterial hypertension.
We search signs of necrosis.

Search similar cases...

Fig. 5. The TETRA user query interface.

Patient id	Examination id	Similar data	Similarity	Conclusion	View
2541708963	1425360	Arterial hypertension, diabetes	85%	Ischemia	Images
8451629730	2518940	Arterial hypertension	45%	Necrosis	Images
1047061409	5058426	Diabetes	35%	None	Images
1526186420	1861512	Arterial hypertension	25%	Ischemia, necrosis	Images

Fig. 6. The TETRA user result interface.

The similarity $sim(c1, c2)$ between two cases c_1 and c_2 is computed as:

$$sim(c_1, c_2) = \frac{\sum_{i=1}^{d_{Sol}} sim_{d_i}(c_1, c_2) \times F(d_i)}{\sum_{i=1}^{d_{Sol}} F(d_i)} \in [0; 1]$$

We assume that the most frequent diagnoses have to be proposed in priority. The reason is that uninteresting diagnosis may appear in the report. However, what interests the physicians is more likely the diagnoses appearing the most of the times rather than the fortuitous ones. For this, the similarity between c_1 and c_2 w.r.t a diagnosis d_i is weighted by $F(d_i)$, the frequency of d_i in the case base.

TETRA user interfaces. Figure 5 illustrates the TETRA query interface. With this interface, the physician can query the TETRA system to search the past image interpretations the most similar to her current interpretation task. The most similar cases are searched according to:

- the patient information, coming from the patient idenfication number;
- the clinical information and diagnosis hypothesis;
- the imaging protocol;
- the vector dimensions (clinical information, biological information) that have to be used in the similarity computation.

The results composed of the most similar image interpretations with their conclusion and statistics, as illustrated in Fig. 6, are returned to the physician who can directly access the images, in order to compare them to the image she has to interpret.

4 Results

Evaluation of the semantic annotation process. 100 cases were chosen randomly to evaluate the TETRA semantic annotation process. A nuclear physician checked the 100 annotation results as being correctly annotated or not. On 100 cases:

- 100 have been correctly parsed into medical background, goal and diagnosis parts;
- 100 have been correctly annoted (i.e. all the information contained in the imaging report were found);
- only 1 error of negation in 1 imaging report occurred.

Methodology for the evaluation of the TETRA prediction capability. TETRA has been evaluated by the Area Under Curve (AUC) of the ROC curve and the sensitivity and specificity according to several thresholds. Figure 7 presents the ROC curve and numerical details about this curve.

Let $T_\theta^+(d) = \{c_i \mid p(d, c_i) \geq \theta\}$ be the set of cases having to find the diagnosis d with a probability greater or equal to θ and $T_\theta^-(d)$ be the set of cases having a probability to find the diagnosis d less than θ with $T_\theta^-(d) = \{c_i \mid p(d, c_i) < \theta\}$. The sensitivity (also called recall in information retrieval context) measures the proportion of cases with a given solution that are correctly identified as such with $Se_\theta(d) = \frac{|C(d)^+ \cap T_\theta^+(d)|}{|C(d)^+|}$. In our case, the sensitivity measures the proportion of cases with a given diagnosis that are correctly identified by TETRA. The specificity measures the proportion of cases without a given solution that are correctly identified as such with $Sp_\theta = \frac{|C(d)^- \cap T_\theta^-(d)|}{|C(d)^-|}$. In our case, the specificity measures the proportion of cases without a given diagnosis that are correctly identified by TETRA.

Result analysis. TETRA offers moderatly good results to predict the presence of ischemia with an AUC of 0.66 [0.62–0.70] and a threshold with a sensitivity of 0.73 [0.61;0.78] and a specificity of 0.56 [0.43;0.60], and very good results to predict the presence of necrosis with an AUC of 0.81 [0.78;0.84] and a threshold with a sensitivity of 0.83 [0.74;0.90] and a specificity of 0.69 [0.64;0.74]. These diagnoses are the only two possible diagnoses in the myocardial scintigraphy. Predictions without extending the vectorial representation with most general concept are less effective. As TETRA is based on clinical manifestations of diseases, the more a disease is severe and/or lasts a long time, the more it is likely to be symptomatic or to require care. So, the more TETRA uses medical informations, the more TETRA will give good forecasts. This is the reason why TETRA is less effective for ischemia, which is often unknown and asymptomatic, than for necrosis, which is more symptomatic and results in several medical events. With a representative case base (8000 cases) and a ROC curve globaly convex and consistent, we can assume that the thresholds found are generalizable to cases outside the case base.

The analysis of the most relevant features for each diagnosis, which are given in Table 1, shows also interesting things:

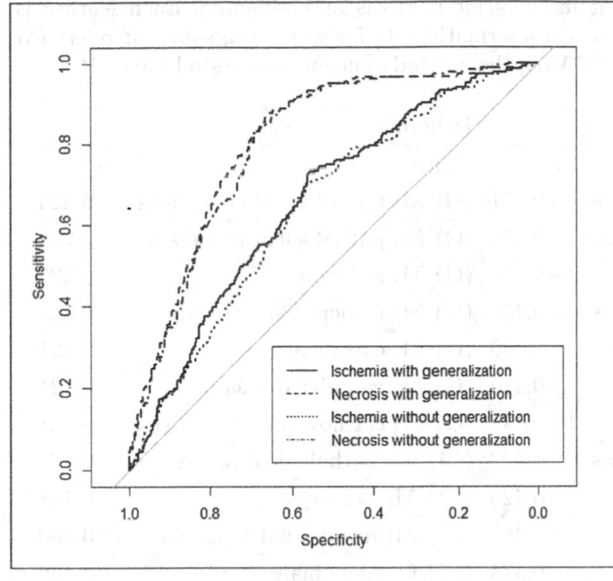

With generalization:

	Ischemia
AUC	0.66 [0.62;0.70]
Se	0.73 [0.61;0.78]
Sp	0.56 [0.43;0.60]
	Necrosis
AUC	0.81 [0.78;0.84]
Se	0.83 [0.74;0.90]
Sp	0.69 [0.64;0.74]

Without generalization:

	Ischemia
AUC	0.65 [0.61;0.68]
Se	0.65 [0.55;0.72]
Sp	0.59 [0.52;0.64]
	Necrosis
AUC	0.80 [0.77;0.84]
Se	0.82 [0.69;0.88]
Sp	0.69 [0.63;0.73]

Fig. 7. ROC curve and numerical results about the TETRA performance on ischemia and necrosis diagnoses.

- the relevance of the twenty necrosis features (presented in the table) is greater than the best relevance of ischemia features, which explains a lower effeciency of TETRA for ischemia, the features being less relevant;
- necrosis and ischemia share a lot of relevant features, which is consistent from a medical point of view since these two diagnoses have the same risk factors;
- all the most relevant features found are officially considered as cardiovascular risk factors, which corroborates the robustness of the model;
- all types of features of the clinical data (MeSH, ICD-10, CCAM) and several features of the goal are represented in the twenty most relevant features, which means that all these types of features play an important role in the construction of the model.

Benefits. TETRA forecasts are reliable and the AUC of the ROC curve is rather high. Its forecasts helps nuclear physicians for image interpretation. It can be used as the summary of the patient medical background for a given diagnosis. As said previously, diagnoses with more clinical manifestations are more able to be correctly predicted. TETRA can easily be adapted to all other protocols of nuclear and non-nuclear imaging, and even to biological or clinical tests. Moreover, in case of examination contraindications, as it is the case in particular in cardiac tests, TETRA provides a good alternative to evaluate the probability for some diagnoses, without or before passing the test. TETRA could also avoid useless tests, if the probability of a diasgnosis is very high or, on the contrary, very low.

Table 1. TETRA most relevant features for necrosis and ischemia. Each feature is decribed by its origin (CI for clinical information, G for goal), the source it refers (M for MeSH, I for ICD-10, C for CCAM), the related concept, and its relevance R.

Necrosis		Ischemia	
Feature	R	Feature	R
(G) M: pathological conditions	−0.451	(G) M: pathological conditions	−0.224
(G) M: pathological processes	−0.451	(G) M: pathological processes	−0.224
(G) M: ischemia	−0.451	(G) M: ischemia	−0.223
(G) M: cardiovascular diseases	−0.450	(G) M: myocardial ischemia	−0.221
(G) M: cardiopathies	−0.450	(G) M: cardiopathies	−0.221
(G) M: vascular diseases	−0.450	(G) M: vascular diseases	−0.221
(G) M: myocardial ischemia	−0.449	(G) M: cardiovascular diseases	−0.221
(CI) M: pathological processes	0.427	(CI) M: pathological processes	0.171
(CI) M: ischemia	0.420	(CI) M: ischemia	0.168
(CI) M: myocardial ischemia	0.387	(CI) M: myocardial ischemia	0.160
(CI) M: necrosis	0.373	(CI) Gender: male	0.160
(CI) I: I20-I25	0.370	(CI) M: cardiopathies	0.159
(CI) M: cardiopathies	0.366	(CI) M: pathological conditions	0.155
(CI) M: pathological conditions	0.364	(CI) I: I25	0.115
(CI) I: I25	0.313	(CI) I: I20-I25	0.108
(CI) I: Z95	0.264	(CI) C: DAQL009	0.103
(CI) I: I30-I52	0.248	(CI) I: Z95	0.102
(CI) I: I00-I99	0.247	(CI) I: I25.1	0.096
(CI) C: 19.01.09.02	0.244	(CI) I: Z95.5	0.088
(CI) I: Z95.5	0.243	(CI) C: 04.01.07	0.087

Limits. Currently, terms of radiological and nuclear semiology are not parts of the MeSH ontology, so radiological signs can be forgotten by the system and it is mainly the clinical interpretation that is recognized. Moreover, stays are sometimes coded by different persons and maybe with missing informations, so it is possible that occasionally some similar cases are not displayed, but the cases displayed are necessarily similar. This lack of precision in the case descriptions directly impacts the sensitivity in the results of TETRA. So, some efforts have to be done to improve the case descriptions.

5 Conclusion

This paper presented TETRA, a case-based decision support system, and shows how it can be used to improve the nuclear image interpretation. TETRA demonstrates that it is possible to calculate the similarity between cases and to displays the most similar cases to physician in order to help him to make a decision. Furthermore, calculating the probability of diagnoses is also possible and permits a thinner interpretation for the nuclear physician.

Currently, TETRA is based on EHR, biological data and imaging reports. It could be interesting to add a Content-Based Image Retrieval (CBIR) system to improve the efficiency of the system by suggesting similar cases based among others directly on the images themselves and using images to forecast more efficiently. Furthermore, it will permit eventually to forecast without the imaging reports [17,18], even if radiology and nuclear medicine images are always to be interpreted depending on the context. Combining TETRA forecasts with an automatic images interpretation seems then to be a good track.

Another improvment could consist in using several thresholds, according to what interests the nuclear physicians. For example, a more sensitive threshold or a more specific threshold can be chosen. Two thresholds, a low and a high, can also be chosen, in order to separate cases in three groups: low, moderate or high probability for each diagnosis. This could improve the sensitivity and the specificity of TETRA.

Finally, the possibilities in which TETRA can be used are beyound the initial objective of helping nuclear physicians to interpret images. For example, TETRA could also be used:

- for a research purpose: by choosing the cases relevant to a study according to the inclusion and exclusion criteria, and determining imaging interpretation or risk factors, or even directly to retrospectively compute risk factors in large amount of population;
- for teaching purpose: by choosing the cases relevant to a particular theme;
- for medical purpose: by giving directly to the clinicians, in some cases, probabilities concerning some diagnoses before or without imaging test.

References

1. Riesbeck, C.K., Schank, R.C.: Inside Case-Based Reasoning. Lawrence Erlbaum Associates Inc., Hillsdale (1989)
2. Reiner, B.I.: Medical imaging data reconciliation, part 2: clinical order entry and imaging report data reconcilia-tion. J. Am. Coll. Radiol. (JACR) **8**(10), 720–724 (2011)
3. Goldzweig, C.L., Orshansky, N.M., Paige, G., Miake-Lye, I.M., Beroes, J.M., Ewing, B.A., Shekelle, P.G.: Electronic health record-based interventions for improving appropriate diagnostic imaging: a systematic review and meta-analysis. Ann. Intern. Med. **162**(8), 557–565 (2015)
4. Cordier, A., Dufour-Lussier, V., Lieber, J., Nauer, E., Badra, F., Cojan, J., Gaillard, E., Infante-Blanco, L., Molli, P., Napoli, A., Skaf-Molli, H.: Taaable: a case-based System for personalized cooking. In: Montani, S., Jain, L.C. (eds.) Successful Case-based Reasoning Applications-2. Studies in Computational Intelligence, vol. 494, pp. 121–162. Springer, Heidelberg (2014)
5. Gøeg, K.R., Cornet, R., Andersen, S.K.: Clustering clinical models from local electronic health records based on semantic similarity. J. Biomed. Inform. **54**, 294–304 (2015)

6. Guyatt, G., Cairns, J., Churchill, D., et al.: Evidence-based medicine: a new approach to teaching the practice of medicine. JAMA **268**(17), 2420–2425 (1992)
7. Perner, P.: Case-based reasoning and the statistical challenges. Qual. Reliab. Eng. Int. **24**(6), 705–720 (2008)
8. Sandefer, R.H., Marc, D.T., Kleeberg, P.: Meaningful use attestations among us hospitals: The growing rural-urban divide. Perspect. Health Inf. Manag. **12** (2015)
9. Schmidt, R., Gierl, L.: The roles of prototypes in medical case-based reasoning systems. In: 4th German Workshop on CBR-System Development and Evaluation, Humbolt University, Informatik-Berichte, Berlin, pp. 207–216 (1996)
10. Bellazzi, R., Montani, S., Portinale, L.: Retrieval in a prototype-based case library: a case study in diabetes therapy revision. In: Smyth, B., Cunningham, P. (eds.) EWCBR 1998. LNCS, vol. 1488, pp. 64–75. Springer, Heidelberg (1998). doi:10.1007/BFb0056322
11. Sharaf-El-Deen, D.A., Moawad, I.F., Khalifa, M.E.: A new hybrid case-based reasoning approach for medical diagnosis systems. J. Med. Syst. **38**(2), 9 (2014)
12. Vallati, M., Gatta, R., De Bari, B., Magrini, S.M.: Clinical similarities: an innovative approach for supporting medical decisions. Stud. Health Technol. Inf. **192**, 1114 (2013)
13. Bratsas, C., Koutkias, V., Kaimakamis, E., Bamidis, P.D., Pangalos, G.I., Maglaveras, N.: KnowBaSICS-M: an ontolo-gy-based system for semantic management of medical problems and computerised algorithmic solutions. Comput. Meth. Programs Biomed. **88**(1), 39–51 (2007)
14. Popescu, M., Arthur, G.: Ontoquest: a physician decision support system based on ontological queries of the hospital database. In: AMIA Annual Symposium Proceedings, pp. 639–643 (2006)
15. Richter, M.M., Wess, S.: Similarity, Uncertainty and Case-Based Reasoning in Patdex. Springer, Dordrecht (1991)
16. Stram, R., Reuss, P., Althoff, K.-D., Henkel, W., Fischer, D.: Relevance matrix generation using sensitivity analysis in a case-based reasoning environment. In: Case-Based Reasoning Research and Development, pp. 402–412
17. Kurtz, C., Beaulieu, C.F., Napel, S., Rubin, D.L.: A hierarchical knowledge-based approach for retrieving similar medi-cal images described with semantic annotations. J. Biomed. Inform. **49**, 227–244 (2014)
18. El-Naqa, I., Yang, Y., Galatsanos, N.P., Nishikawa, R.M., Wernick, M.N.: A similarity learning approach to content-based image retrieval: application to digital mammography. IEEE Trans. Med. Imaging **23**(10), 1233–1344 (2004)

Case-Based Team Recognition Using Learned Opponent Models

Michael W. Floyd[1(✉)], Justin Karneeb[1], and David W. Aha[2]

[1] Knexus Research Corporation, Springfield, VA, USA
{michael.floyd, justin.karneeb}@knexusresearch.com
[2] Navy Center for Applied Research in AI,
Naval Research Laboratory (Code 5514), Washington, DC, USA
david.aha@nrl.navy.mil

Abstract. For an agent to act intelligently in a multi-agent environment it must model the capabilities of other agents. In adversarial environments, like the beyond-visual-range air combat domain we study in this paper, it may be possible to get information about teammates but difficult to obtain accurate models of opponents. We address this issue by designing an agent to learn models of aircraft and missile behavior, and use those models to classify the opponents' aircraft types and weapons capabilities. These classifications are used as input to a case-based reasoning (CBR) system that retrieves possible opponent team configurations (i.e., the aircraft type and weapons payload per opponent). We describe evidence from our empirical study that the CBR system recognizes opponent team behavior more accurately than using the learned models in isolation. Additionally, our CBR system demonstrated resilience to limited classification opportunities, noisy air combat scenarios, and high model error.

Keywords: Beyond-visual-range air combat · Autonomous agents · Team recognition · Opponent modeling

1 Introduction

Beyond-visual-range (BVR) air combat is a modern style of air-to-air combat where teams of aircraft engage each other over large distances using long-range missiles [1]. This differs from the classic dogfighting combat of World Wars I and II, where aircraft used short-range weaponry in fast-paced, close-quarters combat. Whereas dogfighting lends itself well to reactive control strategies, BVR allows for longer-term strategic planning and reasoning. For an agent that engages in air combat, both styles offer similar challenges including an adversarial environment, imperfect information, and real-time performance constraints. While the large distance between aircraft provide BVR agents more time to reason than dogfighting agents, it also increases uncertainty when observing other aircraft.

One significant limitation of long distance observations is that they make it difficult to accurately identify the capabilities of opponent aircraft. Observations are made through various types of long-range sensors rather than being observed directly by a

D.W. Aha and J. Lieber (Eds.): ICCBR 2017, LNAI 10339, pp. 123–138, 2017.
DOI: 10.1007/978-3-319-61030-6_9

pilot, making it difficult to sense opponents with sufficient precision to accurately detect their capabilities (e.g., maximum speed, maneuverability, flying range). For example, at close range it may be possible to visually differentiate the type of aircraft based on shape or defining characteristics (i.e., paint, materials, and engine type) but onboard sensors may be unable to provide information other than the aircraft's position, speed, and heading. Similarly, while it is possible to detect when an opponent fires a missile, it is difficult to determine the exact properties of an opponent's weapons (e.g., range, maximum speed, payload) through long-range sensors alone. An opponent's aircraft type and weapon capabilities could be provided as part of a pre-mission briefing, but given the adversarial nature of air combat, such information may be outdated (e.g., a last-minute aircraft change) or erroneous (e.g., deception by opponents). Having inaccurate opponent information in BVR combat can result in the agent wasting resources (e.g., firing a missile an opponent can easily evade), selecting sub-optimal goals or plans (e.g., based on incorrect assumptions about an opponent's possible actions), or putting itself in dangerous situations (e.g., underestimating an opponent's weaponry). BVR combat scenarios typically involve engaging with a team of opponents, thereby compounding the potential impact of incorrect assumptions about opponents.

Our work has two primary contributions. First, we describe an approach for learning models to predict the movement of aircraft and missiles in BVR scenarios. When encountering an unknown aircraft, these models can be used to classify the type of aircraft and its weapons capabilities. Second, we present a case-based reasoning (CBR) system that can use the classification of individual aircraft to determine the composition of an opposing team. Our approach requires only a small subset of aircraft or missiles to be correctly classified to perform accurate retrieval, making it resilient to classification errors (i.e., due to learning error or unexpected opponent behavior) and limited opportunities to classify opponents (i.e., when only certain observed behaviors can be used for classification).

In the remainder of this paper we describe our approach for opponent model learning and team recognition. Section 2 describes the BVR combat domain and motivates why accurate information about aircraft type and weapons capabilities are necessary. Our approach for learning aircraft and missile models is presented in Sect. 3, with a focus on how the models can be used for classification. Section 4 describes our case-based team recognition system, and how classifications of individual aircraft and missiles can be used to determine the composition of the entire team. In Sect. 5, we report evidence that our system improves team recognition performance in BVR scenarios. Related work is discussed in Sect. 6, followed by conclusions and topics of future work in Sect. 7.

2 Beyond-Visual-Range Air Combat

BVR scenarios occur in large airspaces (i.e., thousands of square kilometers) with opposing aircraft located tens or hundreds of kilometers from each other. Figure 1 shows a graphical representation of a BVR engagement between two opposing teams, each of which has five aircraft. The objective of each team is to destroy their opponents

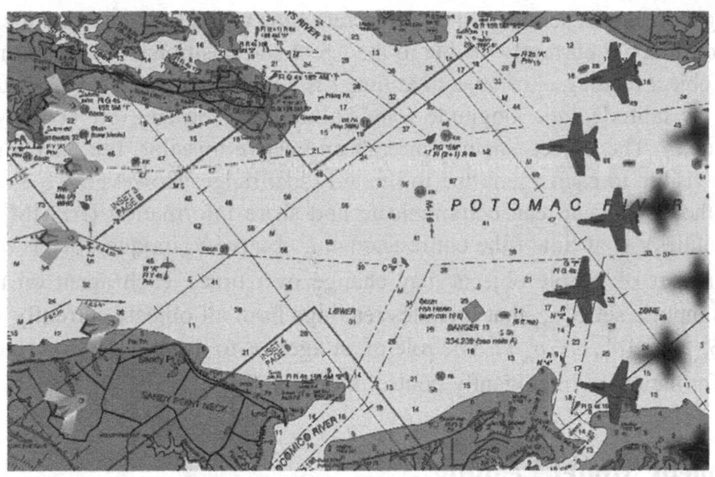

Fig. 1. Graphical representation of two teams of aircraft engaged in a 5 vs 5 beyond-visual-range air combat scenario (aircraft size is not shown to scale)

or force them to retreat. Given the large distances involved, aircraft are equipped with active radar homing missiles that have ranges of approximately 50 km.

We use a high-fidelity BVR air combat simulator for our studies, the Advanced Framework for Simulation, Integration, and Modeling (AFSIM) [2]. AFSIM allows for control of a simulated aircraft using low-level control commands or high-level actions. Additionally, aircraft can be controlled programmatically (e.g., scripts or agents) or by human pilots using physical hardware. In AFSIM, each controller (i.e., script, agent, human) pilots a single aircraft. For the remainder of this paper, we assume that aircraft are controlled by intelligent agents.

At the start of a BVR mission, each agent receives a *mission briefing* that contains information about its teammates and its opponents. This information includes the number of aircraft per team, the type of each aircraft (i.e., the aircraft architecture, maximum speed, maneuverability), and each aircraft's weapons capabilities (i.e., the range and speed of its missiles). For teammates, this information can be assumed to be accurate. However, information about opponents may come from assumptions, intelligence reports, or previous encounters, so there is no guarantee that mission briefing data is accurate. As such, an agent that relies on this information will need to verify and update it during a mission. There are several reasons why information about an opponent's aircraft type and weaponry are vitally important. First, it directly impacts the attack ranges of the agent and its opponents. Underestimating an opponent's aircraft type will cause the agent to fire missiles that the opponent can easily evade, whereas overestimation will prevent the agent from firing in advantageous positions. Similarly, overestimating the opponent's weapons capabilities will cause the agent to engage from longer distances, possibly never entering a reasonable firing range, and underestimating may cause the agent to fly into dangerous positions. Second, an accurate model of each opponent and their capabilities directly influences an agent's ability to perform long-term prediction, select appropriate goals, and generate appropriate plans.

Each agent receives sensory input at discrete time internals. The input includes the set of objects that are currently visible to the agent and positional information for each object. An object reading o_i^t of object i at time t is a tuple $o_i^t = \langle lat_i^t, long_i^t, a_i^t, b_i^t, v_i^t, ac_i^t \rangle$ containing its latitude lat_i^t, longitude $long_i^t$, altitude a_i^t, bearing b_i^t, velocity v_i^t, and acceleration ac_i^t. The objects include aircraft and active missiles, but only a subset of objects are visible to each agent due to limited radar range. However, we assume that agents on the same team can communicate and share information (AFSIM provides such capabilities). If at time t the entire team can observe n_t unique objects $o_1^t, \ldots, o_{n_t}^t$ (i.e., the number of visible objects may change over time), each agent on that team receives as input a set S_{team}^t that includes readings from all objects currently visible to the team ($S_{team}^t = \{o_1^t, \ldots, o_{n_t}^t\}$). The role of an agent is to use the mission briefing and sensory information to intelligently control the aircraft.

3 Opponent Model Leaning

In Sect. 2 we described why agents require accurate models of their opponents to operate efficiently in BVR scenarios but did not address what the models contain or how they are used. Our work focuses on models of an opponent's *maneuverability* and *weapon range*. The maneuverability is based on its aircraft type (e.g., F-16 Fighting Falcon, F/A-18 Super Hornet, Su-27 Flanker, MiG-29 Fulcrum) and incorporates velocity, acceleration, and turning radius. Similarly, the weapon range is based on the type of missiles an aircraft is equipped with and their effective range (e.g., short-range AA-11 Archer, medium-range AIM-120 AMRAAM, long-range AIM-54 Phoenix).

The primary challenge of using aircraft and missile models is that there are limited opportunities to differentiate between the possible models. Aircraft types differ based on their top-end performance but the majority of the time all aircraft will operate similarly. For example, aircraft use cruising speeds that are significantly less that their maximum speed, so all aircraft will appear identical when cruising. It is only when an aircraft operates at their top-end that they show noticeable differences. Similarly, the type of weapons an aircraft is equipped with can be determined only when a missile is fired.

We restrict our models to observations that can reliably differentiate between different aircraft and missiles. The following information is used:

- **Aircraft Models:** The most likely time for an aircraft to display its top-end performance is when it is threatened. As such, observations are collected while an aircraft is evading a missile. If at time t a missile is fired at aircraft i, readings for the evading aircraft are added to the set $\mathcal{A}_i$ during a window of length w_1: $\mathcal{A}_i = \mathcal{A}_i \cup \{o_i^t, \ldots, o_i^{t+w_1}\}$. If the missile is destroyed before the end of the window (i.e., it reaches its maximum range and crashes, or collides with an object), any observations after destruction are not added to the set. This is because the missile is no longer a threat so the aircraft will no longer evade it. Since each aircraft can be attacked multiple times, the set is extended during each attack. There is no guarantee that all observations in the set are of the aircraft actively evading a missile. For example, an aircraft could determine that its current cruising speed is sufficient to evade the missile, or be unaware that a missile has been fired at it. However, we assume that a sufficient number of observations will be of active evasion.

- **Weapons Models:** Missiles do not display the same level of agency as aircraft (i.e., they fly at maximum speed towards their target), so observations are collected as soon as a missile is detected. If at time t missile j is fired by aircraft i, readings for the missile are added to the set $\mathcal{W}_i$ during a window of length w_2: $\mathcal{W}_i = \mathcal{W}_i \cup \{o_j^t, \ldots, o_j^{t+w_2}\}$. As with aircraft, observations are not added after the missile is destroyed (i.e., if the missile is destroyed before w_2). This groups together the observations of all missiles fired by an aircraft and assumes that each aircraft is equipped with a single type of missile (although we would like to relax that assumption in future work).

3.1 Model Training

Training the models requires obtaining a set of training observations for each type of aircraft and missile. However, in adversarial domains this can be challenging. The primary difficulty is collecting observations that represent actual engagements. Engagements are likely rare, so there are limited opportunities to collect training data. There is also the possibility of the opponent developing or deploying new aircraft or missiles (i.e., with no existing model).

To overcome these challenges, our models are trained on observations of friendly aircraft during training missions. The missions are simplified scenarios using simulated missiles (i.e., they will not damage the aircraft) where one aircraft pursues and simulates attack on another. Each training mission ends when the target aircraft is hit or successfully evades. The parameters for a mission are: *target's aircraft type*, *attacker's missile type*, *initial distance between aircraft*, *starting altitude of each aircraft*, *starting velocity of each aircraft*, and *relative heading of each aircraft*. This allows data to be collected for each aircraft and missile type using a variety of initial configurations (e.g., based on expert input or random sampling). Data collection is restricted only by the time and availability of training aircraft.

Uncertainty about possible opponent aircraft and missile types is handled by having friendly aircraft perform synthetic opponent behavior. For aircraft, this involves placing artificial limits on the training aircraft's *turning radius*, *acceleration*, and *maximum velocity*. For missiles, limits are placed on the training missile's *maximum range*, *acceleration*, and *maximum velocity*. Thus, modifying one of more of these parameters effectively creates a synthetic opponent that can be used to train a new model. It is possible that unrealistic models will be learned (i.e., the opponent does not use a similar aircraft or missile) or that it is not possible to replicate a particular aircraft or missile type (e.g., the opponent aircraft's maneuverability exceeds the training aircraft's top-end performance). However, we anticipate the impact of superfluous or unobtainable models is offset by the performance benefits of learning valid models.

If l synthetic aircraft types and k synthetic missile types are used, l aircraft models $M_{air}^1, \ldots, M_{air}^l$ and k missile models $M_{mis}^1, \ldots, M_{mis}^k$ are learned. Each model is trained using all observations of that object type collected during training missions (i.e., the set

$\mathcal{A}_i$ containing all observations of aircraft type i and $\mathcal{W}_j$ containing all observations of missile type j). Input values are current observations (e.g., observed values at time t) and outputs estimate the expect rate of change (e.g., the rate of change between time t and time $t+1$). If an observation is the last in a temporally related sequence (i.e., the last observation of an evasion or missile flight), it does not have a subsequent observation to calculate rate of change so is not used for training. The inputs and outputs are:

- **Aircraft**
  - **Inputs**: *bearing* (degrees), *velocity* (meters per second), *distance to attacking missile* (meters), *velocity of attacking missile* (meters per second)
  - **Outputs**: *rate of altitude change* (meters per second), *rate of separation from attacking missile* (meters per second, with positive values representing the aircraft distancing itself from the missile)
- **Missile**
  - **Inputs**: *altitude* (feet), *flight time* (seconds)
  - **Output**: *acceleration* (meters per second squared)

Models can be learned using any algorithm that can learn a mapping from continuous inputs to continuous outputs. However, for the remainder of this paper we use the M5′ algorithm [3]. M5′ is a decision tree induction algorithm where each leaf node contains a regression model. Training instances are first used to build the tree, and then all training instances that arrive at the same leaf node are used to train a linear regression model for that node. For an input instance, it traverses the tree to a leaf node and its outputs are calculated using the regression model at that node. Since there are two outputs for aircraft models, one decision tree is used per output.

3.2 Model-Based Classification

The learned models are used during scenarios to continuously predict the movement of aircraft and missiles. Since the models use values from time t to predict the rate of change between t and $t+1$, the output of a model can be evaluated at each subsequent time step. During an evasion, all aircraft models $M^1_{air}, \ldots, M^l_{air}$ are used to generate predicated outputs $p^1_{air_t}, \ldots, p^l_{air_t}$ (i.e., each prediction is a tuple containing the rate of altitude change and rate of separation from attacking missiles) at each time t. Similarly, during the flight of a missile, all missile models $M^1_{mis}, \ldots, M^k_{mis}$ are used at each time t to generate predicted outputs $p^1_{mis_t}, \ldots, p^k_{mis_t}$ (i.e., each predication is the acceleration). At time $t+1$, the observed values o_{air_t} and o_{mis_t} are computed.

If the models have been used to predict values between time t and $t+c$, the aircraft or missile is classified based on the model that minimizes the distance between predictions and observations:

$$class_{air} = \underset{i=1\ldots l}{\operatorname{argmin}}(dist^i_{air}), \quad dist^i_{air} = \sum_{j=t}^{t+c} dist(p^i_{air_j}, o_{air_j})$$

$$class_{mis} = \underset{i=1...k}{\operatorname{argmin}}(dist^i_{mis}), \quad dist^i_{mis} = \sum_{j=t}^{t+c} dist(p^i_{mis_j}, o_{mis_j})$$

Although classifications can be made at any time, in practice we use only the classifications obtained by observing the entire sequence (i.e., entire evasion or missile flight). For missiles, the distance function $dist(p_{mis}, o_{mis})$ computes the absolute distance between the predicted and observed values (i.e., $|p_{mis} - o_{mis}|$). The distance function for aircraft $dist(p_{air}, o_{air})$ is slightly more complicated since each value is a tuple containing both the rate of altitude change Δalt and rate of separation from attacking missile Δsep (i.e., $p_{air} = \langle \Delta alt_p, \Delta sep_p \rangle$ and $o_{air} = \langle \Delta alt_o, \Delta sep_o \rangle$). The distance function computes the average absolute distance between the output: (i.e., $\frac{|\Delta alt_p - \Delta alt_o| + |\Delta sep_p - \Delta sep_o|}{2}$). The confidence in each of the i models is also calculated, with values ranging from 0 to 1 (inclusive):

$$conf^i_{air} = \frac{\sum_{j=1...l}(dist^j_{air}) - dis^i_{air}}{\sum_{j=1...l}(dist^j_{air})}, \quad conf^i_{mis} = \frac{\sum_{j=1...k}(dist^j_{mis}) - dis^i_{mis}}{\sum_{j=1...k}(dist^j_{mis})}$$

The confidence values are stored in the sets $\mathcal{CONF}_{air} = \{conf^1_{air}, \ldots, conf^l_{air}\}$ and $\mathcal{CONF}_{mis} = \{conf^1_{mis}, \ldots, conf^k_{mis}\}$. Thus, each classification outputs a class label (i.e., $class_{air}$ or $class_{mis}$) and the confidence in each possible label (i.e., $\mathcal{CONF}_{air}$ or $\mathcal{CONF}_{mis}$).

4 Case-Based Team Recognition

The learned models can be used to classify individual aircraft and missiles but, as we discussed in the previous section, the situations when classification can be performed are limited. When engaging a team of opponents, it is possible that some aircraft will never evade or fire missiles. To overcome the scarcity of classification opportunities, and therefore the scarcity of class labels, we use a case-based team recognition approach.

We assume the availability of a case base containing known compositions of opponent teams. Each case C contains both the team composition T and team properties P: $C = \langle T, P \rangle$. The team composition is a set containing the aircraft type and missile type of each member of the team: $T = \{\langle class'_{air}, class'_{mis} \rangle, \langle class''_{air}, class''_{mis} \rangle, \ldots\}$. The properties include additional information about the team including the team leader, base of operations, and records of previous encounters. The goal of the CBR process is to retrieve a case that is similar to the opponent observations. First, a target team T_{tar} is created by merging the team provided by the mission briefing T_{MB} and the observed team T_{obs}. While T_{MB} contains a full, although possibly incorrect, team, T_{obs} may contain unknown values if only a subset of classifications have been performed (e.g., $class_{air} = \emptyset$, $class_{mis} = \emptyset$, or both are unknown). The method for merging the mission briefing and observations is show in Algorithm 1.

Algorithm 1: Merging mission briefing and observations

Function: *merge*(T_{MB}, T_{obs}) **returns** T_{tar}

1 $T_{tar} \leftarrow \emptyset$;
2 **foreach** $\langle class_{air}, class_{mis}\rangle \in T_{obs}$ **do**
3   **if** $\langle class_{air}, class_{mis}\rangle \in T_{MB}$ **then**
4         $T_{tar} \leftarrow T_{tar} \cup \langle class_{air}, class_{mis}\rangle$;
5         $T_{MB} \leftarrow T_{MB} \setminus \langle class_{air}, class_{mis}\rangle; T_{obs} \leftarrow T_{obs} \setminus \langle class_{air}, class_{mis}\rangle$;
6 **foreach** $\langle class_{air}, class_{mis}\rangle \in T_{obs}$ **do**
7   **foreach** $\langle class'_{air}, class'_{mis}\rangle \in T_{MB}$ **do**
8   **if** $class_{mis} = class'_{mis}$ **then**
9             $T_{tar} \leftarrow T_{tar} \cup \langle class'_{air}, class'_{mis}\rangle$;
10             $T_{MB} \leftarrow T_{MB} \setminus \langle class'_{air}, class'_{mis}\rangle; T_{obs} \leftarrow T_{obs} \setminus \langle class_{air}, class_{mis}\rangle$;
11   **break**;
12 **foreach** $\langle class_{air}, class_{mis}\rangle \in T_{obs}$ **do**
13   **foreach** $\langle class'_{air}, class'_{mis}\rangle \in T_{MB}$ **do**
14   **if** $class_{air} = class'_{air}$ **then**
15             $T_{tar} \leftarrow T_{tar} \cup \langle class'_{air}, class'_{mis}\rangle$;
16             $T_{MB} \leftarrow T_{MB} \setminus \langle class'_{air}, class'_{mis}\rangle; T_{obs} \leftarrow T_{obs} \setminus \langle class_{air}, class_{mis}\rangle$;
17   **break**;
18 **foreach** $\langle class_{air}, class_{mis}\rangle \in T_{obs}$ **do**
19   **foreach** $\langle class'_{air}, class'_{mis}\rangle \in T_{MB}$ **do**
20         $tar_{air} = class_{air}$; $tar_{mis} = clas_{mis}$;
21   **if** $tar_{air} = \emptyset$ **then** $tar_{air} = class'_{air}$;
22   **if** $tar_{mis} = \emptyset$ **then** $tar_{mis} = class'_{mis}$;
23         $T_{tar} \leftarrow T_{tar} \cup \langle tar_{air}, tar_{mis}\rangle$;
24         $T_{MB} \leftarrow T_{MB} \setminus \langle class'_{air}, class'_{mis}\rangle; T_{obs} \leftarrow T_{obs} \setminus \langle class_{air}, class_{mis}\rangle$;
25   **break**;
26 **return** T_{tar};

The algorithm starts with an empty team (line 1) and adds aircraft to the team using a priority-based merging method. First, aircraft are added if both the mission briefing and observations agree on the type of aircraft and missile (lines 2–5). Second, aircraft are added if the mission briefing and observations agree on the missile type (lines 6–11). Third, aircraft are added if there is agreement on aircraft type (lines 12–17). For all three previous merging steps, the aircraft is added using the labels stored in the mission briefing (although for the first merging method the labels are identical). This is done because the observations may be missing labels, so the information from the mission briefing is used to ensure a fully-defined team. Finally, any remaining aircraft that do not have a full or partial match between the mission briefing and the observations are merged (lines 18–25). Priority is given to the observed labels, and only if there is a missing label is information from the mission briefing used (lines 21 and 22).

The method used to fill in unknown values is uninformed; it uses the value from the first available aircraft in the mission briefing. After merging, the number of aircraft stored in T_{tar} is equal to the number that were originally in T_{obs} and T_{MB} (e.g., if T_{obs} and T_{tar} both contained five aircraft, T_{MB} will contain five aircraft).

Consider an example where $T_{MB} = \{\langle 1,B\rangle, \langle 3,A\rangle, \langle 2,C\rangle\}$ and $T_{obs} = \{\langle 2,C\rangle, \langle 2,A\rangle, \langle \emptyset,C\rangle\}$. T_{tar} is initially empty (line 1). The first merger stage (lines 2–5) finds one perfect match $\langle 2,C\rangle$ that is added to T_{tar} and removed from T_{MB} and T_{obs} ($T_{tar} = \{\langle 2,C\rangle\}$, $T_{MB} = \{\langle 1,B\rangle, \langle 3,A\rangle\}$ and $T_{obs} = \{\langle 2,A\rangle, \langle \emptyset,C\rangle)$. The second merger stage (lines 6–11) matches $\langle 3,A\rangle$ and $\langle 2,A\rangle$ because they have identical missile types. They are removed from their respective teams and $\langle 3,A\rangle$ is added to T_{tar} because priority is given to aircraft from the mission briefing ($T_{tar} = \{\langle 2,C\rangle, \langle 3,A\rangle\}$, $T_{MB} = \{\langle 1,B\rangle\}$ and $T_{obs} = \{\langle \emptyset,C\rangle)$. The third merger stage (lines 12–17) does not result in any changes because T_{MB} and T_{obs} no longer contain any aircraft with matching aircraft types. The forth merging stage (lines 18–25) pairs the remaining aircraft $\langle 1,B\rangle$ and $\langle \emptyset,C\rangle$ and merges their class labels. Priority is given to $\langle \emptyset,C\rangle$ because it came from T_{obs}, but its missing value is filled in with the associated label from $\langle 1,B\rangle$. The merged aircraft $\langle 1,C\rangle$ is added to T_{tar}, and the other aircraft are removed from their teams. This results in a final merged team of $T_{tar} = \{\langle 2,C\rangle, \langle 3,A\rangle, \langle 1,C\rangle\}$, with T_{MB} and T_{obs} now empty.

After the mission briefing and observations are merged, the target team is used to retrieve from the case base the case containing the most similar team. Similarity between a target team T_{tar} and a source team T_{src} is computed using Algorithm 2. The similarity function performs a greedy matching where the labels for each aircraft in the source team are matched to the aircraft with the most similar labels in the target team. Since the algorithm is greedy, aircraft in the source case are iterated over based on order of occurrence (line 2) and their best match is determined without considering the optimal global match (lines 3–7). Once an aircraft from the target team has been found as the best match for an aircraft in the source team, it is not considered as a possible match for any other aircraft (line 8). The similarity between the labels of two aircraft (line 5) is calculated using the local similarity function $sim(\ldots)$ (lines 11–13). The local similarity function first retrieves the confidence in each of the possible class labels (lines 11 and 12). Recall that these confidence values are computed after each classification, so any class labels that came as a result of observations will have these confidence values computed (i.e., any parts of T_{tar} that came from T_{obs}). For class labels that originated from the mission briefing, all possible class labels are given an equal confidence. The labels from the source team are used to retrieve the confidence the target team has in those labels, and their average value is returned (line 13). Since the target team's classification labels are chosen by selecting the label with the highest confidence, similarity will be highest when all labels are identical (i.e., $class_{air} = class'_{air}$ and $class_{mis} = class'_{mis}$). However, the similarity function takes into account the relative similarity of class labels by also using the confidence of non-matching labels, although they will result in lower similarity than matching labels.

Algorithm 2: Similarity between teams

```
Function: similarity(T_tar, T_src) returns sim
1  sim = 0;
2  foreach ⟨class_air, class_mis⟩ ∈ T_src do
3      bestMatch = ∅; bestSim = −1;
4      foreach ⟨class'_air, class'_mis⟩ ∈ T_tar do
5          localSim = sim(⟨class_air, class_mis⟩, ⟨class'_air, class'_mis⟩);
6          if localSim > bestSim then
7              bestSim = localSim; bestMatch = ⟨class'_air, class'_mis⟩;
8      T_tar ← T_tar \ bestMatch;
9      sim = sim + bestSim;
10 return sim;

Function: sim(⟨class_air, class_mis⟩, ⟨class'_air, class'_mis⟩) returns sim
11 CONF_air = retrieveConfidence(class'_air);
12 CONF_mis = retrieveConfidence(class'_mis);
13 return (confidence(class_air, CONF_air) + confidence(class_mis, CONF_mis)) / 2;
```

For an example of Algorithm 2, we consider when $T_{tar} = \{\langle A, 1\rangle, \langle B, 2\rangle\}$ and $T_{src} = \{\langle B, 1\rangle, \langle A, 2\rangle\}$. We assume $\langle A, 1\rangle$ came from observations (i.e., merged from T_{obs} in Algorithm 1) and has known confidence values (calculated during classification): $conf_{air}^{A} = 0.7$, $conf_{air}^{B} = 0.3$, $conf_{mis}^{1} = 0.6$, and $conf_{mis}^{2} = 0.4$. We assume $\langle B, 2\rangle$ came from the mission briefing (i.e., merged from T_{MB}) so the confidence values are all equal: $conf_{air}^{A} = conf_{air}^{B} = 0.5$, and $conf_{mis}^{1} = conf_{mis}^{2} = 0.5$. The first iteration (lines 2–9) finds a match for $\langle B, 1\rangle$. The similarity between $\langle B, 1\rangle$ and $\langle A, 1\rangle$ (line 5) is calculated by first retrieving the associated confidence values of $\langle A, 1\rangle$ (lines 11 and 12). As we mentioned previously, the confidence values associated with $\langle A, 1\rangle$ are $conf_{air}^{A} = 0.7$, $conf_{air}^{B} = 0.3$, $conf_{mis}^{1} = 0.6$, and $conf_{mis}^{2} = 0.4$. The confidence in class labels B and 1 are retrieved (i.e., since $\langle A, 1\rangle$ is being compared to $\langle B, 1\rangle$), resulting in the values $conf_{air}^{B} = 0.3$ and $conf_{mis}^{1} = 0.6$. These values are used to compute the similarity: $sim_{B1-A1} = 0.5 \times (conf_{mis}^{B} + conf_{air}^{1}) = 0.5 \times (0.3 + 0.6) = 0.45$. The similarity between $\langle B, 1\rangle$ and $\langle B, 2\rangle$ is calculated in a similar manner, but using the confidence values from $\langle B, 2\rangle$: $sim_{B1-B2} = 0.5 \times (conf_{mis}^{B} + conf_{air}^{1}) = 0.5 \times (0.5 + 0.5) = 0.5$. Thus, $\langle B, 1\rangle$ is matched with $\langle B, 2\rangle$ because it has the higher similarity ($sim_{B1-B2} > sim_{B1-A1}$). During the second iteration $\langle A, 2\rangle$ is matched with $\langle A, 1\rangle$ as they are the only two remaining, resulting in $sim_{A2-A1} = 0.55$. The similarity returned by Algorithm 2 is $sim_{B1-B2} + sim_{A2-A1} = 1.05$.

5 Evaluation

In this section, we evaluate our claim that *our case-based technique improves team recognition*. Our evaluation tests the following hypotheses:

H1: The teams retrieved by the CBR system are similar to the opponent's actual team (i.e., are composed of similar aircraft).

H2: The team retrieved by the CBR system is more accurate than the team defined in the mission briefing.

H3: The team retrieved by the CBR system is more accurate than relying exclusively on observations.

H4: The observed team using the learned models is more accurate than the team defined in the mission briefing.

5.1 Data Collection and Model Training

Our evaluation uses three synthetic aircraft types and five synthetic missile types. As a result, three aircraft models and five missile models are learned. The default aircraft type has similar maneuverability to an F-16 fighter jet. The other two aircraft types are modifications of the default aircraft. One has a 35% increase in maneuverability (i.e., maximum velocity, acceleration and turn radius) and the other has a 35% decrease in maneuverability. The default missile type has similar properties to missiles used by an F-16. The additional missiles are variations of the default missile with their range and maximum velocity modified. The variations are: 20% decrease, 10% decrease, 10% increase, and 20% increase.

The training missions place each aircraft type and missile type in a variety of mission configurations. For collecting aircraft data, the initial configurations use a sampling of values that are expected to be encountered during actual encounters: altitudes of the attacked aircraft (feet) from the set $\{1000, 2000, \ldots, 20000\}$, velocities of the attacked aircraft (meters per second) from the set $\{200, 225, \ldots, 350\}$, bearings of the attacked aircraft (degrees) from the set $\{0, 30, \ldots, 180\}$, and distances between the two aircraft (kilometers) from the set $\{25, 50, 75\}$. Missile data is collected with a similar set of initial configuration values: altitudes of the attacking aircraft from the set $\{1000, 2000, \ldots, 20000\}$, velocities of the attacking aircraft from the set $\{200, 225, \ldots, 350\}$, and distances between the two aircraft from the set $\{25, 50, 75\}$. Aircraft are observed when evading a missile for a maximum of 60 s (i.e., $w_1 = 60$) and missiles are observed for a maximum of 40 s (i.e., $w_2 = 40$).

As we mentioned earlier, models are learned using the M5′ algorithm. Identical settings are used to train each model: a minimum branch size of 20 (i.e., a node must contain at least 20 training instances before branching) and a minimum error reduction of 0.5 (i.e., branching must reduce error by at least 0.5).

5.2 Experimental Setup

Our evaluation scenarios involve two teams of five aircraft engaged in BVR air combat. The *base scenario* arranges each team in a column with teammates spaced 5.5 nautical miles (approximately 10.2 km) from each other and opposing teams at a distance of 40 nautical miles (approximately 74.1 km). The aircraft start at an altitude of 17,000 ft and face in the direction of their enemies (i.e., east or west). The base scenario was used to generate 200 *random scenarios* where each aircraft's position is modified by between –3 and 3 nautical miles (approximately 5.6 km) according to a uniform random distribution in both the north/south and east/west directions. Additionally, each aircraft's altitude is modified between 0 and 2500 ft and its bearing between –15 and 15° (according to a uniform random distribution). Figure 1 shows a graphical representation of one such random scenario. Similar to the training missions, the evaluation scenarios use simulated missiles so no aircraft are damaged or destroyed. Each scenario has a duration of 10 min.

The CBR system uses a case base composed of 10 expert-authored cases, with each of the cases containing a different team composition (i.e., the aircraft type and missile type of each aircraft). Before a scenario is run, each team is assigned a team composition based on a randomly selected case (according to a uniform distribution). This represents each team's *true composition*. Additionally, each team is given a mission briefing containing the assumed composition of their opponents. The *mission briefing composition* is also randomly selected from the teams defined in the case base (according to a uniform distribution). The CBR system operates as an external observer and performs team recognition on one team per run (i.e., either the left team or the right team). Each scenario is repeated twice so that the CBR system has to recognize both teams, resulting in 400 total runs. During each scenario, the models are used to classify the aircraft and those values are merged with the mission briefing (i.e., Algorithm 1) to create an *observed composition*. Both the observed composition and mission briefing composition are used by the CBR system to retrieve the *CBR composition* (i.e., using Algorithm 2).

To measure the effectiveness of team recognition, we use two metrics: *team recognition accuracy* and *average team distance*. Team recognition accuracy measures the percentage of scenarios where a predicted team composition (i.e., mission briefing, observed, or CBR) is identical to the true composition. Average team distance measures the distance between the predicted team and the true team. Since the models are ordered based on how much they differ from the default F-16 model (i.e., –35%, 0%, and 35% for aircraft, and –20%, –10%, 0%, 10%, and 20% for missiles), the distance between two models is measured by how their indexes in the sorted lists differ. Aircraft models have a maximum distance of 2, and missile models have a maximum distance of 4. For example, the default missile model differs from itself by a distance of 0, but a distance of 2 from both the –20% and 20% models. The team distance is the summation of all model distances, and that value is averaged over all scenarios.

5.3 Results and Discussion

Our results are shown in Table 1. The team recognition performance of our CBR system is a statistically significant improvement over mission briefing and observation-based compositions across all metrics (using a paired t-test with $p < 0.001$). This provides strong support for **H2** and **H3**. Additionally, the CBR system was able to identify the correct team nearly 90% of the time and had a low average distance from the team's true composition, providing support for **H1**. The observation-based team composition was a statistically significant improvement over the mission briefing composition using the average team distance metric, but a significant decrease using team recognition accuracy. The reason for this is because the mission briefing and CBR team compositions are guaranteed to be valid (i.e., team compositions are selected from teams contained in the case base). However, the observations are not restricted in such a way, often leading to team configurations that cannot be used as true compositions. Even though this gives the observation-based composition a disadvantage over the mission briefing composition, and results for team recognition accuracy worse than random, its recognized teams are much closer to the true composition. This provides partial support for **H4**.

Table 1. Results of team recognition over 400 experimental runs

Prediction source	Team recognition accuracy	Average team distance		
		Aircraft models	Missile models	Total
Mission briefing	10.0%	3.32	5.84	9.16
Observations	4.8%	2.60	1.72	4.32
CBR	89.8%	0.19	0.31	0.50

Our results also demonstrate that the opportunities to use the learned models for classification are relatively rare. On average, there are 3.6 aircraft and 4.5 missiles per run that performed behaviors that could be used to classify them (i.e., evading or firing a missile). Overall, only 12% of the scenarios had sufficient data to classify all 5 aircraft and missiles in the run. Additionally, the models are learned so there is a possibility of error during learning or classification (i.e., class labels may be incorrect). The CBR process helps reduce the impact of missing information and error by allowing for partial team matches during retrieval, resulting in improved team recognition performance.

6 Related Work

Our previous work related to the BVR domain has primarily focused on discrepancy detection [4] and opponent behavior recognition [5]. Team recognition can be thought of as a form of both discrepancy detection (i.e., a discrepancy in the expected team composition) and behavior recognition (i.e., an aircraft's behavior is based on its aircraft and missile type), but our prior work reasons about opponents at a higher level of abstraction (i.e., actions, plans, and goals) and cannot detect variations in an aircraft's maneuverability or weapons capabilities. Similarly, single and multi-agent behavior recognition [6] has historically focused on identifying agents' actions, activities, and behaviors. Simultaneous Team Assignment and Behavior Recognition (STABR) identifies the behavior of agents in a multi-agent environment and determines the team to which they belong [7]. This differs from our work in that it focuses on team assignment (rather than determining the capabilities of each agent) and allows for dynamic team changes (rather than a static set of teammates and enemies).

Case-based reasoning has been used for multi-agent behavior recognition in soccer [8]. Cases store environmental trigger conditions and behaviors the agents will take when the triggers occur. Similarly, plan recognition has been used as part of a case-based reinforcement learner to identify the plans of opponent teams in American Football [9]. Both of these approaches identify the coordinated behaviors of teams but cannot be used to identify changes in team composition. For example, if an elite player was substituted for a weak player, the systems could not identify the change. CBRetaliate responds to decreased mission performance using case-based reinforcement learning [10]. This allows it to respond to changes in the underlying strategies used by an opposing team. Their approach is similar to our own in that CBRetaliate detects discrepancies between the expected and observed behaviors of an opponent, but differs in that it identifies a team-level strategy rather than the composition of the team. Case-based multi-agent coordination in robotic soccer [11] is similar to our work in that cases are composed (in part) of information about agent teams. While soccer provides many similar challenges to BVR combat (e.g., noise, adversaries, non-deterministic actions), their prior work uses cases to control teammates rather than reason about opponents. Soccer is also similar to BVR combat in that it is a multi-agent environment which requires object matching due to partial observability, with greedy matching often preferable to optimal matching due to real-time constraints [12].

To the best of our knowledge, other applications of AI in BVR air combat have been restricted to expert-authored scripted agents [3] in high-fidelity simulators, and initial flight formation [13] and target assignment [14] in low-fidelity simulators. Unlike our approach, these systems do not consider the possibility that initial assumptions about opponents may be incorrect and should be continually assessed and revised as needed.

7 Conclusions

We presented a technique for case-based team recognition. Our approach uses learned models to classify an opponent's aircraft and missile types and utilizes that information during case retrieval. We tested our CBR system in simulated beyond-visual-range air combat scenarios and reported significantly increased team recognition performance compared to relying on the models or mission briefing data alone.

Our empirical results are promising but several areas of future work remain. We evaluated our CBR system as an external observer of BVR scenarios. We plan to incorporate the capabilities into individual agents so they can use the recognized teams to modify their own behavior. This will require evaluating both team recognition performance and influence on mission performance. Additionally, we plan to extend our approach to allow heterogeneous weapons systems (i.e., each aircraft can be equipped with multiple missile types). Finally, we plan to investigate team recognition countermeasures. A BVR agent could give the appearance of having different capabilities to influence their opponent's tactical decisions.

Acknowledgements. Thanks to OSD ASD (R&E) for supporting this research.

References

1. Shaw, R.L.: Fighter Combat: Tactics and Maneuvering. Naval Institute Press, Annapolis (1985)
2. Clive, P.D., Johnson, J.A., Moss, M.J., Zeh, J.M., Birkmire, B.M., Hodson, D.D.: Advanced Framework for Simulation, Integration and Modeling (AFSIM). In: Proceedings of the 13th International Conference on Scientific Computing, pp. 73–77 (2015)
3. Wang, Y., Witten, I.H.: Inducing model trees for continuous classes. In: Poster Papers of the 9th European Conference on Machine Learning, Prague, pp. 128–137. Springer (1997)
4. Karneeb, J., Floyd, M.W., Moore, P., and Aha, D.W.: Distributed discrepancy detection for BVR air combat. In: Proceedings of the IJCAI Workshop on Goal Reasoning, New York (2016)
5. Borck, H., Karneeb, J., Floyd, M.W., Alford, R., Aha, D.W.: Case-based policy and goal recognition. In: Hüllermeier, E., Minor, M. (eds.) ICCBR 2015. LNCS, vol. 9343, pp. 30–43. Springer, Cham (2015). doi:10.1007/978-3-319-24586-7_3
6. Intille, S.S., and Bobick, A.F.: A framework for recognizing multi-agent action from visual evidence. In: Proceedings of the 16th National Conference on Artificial Intelligence, pp. 518–525. AAAI Press, Orlando (1999)
7. Sukthankar, G., Sycara, K.P.: Simultaneous team assignment and behavior recognition from spatio-temporal agent traces. In: Proceedings of the 21st National Conference on Artificial Intelligence, pp. 716–721. AAAI Press, Boston (2006)
8. Wendler, J., Bach, J.: Recognizing and predicting agent behavior with case based reasoning. In: Proceedings of the RoboCup Robot Soccer World Cup, pp. 729–738 (2003)
9. Molineaux, M., Aha, D.W., Sukthankar, G.: Beating the defense: using plan recognition to inform learning agents. In: Proceedings of the 22nd International Florida Artificial Intelligence Research Society Conference, pp. 337–343. AAAI Press, Sanibel Island (2009)
10. Auslander, B., Lee-Urban, S., Hogg, C., Muñoz-Avila, H.: Recognizing the enemy: combining reinforcement learning with strategy selection using case-based reasoning. In: Althoff, K.-D., Bergmann, R., Minor, M., Hanft, A. (eds.) ECCBR 2008. LNCS, vol. 5239, pp. 59–73. Springer, Heidelberg (2008). doi:10.1007/978-3-540-85502-6_4
11. Ros, R., López de Màntaras, R., Arcos, J.L., Veloso, M.: Team playing behavior in robot soccer: a case-based reasoning approach. In: Weber, R.O., Richter, M.M. (eds.) ICCBR 2007. LNCS, vol. 4626, pp. 46–60. Springer, Heidelberg (2007). doi:10.1007/978-3-540-74141-1_4
12. Floyd, M.W., Esfandiari, B., Lam, K.: A case-based reasoning approach to imitating RoboCup players. In: Proceedings of the 21st International Florida Artificial Intelligence Research Society Conference, pp. 251–256. AAAI Press, Coconut Grove (2008)

13. Luo, D.-L., Shen, C.-L., Wang, B., Wu, W.-H.: Air combat decision-making for cooperative multiple target attack using heuristic adaptive genetic algorithm. In: Proceedings of the 4th International Conference on Machine Learning and Cybernetics, pp. 473–478 (2005)
14. Mulgund, S., Harper, K., Krishnakumar, K., Zacharias, G.: Air combat tactics optimization using stochastic genetic algorithms. In: Proceedings of the IEEE International Conference on Systems, Man, and Cybernetics, pp. 3136–3141 (1998)

The Mechanism of Influence of a Case-Based Health Knowledge System on Hospital Management Systems

Dongxiao Gu[1], Jingjing Li[1], Isabelle Bichindaritz[2], Shuyuan Deng[3], and Changyong Liang[1(✉)]

[1] School of Management, Hefei University of Technology, Hefei, China
{gudongxiao, cyliang}@hfut.edu.cn, 847840274@qq.com

[2] Computer Science Department, SUNY Oswego, Oswego, USA
ibichind@oswego.edu

[3] Sheldon B. Lubar School of Business, University of Wisconsin-Milwaukee, Milwaukee, USA
dengs@uwm.edu

Abstract. In hospitals in China, the rapid development of intelligent knowledge-based systems has been accompanied by the widespread adoption of case-based health knowledge systems (CBHKS). Their implementation has provided a great opportunity for Management promotions in hospitals. However, the impact of the use of CBHKS on the improvement of hospital management efficiency has not been clearly addressed in the literature. In this study, we investigate the role of CBHKS in improving hospital management through group effectiveness and leadership performance-maintenance theory (PM). We developed a conceptual model and empirically tested it. From the theoretical standpoint, the establishment of the model not only enriched the group effectiveness and leadership performance and maintenance, but also played a positive role in promoting CBHKS systems as well as the sustained development of and innovation in group effectiveness and leadership performance-maintenance theory. From the practical standpoint, the validation of the study provides an important reference for the improvement of hospital management level and efficiency. Our findings have important implications for the effective use of CBHKS in hospitals.

Keywords: Case-based systems · Group effectiveness · Hospital managerial performance · Knowledge-based system · Healthcare information management

1 Introduction

With the rapid development of modern information technology (IT) and applications of artificial intelligence, intelligent knowledge-based information systems have been widely implemented throughout the world. Healthcare is no exception and a few large hospitals in China have adopted case-based health knowledge systems (CBHKS) that integrate case-based reasoning and cloud computing and assist diagnostic tasks and decision making. There are three main factors fostering the implementation of CBHKS

D.W. Aha and J. Lieber (Eds.): ICCBR 2017, LNAI 10339, pp. 139–153, 2017.
DOI: 10.1007/978-3-319-61030-6_10

in hospitals: (1) Traditional health information systems (HIS) generally do not provide knowledge support for medical diagnosis or hospital management, which is urgently required for implementing precision medicine in hospitals; (2) Large-scale historical cases accumulated in electronic health records (EHRs) afford a rich source of data to build case-based knowledge systems; (3) By deepening collaborations between hospitals and universities, more and more professors of artificial intelligence and data mining engage in collaborative research and development of CBHKS with hospitals. CBHKS systems can provide knowledge support for many functional areas in healthcare and hospital management, such as medical diagnoses and prognoses, drug inventory management, nursing management, medical equipment management, managerial decision making, and medical material management. Because of their comprehensiveness and efficiency in knowledge-based reasoning, CBHKS are playing an important role in improving hospital management. They effectively facilitate medical knowledge discovery and management's decision making.

Despite the investments made in CBHKS implementation, the mechanisms of their influence on management performance are still unclear. However, understanding these mechanisms is important for the more effective use of CBHKS and for justifying their costs. This study examines the postadoption use of CHBKS, which is closely associated with the long-term performance of hospital management. Using expectation confirmation theory, Bhattacherjee [1] was among the first to investigate the continued use of IT. His work was followed by an array of studies that focused on information systems (IS) post-adoption behavior (ISPAB) [2–4]. ISPAB explains why individuals and organizations continue to use and expand IS functions and proposed a set of factors affecting the continuing utilization of IS and the expansion of individuals and organizations. This stream of research forms the theoretical foundation for the continued use of CHBKS in hospitals. ISPAB research has been conducted at both the individual level and the organizational level. However, studies at the group level are scarce, a group being defined as a formal or informal team within an organization. Group-level factors — group performance, group learning, group satisfaction, and the degree of external satisfaction — may influence the postadoption attitudes and behaviors of individuals within a group as well as postadoption decisions at the organizational level. Thus, examining continued use of IS at the group level will fill an important gap in ISPAB research.

In this study, we focus on how CBHKS systems improve hospital management through group-level factors. Specifically, we believe that these group factors, namely, group performance, group learning, group satisfaction, and the degree of external satisfaction, mediate the effect of CBHKS on hospital management. By applying Misumi's [5] performance-maintenance (PM) theory of leadership, we can effectively measure hospital management. More broadly, this study aims at providing an evaluation of the usefulness of CBR systems in healthcare services and management, through the example of a CBHKS system currently in use in a hospital. This type of study provides very important information for future projects in the same field.

Earlier studies have shown that IT utilization positively affects an organization's performance. For example, Andersen [6] tested the positive impact of Intranet and Internet usage on organizations' sales growth and innovation. In their study of hospital management, Devaraj and Kohli [7] investigated eight hospitals that adopted decision support systems (DSS) and found that a high level of IT applications in hospitals improves their financial performance and quality. At the group level, scholars studied the role of group effectiveness in the interaction between IS utilization and organizational management. Hospitals comprise various departments, institutes, centers, etc. These groups can develop group cohesiveness and teamwork within the confines of their processes and research [8]. Group effectiveness is an important manifestation of an organization's effectiveness as well as a sign of managerial competency in hospitals. It is an antecedent of group performance. It is also lead to group members' satisfaction. Therefore, it is imperative to understand how CBHKS systems improve managerial performance and maintenance in hospitals through group effectiveness.

Focusing on group effectiveness and drawing on the PM theory of leadership, this study investigates the impact of CBHKS on hospital management. Furthermore, we reveal how group effectiveness mediates the effect of the use of CBHKS systems on organizational and managerial performance. Our study addresses these questions:

(1) Does the implementation of CBHKS improve managerial performance and maintenance in hospitals?
(2) How does the implementation of CBHKS affect managerial performance and maintenance in hospitals?
(3) What role does group effectiveness play in mediating the impact of CBHKS on managerial performance and maintenance in hospitals?

Answering these questions is important for bridging related findings at the individual and organizational levels. Our study not only contributes to the literature on hospital management, but from a practical perspective, it will promote the implementation of CBHKS in order to improve the satisfaction of hospital staff and patients and enhance managerial performance and maintenance in hospitals.

2 Theoretical Foundations

2.1 Group Effectiveness

Hackman [9] proposed the earliest definition of group effectiveness. In his study, group effectiveness has been defined from three perspectives: work output, impact on group members (including satisfaction), and the improvement of group problem-solving skills for the future. Later research supplemented and perfected the definition of group effectiveness. Sundstrom and Mclntyre [10] used member satisfaction, performance, external satisfaction, and team learning to measure group effectiveness. Cohen and Baily [11] further noted that the three aspects of group effectiveness are: (1) group performance, namely efficiency, productivity, response speed, quality, customers' satisfaction with service, and innovation; (2) members of the group, namely satisfaction of the members, commitment and trust in management; and (3) members' behavior, namely members' absenteeism, turnover, and safety.

Based on these definitions, we used four dimensions proposed by Sundstrom and Mclntyre [10] to evaluate group effectiveness: group performance, group member satisfaction, group learning, and external satisfaction. Most studies on group effectiveness have focused mostly on constructing a reasonable group management system that enhances group effectiveness. The influence of group effectiveness is relevant to many levels in a management hierarchy, such as individual, team, business unit, and organization [11–13]. In this study, we focus on the impact of group effectiveness on organizational performance, explaining why the adoption of CBHKS in hospitals affects group work and influences management performance through group effectiveness. Measuring these four dimensions of group effectiveness as intermediary variables, we developed a research model providing insights into the impact of CBHKS on hospital performance and maintenance.

2.2 Performance-Maintenance (PM) Theory of Leadership

Misumi [14] proposed the PM theory of leadership in the 1970s. This theory divides leadership behavior into two categories, i.e., Performance (P) and Maintenance (M). The performance function means that a leader must design detailed work plans, task and strictly monitor staff to carry out the plans, with the goal of ensuring that a group's specific goals can be achieved. The maintenance function means that a leader should be concerned about his or her staff, try to build a harmonious relationship with his or her staff members, carefully encourage them, and pay attention to proper authorization to ensure ongoing improvement in the organization's normal operation.

In PM theory, leaders have two main management objectives: organizational goals and organizational relationships. Organizational goals focus on how to improve managerial performance; organizational relationship focuses on how to establish a stable and harmonious inner environment that maintains and sustains the organization. Echoing PM theory, we establish management effectiveness and managerial maintenance as the dual goals when adopting CBHKS in hospitals and when measuring the impact of CBHKS on hospital management.

2.3 CBHKS System Description

A CBHKS system is designed and implemented with practicality, usefulness, and ease of use in mind. The main purpose of this system is to provide a knowledge-based decision-making support for doctors' diagnostic and therapeutic processes, as well as for hospital leaders or managers problem solving process, using the most similar historical administrative cases (i.e. issues and their solutions) for reference. For example, in one of the largest hospitals in East China, the main function of the CBHKS system includes mainly five modules (see Figs. 1 and 2): system administration, EHR cases, case features management and maintenance, historical hospital management cases (issues and solutions), as well as case base maintenance. Its main design principles include: (1) to satisfy the requirements of knowledge acquisition for diagnosis and managerial decision support; (2) ease of use for the doctors and hospital managers; and (3) reliability and expansibility. To satisfy the effective case acquisition requirement, we adopted an object-oriented case representation. The original case base has over 24,700 cases and this number is rapidly growing while CBHKS expands to new

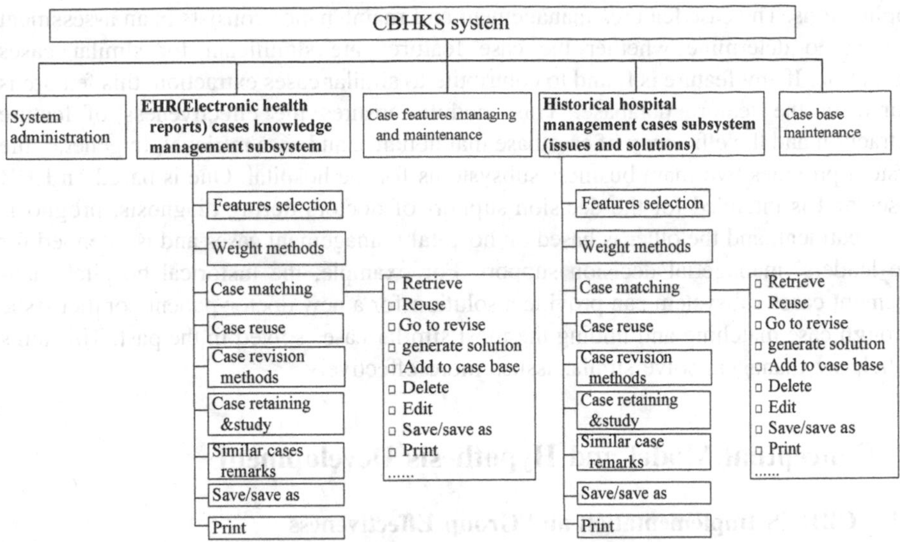

Fig. 1. The main functional structure of a CBRKS system

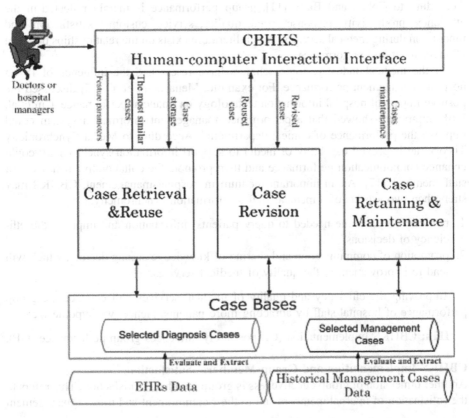

Fig. 2. CBHKS systems' core components and interaction mechanism.

applications. The case features management and maintenance consists in an assessment process to determine whether the case features are significant for similar cases extraction. If any feature is found to contribute to similar cases extraction, this feature is stored in the feature database. This module ensures the effectiveness of feature extraction and the efficiency of case base maintenance and case retrieval. In general, the system provides two main business subsystems for the hospital. One is based on EHR cases and is intended for the decision support of doctors during diagnosis, prognosis and treatment, and the other is based on hospital management cases and is intended for the leaders' managerial decision support. For example, the historical hospital management cases subsystem can provide a solution for a new doctor-patient conflict issue through case matching and finding the most similar cases solved in the past. This aims at helping managers solve similar issues more effectively.

3 Conceptual Model and Hypothesis Development

3.1 CBHKS Implementation and Group Effectiveness

CBHKS Implementation and Group Performance

According to Cohen and Baily [11], group performance is mainly reflected in the efficiency, productivity, response speed, quality, service, customer satisfaction, and innovation during medical services. A rich literature exists on the relationships between IT investment and performance.

In the medical industry, some scholars have researched the influence of IS on hospital management performance. For example, Menachemi et al. [15] discussed the positive impact of hospital information technology on financial performance. Cui et al. [16] empirically showed that a performance management information system could improve the performance of clinical departments. According to Media Synchronicity Theory, an improved capability of media to support information synchronicity could enhance communication performance and then promote the collaborative behaviors of staff members [17]. As an important medium in hospital management, CBHKS may strengthen two aspects of a medical staff's information synchronicity:

1. reduction in the time needed to query patients' information and improve the efficiency of decisions;
2. promotion of communication and sharing of knowledge among doctors, which will lead to improvement in the quality of medical services.

Improving the efficiency and quality of medical services will enhance the group performance of hospital staff by attracting more patients. Hence, we hypothesize:

H1a: CBHKS Implementation (CI) positively influences group performance (GP).

CBHKS Implementation and Group Members' Satisfaction

Another dimension of group effectiveness is group members' satisfaction. This refers to the satisfaction of internal members and to their commitment and trust in management

[11]. In the context of a hospital, it represents the work attitudes of doctors, nurses, and other medical staff.

Considerable research has studied the relationships between IS implementation and group member satisfaction. For example, Ammenwerth et al. [18] believe that IT could be used to improve the sense of responsibility in nursing care, thus, improving the efficiency of nurses as well as increasing the satisfaction of patients and other members of the hospital staff. Langfred [19] posited that team performance depends on an individual's and a team's autonomy, both of which could be improved by IS implementation by encouraging users' initiative and adding to their satisfaction.

Media Synchronicity Theory posits that communication effectiveness may be enhanced by faster information conveyance and higher information convergence [17], which in turn will improve users' satisfaction with access to information and its use. We believe that the implementation of CBHKS improves doctors' autonomy. Doctors can obtain information conveniently, process information efficiently, and reduce redundant work. The improvement of efficiency is more pronounced in medical group collaboration. All of these improve the satisfaction of members of the medical staff, facilitating their completion of clinical tasks. More efficient clinical processes will improve patients' satisfaction, and in turn, improve their relationship with the medical staff. Hence, we propose the following hypothesis:

H1b: CBHKS Implementation positively influences group members' satisfaction (GMS).

CBHKS Implementation and Group Learning

The relationship between IT implementation and group learning has received a lot of attention in the literature. Scholars believe that IT plays an important role in the process of shifting to a "learning organization" by providing technical support for effectively storing, organizing, and modifying information and knowledge. Furthermore, some scholars believe that IT could facilitate organizational learning [20]. Related studies suggest that the implementation of new technologies often requires employees to update their skills [21, 22].

In hospitals, the use of CBHKS also promotes group learning among the staff. In particular, the implementation of such a system requires doctors to learn how to use it correctly, which can be regarded as a technical learning process for a group. Moreover, the implementation of CBHKS will promote information and knowledge sharing among doctors and enhance their clinical problem-solving skills. Last but not least, the implementation of CBHKS can encourage doctors to cooperate during the problem-solving process. Hence, we hypothesize:

H1c: CI positively influences group learning (GL).

CBHKS Implementation and External Satisfaction

External satisfaction is a consequence of group behavior and a dimension of group effectiveness. According to Sundstrom and McIntyre [10], higher group effectiveness leads to higher client satisfaction. Information conveyance and information convergence through media could not only affect a staff's work efficiency, but also improve a customer's perceived information transparency [17], which promotes trust among contracting parties and improves external satisfaction.

The implementation of IS is likely to improve patient satisfaction and hospital management and reduce administrative costs. Aamodt et al. [23] posited that the utilization of IS can optimize inspection processes, standardize operating procedures, and improve patient and staff satisfaction. Aiken [24] concluded that the application of IS in hospitals optimizes the treatment process by shortening treatment time and improving service quality, thus enhancing patient communication and improving patient satisfaction with his or her medical treatment.

The implementation of CBHKS is likely to improve external satisfaction for several reasons. First, CBHKS can save doctors time by improving their efficiency. As a result, doctors will have more time to discuss treatment with other doctors and rethink their work, making treatment more effective. This is likely to enhance patients' satisfaction. Second, implementation of CBHKS will showcase a hospital's IT capability and commitment. This will create a better image of the hospital, potentially improving external satisfaction. Hence, we hypothesize:

H1d: CI positively influences external satisfaction (ES).

3.2 Group Effectiveness and Hospital Management

Group Performance and Hospital Management

Only a few studies have analyzed the relationships between hospital management and group effectiveness. Besstremyannaya et al. [25] believe that hospital managerial performance is a kind of output that reflects hospital input as represented by cost efficiency. Implementation of CBHKS, as a kind of information technology input, may improve the efficiency, productivity, and responsiveness of a group of doctors [11] and as a result improve patients' satisfaction with hospital services. High patient satisfaction leads to more use of medical services from the same hospital, increasing its financial income, an important indicator of managerial performance. Hence, the usage of CBHKS may have a significant impact on managerial performance.

Hospital managerial maintenance means keeping the relationships in a hospital stable and harmonious through appropriate communication while supporting employees' autonomy. It preserves a hospital's social stability. High group performance will improve financial performance [26], which could improve the income of individuals and groups under the generally accepted performance appraisal system and enhance the stability of the organization. Furthermore, high group performance implies harmonious interpersonal relationships and efficient communication in organizations, which could foster group members' job satisfaction and improve managerial maintenance. Hence we hypothesize:

H2a: Group performance (GP) positively influences hospital managerial performance (HMP).

H2b: GP positively influences hospital managerial maintenance (HMM).

Group Members' Satisfaction and Hospital Management

Group members' satisfaction is related to the recognition, responsibilities, supervision, and opportunities offered during their work [27]. A number of researchers have empirically found that satisfaction is related to performance, such as an organization's outcomes, management effectiveness, etc.

In the context of hospital management, many scholars have studied the relationships between group members' satisfaction and hospital performance. Wagner et al. [28] tested the positive effect of professional nurses' communication satisfaction to hospital's performance. Mascia et al. [29] studied how hospital restructuring influences job satisfaction in physicians. The use of CBHKS is likely to improve communication efficiency and accelerate the efficiency of searching for knowledge. It is also likely to trigger the restructuring of a hospital's business processes and shorten diagnosis time. Thus, the presence of a CBHKS may improve both the staff's job satisfaction and the hospital's performance. This is an important indicator of management performance and maintenance. Hence we hypothesize:

H3a: Group members' satisfaction (GMS) positively influences hospital managerial performance (HMP).
H3b: Group members' satisfaction positively influences hospital managerial maintenance (HMM).

Group Learning and Hospital Management

Group learning will be helpful for better communication and coordination in teams. It has been shown to have positive influence on new product quality through the improvement of management. Hassan et al. [30] suggested that team learning behaviors play a significant role in ensuring the success of marketing teams. Argote [31] posited that the occurrence of team learning was mainly manifested in the relatively lasting change in team members' knowledge and performance. He also used organizational and team learning curves to measure team learning performance. Chen [32] found a significant positive relationship between a team's learning ability and its performance.

In hospitals, leaders, department directors, managers, and medical staff can use various functions of CBHKS to conduct group learning and acquire information and knowledge for managerial decision making. All of these could improve the professional skills of a medical group and deliver a high quality of medical service to patients. These gains also could lead to more efficiency that would permit treating more patients as well as earn their trust, which eventually could improve hospital managerial performance. In addition, CBHKS could also help improve the effectiveness of group learning and increase the communication behavior of group members, which in turn could increase team cohesion and promote the hospital's managerial maintenance. Hence, we propose the following two hypotheses:

H4a: Group learning (GL) positively influences hospital managerial performance (HMP).
H4b: Group learning (GL) positively influences hospital managerial maintenance (HMM).

External Satisfaction and Hospital Management

Previous studies have investigated the relationships between external satisfaction and managerial performance and maintenance. External satisfaction is a kind of work attitude from external stakeholders, while managerial maintenance directly embodies organizational effectiveness [33]. Terziovski [34] investigated the relationships among quality management practices, customer satisfaction, and productivity improvement. He suggested that improvement of customers' satisfaction correlates positively with productivity improvement.

External satisfaction in hospitals mainly refers to patients' satisfaction. The implementation of CBHKS improves treatment effectiveness and efficiency, saves patients' time, and then improves external satisfaction. Patients' experiences are vital in determining whether patients will choose a hospital again. Moreover, patients' satisfaction influences the service choices of potential customers through word-of- mouth. As one dimension of group effectiveness, external satisfaction is an important reflection of whether the work of medical groups is effective. As a result, external satisfaction can directly influence hospital performance. In addition, CBHKS can promote medical effectiveness, which also improves the satisfaction of patients. This improves doctor-patient relationships, which indirectly enhance the stability of medical groups. Hence, external satisfaction influences the hospital's continuing development and maintenance. Therefore, we propose the following hypotheses:

H5a: External satisfaction (ES) has a significant positive effect on hospital managerial performance (HMP).
H5b: External satisfaction (ES) has a significant positive effect on hospital managerial maintenance (HMM).

Based on the above hypotheses, we present our research model in Fig. 3.

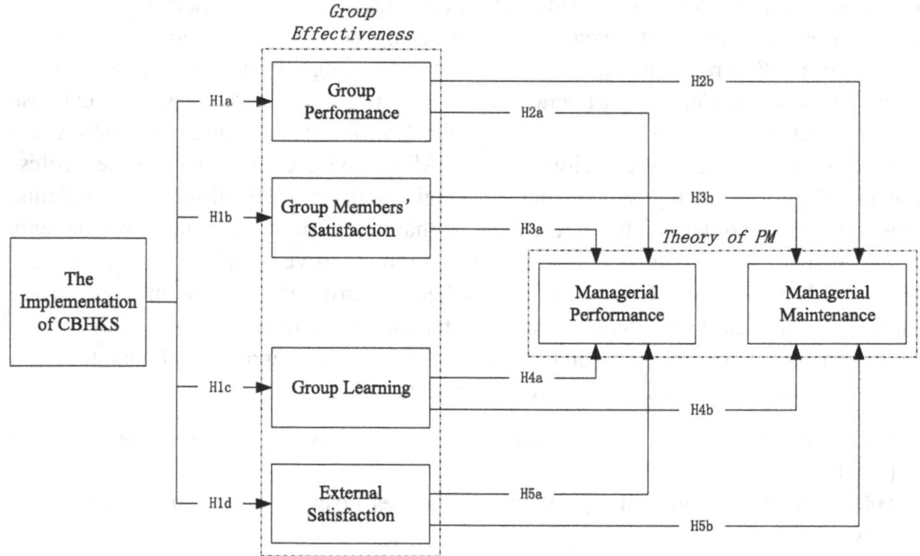

Fig. 3. Research model.

4 Experiment and Results

We cooperated with the hospital to complete the experiments. The participants were randomly picked from the doctors and hospital managers by using their employee ID. The office directors informed the selected participants and explained the purpose of the study. The hospital explained the purpose of the experiment and privacy protection measures to all the participants. We collected data department by department in the hospital. Participation was voluntary. To ensure data quality, a small gift with a value of approximately $10 was provided to each respondent who completed and returned the questionnaire. Besides participants' basic information (age, gender, and education), the questionnaire comprised 41 questions. All of the items were measured with a seven-point Likert scale, ranging from 1 (strongly disagree) to 7 (strongly agree). The survey and experiment lasted for two and a half months. In our questionnaire, all the items are designed to ask questions regarding the CBHKS system that is being used in the hospital we mention in Sect. 2.3. All the participants are the users of this system and can understand the difference between CBRKS and general EHRs. 225 responses were collected, but 11 responses were discarded as incomplete. Hence, we used 214 responses in the analysis. This study adopted SmartPLS 2.0 for data analysis and the validation of the model and its hypotheses. The measurement model analysis mainly involves the assessment of reliability and validity. The validity of construct is commonly measured from two aspects, i.e. convergent validity and discriminant validity. All of the Cronbach's α and composite reliability values exceed 0.8, indicating very good reliability. All of the AVE values exceed 0.5, suggesting good convergence validity. The square root of each construct's AVE value exceeds its correlations with other constructs, indicating strong discriminant validity.

In this study, the relationships among CBHKS implementation, group effectiveness, and the improvement of managerial performance and managerial maintenance in hospital management were thoroughly investigated. The results show that CBHKS implementation has a significant, positive effect on group effectiveness. Moreover, all of the four dimensions of group effectiveness have significant, positive effects on hospital management performance. Group members' satisfaction and external satisfaction also have significant, positive effects on hospital management maintenance. But the effects of group performance and group learning on managerial maintenance are not significant, as show in Table 1.

According to the data analysis results, the positive influences of group performance and group learning on managerial maintenance are not significant ($\beta_{H2b} = 0.0758$ and $\beta_{H4b} = -0.0244$). Table 1 contains a summary of these results. According to the PM Theory of Leadership, the *maintain* function represents efforts to maintain and strengthen group relationships, such as team collaboration spirit, level of satisfaction, and code of conduct. While the *performance* function benefits from visibility and objective evaluation criteria, the assessment of the *maintain* function generally contains some subjective or emotional attributes. Hence, the evaluation of the *maintain* function needs relatively more time than that of the *performance* function. The evaluation

Table 1. Structural parameter estimates - Note. **p < .01

Hypothesized path	Standardized path coefficients (β)	T-value	Results
H1a: CBHKS implementation → group performance	0.5688**	24.3204	Supported
H1b: CBHKS implementation → group members' satisfaction	0.4712**	16.2104	Supported
H1c: CBHKS implementation → group learning	0.5434**	19.883	Supported
H1d: CBHKS implementation → external satisfaction	0.5115**	20.4664	Supported
H2a: group performance → hospital managerial performance	0.1396**	3.1869	Supported
H2b: group performance → hospital managerial maintenance	0.0758	1.8325	Not supported
H3a: group members' satisfaction → hospital managerial performance	0.2025**	4.8783	Supported
H3b: group members' satisfaction → hospital managerial maintenance	0.2793**	8.9865	Supported
H4a: group learning → hospital managerial performance	0.2339**	5.8591	Supported
H4b: group learning → hospital managerial maintenance	−0.0244	0.5187	Not supported
H5a: external satisfaction → hospital managerial performance	0.307**	9.2919	Supported
H5b: external satisfaction → hospital managerial maintenance	0.5342**	16.7912	Supported

duration of the CBHKS system application was probably not long enough to support an accurate evaluation of the *maintain* function in our current study, which could cause hypotheses H_{2b} and H_{4b} to be not supported. CBHKS is an important toolkit for collaborative diagnosis and management, which can improve dynamic group learning ability. Hence its implementation heavily affects group performance and group learning. External satisfaction likely triggers a sense of pride in group members, which could probably usefully arouse their team spirit and make them adhere to behavioral norms. Consequently, external satisfaction can greatly influence hospital managerial maintenance. Based on the above analysis, for a decision maker considering implementing a CBHKS in her hospital, one of the benefits she can anticipate from a successful CBHKS implementation is an improvement in managerial performance. Another benefit she can anticipate from using a CBHKS is to successfully promote managerial maintenance by improving the level of external satisfaction and group members' satisfaction.

5 Conclusions

This study not only enriched the group effectiveness and leadership performance-maintenance theory, but also provided a convenient reference for medical workers and patients. The validation of the study also provides an important reference value on the improvement of a hospital management level and efficiency. This study discussed how implementing and using a CBHKS improves hospital managerial performance and managerial maintenance through group effectiveness and the PM theory. Our results support most of the research hypotheses. Although our example is confined to East China, similar systems start being used more broadly. Hence the results of our current research have the potential to generalize across the country.

Acknowledgements. This research is partially supported in data collection, analysis, and interpretation by the National Natural Science Foundation of China (NSFC) under grants No. 71331002, 71301040, 71273125, and 71503033, as well as by National Social Science of China (Major Program) with No. 12&ZD221.

Appendix A

Group Effectiveness: There are four dimensions proposed by Sundstrom and Mclntyre to evaluate group effectiveness: group performance, group member satisfaction, group learning, and external satisfaction [10].

Group Performance: According to the definition by Cohen and Baily, group performance is mainly reflected in the efficiency, productivity, response speed, quality, service, satisfaction, and innovation during medical services [11].

Group Members' Satisfaction: This refers to the satisfaction of internal members and to their commitment to and trust in management. In a healthcare context, it represents the work attitude of doctors, nurses, and other medical staff.

Group Learning: It represents a team learning process or behavior when learning new things or adapting to a new and dynamic environment.

External Satisfaction: It denotes a kind of positive work attitude from external stakeholders. The implementation of CBHKS improves treatment effectiveness and efficiency, which in turn improves external satisfaction.

Performance-Maintenance (PM) Theory of Leadership: The performance function means that a leader must design detailed work plans, task staff to carry out the plans, and strictly monitor staff, with the aim of ensuring that a group's specific goals can be achieved. The maintenance function means that a leader should be concerned about his or her staff, try to build harmonious relationships with his or her staff members, carefully encourage them, and pay attention to proper authorization to ensure ongoing improvement in the organization's normal operations.

Managerial Maintenance: Hospital managerial maintenance means keeping the relationships in a hospital stable and harmonious through appropriate communication while supporting employees' autonomy.

References

1. Bhattacherjee, A.: Understanding information systems continuance: an expectation-confirmation model. MIS Q. **25**, 351–370 (2001)
2. Wang, E.T.G., Tai, J.C.F., Grover, V.: Examining the relational benefits of improved interfirm information processing capability in buyer-supplier Dyads. MIS Q. **37**, 149–173 (2013)
3. Bhattacherjee, A., Sanford, C.: Influence processes for information technology acceptance: an elaboration likelihood model. MIS Q. **30**, 805–825 (2006)
4. Lin, C.S., Wu, S., Tsai, R.J.: Integrating perceived playfulness into expectation-confirmation model for web portal context. Inf. Manag. **42**, 683–693 (2005)
5. Misumi, J., Peterson, M.F.: The performance-maintenance (PM) theory of leadership: review of a Japanese research program. Adm. Sci. Q. **30**, 198–223 (1985)
6. Andersen, T.J.: Information technology, strategic decision making approaches and organizational performance in different industrial settings. J. Strateg. Inf. Syst. **10**, 101–119 (2001)
7. Devaraj, S., Kohli, R.: Performance impacts of information technology: is actual usage the missing link? Manag. Sci. **49**, 273–289 (2003)
8. Halvorsen, K.: Team decision making in the workplace: a systematic review of discourse analytic studies. J. Appl. Linguist. Prof. Pract. **7**, 273–296 (2010)
9. Hackman, J.R.: The design of work teams, handbook of organizational behavior, prentice-hall. Englewood Cliffs. NJ. **35**, 299–301 (1987)
10. Sundstrom, E., Mcintyre, M.: Measuring Work-Group Effectiveness: Practices, Issues and Prospects, Working paper. University of Tennessee, Department of Psychology, Knoxville, TN (1994)
11. Cohen, S.G., Bailey, D.E.: What makes teams work: group effectiveness research from the shop floor to the executive suite. J. Manag. **23**, 239–290 (1997)
12. Gupta, V.K., Huang, R., Yayla, A.A.: Social capital, collective transformational leadership, and performance: a resource-based view of self-managed teams. J. Manag. Issues **23**, 31–45 (2011)
13. Tesluk, P.E., Mathieu, J.E.: Overcoming roadblocks to effectiveness: incorporating management of performance barriers into models of work group effectiveness. J. Appl. Psychol. **84**, 200–217 (1999)
14. Misumi, J., Shinohara, H., Sato, S.: Effect of P-M leadership behavior patterns upon performance factors in perceptual motor learning (i). Jpn. J. Exp. Soc. Psychol. **14**, 31–47 (1974)
15. Menachemi, N., Burkhardt, J., Shewchuk, R., Burke, D., Brooks, R.G.: Hospital information technology and positive financial performance: a different approach to finding an ROI. J. Healthc. Manag. **51**, 40–58 (2006)
16. Cui, Y., Wu, Z., Lu, Y., Jin, W., Dai, X., Bai, J.: Effects of the performance management information system in improving performance: an empirical study in Shanghai ninth people's hospital. Springerplus **5**, 1785 (2016)

17. Dennis, A.R., Valacich, J.S.: Media, tasks, and communication processes: a theory of media synchronicity. MIS Q. **32**, 575–600 (2006)
18. Ammenwerth, E., Gräber, S., Herrmann, G., Bürkle, T., König, J.: Evaluation of health information systems—problems and challenges. Int. J. Med. Inf. **71**, 125–135 (2003)
19. Langfred, C.W.: Autonomy and performance in teams: the multilevel moderating effect of task interdependence. J. Manag. **31**, 513–529 (2005)
20. Anand, K.J.: Clinical importance of pain and stress in preterm neonates. Biol. Neonate **73**, 1–9 (1998)
21. Carpenter, M.A.: Upper echelons research revisited: antecedents, elements, and consequences of top management team composition. J. Manag. **30**, 749–778 (2004)
22. Kazahaya, G.: Harnessing technology to redesign labor cost management reports. Healthc. Fin. Manag. J. Healthc. Financ. Manag. Assoc. **59**, 94–100 (2005)
23. Aamodt, A.: Knowledge-intensive case-based reasoning in CREEK. In: Funk, P., González Calero, P.A. (eds.) ECCBR 2004. LNCS, vol. 3155, pp. 1–15. Springer, Heidelberg (2004). doi:10.1007/978-3-540-28631-8_1
24. Aiken, L.H., Sermeus, W., Heede, K.V.D., Sloane, D.M., Busse, R., McKee, M., Bruyneel, L., Rafferty, A.M., Griffiths, P., Morenocasbas, M.T.: Patient safety, satisfaction, and quality of hospital care: cross sectional surveys of nurses and patients in 12 countries in Europe and the United States. BMJ **344**, e1717 (2012)
25. Besstremyannaya, G.: Managerial performance and cost efficiency of Japanese local public hospitals: a latent class stochastic frontier model. Health Econ. **20**, 19–34 (2011)
26. Hillman, A.L., Pauly, M.V., Kerstein, J.J.: How do financial incentives affect physicians' clinical decisions and the financial performance of health maintenance organizations? N. Engl. J. Med. **321**, 86–92 (1989)
27. Sharma, S., Cavallaro, G., Rosato, A.: The effect of reward structures on the performance of cross-functional product development teams. J. Mark. **65**, 35–53 (2001)
28. Wagner, J.D., Bezuidenhout, M.C., Roos, J.H.: Communication satisfaction of professional nurses working in public hospitals. J. Nurs. Manag. **23**, 974–982 (2014)
29. Mascia, D., Morandi, F., Cicchetti, A.: Hospital restructuring and physician job satisfaction: an empirical study. Health Policy **114**, 118–127 (2014)
30. Hassan, M., Aksel, I., Yaqub, M.Z., Aldemir, Z.: Team learning and its impact on marketing team performance: an empirical study. Int. Bus. Res. **4**, 124–131 (2011)
31. Argote, L.: Organizational learning: creating, retaining, and transferring knowledge. Adm. Sci. Q. **45**, 622–625 (2000)
32. Chen, G.Q.: Empirical study on relationship among organizational influential factors, (organizational) learning capabilities and organizational performance. J. Manag. Sci. China **8**, 48–61 (2005)
33. Harrison, D.A., Newman, D.A., Roth, P.L.: How important are job attitudes? Meta-analytic comparisons of integrative behavioral outcomes and time sequences. Acad. Manag. J. **49**, 305–325 (2006)
34. Terziovski, M.: Quality management practices and their relationship with customer satisfaction and productivity improvement. Manag. Res. News **29**, 414–424 (2006)

Scaling Up Ensemble of Adaptations for Classification by Approximate Nearest Neighbor Retrieval

Vahid Jalali(✉) and David Leake

School of Informatics and Computing, Indiana University,
Bloomington, IN 47408, USA
{vjalalib,leake}@indiana.edu

Abstract. Acquisition of case adaptation knowledge is a classic challenge for case-based reasoning. A promising response is learning adaptation rules from cases in the case base, using the *case difference heuristic* (CDH). In previous research we presented Ensembles of Adaptations for Regression (EAR), an approach that uses a CDH-based method to generate adaptation rules and then exploits the availability of multiple learned rules to apply ensemble-based adaptation. We extended EAR to classification tasks, with Ensembles of Adaptations for Classification (EAC), which showed promising accuracy results. EAR and EAC are practical for standard case bases, but become computationally expensive for large case bases and large ensembles, primarily due to retrieval cost. This paper presents research on scaling up EAC by integrating it with EACH, a new method for efficient approximate retrieval that extends locality-sensitive hashing retrieval to categorical features. Experimental results support the ability of the EAC with EACH (Ensemble of Adaptations for Classifications Hashing) to maintain accuracy while increasing efficiency. In addition, EACH could be applied as a standalone method to provide scalable approximate nearest neighbor retrieval in other CBR retrieval contexts.

Keywords: Case adaptation learning · Case difference heuristic · Ensemble of adaptations for classification · Locality sensitive hashing · Value difference metric

1 Introduction

How best to acquire case adaptation knowledge has been a longstanding challenge for case-based reasoning (CBR). Its difficulty has prompted much interest in machine learning methods for learning adaptation knowledge. An especially strong research current has focused on learning adaptation knowledge from cases in the case base by the *case difference heuristic* (CDH) approach, proposed by Hanney and Keane [9]. This approach generates adaptation rules from the case base, by comparing pairs of cases in the case base. For each pair it generates a

D.W. Aha and J. Lieber (Eds.): ICCBR 2017, LNAI 10339, pp. 154–169, 2017.
DOI: 10.1007/978-3-319-61030-6_11

rule in two parts, (1) a description of the difference between the problem parts of the two cases, and (2) a description of the difference between their solutions. Given an input problem and retrieved source case to adapt, an adaptation rule applies if the difference between their problem parts is similar to the problem difference of the rule. When the rule applies, the source case is adapted by applying a difference similar to the solution difference in the rule. Adaptation rule generation in the spirit of CDH has been effective in contexts including numerical prediction [13,16,17], classification [12], and adaptation of structured cases [1,19].

A potential issue for automatically-derived adaptation rules is the uncertain reliability of those rules. Our previous work on Ensembles of Adaptations for Regression (EAR) [13] and Ensembles of Adaptations for Classification (EAC) [12] addresses the reliability issue by exploiting the ability of the CDH approaches to generate many adaptation rules, to use ensembles of adaptations (cf. [26]). Given an input problem, EAR and EAC retrieve multiple cases and can adapt them by multiple system-generated adaptation rules, providing multiple candidate solutions to combine. Experiments support their ability to improve adaptation accuracy. However, the accuracy benefits come at the cost of increased computational cost, as ensembles require multiple retrievals and adaptations. When EAR or EAC is configured for lazy adaptation rule generation, generating adaptation rules on the fly rather than in advance, the cost for each adaptation is increased; when large ensembles are used, the total cost is multiplied.

Retrievals are needed at three points in the EAR/EAC process: to retrieve cases from which to build adaptation rules, to retrieve collections of source cases to adapt, and to retrieve stored adaptation rules. For EAR, we addressed the retrieval cost issue in BEAR [11] by applying a big data retrieval method, locality-sensitive hashing, to its retrieval process. LSH (Locality Sensitive Hashing), as an approximate nearest-neighbor method, loses some retrieval accuracy compared to conventional (and more costly) methods. However, experiments showed that on sample data sets, BEAR's use of ensembles enabled numeric prediction accuracy comparable to that of traditional k-NN without LSH, but with much better scalability [11]. Consequently, pursuing such an approach was appealing for increasing the scalability of EAC as well. However, this faced an impediment: the absence of existing locality sensitive hashing schemes suitable for handling symbolic data as required for EAC.

The goal of the work in this paper is to develop methods for large-scale application of the EAC approach [12] for domains with mixed categorical and numeric input features. It addresses this goal by introducing and evaluating a novel locality-sensitive hashing (LSH) algorithm to meet the specific requirements enforced by the EAC method. It tests an implementation of that scheme in the system EACX (**EAC** e**X**tended to scale). Experimental results illuminate the accuracy tradeoff, showing a relatively low drop in accuracy of EACX compared to that of EAC and increased efficiency. This demonstrates the ability of the ensemble-based approach of EACX to compensate for its reliance on approximate nearest neighbors compared to a brute force kNN with higher time complexity.

The paper begins with a perspective on the use of LSH for CBR system scaleup and an overview of the EAC approach. It next discusses LSH and describes the new locality-sensitive hashing approach, EACH, EACH's use in EACX, and its time complexity. It closes with evaluation and future work.

2 Scaling Up CBR with Big Data Retrieval Methods

As big data applications become increasingly prevalent, scaleup concerns take on new importance. The CBR community has long recognized the utility issues arising from case base growth (*e.g.*, [20]), as can arise for large-scale tasks or long-lived CBR systems. The primary response has been research on competence-based deletion and competence-aware case base construction (*e.g.*, [21,22,25,27]). Unfortunately, methods for increasing efficiency by reducing the size of the case base result in information loss for the CBR system, and are commonly expected to reduce accuracy.

An alternative is to apply big data methods to increase the scalability of retrieval [6,11], for example, by applying Locality Sensitive Hashing (LSH). LSH is a family of methods for efficient retrieval of approximate nearest neighbors, without limiting case base size. Thus like case base compression, it sacrifices accuracy for efficiency. However, it offers two benefits in contrast to case compression. First, because no cases need to be deleted, it avoids case information loss. By setting LSH parameters, a user can adjust the level of accuracy retained. Second, it avoids the cost of case-base compression (which, in a large-scale scenario, might need to be done repeatedly).

In previous work we introduced BEAR [11], an approach for scaling up Ensemble of Adaptations for Regression (EAR) to larger case bases by using big data methods (LSH and MapReduce). BEAR uses LSH in Euclidean space to find approximate nearest neighbors for case and adaptation rule retrieval. Conventional instance-based methods using exact nearest neighbors must calculate the distance between the input case and all other cases in the case base to pick the top few cases closest to the input case. In contrast, BEAR applies locality sensitive hashing to partition cases into different buckets, where it is likely that cases hashed to the same bucket are in close vicinity of each other. This enables limiting the search for the nearest neighbors to the cases that are in the same bucket as the input query instead of all case in the case base. This results in significantly improved efficiency. The standard drawback of LSH is a decrease in accuracy, due to use of approximate nearest neighbors. However, experiments showed that the use of EAR's ensemble method largely compensated for the accuracy drop, providing substantially improved scalability with little accuracy penalty. The success of BEAR motivated us to explore an analogous method for EAC.

3 Background

3.1 Ensemble of Adaptations for Classification

EAC [12] is a method for predicting target values in domains with categorical and numeric features by applying an ensemble of adaptations to adjust the solutions of the input query's nearest neighbors. To calculate the distance between a pair of cases, EAC uses the *Case Value Distance Heuristic Metric* (CVDHM) inspired by Wilson and Martinez's Heterogeneous Value Difference Metric [24]. For a pair of cases (c_1, c_2), we defined EAC's distance metric as:

$$\text{casedist}_f(c_1, c_2) \equiv (\sum_{i=1}^{n} \mid CVDHM_i(c_{1,i}, c_{2,i}) \mid^p)^{\frac{1}{p}} \tag{1}$$

where n is the number of features, $c_{1,i}$, $c_{2,i}$ are the values of the ith feature of cases c_1 and c_2 respectively, and p is a real value that is greater or equal to 1; p which is set to 2 in EAC. For any feature f taking on values "a" and "b" in cases c_1 and c_2, respectively, CVDHM calculates their feature distance by:

$$CVDHM_f(a,b) \equiv \begin{cases} 1, & \text{if a or b is unknown} \\ 1, & \text{if f is categorical and a or b is not observed in the training data} \\ \text{diffc}_f(a,b), & \text{if f is categorical} \\ min(1, \text{diffn}_f(a,b)), & \text{if f is numeric} \end{cases} \tag{2}$$

The function diffc, used for categorical features, is Stanfill's *Value Difference Metric* [23], a supervised learning method used to quantify the difference between two categorical features for domains with categorical target values. For regression tasks diffc is the unpaired t-test statistic of the target values corresponding to values "a" and "b" of feature "f", as explained below. The function diffn is the distance between two numeric features, defined by:

$$\text{diffn}_f(a,b) \equiv \frac{\mid a - b \mid}{4\sigma_f} \tag{3}$$

where σ_f represents the standard deviation of feature f. When adaptation rules are indexed by the differences they address, as in EAC, the CVDHM can be applied to adaptation rule retrieval as well. For reasons of space, we refer readers to Jalali and Leake [12] for more details on Case Value Distance Heuristic Metric.

If the target values of cases for values "a" and "b" of feature "f" are roughly normally distributed (as ensured, for large numbers of cases, by the central limit theorem), and their variances are roughly equal, then the unpaired t-test statistic is calculated as:

$$\text{diffc}_f(a,b) \equiv normalize(\frac{\bar{c_a} - \bar{c_b}}{SE(\bar{c_a} - \bar{c_b})}) \tag{4}$$

where $\bar{c_a}$ and $\bar{c_b}$ represent the sample mean of the target values associated with feature values "a" and "b" respectively. SE is the standard error of the difference between the means of the target values associated with values "a" and "b" which is calculated based on *pooled standard deviation.* The function *normalize* is a method that normalizes the diffc_f values by subtracting their mean from them and dividing the result by the standard deviation of the diffc_f for feature "f".

3.2 Locality-Sensitive Hashing

Locality Sensitive Hashing (LSH) [10] is a method for rapid approximate retrieval of nearest neighbors. LSH maps data points to partitions/buckets, with the goal of placing approximate nearest neighbors in the same buckets/partitions. The core module of any LSH algorithm is the hashing family used for partitioning the data points. LSH methods can be categorized on several dimensions, such as characteristics of the underlying domain, the similarity measure used to find the nearest neighbors, whether a data-driven approach is used to construct the hashing function, and if so, whether the construction takes a supervised or unsupervised approach. For example, Datar et al. [7] introduce a p-stable distribution-based LSH for Euclidean space; Simhash [5] uses angle-based distance, and minhash [3] uses set/Jaccard similarity.

Applying LSH to Ensembles of Adaptation for Classification requires LSH to handle domains with categorical or mixed categorical and numeric features. However, little work has attempted to apply LSH to such domains. Lee [14] proposed a hashing method for such domains which separates the categorical and numeric features and applies different families of hashing methods for each group. It then hashes the data points to two sets of buckets for categorical and numeric feature values. For categorical values, Lee uses three distance measures derived from the training data without considering labels. At retrieval, a point is considered a nearest neighbor when the query and the case are in the same bucket based on both their categorical and numeric features.

A prerequisite for applying LSH in any domain is that the cases' input features should be in a format consumable by the LSH techniques to be used. For EAC, this means that categorical features should be converted to a numeric representation. For this conversion, *one hot encoding* is a widely used method. One hot encoding converts a categorical feature with m distinct values to a vector of m bits where for every categorical value, all bits except one will be zero. For example, a categorical feature with possible values *high* and *low* will be converted to a vector of two bits where *01* and *10* can represent *high* and *low* respectively. Unfortunately for the similarity-based matching required by CBR, the one hot encoding approach is essentially equivalent to relying on exact matching when searching for similar cases. Consequently, in this work we introduce a novel technique that transforms cases rather than their categorical features to a new domain and applies LSH to find approximate nearest neighbors for the transformed cases. Our new hashing scheme transforms cases to their new representation by using the value difference metric for learning the similarities between categorical features, enabling searching for similarity in presence of categorical features, rather than relying on exact matching.

In the next section we introduce our new hashing scheme. In contrast to Lee's method, our method considers numeric and categorical feature values together, using the *heterogeneous value difference metric* [24]. Also, instead of using a data-driven similarity measure that only takes the distribution of the feature values into consideration, we use a supervised method dictated by the underlying similarity measure. This method considers the relation between feature values and target values as well.

3.3 EACH: A Locality Sensitive Hashing Method for Domains with Categorical Features

To facilitate processing in the context of EAC, a hashing scheme must be able to deal with both categorical and numeric input features. In addition, the hashing scheme should serve for both case and adaptation rule retrieval purposes. Our scheme, EACH, (Ensemble of Adaptations for Classification Hashing) uses a distance-based hashing scheme inspired by random hyperplane-hashing followed by a classic LSH method (*e.g.,* min-hash, a set similarity-based LSH method that is an approximation of the Jaccard coefficient, or an angle or distance-based hashing scheme).

Algorithm 1 summarizes the EACH hashing method for case retrieval. At the first step EACH generates M random points p with a set of predefined dimensions (at most as many dimensions as the number of dimensions in the underlying domain). Next, the distance between every case in the case base and each of these randomly generated points is calculated and compared with a given threshold τ. For each case, if the distance from the case to point "p" is less than τ, then a "0" bit will be generated in its hashing sequence, otherwise a "1" will be put in that position. By the end of this step, all cases in the case base are converted to sequences of M bits. The *GenerateRandomPoint* performs this conversion. For a case base with N cases, N sequences of "M" bits will be generated. *getDistance* returns the distance between two points according to Eq. 1. In the last step, the bit sequences of the cases are partitioned by using an LSH method, which can be selected based on the domain. For example, suppose *minhash* is used as the underlying hashing scheme. Because minhash approximates the Jaccard similarity between these bit sequences, cases that have more similar sequences are more likely to be partitioned to the same buckets, which in turn means that cases that are in close vicinity of each other are more likely to be partitioned to the same buckets.

Algorithm 2 describes the process of *GenerateRandomPoint*. In the algorithm, *getDistinctValues* is a method that takes a list of values and returns the distinct values in the list, and *Random* generates a random number between the input arguments (inclusive) passed to the method.

A good hashing scheme maximizes the total entropy of the sequences of bits that represent the distance between cases in the case base and the generated points by *GenerateRandomPoint*. Maximizing this requires an appropriate

Algorithm 1. EACH's basic algorithm

```
Input:
CB: case base
dims: the set of the dimensions to keep
M: number of random points to generate
τ: distance threshold
lsh_conf: LSH configuration settings
Output: hashed version of the case base
  RandomPoints ∈ ℝ^{M×|dims|}
  for m ∈ {1,...,M} do
    RandomPoints.add(GenerateRandomPoint(dims, CB))
  end for
  CaseBaseBitSequence ∈ ℝ^{M×|CB|}
  for each case ∈ CB do
    BitSequence ∈ ℝ^M
    for each point ∈ RandomPoints do
      distance ← getDistance(case, point)
      if distance ≤ τ then
        BitSequence.append(1)
      else
        BitSequence.append(0)
      end if
    end for
    CaseBaseBitSequence.add(BitSequence)
  end for
  return LSH(CaseBaseBitSequence, lsh_conf)
```

Algorithm 2. GenerateRandomPoint's algorithm

```
Input:
CB: case base
dims: the set of the dimensions to keep
Output: A point with |dims| features
  RandomPoint ∈ ℝ^{|dims|}
  for each dim ∈ dims do
    distinctValues ← getDistinctValues(CB_dim)
    r ← Random(1, |distinctValues|)
    RandomPoint_dim ← distinctValues_r
  end for
  return RandomPoint
```

value for τ in EACH. We propose two candidate methods for selecting suitable values. First, a hill climbing search could be used to approximate the optimal value of τ, where the objective function is the total entropy of the bit sequences generated for all cases in the case base. Second, a heuristic could calculate the distance between every pair of cases in the case base and then choose the median of these distances as τ. To keep EACH applicable to larger case bases, sampling techniques [8] can be used in conjunction with the above methods.

4 EACX: EAC at Scale

Given EACH, it becomes possible to scale up EAC with an approach similar to that taken in our previous work on making Ensemble of Adaptation for Regression scalable in BEAR [11]. BEAR used p-stable hashing in the Euclidean space to find approximate nearest neighbors for case and rule retrieval. However, BEAR was only applicable to regression tasks in domains with solely numeric features, due to the limitations of its p-stable distribution-based LSH. Unlike BEAR, EACX can be applied to both regression and classification tasks in domains with numeric and categorical input features.

4.1 Where EACH is Used in EACX

EACX uses EACH to find nearest neighbors at each of three steps in the process for Ensembles of Adaptations for Classification. First, it uses EACH for rule generation, when retrieving the nearest neighbors of every case in the case base so that a set of adaptation rules can be generated by comparing each case with its top few nearest neighbors. EACX generates adaptation rules based on the *Case Difference Heuristic* according to the process explained by Jalali, Leake, and Forouzandehmehr [12]. Second, it uses EACH to find the nearest neighbors to the input problem to use their solutions in predicting the target value of the input problem. The values of the nearest neighbors of the input query are adjusted by applying an ensemble of adaptations and the final solution is built by combing these adjusted values. Third, it uses EACH to retrieve the ensemble of adaptations for individual pairs of the input problem and its nearest neighbor cases. For each pair, EACX retrieves adaptation rules that address differences similar to those observed between the input query and the neighbor under consideration.

4.2 EACX's Architecture

Figure 1 depicts the general process of the EACX method. First, the Case Value Distance Heuristic Metric is trained based on the cases in the case base and used in EACH to partition the cases in the case base and to find nearest neighbors to the input query. At this stage EACH serves two purposes: First, finding the nearest neighbors of the cases in the case base so that a set of adaptation rules can be generated by comparing every case with its nearest neighbors; Second, finding the nearest neighbors of the input query in the case base. Next, another round of training the Case Value Distance Heuristic Metric is conducted to enable similarity assessment for the generated rules. This rule-based similarity measure is used in partitioning the rule base, by training a rule-based version of EACH; that in turn will be used to find the nearest neighbor rules for the pairs of the input query and the retrieved cases (*i.e.*, the input query's nearest neighbors), in the previous step.

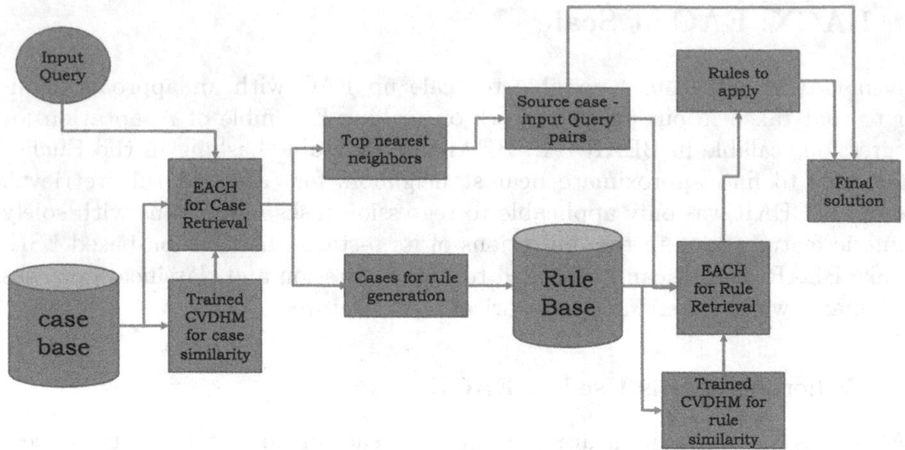

Fig. 1. Illustration of the generic process of EACX

4.3 Time Complexity of EAC and EACX

As a key motivation for EACX is scaleup, we consider its effect on computational complexity. Calculating exact nearest neighbors is O(n), where n is the number of cases in the case base. (Note that more efficient standard methods, such as k-d trees, are not applicable to categorical features.) In principle, $2\binom{n}{2}$ rules could be generated from a case base of n cases, but EACX limits the set of generated rules at $O(n \times c)$ rules, where c is a positive constant integer. Thus computational complexity of EAC is O(n); processing time depends on a constant factor of number of neighbors to consider (k) × a constant number of rules to apply (c). As the number of cases in the case base grows, the time required for EAC, even for small values of c, and k becomes very noticeable in practice.

In contrast, using EACH the time complexity of nearest neighbor search can be decomposed into two different components: hashing and retrieval. The time complexity of the hashing stage is equal to the time complexity of converting cases/rules to a sequence of bits, which is equal to $O(n \times d)$ and $O(n \times c \times l)$ for case base and rule base hashing respectively, where d is the number of dimensions in the underlying domain and l is the number of randomly generated points in EACH. However, we note that this is a one-time cost and once the cases and rules are hashed there is no need to repeat this process for future queries. In other words, once hashing is done, the time complexity of retrieving nearest neighbors is decoupled from the case/rule base size. This is the advantage of EACX over EAC. The time complexity of using LSH to find the nearest neighbors depends on the LSH method used. For example, if min-hash is used as the underlying LSH method the time complexity depends on the number of hash tables, bands and the signature length. However, for a sufficiently large case base this is guaranteed to be orders of magnitude less than brute-force nearest neighbor search.

5 Instantiations of EACH and EACX

As EACH and EACX are both general approaches which can be defined with different configurations, we define specific instantiations of both, respectively called EACH1 and EACX1, which will be used in the evaluation.

5.1 EACH1: An Instantiation of EACH

EACH1 uses the Case Value Distance Heuristic Metric (CVDHM) as the *getDistance* method. CVDHM calculates the distance between values "a" and "b" of feature "f" according to Eq. 1. It uses min-hash its LSH method. The implementation of min-hash in EACH1 follows Broder [3,4] and is from the *sparkneighbors* package[1]. To prepare *sparkneighbors* to calculate the min-hash of a case, it is provided with the number of dimensions in the domain, the number of hash tables, signature length, a prime module, and the number of bonds.

5.2 EACX1: An Instantiation of EACX

EACX depends on EAC and EACH, each of which depends on its own parameters; this results in a wide range of possible customizations of EACX. For EAC, choices include the choice of source cases to adapt to build the final solution (*e.g.,* how many source cases, and how those cases are selected), the choice of cases to generate the adaptation rules, and the method to derive adaptation rules from pairs of cases; each set of choices for these yields different variations of the method.

In EACX1, source case selection is "Local", meaning that EACX1 relies on the top nearest neighbors of the input query to build the final solution. The choice of case selection to generate adaptation rules in EACX1 is "Global cases - Local neighbors", meaning that every case in the case base participates in the rule generation process and adaptation rules are generated by comparing every case with its top few nearest neighbors. To read more about these choices and other alternatives we refer the readers to our initial work on Ensemble of Adaptations for Regression [13]. The method to generate adaptation rules is the *Generalized Case Difference Heuristic* (GCDH) introduced in Jalali, Leake and Forouzandehmehr [12].

The Generalized Case Difference Heuristic approach extends the Case Difference Heuristic so that it can be applied to domains with categorical target values. The Case Difference Heuristic relies on exact matches for categorical input features; this is enhanced by using a more general method, the Case Value Distance Heuristic Metric in GCDH. The main idea of GCDH is to represent adaptation rules as:

$$(\Delta_{f_1}, \ldots, \Delta_{f_k}) \Rightarrow \Delta_t \tag{5}$$

[1] https://github.com/karlhigley/spark-neighbors.

where Δ_f's are calculated according to Eq. 2 and Δ_t is a numeric value when the target value is numeric and is a pair of categorical values when the target value is categorical. EACX1 uses EACH1 for locality sensitive hashing.

6 Evaluation

We tested the effectiveness of EACH1 and EACX1 with experiments testing their accuracy in four standard domains. Because EACX uses approximate rather than exact search we expected its overall accuracy to be lower than that of EAC. However, one of the goals of our experiments was to investigate how much quality can be retained while achieving the LSH-based speedup and to what extent EACX's ensemble-based approach mitigates the inaccuracies stemming from LSH.

Our experiments address the following questions:

1. How does the accuracy of EACX compare to that of k-NN with LSH?
2. How does the accuracy of EACX compare to other state of the art machine learning methods for classification?
3. How much does the ensemble-based approach of EACX contribute to its accuracy, compared to a non-ensemble method?

6.1 Experimental Design

We tested EACX1 for four classification domains from the UCI repository [15]:

1. Activity Recognition (Activity): Detecting human activity based on sensory data from smart phones and smart watches.
2. Balance Scale (Balance): Predict whether scale will tip right, left, or stay balanced
3. Car Evaluation (Car): Predict the acceptability of an automobile (unaccepted, accepted, good, and very good)
4. Breast Cancer Detection (BC): Predicting whether a tumor is benign or malignant.

For the activity domain 1000 cases were drawn randomly from the original data set. For the rest of the domains, all cases were used after removing cases with missing values. (We note that EACX could be applied to domains with missing values after pre-processing for handling missing features (*e.g.,* [2])). Table 1 summarizes the characteristics of the test domains.

EACX1 is implemented in Apache Spark MLlib [18]. Spark MLlib provides built-in support for cross validation and grid search which we used for tuning the parameters of the tested classification methods. Our experiments measure the accuracy (*i.e.,* the percent of correctly classified tests out of the entire test population) of kNN, EAC, Random Forest (RF), Logistic Regression (LR), kNN using EACH1 (kNN-LSH) and EAC using EACH (*i.e.,* EACX1) and an ablated version of EACX which uses one adaptation per source case (EACX1-a).

Table 1. Characteristics of the test domains

Domain name	# Categorical features	# Numeric features	# Cases	# Unique combination of categorical features	# Unique solutions
Activity	3	3	1000	49	4
Balance	4	0	625	625	3
Car	6	0	1728	1728	4
BC	0	9	653	0	2

The parameters tuned for RF are the number of trees and maximum tree depth of the tree. We use *entropy* as the impurity measure. For kNN and kNN-LSH the number of source cases used for building the solution was the parameter tuned. The number of source cases to use, the number of adaptations to apply for adjusting the value of each source case and the number of nearest neighbors of every case in the case base to compare with itself for the adaptation rule generation purpose were the values tuned for EAC and EACX1. In the case of EACX1 and EACX1-a, the value of τ is determined by taking the median of the distances from cases to the randomly generated points by EACH1. Models were trained on 80% of the case base and tested on the remaining 20%. The testing process was repeated 10 times for each domain and test/control groups are decided in a random fashion.

6.2 Results on Questions 1 and 2: Comparative Accuracy of EACX1

We conducted experiments to compare the accuracy of kNN-LSH and EACX1 with Random Forest, Logistic Regression, kNN and EAC. Because kNN-LSH and EACX1 are basically kNN and EAC augmented by EACH1, their comparison with kNN and EAC sheds light on their expected accuracy loss due to using approximate nearest neighbors. In addition, the comparison between kNN-LSH and EACX1 with Random Forest and Logistic Regression can determine the competence of our EACH-powered classifiers compared to other machine learning techniques.

Table 2 shows the accuracy of the tested classifiers in four sample domains. EAC has the highest accuracy in Car and BC, while in the Balance and Activity domains, EACX1 and RF show higher accuracy respectively.

Figure 2 depicts the percentage of improvement in accuracy of kNN over kNN-LSH and EAC over EACX1 in the test domains. In three out of four domains (all except Car), kNN's performance has a higher margin compared to kNN-LSH than that of EAC compared to EACX1. We attribute this to the fact that the ensemble-based approach of EACX1 gives it more opportunity to compensate the inaccuracies introduced by using approximate neighbors. We also note that in all domains EACX1 shows higher accuracies compared to kNN, despite the fact that—unlike kNN—EACX1 relies on approximate nearest neighbors.

Table 2. Estimation error of LR, RF, kNN, EAC, kNN-LSH and EACX1 methods in four sample domains

Domain name	LR	RF	kNN	EAC	kNN-LSH	EACX1
Car	26.28%	4.29%	5.14%	**2.57%**	7.37%	4.74%
Balance	13.05%	16.33%	14.56%	12.96%	17.72%	**2.80%**
BC	7.42%	3.28%	2.62%	**1.66%**	6.26%	1.98%
Activity	1.51%	**1.33%**	2.50%	1.50%	5.74%	1.57%

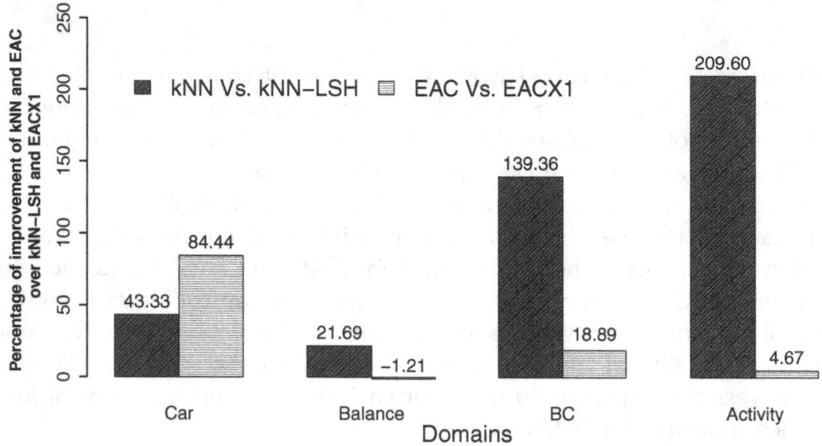

Fig. 2. Percentage of improvement of kNN and EAC over kNN-LSH and EACX1

6.3 Results on Question 3: Effect of Applying Ensembles of Adaptations

To assess the benefit of using ensemble of adaptations in EACX1 to compensate for the use of approximate nearest neighbors (and automatically generated adaptation whose quality cannot be guaranteed) we implemented an ablated version of EACX1 (EACX1-a) that uses one adaptation per source case.

Figure 3 depicts the percentage of improvement of EACX1 and EACX1-a over kNN-LSH. In all domains, EACX1 shows higher percentage of improvement in the estimation error over kNN-LSH compared to EACX1-a which supports the benefit of the ensemble-based approach of EACX-1. In three domains, EACX1-a shows a relatively narrow margin (for BC domain the margin is relatively larger) over kNN-LSH and in the Activity domain it shows slightly worse performance than kNN-LSH.

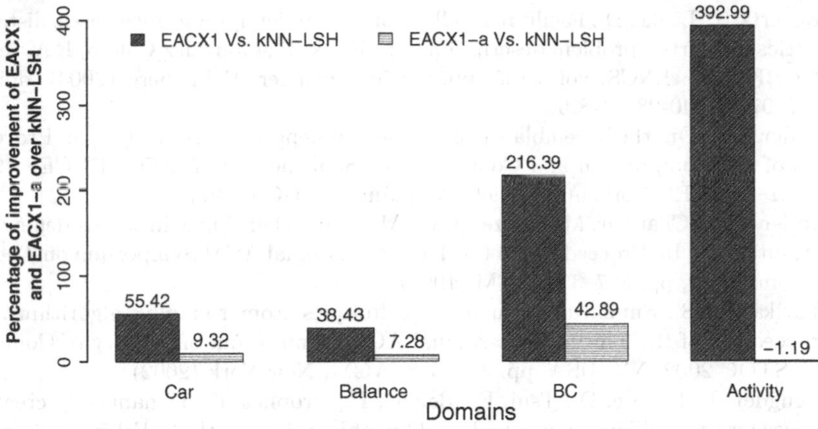

Fig. 3. Percentage of improvement of EACX1 and EACX1-a over kNN-LSH

7 Conclusion and Future Work

This paper has presented an approach to scaling up Ensembles of Adaptations for Classification based on EACH, a new Locality Sensitive Hashing method for domains with numeric and categorical input features and target values. It also introduced EACX, a scalable case-based classification approach that applies an ensemble of adaptions to adjust the predictions derived from the source cases. EACX uses EACH to find the nearest neighbors in source case and adaptation retrieval. Using Locality Sensitive Hashing decreases the time complexity of EACX by using approximate nearest neighbors, enabling its application on larger case bases. We introduced EACH1 and EACX1 instantiations of EACH and EACX and evaluated their performance on four sample domains. Results showed that EACX can provide accuracy as good as the brute-force alternatives.

As the next step, we plan to test EACX on domains with tens of millions of cases to directly evaluate it in big data settings. Other potential future directions include comprehensive study of the effect of variations in EACH, by using different similarity measures such as Jaccard, Hamming, Euclidean and Cosine and exploring similarity measures other than the Case Value Distance Heuristic Metric for distance between categorical feature values, and comparing alternative random point generation methods such as density-based random point generation.

References

1. Badra, F., Cordier, A., Lieber, J.: Opportunistic adaptation knowledge discovery. In: McGinty, L., Wilson, D.C. (eds.) ICCBR 2009. LNCS (LNAI), vol. 5650, pp. 60–74. Springer, Heidelberg (2009). doi:10.1007/978-3-642-02998-1_6

2. Bogaerts, S., Leake, D.: Facilitating CBR for incompletely-described cases: distance metrics for partial problem descriptions. In: Funk, P., González Calero, P.A. (eds.) ECCBR 2004. LNCS, vol. 3155, pp. 62–76. Springer, Heidelberg (2004). doi:10.1007/978-3-540-28631-8_6
3. Broder, A.: On the resemblance and containment of documents. In: Proceedings of the Compression and Complexity of Sequences 1997, SEQUENCES 1997, pp. 21–29. IEEE Computer Society, Washington, DC (1997)
4. Broder, A.Z., Charikar, M., Frieze, A.M., Mitzenmacher, M.: Min-wise independent permutations. In: Proceedings of the Thirtieth Annual ACM Symposium on Theory of Computing, pp. 327–336. ACM (1998)
5. Charikar, M.S.: Similarity estimation techniques from rounding algorithms. In: Proceedings of the Thiry-fourth Annual ACM Symposium on Theory of Computing, STOC 2002, NY, USA, pp. 380–388. ACM, New York (2002)
6. Daengdej, J., Lukose, D., Tsui, E., Beinat, P., Prophet, L.: Dynamically creating indices for two million cases: a real world problem. In: Smith, I., Faltings, B. (eds.) EWCBR 1996. LNCS, vol. 1168, pp. 105–119. Springer, Heidelberg (1996). doi:10.1007/BFb0020605
7. Datar, M., Immorlica, N., Indyk, P., Mirrokni, V.S.: Locality-sensitive hashing scheme based on p-stable distributions. In: Proceedings of the Twentieth Annual Symposium on Computational Geometry, SCG 2004, NY, USA, pp. 253–262. ACM, New York (2004)
8. Gilks, W.R., Richardson, S., Spiegelhalter, D.J.: Introducing markov chain monte carlo. Markov Chain Monte Carlo Pract. **1**, 19 (1996)
9. Hanney, K., Keane, M.T.: Learning adaptation rules from a case-base. In: Smith, I., Faltings, B. (eds.) EWCBR 1996. LNCS, vol. 1168, pp. 179–192. Springer, Heidelberg (1996). doi:10.1007/BFb0020610
10. Indyk, P., Motwani, R.: Approximate nearest neighbors: towards removing the curse of dimensionality. In: Proceedings of the Thirtieth Annual ACM Symposium on Theory of Computing, STOC 1998, NY, USA, pp. 604–613. ACM, New York (1998)
11. Jalali, V., Leake, D.: CBR meets big data: a case study of large-scale adaptation rule generation. In: Hüllermeier, E., Minor, M. (eds.) ICCBR 2015. LNCS (LNAI), vol. 9343, pp. 181–196. Springer, Cham (2015). doi:10.1007/978-3-319-24586-7_13
12. Jalali, V., Leake, D., Forouzandehmehr, N.: Ensemble of adaptations for classification: learning adaptation rules for categorical features. In: Goel, A., Díaz-Agudo, M.B., Roth-Berghofer, T. (eds.) ICCBR 2016. LNCS (LNAI), vol. 9969, pp. 186–202. Springer, Cham (2016). doi:10.1007/978-3-319-47096-2_13
13. Jalali, V., Leake, D.: Extending case adaptation with automatically-generated ensembles of adaptation rules. In: Delany, S.J., Ontañón, S. (eds.) ICCBR 2013. LNCS (LNAI), vol. 7969, pp. 188–202. Springer, Heidelberg (2013). doi:10.1007/978-3-642-39056-2_14
14. Lee, K.M.: Locality-sensitive hashing for data with categorical and numerical attributes using dual hashing. Int. J. Fuzzy Log. Intell. Syst. **2**(2), 98–104 (2014)
15. Lichman, M.: UCI machine learning repository (2013). http://archive.ics.uci.edu/ml
16. McDonnell, N., Cunningham, P.: A knowledge-light approach to regression using case-based reasoning. In: Roth-Berghofer, T.R., Göker, M.H., Güvenir, H.A. (eds.) ECCBR 2006. LNCS (LNAI), vol. 4106, pp. 91–105. Springer, Heidelberg (2006). doi:10.1007/11805816_9

17. McSherry, D.: An adaptation heuristic for case-based estimation. In: Smyth, B., Cunningham, P. (eds.) EWCBR 1998. LNCS, vol. 1488, pp. 184–195. Springer, Heidelberg (1998). doi:10.1007/BFb0056332
18. Meng, X., Bradley, J.K., Yavuz, B., Sparks, E.R., Venkataraman, S., Liu, D., Freeman, J., Tsai, D.B., Amde, M., Owen, S., Xin, D., Xin, R., Franklin, M.J., Zadeh, R., Zaharia, M., Talwalkar, A.: MLlib: machine learning in apache spark. CoRR abs/1505.06807 (2015). http://arxiv.org/abs/1505.06807
19. Müller, G., Bergmann, R.: Learning and applying adaptation operators in process-oriented case-based reasoning. In: Hüllermeier, E., Minor, M. (eds.) ICCBR 2015. LNCS (LNAI), vol. 9343, pp. 259–274. Springer, Cham (2015). doi:10.1007/978-3-319-24586-7_18
20. Smyth, B., Cunningham, P.: The utility problem analysed. In: Smith, I., Faltings, B. (eds.) EWCBR 1996. LNCS, vol. 1168, pp. 392–399. Springer, Heidelberg (1996). doi:10.1007/BFb0020625
21. Smyth, B., Keane, M.: Remembering to forget: a competence-preserving case deletion policy for case-based reasoning systems. In: Proceedings of the Thirteenth International Joint Conference on Artificial Intelligence, pp. 377–382. Morgan Kaufmann, San Mateo (1995)
22. Smyth, B., McKenna, E.: Building compact competent case-bases. In: Althoff, K.-D., Bergmann, R., Branting, L.K. (eds.) ICCBR 1999. LNCS, vol. 1650, pp. 329–342. Springer, Heidelberg (1999). doi:10.1007/3-540-48508-2_24
23. Stanfill, C., Waltz, D.L.: Toward memory-based reasoning. Commun. ACM **29**(12), 1213–1228 (1986)
24. Wilson, D.R., Martinez, T.R.: Improved heterogeneous distance functions. J. Artif. Int. Res. **6**(1), 1–34 (1997)
25. Wilson, D., Martinez, T.: Reduction techniques for instance-based learning algorithms. Mach. Learn. **38**(3), 257–286 (2000)
26. Wiratunga, N., Craw, S., Rowe, R.: Learning adaptation knowledge to improve case-based reasoning. Artif. Intell. **170**, 1175–1192 (2006)
27. Zhu, J., Yang, Q.: Remembering to add: competence-preserving case-addition policies for case base maintenance. In: Proceedings of the Fifteenth International Joint Conference on Artificial Intelligence, pp. 234–241. Morgan Kaufmann (1999)

A CBR System for Efficient Face Recognition Under Partial Occlusion

Daniel López-Sánchez(✉), Juan M. Corchado, and Angélica González Arrieta

C/ Espejo, s/n, Edificio I+D+i, BISITE Research Group, Salamanca 37007, Spain
{lope,angelica}@usal.es

Abstract. This work focuses on the design and validation of a CBR system for efficient face recognition under partial occlusion conditions. The proposed CBR system is based on a classical distance-based classification method, modified to increase its robustness to partial occlusion. This is achieved by using a novel dissimilarity function which discards features coming from occluded facial regions. In addition, we explore the integration of an efficient dimensionality reduction method into the proposed framework to reduce computational cost. We present experimental results showing that the proposed CBR system outperforms classical methods of similar computational requirements in the task of face recognition under partial occlusion.

Keywords: Face recognition · Partial occlusion · Dimensionality reduction

1 Introduction

This work focuses on the design and implementation of a Case-Based Reasoning (CBR) system for efficient face recognition, with a special focus on robust face recognition under partial occlusion conditions[1]. Although the problem of face recognition has been extensively addressed in the available literature, most state-ot-the-art proposals impose a series of constraintsthat limit their applicability in real world scenarios, where only a limited amount of computational power and training information is available.

The CBR method proposed in this paper seeks to cover the full recognition process (i.e., face detection, normalization, and identity prediction). In addition, we focused on methods which are able to work under the constraints of low computational power and little training information. As opposed to other occlusion-robust face recognition systems, the proposed CBR framework does not make any assumption about the nature of occlusion that it will have to face at test time. We also studied the possible integration of an efficient dimensionality reduction method in the proposed framework to reduce computational

[1] In the context of face recognition, partial occlusion refers to the situation where some parts of the faces the system must identify are covered by some artefact.

D.W. Aha and J. Lieber (Eds.): ICCBR 2017, LNAI 10339, pp. 170–184, 2017.
DOI: 10.1007/978-3-319-61030-6_12

cost. The experimental results presented in this paper show that the proposed method outperforms traditional face recognition methods in the task of partially occluded face recognition.

The rest of this paper is structured as follows. Section 2 reviews some of the most relevant works in the field of face recognition, with special attention to approaches robust to face occlusion. The proposed CBR system and the different preprocessing methods are described in detail in Sect. 3. Section 4 empirically compares the proposed CBR system with some alternative classical methods, with special emphasis on partial occlusion scenarios. Finally, Sect. 5 summarizes the conclusions of this work and outlines some promising future research lines.

2 Related Work

In this section, we summarize some of the most relevant works on the topic of face recognition under partial occlusion. Ekenel [5] hold the idea that most of the accuracy loss registered by face recognition systems when dealing with partially occluded images is due to alignment errors, rather than information corruption by the occlusion. To address this problem, they proposed a method which seeks to minimize the distance between each sample in the training set and a new observation by evaluating a number of different alignment variations. As a consequence, searching for the best alignment variation requires hundreds of comparisons for each training sample. Although this method achieved notable accuracy rates, the computational cost supposes a major drawback.

Other authors [12,16] divide facial images into a number of delimited regions. After this, they seek to model those occlusion areas by using Principal Component Analysis or a Self Organized Map. Nevertheless, most occlusion-robust face recognition systems include a previous step to identification where they determine which parts of the images are affected by occlusion. Some studies used manually annotated occluded/non-occluded facial image patches to explicitly train a classifier [13]. However, this approach has the drawback of needing occluded face images during the training stage. As a consequence, if the nature of occlusion faced by the system in production is not the same as during the training stage, the accuracy of the occlusion detector might be affected.

Using color-based segmentation methods to detect occluded facial regions has also been proposed in the literature [7]. However, these methods are very sensitive to lighting conditions and assume that the occlusion is not caused by artefacts with human-skin color. More recently, several authors have tried to apply the recent advances in the field of *deep learning* to the task of face recognition. Nowadays, the state-of-the-art on one of the most widespread face recognition datasets, namely the *Labeled Faces in the Wild* (LFW), is held by a deep neural network trained by the scientists at Baidu [9]. The major drawbacks of this approach are the computational costs and the need for a large training dataset.

Finally, it is worth noting the CBR methodology has been applied in the literature to acquire emotional context about the users of recommender systems, based on their facial expressions [10]. However, to the best of our knowledge the CBR methodology has not yet been applied to the task of facial recognition under partial occlusion.

3 Proposed Framework

In this section, we describe in detail both the proposed CBR framework and the selected pre-processing steps needed for face recognition. At test time, when the system is presented with an image that contains a human face in it, the following processing stages are executed: (1) A region of interest is determined for the human face in the image; (2) the detected face is aligned[2]; (3) the image is pose-normalized, rotating and scaling the face to a standard size and orientation; (4) the lighting conditions of the image are normalized; (5) a feature extraction method is applied; and finally (6) the proposed retrieval and reuse stages are executed to emit a prediction regarding the identity of the person in the original image (Fig. 1).

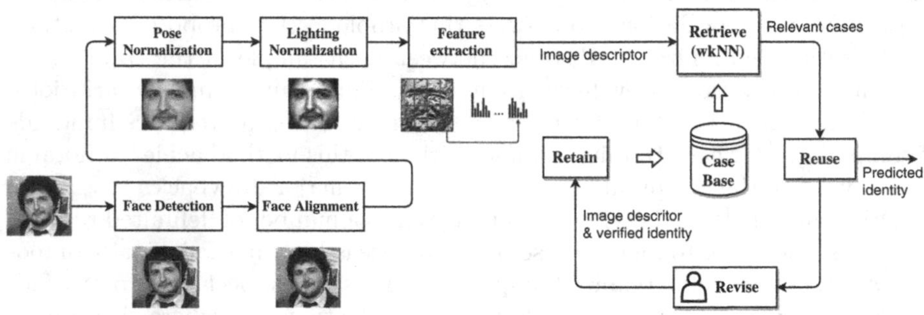

Fig. 1. Architecture of the proposed CBR framework and preprocessing steps.

3.1 Preprocessing

This section describes the successive preprocessing stages executed before the actual retrieval and reuse stages in the proposed CBR framework.

Face Detection. The face detection stage is in charge of finding a preliminary Region of Interest (ROI) for the human face present in the input image. One of the most widespread face detection methods is based on Histogram of Oriented Gradients (HOG) descriptors. This descriptor counts the number of occurrences of each gradient orientation in localized regions of the image. The face detector is then build using a linear classifier with a sliding window over the HOG descriptor of the image. For our experiments, we used the HOG face detector provided by the Dlib C++ library [8].

[2] In the context of face recognition, face alignment refers to the task of locating a series of facial key-points in an image, such as eyes, nose, mouth corners, etc.

Face Alignment. Face alignment consist of automatically predicting the location of a series of facial key-points in the input image. Some of the most popular methods are based on the idea of cascade regression, which provides a greater accuracy and faster processing times than classical methods. In particular, our framework leverages the face alignment algorithm proposed by V. Kazemi in 2014 [17]. Here, the author proposes using a cascade regression model where each successive level refines the alignment coordinates proposed by the previous level. In particular, the base regression models used by V. Kazemi consisted of regression-tree ensembles. For the experiments with automatic face alignment in this paper we used the pretrained model provided by Dlib C++ [8].

Pose Normalization. Once face detection and alignment have been performed, the estimated position of facial key-points in the input image is available. A pose-normalized image is then generated with these facial key-points as a basis by rotating and cropping the image to display the aligned face in its center, in a vertical pose. In addition, the resulting image is resized to a standard size.

Light Normalization. Light normalization algorithms seek to reduce the amount of intra-class variance exhibited by images from face recognition tasks with unconstrained lighting conditions. Histogram Equalization (HE) is arguably the simplest option for light normalization. This method maps the histogram of the original image $H(i)$ to a more uniform distribution. To achieve this, the so called cumulative distribution function $H'(i)$ is used:

$$H'(i) = \sum_{j=0}^{i} H(j) \tag{1}$$

Once $H'(i)$ has been computed, it is normalized to ensure that its maximum value corresponds to the maximum valid pixel value in the desired image format. Next, the following function is used to calculate pixel intensities in the resulting image:

$$equalizada(x, y) = H'(original(x, y)) \tag{2}$$

Due to its simplicity, efficiency and good performance, HE was used to normalize the lighting conditions in all experiments in this paper.

3.2 Feature Extraction: Local Binary Patterns

Using raw pixel values as features to directly train some classification algorithm is not very practical. The main reason for this being that such representation of images often contains undesired information such as noise or lighting variations. In addition, the number of pixels in images is usually too big to train a classifier efficiently. In this paper, we focus on a specific family of feature descriptors known as *Local Binary Patterns* (LBP) [14]. As described in the following sections, the localized nature of this descriptor will allow us to maintain features from occluded regions isolated from those extracted from visible parts of the face.

The LBP descriptor labels pixels in an image by considering value differences with their neighbors. This label is treated afterwards as a binary number. The use of a circular neighborhood and bilinear interpolation over non-integer pixel coordinates enables the use of this descriptor for an arbitrary neighbor number and neighborhood radio [15]. The notation $LBP_{P,R}$ is often used to refer to the LBP descriptor with P neighbors and a radio of value R. It has been proved that, using the $LBP_{8,1}$ descriptor, almost 90% of extracted labels are uniform (i.e., its binary representation contains two transitions at most) [15]. For this reason, a variant of LBP was designed where non-uniform patterns are merged together in a single label. This variant of the descriptor is known as uniform LBP ($LBP^{u}_{P,R}$).

Before training a classifier, the LBP representation is often refined by dividing the image in a number of blocks (arranged in a grid structure) and counting the number of concurrences of patterns in each block. After this, the corresponding histograms of each block are concatenated to form the final descriptor. This process in known as *Local binary pattern histograms* (LBPH).

3.3 Identification: Occlusion-Robust Retrieval and Reuse

The core proposal of this paper consists of a novel dissimilarity function which dynamically inhibits the use of corrupted features while retrieving the most relevant cases from the Case-Base. This section describes how this dissimilarity function is computed and its usage in the context of the proposed CBR framework.

Retrieval and Reuse. First, we introduce a method to detect partial occlusion in LBPH blocks. Conveniently, our method requires no *a priori* knowledge about the nature of occluded blocks. We define the *minimum local distance* for the histogram of an LBP block as the minimum squared Euclidean distance obtained when comparing this histogram with the LBP histograms corresponding to the same facial region in the descriptors stored in the Case-Base. Then, the only assumption made by our method is that minimum local distances of occluded blocks are usually larger than those of unoccluded blocks. To provide insight into the veracity of this assumption, we calculated the distribution of minimum local distances for occluded and unoccluded blocks in the ARFace database[3]; the resulting distributions are shown in Fig. 2. Although some overlapping exists among the two distributions, it might be possible to define a conservative threshold to discard most occluded blocks. More details on how an appropriate value for this threshold is determined can be found in Sect. 4.

Formally, the Case-Base of our framework is defined as a set of identity label y and LBPH descriptor x pairs:

$$CB = \{(y^{(i)}, x^{(i)}), i = 1, 2, \ldots, n\} \tag{3}$$

[3] See Sect. 4 for details about the evaluation database.

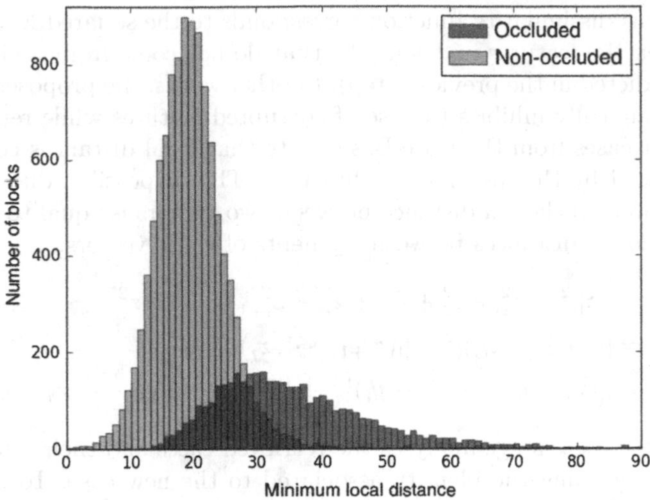

Fig. 2. Differential distribution of minimum local distances for occluded and unoccluded blocks in the ARFace database.

When an unlabeled image I is presented to the system, it is first transformed by the successive preprocessing steps defined in the previous section. Afterwards, the LBPH descriptor $x \in \mathbb{R}^d$ of image I is generated and the retrieval stage begins. Our proposed retrieval stage begins by computing the $n \times d/p$ local distance matrix L, where p is the size of each histogram concatenated to form the LBPH descriptors. Each entry L_{ij} in this matrix corresponds to the local distance between the j-th histogram in x and the j-th histogram in the i-th descriptor in the Case-Base; formally:

$$L_{i,j} = ||(x_{p(j-1)+1}, \cdots, x_{pj}) - (x^{(i)}_{p(j-1)+1}, \cdots, x^{(i)}_{pj})||^2 \quad \text{for} \quad i = 1, 2, \cdots, n \,\text{and}\, j = 1, 2, \cdots, d/p \tag{4}$$

Based on this matrix and the desired *threshold* value for occlusion detection, the retrieve stage computes an occlusion mask $M \in \{0, 1\}^{d/p}$ that determines which of the histograms that conform descriptor x are considered as occluded:

$$M_j = T_h(min(col_j L)) \qquad T_h(x) = \begin{cases} 1 \; if \; x < threshold \\ 0 \; if \; x > threshold \end{cases} \tag{5}$$

Using this occlusion mask, the retrieval stage of the proposed CBR framework finds the k most similar cases to x in the Case-Base, according to the following dissimilarity function:

$$d(x, x^{(i)}) = \sum_{j=1}^{j=d/p} M_j \cdot L_{i,j} \tag{6}$$

Intuitively, this dissimilarity function corresponds to the squared Euclidean distance between the features in x and $x^{(i)}$ that do not come from occluded facial zones (as predicted in the previous step). In other words, the proposed similarity measure dynamically inhibits the use of corrupted features while retrieving the most relevant cases from the Case-Base. Note that local distances computed in Eq. 4 are reused by the dissimilarity function. This is possible due to the fact that the squared Euclidean distance between two vectors is equal to the sum of squared Euclidean distances between segments of those vectors:

$$\begin{aligned} ||x-y||^2 &= ||z||^2 = z_1^2 + z_2^2 + \cdots + z_i^2 + z_{i+1}^2 + \cdots + z_j^2 \\ &= ||(z_1, \cdots, z_i)||^2 + ||(z_{i+1}, \cdots, z_j)||^2 \\ &= ||(x_1 - y_1, \cdots, x_i - y_i)||^2 + ||(x_{i+1} - y_{i+1}, \cdots, x_j - y_j)||^2 \end{aligned} \tag{7}$$

Afterwards, the reuse stage analyses the retrieved cases and their labels to emit a prediction regarding the identity associated to the new case. To this extent, we use the weighted voting scheme proposed in [6]. First, the dissimilarities are used to compute the weight vector:

$$w_j = \frac{1}{d(x, x^{(j)})} \text{ for } j = 1, \cdots, k \tag{8}$$

As explained in [6], the weight vector can be used to estimate the probability that sample x belongs to class c by:

$$P(y = c \mid x) = \frac{\sum_{j=1}^{k} w_j \cdot I(y_j = c)}{\sum_{j=1}^{k} w_j} \tag{9}$$

where $I(y_j = c)$ returns a value of one if the j-th retrieved case belongs to class c, and zero otherwise. Finally, the reuse module obtains the predicted class label as follows:

$$y = arg \max_{c \in C} P(y = c \mid x) \tag{10}$$

where C is the set of all possible class labels (identities).

Computational Complexity. The computational complexity of classical case-retrieval methods (i.e., nearest neighbour search) mainly depends on the method chosen to store the Case-Base. The simplest storage and search method, known as *Naive search*, stores the cases of the Case-Base without any particular order and performs a sequential search over the complete Case-Base in test time. As a consequence, the computational complexities of training and test phases are $\mathcal{O}(1)$ and $\mathcal{O}(nd + nk)$ respectively[4], where n is the number of training cases, d their dimension and k the desired number of nearest neighbours to be considered [18].

[4] This complexity corresponds to the version of the algorithm which computes and stores dissimilarities in a vector of dimension n. If distances are re-computed to find each nearest neighbour, the complexity is $\mathcal{O}(knd)$.

The hyperparameter k is usually considered as a constant. Hence, the complexity of test stage simplifies to $\mathcal{O}(nd)$.

Regarding the proposed method, the training stage has a constant computational cost $\mathcal{O}(1)$ as no computation is performed. For test stage, the computations defined by Eqs. 4 and 5 can be done at the cost of time $\mathcal{O}(n(\frac{d}{p} \cdot p + \frac{d}{p}))$, which simplifies to $\mathcal{O}(nd)$ given that $p > 1$. Finding the most similar cases in the Case-Base according to Eq. 6 takes $\mathcal{O}(n \cdot \frac{d}{p} + kn)$ time; which can be simplified to $\mathcal{O}(nd)$ by considering hyperparameters p and k as constants. Finally, the remaining computations which correspond to the voting process have a complexity of $\mathcal{O}(k)$. Therefore, the computational complexity of the complete test stage is $\mathcal{O}(nd) + \mathcal{O}(nd) + \mathcal{O}(k)$, which simplifies to $\mathcal{O}(nd)$ given that $n >> k$. Hence, we can conclude that the scalability of the proposed retrieval and reuse stages is equivalent to that of classical nearest neighbour search methods.

Revise and Retain. The Revise and Retain stages enable the over-time learning capabilities of the CBR methodology. In the context of the proposed CBR system, the revision should be carried out by a human expert who determines whether an image has been assigned the correct identity. The proposed method can be categorized as a *lazy learning* model, as the generalization beyond training data is delayed until a query is made to the system. As a consequence, the proposed system does not involve training any classifier or model apart from the storage of cases in the Case-Base (as opposed to other occlusion-robust face recognition approaches [12, 13, 16]). For this reason, retaining revised cases only involves storing their case representation into the Case-Base. In addition, this mechanism can also be applied to provide the CBR system with knowledge of previously unseen individuals, thus extending the number of possible identities predicted by the system.

3.4 Multi-scale Local Binary Pattern Histograms

Several studies have found that higher recognition accuracy rates can achieved by combining LBPH descriptors extracted form the same image at various scales [2, 3]. In spite of containing some redundant information, the high-dimensional descriptors extracted in this manner are known to provide classification methods with additional information which enables higher accuracy rates. Unfortunately, the computational costs derived from using such a high-dimensional feature descriptor suppose a serious problem. Apart from that, this image descriptor is perfectly compatible with the proposed method. The only requirement is that histograms corresponding to the same image region are placed next to each other when forming the final descriptor. Then, selecting the correct value for p, the corresponding histograms for a specific face region will be treated as a single occlusion unit (i.e., a set of features which our method considers as occluded or non-occluded as a whole).

Local Dimensionality Reduction with Random Projection. This section tries to address the problem of high-dimensionality of multi-scale LBPH descriptors. In the literature, Chen *et al.* [3] proposed using an efficient dimensionality reduction algorithm to reduce the size of multi-scale LBP descriptors. However, this approach is not directly compatible with the method proposed in the previous section. The reason for this being that we need to keep features from different occlusion units (i.e., facial regions) isolated form each other, so we can later detect and inhibit features coming from occluded facial areas. Classical dimensionality reduction methods such as Principal Component Analysis and Linear Discriminant Analysis produce an output feature space were each component is a linear combination of input features, thus being incompatible with our occlusion detection method. To overcome this limitation, we propose performing dimensionality reduction at a local level. To this extent, the histograms extracted from a specific facial region (at various levels) are considered a single occlusion unit. Then, the Random Projection [1] (RP) algorithm is applied locally to each occlusion unit. As opposed to other dimensionality reduction methods, RP generates the projection matrix from a random distribution. As a consequence, the projection matrix is data-independent and cheap to build.

The main theoretical result behind RP is the Johnson-Lindenstrauss (JL) lemma. This result guarantees that a set of points in a high dimensional space can be projected to a Euclidean space of much lower dimension while approximately preserving pairwise distances between points [4]. Formally, given $0 < \epsilon < 1$, a matrix X with n samples from $\mathbb{R}^p$, and $k > 4 \cdot ln(n)/(\epsilon^2/2 - \epsilon^3/3)$ a linear function $f : \mathbb{R}^p \rightarrow \mathbb{R}^k$ exists such that:

$$(1-\epsilon)||u-v||^2 \leq ||f(u)-f(v)||^2 \leq (1+\epsilon)||u-v||^2 \quad \forall u,v \in X \tag{11}$$

In particular, the map $f : \mathbb{R}^p \rightarrow \mathbb{R}^k$ can be performed by multiplying data samples by a random projection matrix R drawn from a Gaussian Distribution:

$$f(x) = \frac{1}{k} x \cdot R \tag{12}$$

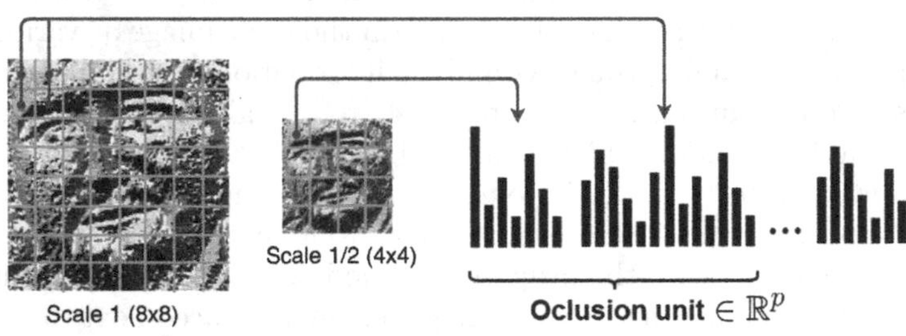

Fig. 3. Features are grouped together in the final descriptor according to the facial region they come from.

As previously said, in order to apply RP in the context of the proposed method, me must first ensure that histograms coming from the same face region are placed together in the final descriptor[5] (see Fig. 3). Afterwards, we can apply the RP method locally to each occlusion unit. Formally, let $x \in \mathbb{R}^d$ be a multi-scale LBPH descriptor where each occlusion unit consists of p features, k a natural number such that $k < p$, and R a $p \times k$ random matrix whose entries have been drawn from $\mathcal{N}(0,1)$. The reduced version of descriptor x is computed as follows:

$$x' = f((x_1, \cdots, x_p)) \ || \ f((x_{p+1}, \cdots, x_{2p})) \ || \cdots || \ f((x_{d-p+1}, \cdots, x_d)) \quad (13)$$

where $||$ denotes vector concatenation. Note that, thanks to the JL-lemma, for a sufficiently large k value the result of applying the proposed retrieval and reuse stages over reduced descriptors is approximately the same as doing it over the original high-dimensional descriptors. To prove this, it suffices to consider the different computations carried out by the proposed retrieval and reuse stages. First, the local distance matrix is computed according to Eq. 4. If we reduce both the new case x and the descriptors $x^{(i)}$ in the Case-Base as described in Eq. 13, and set hyperparameter p to the new size of occlusion units (i.e., $p = k$), Eq. 4 can be rewritten as follows:

$$\begin{aligned} L'_{i,j} = ||(x'_{k(j-1)+1}, \cdots, x'_{kj}) - (x'^{(i)}_{k(j-1)+1}, \cdots, x'^{(i)}_{kj})||^2 \\ \text{for} \quad i = 1, 2, \cdots, n \ \text{ and } \ j = 1, 2, \cdots, d'/k \end{aligned} \quad (14)$$

where $x' \in \mathbb{R}^{d'}$. Then, applying the JL-lemma, for a sufficiently large k value we can ensure that:

$$\begin{aligned} (1 - \epsilon)\ L_{i,j} \leq L'_{i,j} \leq (1 + \epsilon)\ L_{i,j} \\ \text{for} \quad i = 1, 2, \cdots, n \ \text{ and } \ j = 1, 2, \cdots, d'/k \end{aligned} \quad (15)$$

In other words, the distortion induced in matrix L' with respect to L is bounded. The following steps of the proposed method are based on L'. Therefore, if the difference between L' and L is small enough, the proposed retrieval and reuse stages will provide the same results when executed over the reduced descriptors. Section 4 reports on several experiments where the descriptors were reduced with this approach.

4 Experimental Results

This section reports on a series of experiments carried out to assess the performance of the proposed CBR framework in the task of face recognition under partial occlusion. We evaluated the proposed system over a database of facial images with different types of occlusion and using different image descriptors. In addition, we evaluate how much accuracy is lost by using an automated face alignment method as compared to manual human annotations. In particular,

[5] To ease this, we always select gird sizes such that occlusion units defined as cells in the smallest grid contain an integer number of cells from the bigger grids.

the evaluation dataset is the *ARFace database* [11]. This dataset contains about 4,000 color images corresponding to 126 individuals (70 men and 56 women). The images display a frontal view of individuals' faces with different facial expressions, illumination conditions and partial occlusions. The dataset also includes annotations with the exact bounding boxes of faces inside images.

We used the images in the ARFace dataset to create several subsets for our experiments. In particular, we arranged a training set, a validation set, and several test sets with different characteristics:

- *Training set*: one image per individual (neutral, uniform lighting, first session).
- *Validation set*: almost one image per individual[6] (neutral, uniform lighting, second session).
- *Lighting test set*: almost four images per individual (neutral, illumination left/right, first and second sessions).
- *Glasses test set*: almost two images per individual (sunglasses, uniform lighting, first and second session).
- *Scarf test set*: almost two images per individual (scarf, uniform lighting, first and second session).

We evaluate the proposed method against other common classification methods used in the field of face recognition, namely Logistic Regression (LR), Support Vector Machine (SVM) and Naive Bayes (NB). In the case of the proposed method, several hyperparameters need to be adjusted. Hyperparameter p determines the size of the occlusion unit, and is fully determined by the parametrization of the LBP descriptor. The remaining hyperparameter is the *threshold* for occlusion detection. In an ideal scenario, a set of images with partial occlusion would be available to adjust this value. However, one of the goals of this work was to design a method which could operate without any information on the nature of occlusion during training time. Fortunately, it is possible to find a suitable *threshold* value with a validation set of images without occlusion, even if such validation set contains less than one image per individual. This can be achieved by following these steps:

1. The *threshold* is initialized to a sufficiently large value (for large *threshold* values, the proposed method behaves exactly like wkNN. We can use this to determine whether the *threshold* was initialized to a sufficiently large value).
2. The proposed CBR framework is trained over the *Training set* and evaluated over the *Validation set*.
3. The *threshold* value is decreased. Steps 2 and 3 are repeated until a significant loss in the accuracy is registered. This will indicate that some non-occluded blocks in the validation set have been misclassified as occluded, so the *threshold* value is set to the previous value.

The evaluation protocol for all our experiments has been the following: (1) the classifier under evaluation is trained over the training set; (2) the validation set

[6] Second session images are not available for all individuals in the dataset.

is used to perform hyperparameter selection; (3) the classifier, parametrized as determined in the previous step, is re-trained over the union of the training set and the validation set; (4) the trained classifier is evaluated over the different test datasets available.

Experimental Results with Automatic Face Alignment. Table 1 presents the results obtained by using the automatic face detection and alignment methods explained in Sect. 3.1. For single scale descriptors, we used $LBP^u_{8,2}$ histograms over a 8×8 grid, thus obtaining a descriptor of 3,776 dimensions. In the case of multi-scale descriptors, we used $LBP^u_{8,2}$ histograms over 12×12 and 6×6 grids. The resulting descriptor dimension was therefore 10,620. Finally, for our experiments with local RP, each 295-dimensional occlusion unit in the high-dimensional multi-scale descriptor was reduced to 150 features. Therefore, the complete descriptor ended up having a dimension of 5,400 (i.e., approximately half the original dimension).

Table 1. Experimental results with automatic face alignment.

Features	Classifier	Lighting	Scarf	Glasses
$\text{LBP}^u_{8,2}$	wkNN [6]	81.4%	39.4%	30.0%
$\text{LBP}^u_{8,2}$	Proposed CBR $p = 59$; *threshold* $= 27$	96.2%	83.6%	50.2%
$\text{LBP}^u_{8,2}$	SVM (poly kernel)	78.1%	36.9%	25.1%
$\text{LBP}^u_{8,2}$	Logistic regression	84.8%	45.0%	23.4%
$\text{LBP}^u_{8,2}$	Naive Bayes	82.5%	43.7%	20.1%
multi-scale $\text{LBP}^u_{8,2}$	wkNN [6]	98.8%	73.3%	34.9%
Multi-scale $\text{LBP}^u_{8,2}$	Proposed CBR $p =$ 295; *threshold* $= 17$	99.2%	89.2%	50.6%
Multi-scale $\text{LBP}^u_{8,2}$	SVM (poly kernel)	88.1%	59.6%	27.9%
Multi-scale $\text{LBP}^u_{8,2}$	Logistic regression	96.2%	75.1%	28.3%
Multi-scale $\text{LBP}^u_{8,2}$	Naive Bayes	86.29%	72.1%	37.03%
Multi-scale $\text{LBP}^u_{8,2}$ + RP	wkNN [6]	98.5%	66.5%	31.2%
Multi-scale $\text{LBP}^u_{8,2}$ + local RP (see Sect. 3.4)	Proposed CBR $p =$ 150; *threshold* $= 100$	98.8%	90.5%	51.0%
Multi-scale $\text{LBP}^u_{8,2}$ + RP	SVM (poly kernel)	85.5%	55.3%	20.9%
Multi-scale $\text{LBP}^u_{8,2}$ + RP	Logistic regression	93.3%	69.0%	25.5%
Multi-scale $\text{LBP}^u_{8,2}$ + RP	Naive Bayes	84.0%	54.5%	27.1%

Experimental Results with Manual Face Alignment. Table 2 compiles the results obtained by using the manual face annotations provided by the authors of the ARface database. Again, for single scale descriptors we used $LBP^u_{8,2}$ histograms over a 8×8 grid, thus obtaining a descriptor of 3,776 dimensions. In the case of multi-scale descriptors, we used $LBP^u_{8,2}$ histograms with 12×12 and 6×6 grid sizes. Hence, the resulting descriptor dimension was $10,620$. Finally, for our experiments with local RP, each 295-dimensional occlusion unit in the high-dimensional multi-scale descriptor was reduced to 150 features. Therefore, the complete descriptor had a dimension of $5,400$.

Table 2. Experimental results with manual face alignment.

Features	Classifier	Lighting	Scarf	Glasses
$\text{LBP}^u_{8,2}$	wkNN [6]	95.5%	76.5%	69.5%
$\text{LBP}^u_{8,2}$	Proposed CBR $p = 59$; $threshold = 30$	99.5%	91.5%	83.5%
$\text{LBP}^u_{8,2}$	SVM (poly kernel)	96.5%	75.0%	61.0%
$\text{LBP}^u_{8,2}$	Logistic regression	98.5%	81.0%	68.0%
$\text{LBP}^u_{8,2}$	Naive Bayes	94.0%	76.5%	69.0%
Multi-scale $\text{LBP}^u_{8,2}$	wkNN [6]	100%	92.0%	86.0%
Multi-scale $\text{LBP}^u_{8,2}$	Proposed CBR $p = 295$; $threshold = 111$	99.5%	97.0%	92.0%
Multi-scale $\text{LBP}^u_{8,2}$	SVM (poly kernel)	100.0%	92.0%	84.5%
Multi-scale $\text{LBP}^u_{8,2}$	Logistic regression	100.0%	93.0%	89.5%
Multi-scale $\text{LBP}^u_{8,2}$	Naive Bayes	95.5%	93.5%	89.0%
Multi-scale $\text{LBP}^u_{8,2}$ + RP	wkNN [6]	100%	92.0%	86.0%
Multi-scale $\text{LBP}^u_{8,2}$ + Local RP (see Sect. 3.4)	Proposed CBR $p = 150$; $threshold = 111$	99.5%	97.0%	92.0%
Multi-scale $\text{LBP}^u_{8,2}$ + RP	SVM (poly kernel)	100.0%	92.0%	84.5%
Multi-scale $\text{LBP}^u_{8,2}$ + RP	Logistic regression	100.0%	93.0%	89.5%
Multi-scale $\text{LBP}^u_{8,2}$ + RP	Naive Bayes	95.5%	93.5%	89.0%

5 Discussion and Future Work

This work proposed a novel CBR framework for occlusion-robust face detection. The retrieval and reuse stages of the system use a modified version of the weighted k-Nearest Neighbour [6] algorithm to dynamically inhibit features from occluded face regions. This is achieved by using a novel similarity function which discards local distances imputable to occluded facial regions. As opposed to recent deep learning-based methods, the proposed system can operate in domains where only a small amount of training information is available, and does not require any specialized computing hardware to run.

Our theoretical analysis showed that the scalability of the proposed method is equivalent to that of classical Nearest Neighbour retrieval methods. In addition, we proved that the Random Projection algorithm can be applied in a local manner to reduce the dimension of multi-level LBPH descriptors, while ensuring that the proposed retrieval and reuse stages will perform approximately as well as they do over the original high-dimensional descriptors.

Experimental results carried out over the ARFace database show that, in most cases, the proposed method outperforms classic classification algorithms when using LBPH features to identify facial images with partial occlusion. In addition, the proposed framework exhibits a better performance under uncontrolled lighting conditions.

Our experimental results also suggest that much of the accuracy loss registered when working with occluded images is imputable to automatic-alignment errors. In this regard, investigating how automatic face alignment methods can be made more robust to partial facial occlusion emerges as very interesting future research topic. In addition, we intend to evaluate the compatibility of the proposed CBR framework with other local feature descriptors rather than LBPH and other dimensionality reduction methods, and assess the effectiveness of our method on other datasets.

References

1. Achlioptas, D.: Database-friendly random projections. In: Proceedings of the Twentieth ACM SIGMOD-SIGACT-SIGART Symposium on Principles of Database Systems, pp. 274–281. ACM (2001)
2. Chan, C.-H., Kittler, J., Messer, K.: Multi-scale local binary pattern histograms for face recognition. In: Lee, S.-W., Li, S.Z. (eds.) ICB 2007. LNCS, vol. 4642, pp. 809–818. Springer, Heidelberg (2007). doi:10.1007/978-3-540-74549-5_85
3. Chen, D., Cao, X., Wen, F., Sun, J.: Blessing of dimensionality: high-dimensional feature and its efficient compression for face verification. In: Proceedings of the IEEE Conference on Computer Vision and Pattern Recognition, pp. 3025–3032 (2013)
4. Dasgupta, S., Gupta, A.: An elementary proof of a theorem of johnson and lindenstrauss. Random Struct. Algorithms **22**(1), 60–65 (2003)
5. Ekenel, H.K.: A robust face recognition algorithm for real-world applications. Ph.D. thesis, Karlsruhe, University, Dissertation, 2009 (2009)
6. Hechenbichler, K., Schliep, K.: Weighted k-nearest-neighbor techniques and ordinal classification. Technical report, Discussion paper//Sonderforschungsbereich 386 der Ludwig-Maximilians-Universität München (2004)
7. Jia, H., Martinez, A.M.: Face recognition with occlusions in the training and testing sets. In: 8th IEEE International Conference on Automatic Face & Gesture Recognition, 2008, FG 2008, pp. 1–6. IEEE (2008)
8. King, D.E.: Dlib-ml: a machine learning toolkit. J. Mach. Learn. Res. **10**, 1755–1758 (2009)
9. Liu, J., Deng, Y., Huang, C.: Targeting ultimate accuracy: face recognition via deep embedding. arXiv preprint arXiv:1506.07310 (2015)
10. Lopez-de-Arenosa, P., Díaz-Agudo, B., Recio-García, J.A.: CBR tagging of emotions from facial expressions. In: Lamontagne, L., Plaza, E. (eds.) ICCBR 2014. LNCS, vol. 8765, pp. 245–259. Springer, Cham (2014). doi:10.1007/978-3-319-11209-1_18
11. Martinez, A.M.: The AR face database. CVC Tech. Rep. **24** (1998)

12. Martínez, A.M.: Recognizing imprecisely localized, partially occluded, and expression variant faces from a single sample per class. IEEE Trans. Pattern Anal. Mach. Intell. **24**(6), 748–763 (2002)
13. Min, R., Hadid, A., Dugelay, J.-L.: Improving the recognition of faces occluded by facial accessories. In: 2011 IEEE International Conference on Automatic Face & Gesture Recognition and Workshops (FG 2011), pp. 442–447. IEEE (2011)
14. Ojala, T., Pietikäinen, M., Harwood, D.: A comparative study of texture measures with classification based on featured distributions. Pattern Recognit. **29**(1), 51–59 (1996)
15. Ojala, T., Pietikäinen, M., Mäenpää, T.: Multiresolution gray-scale and rotation invariant texture classification with local binary patterns. IEEE Trans. Pattern Anal. Mach. Intell. **24**(7), 971–987 (2002)
16. Tan, X., Chen, S., Zhou, Z.-H., Zhang, F.: Recognizing partially occluded, expression variant faces from single training image per person with SOM and soft k-NN ensemble. IEEE Trans. Neural Netw. **16**(4), 875–886 (2005)
17. Tzimiropoulos, G.: Project-out cascaded regression with an application to face alignment. In: 2015 IEEE Conference on Computer Vision and Pattern Recognition (CVPR), pp. 3659–3667. IEEE (2015)
18. Weber, R., Schek, H.-J., Blott, S.: A quantitative analysis and performance study for similarity-search methods in high-dimensional spaces. VLDB **98**, 194–205 (1998)

Time Series and Case-Based Reasoning for an Intelligent Tetris Game

Diana Sofía Lora Ariza([✉]), Antonio A. Sánchez-Ruiz, and Pedro A. González-Calero

Dep. Ingeniería del Software e Inteligencia Artificial, Universidad Complutense de Madrid, Madrid, Spain
{dlora,antsanch,pagoncal}@ucm.es

Abstract. One of the biggest challenges when designing videogames is to keep a player's engagement. Designers try to adapt the game experience for each player defining different difficulty levels or even different sets of behaviors that the non-player characters will use depending on the player profile. It is possible to use different machine learning techniques to automatically classify players in broader groups with distinctive behaviors and then dynamically adjust the game for those types of players.

In this paper, we present a case-based approach to detect the skill level of the players in the Tetris game. Cases are extracted from previous game traces and contain time series describing the evolution of a few parameters during the game. Once we know the player level, we adapt the difficulty of the game dynamically providing better or worse Tetris pieces. Our experiments seem to confirm that this type of dynamic difficulty adjustment improves the satisfaction of the players.

Keywords: Dynamic difficulty adjustment · Time series · Video games · Tetris · Case-based reasoning · K-nearest neighbor

1 Introduction

Entertainment is the main goal of computer games. The player's level of entertainment can be measured by the level of engagement or immersion within the game [27,28]. Immersion is a state of consciousness where awareness of physical self is lost by being in an artificial environment [20]. To maintain the interest, players should perceive challenges that enhance their skills, clear goals to achieve and receive immediate feedback. In particular, challenges proposed in the game should be at the "right" difficulty level, that is, players should feel challenged but not overwhelmed [3,5].

In order to achieve complex behaviors in realistic environments, computer games need to provide appropriate responses to changing circumstances [6]. Nowadays, the intelligence of non-player characters in most games is the result of discovering what the player will do in advance, while the video game is being made, so that game developers can implement during the production stage a

D.W. Aha and J. Lieber (Eds.): ICCBR 2017, LNAI 10339, pp. 185–199, 2017.
DOI: 10.1007/978-3-319-61030-6_13

standard set of behaviors to respond appropriately [19]. This way, the player can perceive a sense of "intelligence" from the game on certain circumstances. Unfortunately, these approaches are very expensive to develop and test, and they do not always produce the expected result.

Different player modeling techniques can be used to classify and recognize typical user behaviors [4] and personalize the game experience for different player profiles. For example, Dynamic Difficulty Adjustment (DDA) can be used to automatically alter the game difficulty meet the player's expectations [14]. One popular approach to DDA is *Rubber Band AI* [18,30], which means that the player is "bound" together with their enemies by a rubber band, in such a way that, if the player "pulls" in one direction (exhibit more or less skills), their opponents will go that same direction (show more or less complex behavior).

For these reasons, several researchers have been trying to use different machine learning techniques in videogames to automatically model different types of players based on their behaviors [7,8,17,24]. In this paper, we propose a case-based reasoning [1] (CBR) approach to model and detect different types of players in the popular game of Tetris. In particular, we try to predict the player's skill level looking at the evolution of the game during a time window and comparing that evolution with a set of cases extracted from previous games. Cases contains time series describing the progression of a few features for a certain period and are labeled with the player's level, which is automatically obtained from the final score of the game.

Then, we dynamically customize the difficulty of the game for the current player by providing more or less difficult pieces to place in the current game board. Additionally, we have performed an experiment with real players and our results seem to confirm that this type of DDA improves their satisfaction.

The rest of the paper is organized as follows. Section 2 describes the specific version of Tetris used in our experiments. Section 3 explains our CBR approach to model and dynamically detect the skill level of the player. Next, Sect. 4 describes how we change the difficulty of the game dynamically according to the player profile. Section 5 explains the experiment performed with real players and the different effects observed with and without DDA. Finally, the paper closes with related work, conclusions and future work.

2 Tetris Analytics

Tetris is a very popular video game in which the player must place different tetromino pieces that fall from the top of the screen in a rectangular game board. When a row of the game board is complete, i.e. it has no holes, the row disappears, all pieces above drop one row, and the player is rewarded with some points. As the game progresses, the new pieces fall faster and faster, gradually increasing the difficulty, until a piece is placed such that exceeds the top of the board and the game ends. The goal of the game, therefore, is to complete as many lines as possible to increase the score and make room for the next pieces. Although the game is quite simple to play, it is also very difficult to master and hard to solve from a computational point of view [2].

Fig. 1. Tetris Analytics screen capture.

In our experiments, we used Tetris Analytics (Fig. 1), a version of the game implemented in Java that looks like an ordinary Tetris from the player's point of view, but internally provides extra functionality to extract, store and reproduce game traces. From these traces, we can select the most significant features to characterize the playing style of each player, determine their skill level and dynamically adjust the difficulty of the game.

Note that in this game, each time a new piece appears on the top of the board, players has to make two different decisions. The first one, that we call *tactical*, is to decide the final location. The second decision involves all the *moves* required to lead the piece to its final location. In other words, all the game states generated while the piece is going down are considered *moves*, and the last one, which is when the piece settles, is considered a *tactical* one.

In this paper, we consider only the tactical decisions in order to define the skill level of the player, but we are aware that we could also extract valuable information from the concrete moves (looking at parameters like speed, cadence and moves undone) to detect the player's skill level and we plan to extend our work to consider them in the future.

3 A CBR Approach to Player Modeling

In order to obtain a dataset with different styles of play, we asked 10 people with diverse skill levels to play a total of 60 games of Tetris Analytics [15,16] and send us the games traces that were automatically generated. These traces

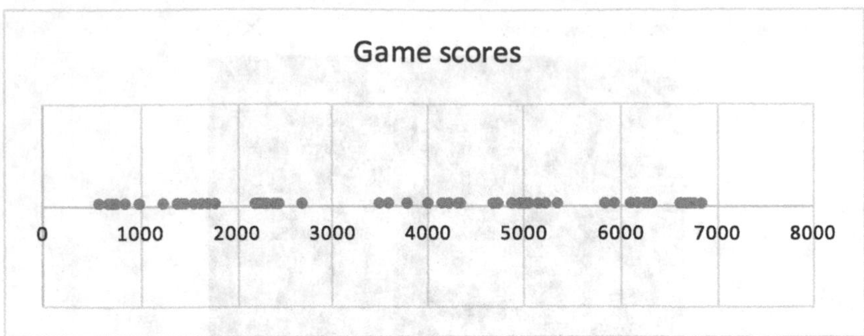

Fig. 2. Total scores of games in train set.

are very useful because they allow us to replay the exact same games they played and extract as much information as we require for our analysis.

Next, we classified the games in 3 different categories each one representing skill levels according to their final scores. Figure 2 shows the total score of each game plotted on the x axis. We notice groups of points close to each other in some intervals and gaps between them. Those gaps are good candidates to split the different skill levels:

- Newbie: total score between 0 and 2999.
- Average: total score between 3000 and 5499.
- Expert: total score more than 5500 points.

In the dataset, 21.6% of the games were classified in the newbie category, 46.2% in the average and 32.2% in the expert profile.

Next, we analyzed the game traces and extracted the following features for every tactical decision the players made in the game, that is, every time they placed a new piece in its final location:

- *Number of piece*: the number of the current piece from the beginning of the game. Since we only consider tactical decisions, the number of piece corresponds to the moment of the game in which this sample is extracted.
- *Current score*: the accumulative game score obtained by the player after placing the current piece. We expect to see a clear correlation between the score during the game and the skill level of the player.
- *Number of holes*: the Tetris board is a matrix of 20 rows and 10 columns. A *hole* is one cell of that matrix. If the pieces are not placed carefully in the board, it is common to leave empty spaces under other pieces in the same column. Good players tend to compact the pieces in the board without leaving holes so that will be easier to complete lines with the next pieces.
- *Height of the board*: measured as the highest row occupied by a piece. Good players tend to play most of the game in the lowest half of the board because each time they complete a line, the height of the board decreases by one unit.

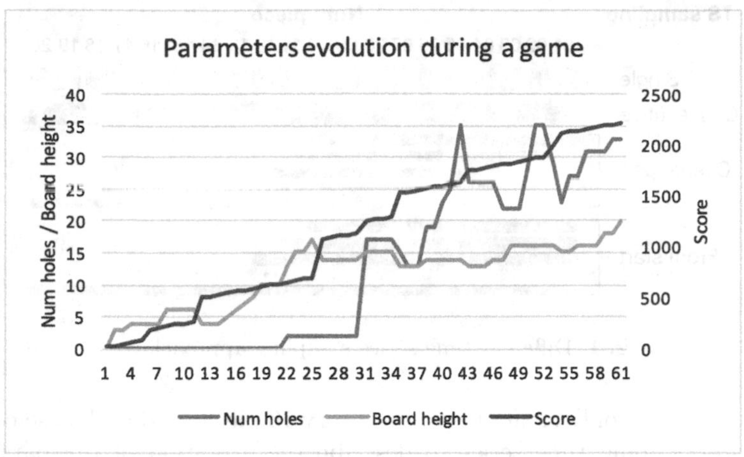

Fig. 3. Evolution of the parameters during a game.

For example, the game state in Fig. 1 correspond to the 8th piece of the game. Game score will be 105 plus the points obtained for placing the current piece. Height of the board is 6 because that is the highest occupied row in the last column on the right side of the board. In this way, a game can be seen as a sequence of tuples describing these values for each piece placed.

It is interesting to note that we characterize the game evolution without a explicit representation of the actions or decisions the player made, we only store some parameters describing the game board progression. We think this is possible in a simple game like Tetris, but might not be enough in more complex games.

The score, number of holes and height of the board summarizes the state of the game after the player places each new piece in the board. In general, it is not easy to predict the skill level of the player from a static picture of the game, it is much easier if we consider the evolution of the game over a time interval. With this idea in mind, we decided to group the data to create time series describing the evolution of each parameter.

For example, Fig. 3 shows the evolution of these parameters during a particular game. The final score is less than 3000 points so this game corresponds to the newbie profile. We can observe that, during the first third of the game, the player was doing a good job compacting the pieces in the board and not leaving holes but then, both the board height and the number of holes began to increase. The game ended when the board height reached the 20th row.

Next, we created a case base where each case describes the evolution of features described above during a given time window. We tried different approaches to build the case base (Fig. 4):

- *Single*: each case only describes the current game state and there is a case for each piece in the game. We do not use time series in this first approach (or equivalently the time window size is 1).

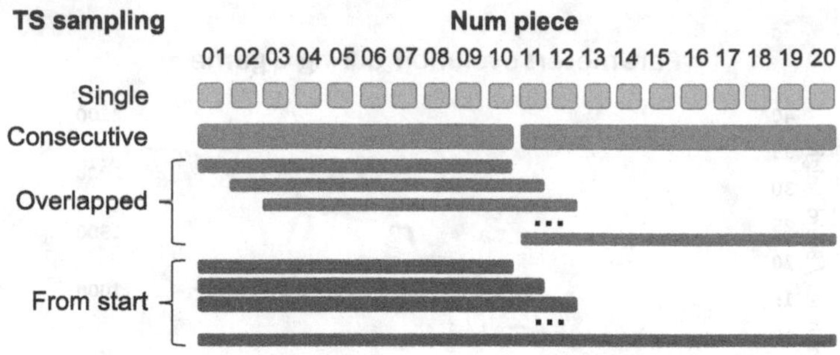

Fig. 4. Different time series sampling approaches.

- *Consecutive*: we only create new cases every 10 pieces and each case contains 3 non-overlapping time series to describe the evolution of the parameters during last 10 pieces. In other words, from tactical move 1 till 10, is one case, from 11 till 20, is another case, and so on. A disadvantage of this approach is that the player can only be classified again every 10 pieces.
- *Overlapped*: it's the same idea but now the time series overlap in time, so we create a case describing the evolution of the parameters during the last 10 pieces.
- *From start*: we create a new case for every piece describing the evolution of the parameters from the beginning of the game to the current moment. In this approach, time series have different length depending of the moment of the game when they are sampled.

Each one of the previous approaches resulted in a different case base. To predict the player profile in an ongoing game, we build a new query case describing the evolution of the parameters during the last time window, and use k-nearest neighbor classifier (k-NN). The current player's skill level is decided by a majority vote among the retrieved cases. It is interesting to note that when we look for similar cases in the case base we only consider those that were created at the same moment in previous games. For example, in order to predict the skill level of the player at piece number 20 we only consider cases created at piece 20 in previous games. This filter reduces significantly the number of cases to consider and let us to compare time series with the same length.

In previous work [16], we built a classifier using clustering analysis. But it did not perform well predicting all player's profile in an ongoing game. It was capable of distinguish extreme cases, like a really bad or good player, but many average players were mistaken as newbies. Average score and number of players with better game experience raised with DDA, but it was not substantial. The mean score of the post-game survey in normal games was 3.25 and improved to 3.46 in games with DDA. With Student's t-test, we were able to confirm that the mean scores are not statistically significant with $p - value = 0.1$. CBR approaches, on the contrary, obtain better results because they are based on a reduced number of features and we end up with a simpler and less noisy dataset. Plus, we can tune the similarity measure to provide different importance to each feature.

Table 1. Precision retrieving k cases to predict the skill class.

	k = 1	k = 3	k = 5	k = 10
CBR single	57.83	58.98	59.82	59.47
CBR consecutive	60.68	63.64	63.86	64.09
CBR overlapped	**63.51**	**64.71**	**65.07**	**65.42**
CBR from start	63.46	64.49	62.93	60.53

Table 1 shows the precision of each case based approach varying the number of cases retrieved (k) to predict players profile using 10-fold cross validation. Note that all approaches based on time series are better than the *Single* classifier. In particular, the approach based on overlapped time series obtains the highest precision (65.42%) with $k = 10$ and has the advantage that can be used at any moment of the game after piece number 10 is settled.

The similarity between two cases was computed as a linear combination of the similarities between their time series. To compare time series we use a simple similarity measure based on the Euclidean distance, but we would like to test other time series similarities in the future [11,22,29]:

$$\begin{aligned} sim_c(c_1, c_2) = {} & 0.70 \times sim_{ts}(c_1.score, c_2.score) + \\ & 0.25 \times sim_{ts}(c_1.holes, c_2.holes) + \\ & 0.05 \times sim_{ts}(c_1.height, c_2.height) \end{aligned}$$

$$sim_{ts}(r, s) = 1 - \sqrt{\sum_{i=1}^{n} (r_i - s_i)^2}$$

The weights shown in the previous formula are the optimal weights for the *Overlapped* approach and were computed using an exhaustive grid search with increments of 0.05 in each parameter. Unsurprisingly, the most important time series is the score (70%), followed by the number of holes in game board (25%) and finally the board height (5%). Probably the board height is not very important because it is only problematic when it is close to 20 and, usually, at that time of the game the number of holes is also high.

4 Dynamic Difficulty Adjustment

There are 2 obvious ways to adjust the difficulty of the game in Tetris: either we can change the speed of the falling pieces or we can select a specific type of piece to appear next. We have decided to use the second approach because it is more difficult to detect and we do not want the players to know that we are making the game easier for them.

The CBR classifier described in the previous section makes predictions about the current player skill level every time the player places a new piece after the

10th piece. From that time, we obtain a new prediction based on the last 10 board states.

Based on those predictions, we have defined a different probability to provide a "good" piece depending on the player profile. If the user is categorized as a *newbie*, then there is 50% probability that the game will help. If categorized as *average*, then there is 30% probability and if it is categorized as *expert*, no help is given and the pieces are generated using the default probability distribution for the pieces in the game.

In order to select a good piece for the player we compute all the possible ways to place each type of piece and use a heuristic to decide how good those final possible boards are. The heuristic tries to maximize the score (i.e. complete new lines) and minimize the number of holes and height of the board. Then we assign to each piece the score of the best game board that can be reached placing the piece optimally. Finally, we randomly choose one of the three best pieces and use it as the next game piece.

Why one of the best 3 pieces and not always the best one? That is what we tried at first in our experiments, but the results were not very good. The problem is that some pieces are very easy to place (like the O-block or the I-block), especially at the beginning of the game, and they were selected very often. In those games, it was evident for the player that something was not right, that we were cheating, and it produced a negative reaction in the players.

To avoid extreme profile changes, we use an inertia function that eases the transition from newbie to expert and vice versa. This way, we only allow the game difficulty level to change to the immediate inferior or superior difficulty level. The inertia function is applied every time a profile is predicted, and it compares the current player profile with the new one. For example, if the current player is really good and he is classified as an expert but for a period he does not play so well, his profile will change to average. Then, if he keeps playing bad, his profile will eventually change to newbie.

We think this smooth transition is a good idea for several reasons. First, it avoids abrupt changes that could be perceived as strange game behaviors. Second, the CBR classifier has a precision of 65.42% predicting the ability of the player, so it is not perfect. Besides, it makes more mistakes at the beginning of the game and requires some time to stabilize. Smooth difficulty changes partially disguise those errors, specially at the beginning of the game. Finally, players do not always play at full level and may have lucky streaks, causing confusion to the classifier.

5 Experiments and Results

We performed an experiment to verify whether our approach to DDA really improves the player's satisfaction. The level of satisfaction during a game is a very complex and subjective feeling that depends on several factors, some of them external to the game itself. For these reasons, the results of this small experiment should be taken carefully and will need to be confirmed with additional experiments.

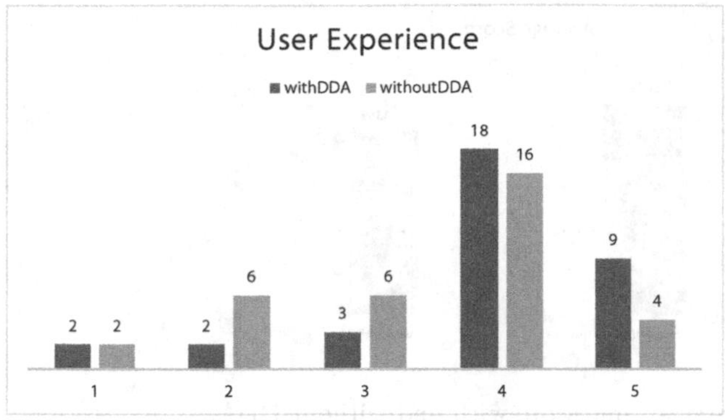

	DDA	no-DDA
Mean	3.92	3.47
σ	1.03	1.09

Fig. 5. Distribution, mean and standard deviation of the players' satisfaction scores with and without DDA.

We asked 18 people with different levels of experience to play 4 games of Tetris. Fortunately, Tetris is a very famous game and most of the people already know how to play. After each game, the players had to evaluate a very simple sentence, "It has been a good Tetris game", with a Likert scale of 5 values were 1 means "strongly disagree" and 5 means "strongly agree". We use this simple question to evaluate the player's level of satisfaction and, additionally, we collect some objective data regarding the game evolution. Games 1 and 3 were normal Tetris games while games 2 and 4 used DDA to help novice and average players. Players did not know the specific goal of our experiment, they only knew that we were interested in measuring their level of satisfaction in different games.

Figure 5 shows the satisfaction results in games with and without DDA. The mean score in normal games was 3.47 and improved to 3.92 in games with DDA. This difference was confirmed to be statistically significant with a Student's t-test (p-value = 0.1). Looking at the distribution of the scores, we can see that most players evaluate better games with DDA. In particular, 8 normal games were evaluated as "bad games" (scores 1 and 2) and 20 normal games were evaluated as "good games" (scores 4 and 5). Using DDA, only 4 games were evaluated as bad and, however, 27 were evaluated as good.

In addition to increasing the player's satisfaction, DDA had other effects in the game metrics. This was expected because, basically, DDA is helping newbie and average players during the game by means of better pieces to place on the board. Figure 6 shows the average score obtained in games with and without DDA. The score with DDA was 14.5% higher than in normal games.

Figure 7 shows the evolution in time of the board height for each of the games played by a random player. The games end when the height reaches the line 20 that corresponds to the top of the board. As we can see, the effect of DDA (games 2 and 4) is remarkable in the second half of the game and increases the game duration for several more pieces.

Another interesting effect of DDA is that players play better because the game is easier, and therefore the CBR classifier categorize them in a higher

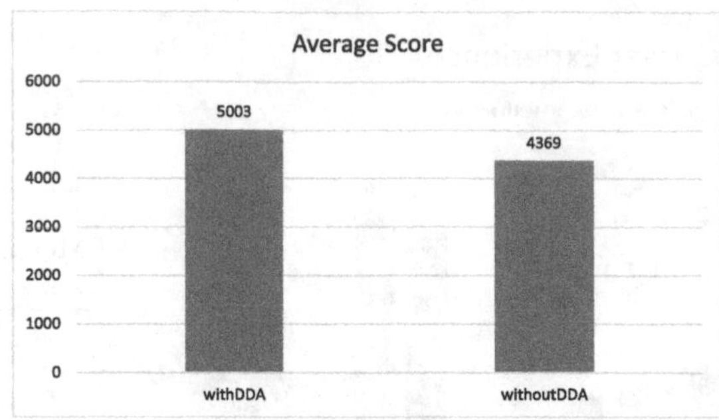

Fig. 6. Average game score with and without DDA.

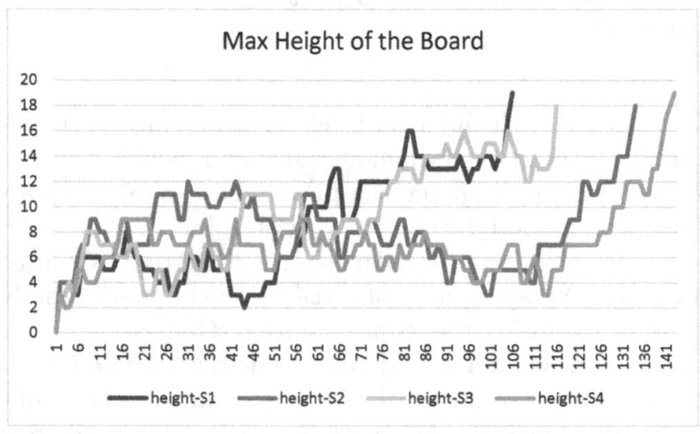

Fig. 7. Evolution of the board height during the 4 games of a random player. Games 1 (height-S1) and 3 (height-S3) are original Tetris games while games 2 (height-S2) and 4 (height-S4) use DDA.

Table 2. Number of games classified in each profile according to their final score with and without DDA.

Profile	With DDA	Without DDA
EXPERT	18	12
AVERAGE	13	13
NEWBIE	5	11

profile. Table 2 shows the number of games that would be classified in each skill profile according to their final score. While the number of average games remains unchanged, the number of games with a low score decreases and the number of games with a high score increases considerably.

We are able to determine if a tactical move is an optimal one from the game state created after the piece is settled. This rate is computed based on an

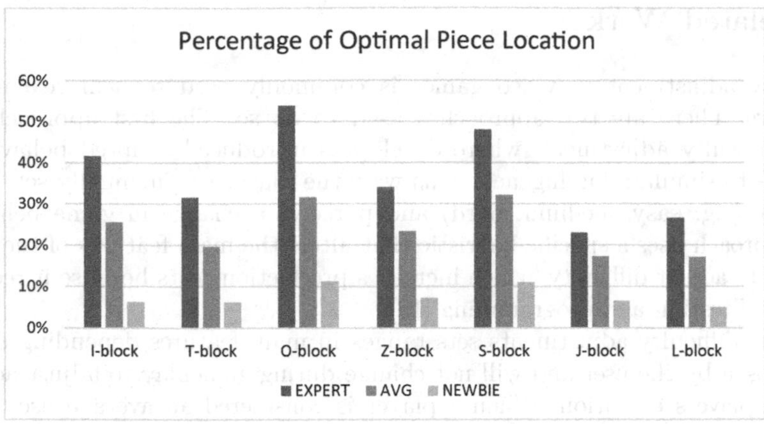

Fig. 8. Percentage of optimal piece location by player profile.

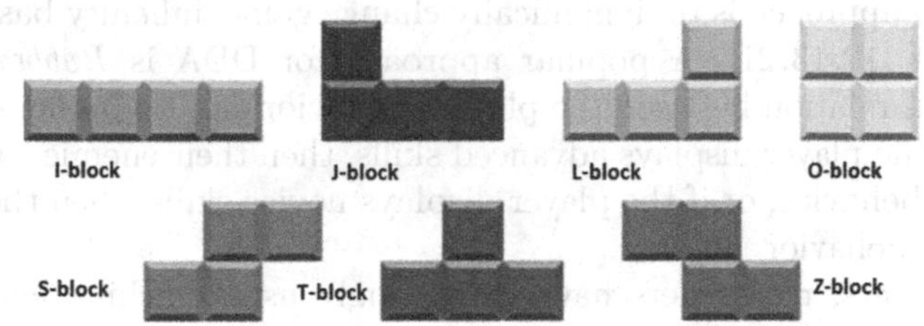

Fig. 9. Tetris pieces names.

heuristic that minimizes height of pieces and number of holes in the board. A move is considered optimal when the piece is placed in such a way that fulfills the heuristic. When no help is given, 66.45% of the times users place the piece in an optimal spot. And this value increases to 76.78% when DDA is active.

Figure 8 shows the percentage of times users placed a piece in an optimal location filtered by player profile and piece name (Fig. 9 shows the names of each piece). Normally, people assume that the easiest piece to locate would be I-block, but as we can see for expert and newbie profiles is O-block. For average profile, the S-block and the O-block are the easiest pieces to locate. Although, the I-block is second easiest piece to locate for experts. This piece analysis will help us to make a better difficult adjustment.

Although this is a small experiment with a few users, the results seem to confirm that our approach to DDA in Tetris really have a positive impact in the player's level of satisfaction. The CBR approach based on time series is able to classify the players dynamically, during the game evolution, according to their level of skill, and the decision to help newbie and average players by providing "good" new pieces from time to time is subtle enough so player do not notice.

6 Related Work

Difficulty adjustment in video games is commonly used to achieve complex behaviors. There are two approaches used to do so. The first approach is a static difficulty adjustment, where developers introduced scripted behavior in the code to simulate intelligence. This way, the player can manually set games difficulty (e.g. easy, medium, hard) and perceive a change in game behavior. This approach uses a specific heuristic that alters the main features of the game in order to adjust difficulty, which increases production costs because it requires additional testing and programming [19].

Static difficulty adjustments sets ranges in main features depending on the level chosen by the user and will not change during gameplay to adjust accordingly to players evolution. When a player is considered an average user or an expert one, it does not mean that all the skills will be above a expected value. In general, users do not learn all the skills at same pace. That is why, this approach is not as accurate as desired.

The second approach is to dynamically change game difficulty based on players interaction [12,13,21]. A popular approach for DDA is *Rubber Band AI*, which creates a relation between the players behavior and their enemies; in such a way that, if the player displays advanced skills, then their enemies will respond with complex behavior, or if the player displays novice skills, then the game will exhibit simple behavior [30].

In recent years, researchers have increasingly used machine learning techniques to create complex behaviors in video games [4,9]. Missura and Gärtner [18] used a simple game where the player shoots down aliens spaceships while those shoot back. They aimed to employ dynamic difficulty adjustments by grouping players into different profiles and supervised prediction from short traces of gameplay. Each game had a limit of 100 s. The first 30 s were used to acquire data, and the rest of the game, they adjust the aliens spaceships speed based on players performance.

Also, there has been several researches that employs CBR to adapt the game to player's profile [24,26]. Sharma et al. [25] present an approach to create a story-based game where the players has an active role in the game narrative. They use a Drama Manager or Director that employs a case-based reasoning approach in its player modeling module to guide the players towards more enjoyable story-lines. Futhermore, CBR has been used in Tetris games with several goals. Floyd and Esfandiari [10] describe a framework called jLOAF for developing case-based reasoning agents that learn by observation. One of their case studies involve an agent learning how to play a Tetris game by observing how an expert plays. The sensory input in Tetris contains the current state of the game region and the piece that needs to be settled. Unfortunately, the agent performs poorly because of the large state-space of the game. Romdhane and Lamontagne [23] investigates how reinforcement learning can improve the management of a legacy case base in Tetris. Cases represent local patterns describing columns where the pieces could be placed. Two of their CBR approaches include retrieving local patterns using a similarity function, which shows a significant increase in the number of lines being removed from the board. It is interesting to note, that they also evaluate future moves in order to seek maximum payoff.

7 Conclusion and Future Work

In this paper, we present a time series and CBR approach to create an intelligent Tetris game. To achieve this, we use a game called "Tetris Analytics", which is like a simple Tetris game, but allow us to extract both game state and player interaction. We tested different approaches to build our train set. CBR Overlapped approach gave higher precision than others. So we used it to build a train set that helps us predict player profile in an ongoing game. Once a profile is computed, we used probability to decide when to help them. The lower the skill level of the player, the higher the help given by the game. To adjust the difficulty of a game, we deliver the best possible piece in terms of increasing the probability of the player doing lines without leaving holes.

Our experimental results indicate that, indeed, user experience and perceived skill level increases when employing dynamic difficulty adjustment. Combining case based reasoning with time series, we are able to analyze the evolution of several features from both game state and player interaction in order to predict player's profile during gameplay. Not only the satisfaction of the player is improved, but also their average performance. For games with dynamic difficulty adjustment, Tetris Analytics delivers an optimal piece computed with the current game state. Average scores from games with DDA are much higher than others. Games categorized as newbies were reduced, while games classified as experts increases considerably. Furthermore, players were able to decrease the max height of the board variable when the system was helping them, meaning that they were able to settle more pieces on the board.

As future work, we would like to improve classifier accuracy, by first trying to increase the size of the case base by collecting more game traces. Tetris Analytics is implemented in Java and was initially conceived for being executed as an applet inside of a web browser. Unfortunately for our system, Java support in web browsers is disappearing and therefore we are considering porting Tetris Analytics to JavaScript or some other browser supported language. This way we could greatly facilitate data collection. Having a larger dataset could also allow us to apply other machine learning techniques and compare the results with the CBR approach. This new version should be released open source along with the datasets in order to allow others to experiment with our data.

We plan also to adapt the selection of the best next piece based on data collected from the players. Given that different types of players can find as easier different pieces we can further adapt the game by providing the easy piece for this particular type of player.

Finally, the good results so far in terms of increasing user satisfaction encourages us to try these techniques with more complex games where more parameters can be tweaked to influence the player experience.

References

1. Aamodt, A., Plaza, E.: Case-based reasoning: foundational issues, methodological variations, and system approaches. AI Commun. **7**(1), 39–59 (1994)
2. Breukelaar, R., Demaine, E.D., Hohenberger, S., Hoogeboom, H.J., Kosters, W.A., Liben-Nowell, D.: Tetris is hard, even to approximate. Int. J. Comput. Geom. Appl. **14**(1–2), 41–68 (2004)
3. Buro, M., Furtak, T.: RTS games as test-bed for real-time AI research. In: Proceedings of the 7th Joint Conference on Information Science (JCIS 2003), vol. 2003, pp. 481–484 (2003)
4. Charles, D., Black, M.: Dynamic player modelling: a framework for player-centred digital games. In: Proceedings of 5th International Conference on Computer Games: Artificial Intelligence, Design and Education (CGAIDE 2004), pp. 29–35 (2004)
5. Charles, D., Kerr, A., McNeill, M.: Player-centred game design: Player modelling and adaptive digital games. In: Proceedings of the Digital Games Research Conference, vol. 285, pp. 285–298 (2005)
6. Charles, D., Kerr, A., McNeill, M., McAlister, M., Black, M., Kcklich, J., Moore, A., Stringer, K.: Player-centred game design: Player modelling and adaptive digital games. In: Proceedings of the Digital Games Research Conference, vol. 285, p. 00100 (2005)
7. Drachen, A., Thurau, C., Sifa, R., Bauckhage, C.: A comparison of methods for player clustering via behavioral telemetry. arXiv preprint arxiv: https://arxiv.org/abs/1407.3950 (2014)
8. Drachen, A., Sifa, R., Bauckhage, C., Thurau, C.: Guns, swords and data: clustering of player behavior in computer games in the wild. In: 2012 IEEE Conference on Computational Intelligence and Games (CIG 2012), pp. 163–170 (2012)
9. Fagan, M., Cunningham, P.: Case-based plan recognition in computer games. In: Ashley, K.D., Bridge, D.G. (eds.) ICCBR 2003. LNCS, vol. 2689, pp. 161–170. Springer, Heidelberg (2003). doi:10.1007/3-540-45006-8_15
10. Floyd, M.W., Esfandiari, B.: A case-based reasoning framework for developing agents using learning by observation. In: 2011 23rd IEEE International Conference on Tools with Artificial Intelligence (ICTAI), pp. 531–538 (2011)
11. Fu, T.C.: A review on time series data mining. Eng. Appl. Artif. Intell. **24**, 164–181 (2011)
12. Hunicke, R.: The case for dynamic difficulty adjustment in games. In: Proceedings of the 2005 ACM SIGCHI International Conference on Advances in Computer Entertainment Technology (ACE 2005), pp. 429–433 (2005)
13. Hunicke, R., Chapman, V.: AI for dynamic difficulty adjustment in games. In: Challenges in Game Artificial Intelligence AAAI, pp. 91–96 (2004)
14. Jennings-Teats, M., Smith, G., Wardrip-Fruin, N.: Polymorph: dynamic difficulty adjustment through level generation. In: Proceedings of the 2010 Workshop on Procedural Content Generation in Games, p. 11. ACM (2010)
15. Lora, D., Sánchez-Ruiz, A.A., González-Calero, P.A.: Difficulty adjustment in tetris with time series (2016)
16. Lora, D., Sánchez-Ruiz, A.A., González-Calero, P.A., Gómez-Martín, M.A.: Dynamic difficulty adjustment in Tetris (2016)
17. Menéndez, H.D., Vindel, R., Camacho, D.: Combining time series and clustering to extract gamer profile evolution. In: Hwang, D., Jung, J.J., Nguyen, N.-T. (eds.) ICCCI 2014. LNCS, vol. 8733, pp. 262–271. Springer, Cham (2014). doi:10.1007/978-3-319-11289-3_27

18. Missura, O., Gärtner, T.: Player modeling for intelligent difficulty adjustment. In: Gama, J., Costa, V.S., Jorge, A.M., Brazdil, P.B. (eds.) DS 2009. LNCS, vol. 5808, pp. 197–211. Springer, Heidelberg (2009). doi:10.1007/978-3-642-04747-3_17
19. Missura, O., Gärtner, T.: Predicting dynamic difficulty. In: Advances in Neural Information Processing Systems, pp. 2007–2015 (2011)
20. Nechvatal, J.: Immersive ideals/critical distances. LAP Lambert Acad. Pub. **2009**, 14 (2009)
21. Ram, A., Ontañón, S., Mehta, M., Ontanón, S., Mehta, M.: Artificial intelligence for adaptive computer games. In: Proceedings of the 8th International Conference on Intelligent Games and Simulation (GAMEON 2007), vol. 8, pp. 1–8 (2007)
22. Rani, S., Sikka, G.: Recent techniques of clustering of time series data: a survey. Int. J. Comput. Appl. **52**(15), 1–9 (2012)
23. Romdhane, H., Lamontagne, L.: Reinforcement of local pattern cases for playing Tetris. In: FLAIRS Conference (2008). http://www.aaai.org/Papers/FLAIRS/2008/FLAIRS08-066.pdf
24. Sharma, M., Mehta, M., Ontanón, S., Ram, A.: Player modeling evaluation for interactive fiction. In: Proceedings of the AIIDE 2007 Workshop on Optimizing Player Satisfaction, pp. 19–24 (2007)
25. Sharma, M., Ontañón, S., Mehta, M., Ram, A.: Drama management and player modeling for interactive fiction games. Comput. Intell. **26**(2), 183–211 (2010)
26. Sharma, M., Ontanón, S., Strong, C.R., Mehta, M., Ram, A.: Towards player preference modeling for drama management in interactive stories. In: FLARIS, pp. 571–576. Association for the Advancement of Artificial Intelligence (AAAI) (2007)
27. Sweetser, P., Wyeth, P.: Gameflow: a model for evaluating player enjoyment in games. Comput. Entertain. (CIE) **3**(3), 3 (2005)
28. Taylor, L.N.: Video games: perspective, point-of-view, and immersion. Ph.D. thesis, University of Florida (2002)
29. Liao, T.W.: Clustering of time series data - a survey. Pattern Recogn. **38**(11), 1857–1874 (2005)
30. Yannakakis, G.N., Hallam, J.: Real-time game adaptation for optimizing player satisfaction. IEEE Trans. Comput. Intell. AI Games **1**(2), 121–133 (2009)

Case-Based Reasoning for Inert Systems in Building Energy Management

Mirjam Minor(✉) and Lutz Marx

Business Information Systems, Goethe University,
Robert-Mayer-Str. 10, 60629 Frankfurt, Germany
minor@cs.uni-frankfurt.de, wirtschaftsinformatik@lmarx.de

Abstract. Energy management systems are a typical example for inert systems where an event or action causes an effect with a delay. Traditional solutions for energy management, such as PID controllers (PID = proportional-integral-derivative loops), control target values efficiently but are sub-optimal in terms of energy consumption. The paper presents a novel, case-based reasoning approach for inert energy management systems that aims to reduce energy wastage in over heating and over cooling for buildings. We develop a case representation based on time series data, taking environmental impact factors into consideration, such as weather forecast data. This includes a post-mortem assessment function that balances energy consumption with comfort for the users. We briefly discuss retrieval and reuse issues. We report on an experimental evaluation of the approach based on a building simulation, including 35 years of historical weather data.

Keywords: Case-based reasoning · Reasoning over time · Energy management

1 Introduction

A system can change its state by the influence of an impact factor. In physics or in biology, we necessarily have a time delay between impact and state change. *Inert Systems* are systems where an event or action takes effect with a delay and over a period of time. The system may either abruptly switch from one discrete state to another after a delay time or cumulate the impact factor and change the state continuously. In the following, we will focus on a case-based approach for the latter. An example of an inert system in nature is the human body where the injection of a drug changes the insulin level for a couple of hours. A technical sample is an energy management system (EMS) for buildings. The movement of a weather front has an impact on the room temperature with a time lag. We have chosen energy management for buildings as a sample application area for controlling inert systems.

Traditionally, Control System Engineering (CSE) [8] is used for this class of problems. It is well understood and widely used in EMSs and other domains.

D.W. Aha and J. Lieber (Eds.): ICCBR 2017, LNAI 10339, pp. 200–211, 2017.
DOI: 10.1007/978-3-319-61030-6_14

PID controllers (proportional-integral-derivative loops) and switching rules are the industry standard for the control of building EMSs [6]. This logic responds to setpoints and schedules for building components, such as heating circuits, radiators, or air handling units. That means that the temporal delays between causes and effects that are characteristic for inert systems are only considered in a reactive manner by the controller. The temporal dependencies are hidden in setpoints and schedules. For instance, the time for pre-heating to change a heating circuit into an 'enabled' mode is expressed by higher setpoints in the early morning schedule. The basic control logic largely ignores forward planning based on weather forecasts, expected occupancy, or renewable energy availability.

More advanced, model-based decision systems aim to optimize the system operation based on modeling, feedback, and forecasts [6]. They use an explicit time model and forward planning in order to optimize the energy consumption at a system level. However, solving an optimization problem has two challenges [6] in comparison to the basic control logic, such as PIDs: It requires analytical building models at design time and, second, it is computationally intensive, i.e. it requires powerful computational units. A novel EMS is desirable with a lower energy consumption than a PID controller but that is easy to deliver, and easy to operate.

As an alternative solution to costly optimization, *Case-based reasoning* (CBR) provides methods for experience reuse. In this paper, we propose a CBR approach for energy management in buildings where experience in operating the EMS is to be reused. The traditional PID controllers of the EMS are replaced by a case-based control unit for the energy supply. Since EMSs for buildings are inert systems, the cases need to be equipped with a concept of time. The core idea is to observe and record the inert behavior of a system by time series of impact factors and state variables. The context description for a case comprises further time series, such as measured values of the building, recent metereological data and weather forecasts. Corrective actions that have led to good results in similar situations in the past are reused to manipulate the system state in the next time step, i.e. to achieve a system state that is close to the setpoint values. For instance, if a setpoint in an EMS specifies a desired room temperature a corrective action is an amount of energy to be supplied to or dissipated from the room. Like a PID controller that provides a corrective action as an output of each control cycle in order to maintain a desired setpoint value, our CBR approach provides a corrective action as an output of each reasoning cycle.

In comparison to the optimization approaches, the case-based approach uses a shallower model. The analytical model is built on similarity functions for cases. We claim that the CBR approach outperforms the basic control approach for building EMSs in terms of energy consumption while providing the same comfort for the occupants.

The remainder of the paper is organized as follows. Related work is discussed in Sect. 2. The case representation is introduced in Sect. 3. A similarity function is presented in Sect. 4. An adaptation rule is specified in Sect. 5. Section 6 addresses the experimental setup while the results are reported in Sect. 7. Finally, a conclusion is drawn in Sect. 8.

2 Related Work

CBR has been used for energy prediction in the recent literature [9,11]. Like our approach, the work uses a notion of time series. The case-based energy prediciton approaches forecast the energy consumption based on energy values from cases with time series data that is similar to the recent situation. The prediction task is different but quite related to the control task that we address in our work. Like in prediciton, we reuse cases with similar time series. In contrast to reusing energy values, we reuse the control actions that have an impact to the inert system and observe the resulting energy consumption.

Temporal context plays a major role in many CBR approaches [5,10]. Recent work on CBR on time series data is reported in Gundersen's survey [2] as well as in the series of RATIC workshops [3,4]. This work was a major source of inspiration for our case representation and recent similarity function.

3 Case Representation for Inert Systems

The CBR approach aims at reusing experiences in improving the settings for the inert system. A case records the experience in corrective actions to keep a target value within assigned limits around a setpoint value, for instance the room temperature within a corridor of 19.5 to 20.5 °C. The case $Case = (P, S, A)$ comprises a problem description P, a solution description S and a quality assessment A of the proposed solution.

P – The problem description records the state of the system and its environment, including the recent settings.
S – The solution description addresses a revision of the settings by corrective actions.
A – The quality assessment contains the results of a post-mortem analysis of the suggested solution.

Time series for *setpoint values*, *measured process values*, and *disturbance values* are recorded for the problem description P. Setpoint values describe the desired state of the inert system. Measured process values are the actual values that might deviate from the setpoint values. Disturbance values are values that have an impact on the measured process variables in addition to the corrective actions. From the point of view of the controller system, they "disturb" the control processes. From the point of view of the users, they are key determinants on the inert system. The values for the time series are recorded at equidistant time points $t_{-m}, \ldots, t_{-2}, t_{-1}, t_0$ with t_0 denoting the current time point at reasoning time. The continuation of some of them might be predicted for the equidistant time points $t_1, t_2, \ldots, t_n$, estimated at time point t_0. An example for the latter is wheather forecasting data.

The solution S records *corrective actions* that are taken to keep the measured process values within a corridor of values around the setpoint values. For the sake of simplicity, we have chosen that the solution is a single corrective action for the setting of the inert system in the next time step initially. We assume that the time interval until the next time point is large enough to measure a first impact of the corrective action. Alternatively, the solution can be described as time series of corrective actions over a number of time steps. Even interleaved phases of retrieve and reuse are possible in principle. However, the latter would lead to concurrent processes which are difficult to handle.

Table 1 illustrates a sample case in an EMS. The distance between time points is one hour. Disturbance values are the solar radiation in minutes per hour (Sun) and the temperature outside the building ($T_{outside}$). The measured process values are the measured room temperature (T_{inside}). The setpoint values are the desired room temperature (T_{target}). The corrective action is the energy (E_{in}) supplied or dissipated via the EMS during the next time step t_1. The case has been recorded at time point t_0. The disturbance values have been measured until t_0. The values for t_1 until t_n are forecast data.

The assessment A is taken when the time frame is over. When time point t_n has passed, the updated values for $t_1, \ldots, t_n$ are used to assess the case. The predicted disturbance values have been replaced by the measured disturbance values. The time series for the measured process values and the corrective actions have been continued. The assessment considers the deviation between setpoint and measured process values by an error function e as well as the corrective actions by an energy consumption function u for $t_1, \ldots, t_n$. It is computed by the assessment function f for a case c by a weighted sum as follows:

$$f(c) = \sum_{i=1}^{n} w_1 \cdot e(i) + w_2 \cdot u(i)$$

The error function e measures the deviation of the actual room temperature T_{inside} from the setpoint value T_{target}:

$$e(i) = |T_{inside}(i) - T_{target}(i)|$$

The energy consumption function u measures the heating or cooling energy of the corrective action. Since the production of cooling energy consumes nearly twice the energy of heating [7] , we multiply cooling energy with the factor 2:

$$u(i) = \begin{cases} E_{in}(i), & E_{in}(i) \geq 0 \\ E_{in}(i) \cdot 2, & E_{in}(i) < 0 \end{cases}$$

The weights w_1 and w_2 specify the balance between reaching the target temperature and saving energy.

4 Case Retrieval

A time event triggers a reasoning cycle starting with the retrieve phase. We have chosen hourly time events. The query describes the current situation of the EMS,

Table 1. The problem description of a sample case.

Time point	Time stamp [yyyymmddhh]	Sun $[\frac{min}{h}]$	$T_{outside}$ [°C]	T_{inside} [°C]	T_{target} [°C]	E_{in} [Wh]
t_{-m-2}	1981083016	60	20.3	19.94	20	−3500
t_{-m-1}	1981083017	42	20.1	20.06	20	−750
t_{-m}	1981083018	0	18.8	21.43	20	0
...	1981083019	0	16.9	21.12	20	0
	1981083020	0	15.4	20.63	20	0
	1981083021	0	13.8	20.02	20	1000
	1981083022	0	13.2	20.07	20	1000
	1981083023	0	11.1	20.04	20	1250
	1981083100	0	10.4	19.96	20	1500
	1981083101	0	10.1	20.00	20	1500
	1981083102	0	9.1	20.00	20	1750
...	1981083103	0	8.0	20.08	20	1750
t_{-2}	1981083104	0	8.1	20.02	20	1750
t_{-1}	1981083105	0	8.3	19.98	20	0
t_0	1981083106	48	9.1	18.62	20	0
t_1	1981083107	60	11.5	–	20	?
...	1981083108	60	14.0	–	20	–
	1981083109	60	16.0	–	20	–
	1981083110	60	17.8	–	20	–
	1981083111	60	19.1	–	20	–
	1981083112	54	20.0	–	20	–
	1981083113	42	20.3	–	20	–
	1981083114	60	20.9	–	20	–
	1981083115	42	21.4	–	20	–
	1981083116	0	21.2	–	20	–
...	1981083117	0	21.0	–	20	–
t_n	1981083118	0	19.7	–	20	–
t_{n+1}	1981083119	0	18.2	–	20	–
t_{n+2}	1981083120	0	17.1	–	20	–

including the setpoint values, measured process values, and disturbance values. The case depicted in Table 1 can serve as a sample query. The retrieval uses a composite similarity measure for a query and a case that aggregates the local similarity measures by a function F:

$$\begin{aligned} sim = F(& sim_{Time\_stamp}, \\ & + sim_{Sun}, \\ & + sim_{T\_outside}, \\ & + sim_{T\_inside}, \\ & + sim_{T\_target}) \end{aligned}$$

F is a weighted sum. sim_{Time_stamp} considers the annual date *date* and the time of day *hour* when the query and the case were recorded each:

$$sim_{Time\_stamp}(query, case) = \frac{1}{1 + |date_{query}(t_0) - date_{case}(t_0)| \cdot |hour_{query}(t_0) - hour_{case}(t_0)|}$$

We assume the values of the time series Sun, $T_{outside}$, T_{inside}, and T_{target} as vectors. The local similarity measures for the time series are computed by means of the City Block Metric [1]. The size of the vectors Sun, $T_{outside}$, and T_{target} is $m+n+1$ and $m+1$ for T_{inside}, since T_{inside} data only exists for the past.

As a starting point, we have chosen straight forward similarity measures. We will investigate further, more sophisticated similarity measures for time series, such as dynamic time warping [10], as a part of our future work.

5 Case Reuse

The solution of the best matching case is reused for the current situation. The solution describes the corrective action for the settings of the system by the amount of energy $E_{in\_case}(t_1)$ to be infused into the building next. However, E_{in_case} has to be adapted to the recent situation. The impact of an energy infusion depends not only on the bare amount of energy supplied or distracted but also on the current room temperature and on the heat capacity of the building. The latter can be specified by a constant $c_{building}$.

The amount of energy is adapted as follows:

$$e_{ad} = E_{in\_case}(t_1) + c_{building} \cdot (T_{inside\_case}(t_0) - T_{inside\_query}(t_0))$$

The difference between the room temperature of the reference case and the current temperature in the building is multiplied with $c_{building}$. In case the current room temperature is lower, more energy is required for heating than in the case (or the energy that is required for cooling can be reduced, i.e. the negative value of E_{in_case} increases). In case the current room temperature is higher than in the case, E_{in_case} decreases analogously. The adaptation could lead to amounts of energy that are not available for heating and cooling in our building. Thus, we introduce the limits E_{min} for cooling and E_{max} for heating. The final amount of energy to be infused is determined by the following clipping function:

$$E_{ad}(t_1) = \begin{cases} E_{min}, & e_{ad} < E_{min} \\ e_{ad}, & E_{min} < e_{ad} < E_{max} \\ E_{max}, & else \end{cases}$$

In future work, the approach might be extended to reuse a sequence of corrective actions $E_{in\_case}(t_1) \ldots E_{in\_case}(t_k)$.

6 Experimental Setup

We have implemented the case-based approach for inert systems and conducted an experimental evaluation with an EMS scenario. The results of the CBR system have been compared to a traditional PID controller with respect to energy consumption and comfort for the occupants.

Ideally, the experiments would be executed in a real building measuring the energy consumption by sensors at the valves and measuring the comfort by acquiring feedback from the real occupants. Since these resources are difficult to obtain, the experiments have been conducted in silico. They involve an energy simulation of a building to approximate the impact of both, the energy infusion by the system as a corrective action and the two metereological parameters sun duration and outside temperature as disturbance variables. A seeding case base has been constructed from real weather data for the time period from 1981 to 2014. The experiments on the behavior of the CBR system were then simulated with the weather data for the year 2015.

In our example we use a grid of one hour for all of our calculations. On the one hand this reflects the inertness of a building. On the other hand this decision is taken to limit the computational complexity.

The energy model of the building assumes a single cubic room with an edge length of 10 m that has one side with glass windows. The relative position of sun is not taken into account. Basically, the temperature of the air in a building depends on the energy flow into and out of the building through walls and windows. The loss or gain of the energy ΔE through walls and windows can be calculated by using the thermal transmittance (also known as U-value or k-value):

$$\Delta E = U \cdot \Delta T \cdot A \cdot \Delta t$$

U characterizes the isolation value of the wall or the window. ΔT is the temperature difference between $T_{outside}$ and T_{inside}. A is the surface that divides inside and outside and Δt is the time span.

The dynamic simulation is done by the iteration:

$$E_{t+1} = E_t + \sum_{i=1}^{n} \Delta E_n$$

where ΔE_n represents the different sources and drains of energy. So far we use ΔE_1 for the energy flow through walls, ΔE_2 for the energy flow through windows, ΔE_3 for the energy of the sun through the windows and ΔE_4 for the energy of the heating and cooling system.

As the base for the dynamic simulation of the building we use historical weather data of Frankfurt a.M./Germany ranging back to 1981 in an one hour resolution containing air temperature, air speed and direction, humidity and minutes of sun per hour. It is important to note that the initial case base is built for that climate and applies only to regions within the same climate classification.

In a post-mortem analysis, we created a case for each hour of the historical weather data as t_0. The setpoint values are fixed to 20.0 °C. We calculated optimal energy values to be infused into the building for four hours, i.e. $t_n = t_4$ regarding the development of the weather and the (simulated) state of the building. In a brute-force approach, we explored the full solution space with energy amounts between $E_{min} = -4\,\text{kWh}$ and $E_{max} = 4\,\text{kWh}$ in a grid of granularity $g = 0.1\,\text{kWh}$. The assessment function f (compare Sect. 3) serves as fitness function to optimize the E_{in} values. For our example we weighted the energy consumption function with zero ($w_2 = 0$) to force the system to generate seeding cases that keep the given temperature as good as possible. The complexity of generating an optimum initial case follows $\mathcal{O}((\frac{E_{max}-E_{min}}{g})^n)$.

Instead of using the simple approach for creating a seeding case base as described above, a wide variety of modifications for a real building is possible and desirable. Changing demands on the target temperature regarding the comfort of the inhabitants of the building are one example. Another opportunity for an extension is to use sliding frames of acceptable min/max temperatures to preserve a maximum of energy.

We designed two variants of an experiment to evaluate our system. The first experiment tests the ability of our system to compete with a common PID controller if it has access to future weather data. The CBR system uses the future weather data to find the best matching case. The similarity of two cases is calculated with a time span of ±12 h where we assume the existence of a high-quality weather prediction for the next 12 h. The second experiment explores the behavior of our system if no future weather data is available. It acts on the same seeding case base as the first experiment, but the best matching case must be retrieved without knowledge of the future weather development. The similarity of two cases is based on the data for the previous 24 h.

We compared our results against a hand optimized PID controller. To keep the computational effort low, this PID controller works as all of our calculations in the one hour grid that we use for the CBR system. On the one hand this decision is debatable, since a real PID controller works in a grid of seconds or minutes and can thus adapt much faster to changes of the impact factors. On the other hand, the delay between demand and delivery of hot and cold water (or air) in a real building lies between 15 min an 30 min. Arguably, our one hour computation grid is not as precise as a real system but it acts similarly. The PID controller uses the difference between inside and outside temperature as the base for its calculations.

7 Experimental Results

For both experiments, we measured the comfort and the entire energy consumption for the year 2015, comparing the PID controller and CBR system. The comfort is measured by the Root Mean Square Error (RMSE) of the deviations from the desired room temperature of 20 °C within the simulated building. Second, we measured the entire energy consumption for the year 2015 at large.

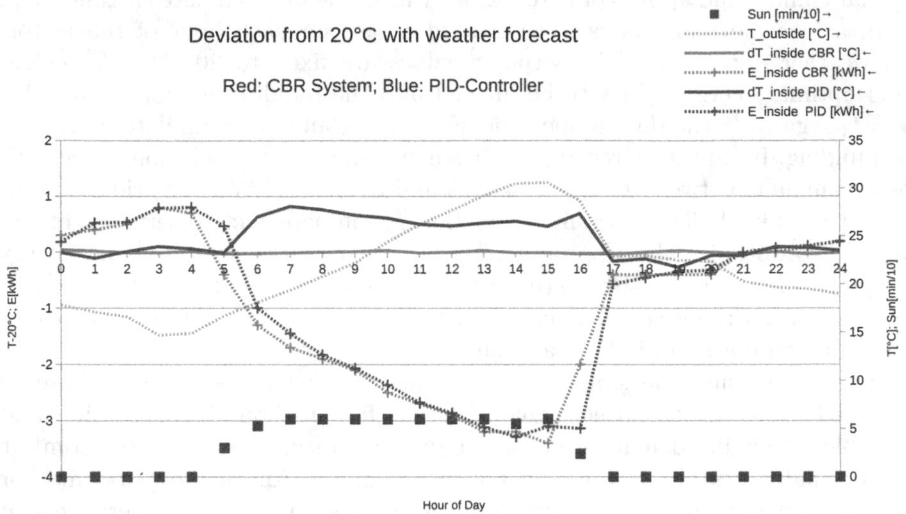

Fig. 1. Comparison of CBR (bright red) and PID (dark blue) with the objective to maintain exactly 20 °C. Future weather data is available. (Color figure online)

For our first experiment, that considers future weather data, an interesting example is the situation for the 12th of June 2015 as depicted in Fig. 1. The outside temperature increases until the early afternoon where the sun seems to be hidden by clouds. During the sunny period, the PID controller results in a room temperature that is slightly higher than 20.5 °C which is the upper boundary of the comfort corridor. The CBR system maintains the 20 °C nearly perfectly. The RMSE of the deviation for the whole year 2015 of the temperature inside the building is 0.035 °K for our CBR system and 0.32 °K for the PID controller. A more palpable metric is the added up deviation of the temperature: $\sum_{2015010100}^{2015123123} |20.0\,°C - T|$ which amounts to 178 °K for the CBR system and 1595 °K for the PID controller for the entire year.

The energy infusion values depicted in the sample in Fig. 1 seem very similar for both, the CBR system and the PID controller. The cumulative annual values confirm this observation. The CBR system used 13.4 MWh of energy for heating and cooling of the building. The PID used 13.7 MWh of energy. This surprising coincidence can be explained simply by the fact that the PID controller infuses too much energy (overprovisioning) about as frequently as too little energy (underprovisioning).

The second experiment without considering future weather data leads as expected to slightly worse (but still reasonable) results for the CBR system.

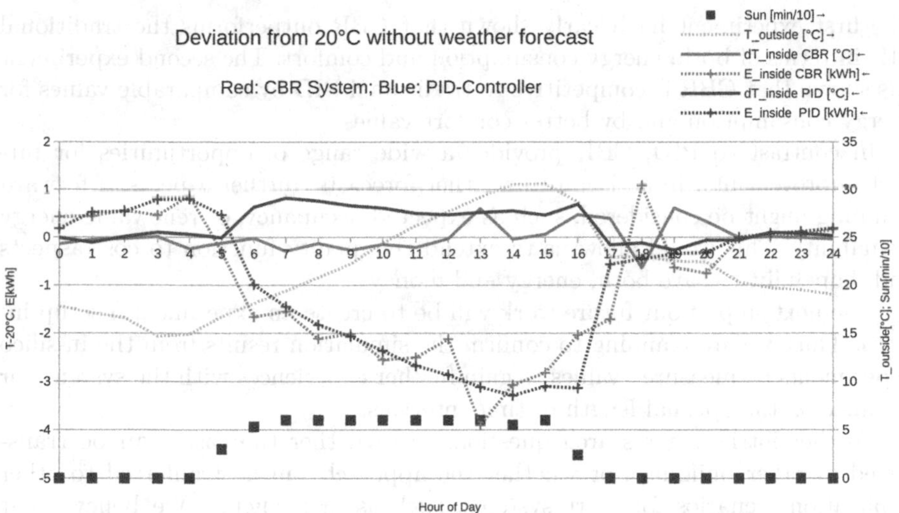

Fig. 2. Comparison of CBR (bright red) and PID (dark blue) with the objective to maintain exactly 20 °C. Future weather data is not available. (Color figure online)

Obviously, the values for the PID controller remain the same as in the first experiment. The results for the same sample day as above is depicted in Fig. 2. It can be seen that both curves for room temperature and energy infusion by the CBR system are less smooth than in the first experiment. The annual RMSE for the CBR system amounts to 0.26 °K. The added up deviation for the cbr system is 1505 °K. This value is still slightly better than the 1595 °K for the PID controller for the entire year. The CBR system used 13.8 MWh of energy for heating and cooling of the building. This is slightly worse than the 13.7 MWh of the PID.

8 Discussion and Conclusion

We have introduced a novel CBR approach for the experience-based control of inert systems and demonstrated the feasibility of the approach in the field of energy management.

We have presented a case representation with time series of impact factors and state descriptors, including setpoint values, measured process values, disturbance values and corrective actions. A straightforward similarity measure has been specified. An adaptation rule considering physical properties such as the heat capacity of a building has been proposed. Our experimental results for a simulated building under real weather conditions provide a proof of concept for using CBR for building EMSs. The experimental results are quite promising in comparison to a traditional PID controller. The first experiment compares the CBR system with a common PID controller if CBR has access to future weather data. The second experiment compares the systems if there is no such access.

The first experiment has clearly shown that CBR outperforms the traditional PID in terms of both, energy consumption and comfort. The second experiment has shown that CBR is competitive to traditional PID by comparable values for energy consumption and by better comfort values.

In contrast to PID, CBR provides a wide range of opportunities for further improvements. In addition to weather forecasts, further aspects of forward planning might be considered, such as expected occupancy, or renewable energy availability. This would allow us to extend the quality function to cost aspects and, hopefully, to save both, energy and money.

The next step of our future work will be to create an experimental setup in-vitro. Thus, we are planning to confirm the simulation results from the in-silico experiments by measured values to gain further experiences with the system, for instance on the optimal length of time intervals.

Further intriguing research questions are whether the cases can be transferred to other buildings, or whether the approach can be tranferred to other application scenarios for inert systems, such as in medicine. We believe that CBR is capable of providing significant benefits for the control of inert systems, especially in reducing modelling efforts and energy consumption.

Acknowledgements. The authors would like to thank Prof. Alfred Karbach and his team at the THM University of Applied Sciences for providing valuable insights to energy systems and many fruitful discussions on user comfort. Further, we thank Dipl.-Ing. Beate Massa whose expert advice on facility management issues we appreciate very much.

References

1. Bergmann, R.: Experience Management: Foundations, Development Methodology, and Internet-Based Applications. Springer, Heidelberg (2002)
2. Gundersen, O.E.: An analysis of long term dependence and case-based reasoning. In: Workshop Proceedings of ICCBR 2014, pp. 129–138 (2014)
3. Gundersen, O.E., Bach, K.: RATIC 2016: workshop on reasoning about time in CBR. In: Workshop Proceedings of ICCBR 2016 (2016)
4. Gundersen, O.E., Montani, S.: RATIC 2014: workshop on reasoning about time in CBR. In: Workshop Proceedings of ICCBR 2014, pp. 125–164 (2014)
5. Leake, D., Wilson, D.: Taking time to make time: types of temporality and directions for temporal context in CBR. In: Workshop Proceedings of ICCBR 2014, pp. 149–153 (2014)
6. Ma, Y.: Model predictive control for energy efficient buildings. Ph.D. thesis, UC Berkeley (2012)
7. Massa, B.: Personal interview on energy cost calculations in facility management, 25-04-2017 at Frankfurt a.M., Germany (2017)
8. Nise, N.S.: Control Systems Engineering, 6th edn. Wiley, Hoboken (2011)
9. Platon, R., Martel, J., Zoghlami, K.: CBR model for predicting a building's electricity use: on-line implementation in the absence of historical data. In: Hüllermeier, E., Minor, M. (eds.) ICCBR 2015. LNCS, vol. 9343, pp. 306–319. Springer, Cham (2015). doi:10.1007/978-3-319-24586-7_21

10. Richter, M.M., Weber, R.: Case-Based Reasoning: A Textbook. Springer, Heidelberg (2013)
11. Shabani, A., Paul, A., Platon, R., Hüllermeier, E.: Predicting the electricity consumption of buildings: an improved CBR approach. In: Goel, A., Díaz-Agudo, M.B., Roth-Berghofer, T. (eds.) ICCBR 2016. LNCS (LNAI), vol. 9969, pp. 356–369. Springer, Cham (2016). doi:10.1007/978-3-319-47096-2_24

Semantic Trace Comparison at Multiple Levels of Abstraction

Stefania Montani[1(✉)], Manuel Striani[2], Silvana Quaglini[3], Anna Cavallini[4], and Giorgio Leonardi[1]

[1] DISIT, Computer Science Institute, University of Piemonte Orientale, Alessandria, Italy
stefania.montani@uniupo.it
[2] Department of Computer Science, University of Torino, Turin, Italy
[3] Department of Electrical, Computer and Biomedical Engineering, University of Pavia, Pavia, Italy
[4] Istituto di Ricovero e Cura a Carattere Scientifico Fondazione "C. Mondino" - on Behalf of the Stroke Unit Network (SUN) Collaborating Centers, Pavia, Italy

Abstract. Event logs constitute a rich source of information for several process analysis activities, which can take advantage of similar traces retrieval. The capability of relating semantic structures such as taxonomies to actions in the traces can enable trace comparison to work at different levels of abstraction and, therefore, to mask irrelevant details, and make the identification of similar traces much more flexible. In this paper, we propose a trace abstraction mechanism, which maps actions in the log traces to instances of ground concepts in a taxonomy, and then allows to generalize them up to the desired level. We also show how we have extended a trace similarity metric we defined in our previous work, in order to allow abstracted trace comparison as well. Our framework has been tested in the field of stroke management, where it has allowed us to cluster similar traces, corresponding to correct medical behaviors, abstracting from details, but still preserving the capabilities of identifying outlying situations.

1 Introduction

Many commercial information systems routinely adopted by organizations and companies worldwide record information about the executed business process instances in the form of an *event log* [14]. The event log stores the sequences (*traces* henceforth [8]) of actions that have been executed at the organization, typically together with execution times.

Event logs constitute a very rich source of information for several process management activities. Indeed, the experiential knowledge embedded in traces is directly resorted to not only in process model discovery [9,14], but also in other tasks, which can rely on Case Based Reasoning [1], and specifically on the functionalities of the retrieval step. For instance, operational support [8] and agile workflow tools [25,29] can take advantage of trace comparison and similar

D.W. Aha and J. Lieber (Eds.): ICCBR 2017, LNAI 10339, pp. 212–226, 2017.
DOI: 10.1007/978-3-319-61030-6_15

trace retrieval. In fact, operational support assists users while process instances are being executed, by making predictions about the instance completion, or recommending suitable actions, resources or routing decisions: to this end, it can exploit similar, already completed instances, retrieved from the log. The agile workflow technology, on the other hand, deals with adaptation and overriding needs in response to expected situations as well as to unanticipated exceptions in the operating environment: in order to provide an effective and quick adaptation support, many agile workflow systems share the idea of recalling and reusing concrete examples of changes adopted in the past, recorded as traces in the event log.

In the currently available tools, however, these activities are supported through a purely syntactical analysis, where actions in the event log are compared and processed only referring to their names. On the other hand, the capability of relating semantic structures such as taxonomies to actions in the traces can enable trace comparison to work at different levels of abstraction (i.e., at the level of instances and/or concepts) and, therefore, to mask irrelevant details, to promote reuse, and, in general, to make trace analysis much more flexible, and closer to the real user needs.

In this paper, we propose a trace abstraction mechanism which maps actions in the log traces to instances of ground concepts (leaves) in a taxonomy, so that they can be converted into higher-level concepts by navigating the hierarchy, up to the desired level.

We also present how we have extended a similarity metric we defined in our previous work [19,20], in order to allow abstracted trace comparison as well.

Our framework has been tested in the field of stroke management, where we have adopted multi-level abstraction and trace comparison to cluster event logs of different stroke units, in order to highlight correct and incorrect behaviors, abstracting from details (such as local resource constraints or local protocols). In our study, the application of the abstraction mechanism allowed us to obtain more homogeneous and compact clusters (i.e., able to aggregate closer examples); however, outliers were still clearly identifiable, and isolated in the cluster hierarchy.

The paper is organized as follows: in Sect. 2 we present the multi-level semantic trace abstraction procedure; in Sect. 3, we discuss how we have extended similarity calculation to deal with abstracted traces; in Sect. 4 we present experimental results; in Sect. 5 we discuss related works, while Sect. 6 is devoted to conclusions and future work.

2 Trace Abstraction

In our framework, trace abstraction has been realized as a two-step procedure.

As a **first step**, every action in the trace to be abstracted is mapped to a ground term of a **taxonomy**.

We have currently applied the framework in the stroke management domain. An excerpt of our stroke management taxonomy is reported in Fig. 1.

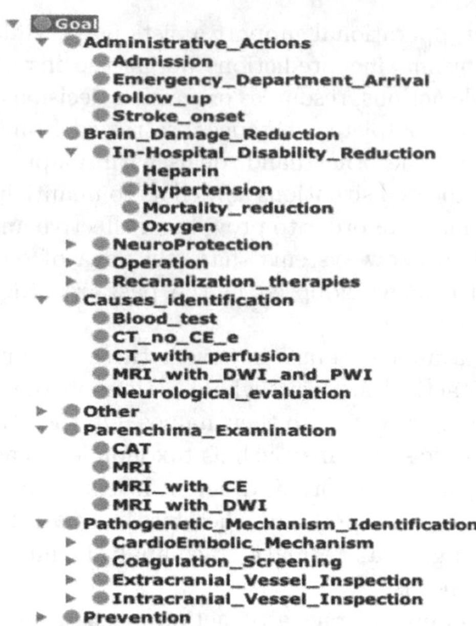

Fig. 1. An excerpt from the stroke domain taxonomy

The taxonomy, which has been formalized by using the Protègè editor, has been organized by goals. Indeed, a set of classes, representing the main goals in stroke management, have been identified, namely: "Administrative Actions", "Brain Damage Reduction","Causes Identification", "Parenchima Examination", "Pathogenetic Mechanism Identification", "Prevention", and "Other" (i.e., none of the previous ones). These main goals can be further specialized into subclasses, according to more specific goals (e.g., "CardioEmbolic Mechanism" is a subgoal of "Pathogenetic Mechanism Identification"), down to the ground actions, that will implement the goal itself (e.g., Computer Aided Tomography (CAT) implements "Parenchima Examination"). Overall, our taxonomy is composed by 136 classes, organized in a hierarchy of four levels.

As a **second step**, actions in the trace are abstracted according to the goals they implement. The level in the taxonomy to be chosen for abstraction (e.g., a very general goal, such as "Prevention", or a more specific one, such as "Early Relapse Prevention"), has to be specified as an input by the user.

When a set of consecutive actions in the trace abstract as the same goal in the taxonomy (at the specified level), our approach merges them into the same abstracted **macro-action**, labeled as the common goal at hand. This procedure requires a proper treatment of *delays* (i.e., periods of time where no action takes place), and of actions in-between that implement a different goal (*interleaved actions* henceforth).

Specifically, the procedure to abstract a trace operates as follows:

- for every action i in the trace:
  - i is abstracted as the goal it implements (i.e., as its ancestor at the taxonomy level selected by the user); the macro-action m_i, labeled as the identified ancestor, is created;
  - for every element j following i in the trace:
    - if j is a delay, its length is added to a variable $tot - delay$, that stores the total delay duration accumulated so far during the creation of m_i;
    - if j is an interleaved action, its length is added to a variable $tot - inter$, that stores the total interleaved actions durations accumulated so far during the creation of m_i;
    - if j is an action that, according to domain knowledge, abstracts as the same ancestor as i, m_i is extended to include j, provided that $tot - delay$ and $tot - inter$ do not exceed domain-defined thresholds. These threshold values are set to limit the total admissible delay time within a macro-action and the total duration of interleaved actions, since it would be hard to justify that two ground actions share the same goal (and can thus be abstracted to the same macro-action), if they are separated by very long delays, or if they are interleaved by many/long different ground actions, meant to fulfill different goals; j is then removed from the actions in the trace that could start a new macro-action, since it has already been incorporated into an existing one;
  - the macro-action m_i is appended to the output abstracted trace which, in the end, will contain the list of all the macro-actions that have been created by the procedure.

The variables $tot - delay$ and $tot - inter$, accumulated during abstraction, are also provided as an output attribute of each macro-action. As discussed in Sect. 3, they will be used as a penalty in abstracted trace similarity calculation.

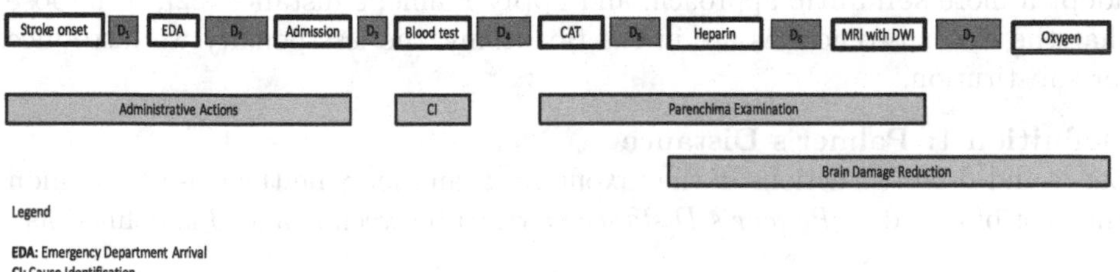

Fig. 2. Abstraction example

Figure 2 shows a trace abstraction example. The three ground actions "Stroke Onset", "EDA" and "Admission" are abstracted to the macro-action "Administrative Actions", when abstracting up to level 1 in the taxonomy of Fig. 1 (where level 0 is the root). While creating this macro-action, the lengths of the two delays D_1 and D_2 are accumulated in $tot - delay$. If $D_1 + D_2$ had exceed the delay threshold, "Admission" would not have added to the macro-action started by "Stroke Onset". On the other hand, "CAT" and "MRI with DWI" are abstracted to the macro-action "Parenchima Examination", and the length of the interleaving action "Heparin" is accumulated in $tot - inter$. Similarly, "Heparin" and "Oxygen" are abstracted to the macro-action "Brain Damage Reduction", and the length of the interleaving action "MRI with DWI" is accumulated in $tot - inter$.

3 Semantic Trace Comparison

In our framework, we have extended a metric we described in [19,20], which worked on ground traces, in order to permit the comparison of abstracted traces as well.

In the current, more general approach, every trace is a sequence of actions (whether ground actions or abstracted macro-actions), each one stored with its execution starting and ending times. Therefore, an action is basically a symbol (plus the temporal information). Starting and ending times allow to get information about action durations, as well as qualitative (e.g., Allen's *before*, *overlaps*, *equals* etc. [2]) and quantitative temporal constraints (e.g., delay length, overlap length [17]) between pairs of consecutive actions/macro-actions.

The main features of the metric published in [19,20] are summarized below. The extensions needed to deal with abstracted traces are also discussed later in this section.

In the metric in [19,20], we first take into account action types, by calculating a modified edit distance which we have called **Trace Edit Distance** [19,20]. As the classical edit distance [18], Trace Edit Distance tests all possible combinations of editing operations that could transform one trace into the other one. However, the cost of a *substitution* is not always set to 1. Indeed, as already observed, we have organized actions in a taxonomy: we can therefore adopt a more **semantic** approach, and apply Palmer's distance [22], to impose that the closer two actions are in the taxonomy, the less penalty we introduce for substitution.

Definition 1: Palmer's Distance
Let α and β be two actions in the taxonomy t, and let γ be the closest common ancestor of α and β. *Palmer's Distance* $dt(\alpha, \beta)$ between α and β is defined as:

$$dt(\alpha, \beta) = \frac{N_1 + N_2}{N_1 + N_2 + 2 * N_3}$$

where N_1 is the number of arcs in the path from α and γ in t, N_2 is the number of arcs in the path from β and γ, and N_3 is the number of arcs in the path from the taxonomy root and γ.

Trace Edit Distance $trace_{NGLD}(P, Q)$ is then calculated as the Normalized Generalized Levenshtein Distance (NGLD) [31] between two traces P and Q (interpreted as two strings of symbols). Formally, we provide the following definitions:

Definition 2: Trace Generalized Levenshtein Distance
Let P and Q be two traces of actions, and let α and β be two actions. The Trace Generalized Levenshtein Distance $trace_{GLD}(P, Q)$ between P and Q is defined as:

$$trace_{GLD}(P, Q) = min\{\sum_{i=1}^{k} c(e_i)\}$$

where $(e_1, \ldots, e_k)$ transforms P into Q, and:

- $c(e_i) = 1$, if e_i is an action insertion or deletion;
- $c(e_i) = dt(\alpha, \beta)$, if e_i is the substitution of α (appearing in P) with β (appearing in Q), with $dt(\alpha, \beta)$ defined as in Definition 1 above.

Definition 3: Trace Edit Distance (Trace Normalized Generalized Levenshtein Distance)
Let P and Q be two traces of actions, and let $trace_{GLD}(P, Q)$ be defined as in Definition 2 above. We define Trace Edit Distance $trace_{NGLD}(P, Q)$ between P and Q as:

$$trace_{NGLD}(P, Q) = \frac{2 * trace_{GLD}(P, Q)}{|P| + |Q| + trace_{GLD}(P, Q)}$$

where $|P|$ and $|Q|$ are the lengths (i.e., the number of actions) of P and Q respectively.

As already observed, the minimization of the sum of the editing costs allows to find the optimal alignment between the two traces being compared.

Given the optimal alignment, we can then take into account temporal information. In particular, we compare the durations of aligned actions by means of a metric we called **Interval Distance** [19,20]. Interval distance calculates the normalized difference between the length of two intervals (representing action durations in this case).

Moreover, we take into account the temporal constraints between two pairs of subsequent aligned actions on the traces being compared (e.g., actions A and B in trace P; the aligned actions A' and B' in trace Q). We quantify the distance between their qualitative constraints (e.g., A and B overlap in trace P; A' meets B' in trace Q), by resorting to a metric known as **Neighbors-graph Distance** [19,20]. If Neighbors-graph Distance is 0, because the two pairs of actions share the same qualitative constraint (e.g., A and B overlap in trace P; A' and B' also overlap in trace Q), we compare quantitative constraints by properly applying Interval Distance again (e.g., by calculating Interval Distance between the two overlap lengths).

In the metric in [19,20], these three contributions (i.e., Trace Edit Distance, Interval Distance between durations, Neighbors-graph Distance or Interval Distance between pairs of actions) are finally put in a linear combination with non-negative weights.

When working on macro-actions, the three contributions still apply (in particular, Palmer's distance allows to compare macro-actions as well, as they appear in the taxonomy). However, the metric in [19,20] needs to be extended, by considering, given the optimal macro-actions alignment, two additional contributions:

- a penalty due to the different length of the delays incorporated into the two aligned macro-actions;
- a penalty due to the different length of interleaved actions in the two aligned macro-actions being compared.

Delay penalty is defined as follows:

Definition 4: Delay Penalty
Let A and B be two macro-actions, that have been matched in the optimal alignment. Let $delay_A = \sum_{i=1}^{k} length(i)$ be the sum of the lengths of all the k delays that have been incorporated into A in the abstraction phase (and let $delay_B$ be analogously defined). Let $maxdelay$ be the maximum, over all the abstracted traces, of the sum of the lengths of the delays incorporated in an abstracted trace. The Delay Penalty $delay_p(A, B)$ between A and B is defined as:

$$delay_p(A, B) = \frac{|delay_A - delay_B|}{maxdelay}$$

As for interleaved actions penalty, we operate analogously to delay penalty, by summing up the lengths of all interleaved actions that have been incorporated within a single macro-action in the abstraction phase.

Definition 5: Interleaving Length Penalty
Let A and B be two macro-actions, that have been matched in the optimal alignment. Let $inter_A = \sum_{i=1}^{k} length(i)$ be the sum of the lengths of all the k interleaved actions that have been incorporated into A in the abstraction phase (and let $inter_B$ be analogously defined). Let $maxinter$ be the maximum, over all the abstracted traces, of the sum of the lengths of the interleaved actions incorporated in an abstracted trace. The Interleaving Length Penalty $interL_p(A, B)$ between A and B is defined as:

$$interL_p(A, B) = \frac{|inter_A - inter_B|}{maxinter}$$

The extended metric working on abstracted traces includes in the linear combination these two penalties as well.

4 Experimental Results

In this paper, we have tested the impact of our abstraction mechanism on the calculation of trace similarity by means of some clustering experiments.

The available event log was composed of more than 15000 traces, collected at the 40 Stroke Unit Network (SUN) collaborating centers of the Lombardia region, Italy. Traces were composed of 13 actions on average.

In our study, we considered the traces of every single stroke unit (SU) separately, and compared clustering results on ground traces with respect to those on abstracted traces. In particular, abstraction was conducted at different levels: at level 1 (*abs*1 henceforth - when actions were abstracted up to the most general medical goals, since the root represents level 0 in the taxonomy of Fig. 1); and at level 2 (*abs*2 henceforth - when actions were abstracted to less general subgoals).

For the sake of brevity, in the following we will show the results on a specific SU, but the other outcomes were very similar.

The metric defined in Sect. 3 has been adopted in the experiments, setting all the linear combination weights to the same value: one would expect the Trace Edit Distance to be the the main component of the similarity measure, but in our specific application domain temporal information is critical; moreover, we wanted to focus on abstraction penalties as well. Abstraction thresholds were common to all traces in the log, and set on the basis of medical knowledge. Interestingly, we also made tests with different thresholds (making changes of up to 10%), but results (not reported due to lack of space) did not differ significantly.

Specifically, we resorted to a hierarchical clustering technique, known as Unweighted Pair Group Method with Arithmetic Mean (UPGMA) [27]. UPGMA operates in a bottom-up fashion. At each step, the nearest two clusters are combined into a higher-level cluster. The distance between any two clusters A and B is taken to be the average of all distances between pairs of objects "x" in A and "y" in B, that is, the mean distance between elements of each cluster. After the creation of a new cluster, UPGMA properly updates a pairwise distance matrix it maintains. UPGMA also allows to build the phylogenetic tree of the obtained clusters.

In the experiments, the hypothesis we wished to test was the following: the application of the abstraction mechanism allows to obtain more *homogeneous* and compact clusters (i.e., able to aggregate closer examples); however, outliers are still clearly identifiable, and isolated in the cluster hierarchy.

Homogeneity is a widely used measure of the quality of the output of a clustering method (see e.g., [10,11,26,30]). A classical definition of cluster homogeneity is the following [30]:

$$H(C) = \frac{\sum_{x,y \in C}(1 - dist(x,y))}{\binom{|C|}{2}}$$

where $|C|$ is the number of elements in cluster C, and $1 - dist(x,y)$ is the similarity between any two elements x and y in C. Note that, in the case of singleton (one-trace) clusters, homogeneity is set to 1 (see e.g., [11]).

The higher the homogeneity value, the better the quality of clustering results.

The average of the homogeneity H of the individual clusters can be calculated on (some of) the clusters obtained through the method at hand, in order to assess clustering quality. We computed the average of cluster homogeneity values level by level in the hierarchies/trees. In order to avoid biases, to calculate homogeneity we resorted to the classical normalized Levensthein's edit distance [18], with no use of semantic, temporal or abstraction information (indeed, if homogeneity is calculated resorting to the metric defined in this paper, it obviously increases when working on abstracted traces, since Palmer's distance [22] decreases when operating at higher levels of the hierarchy; see the results in [21]).

We report on the results of applying UPGMA to the 200 traces of an example SU. Figure 3 shows the top levels of the cluster tree obtained when working on ground traces. Analogously, Figs. 4 and 5 report the top levels of the cluster trees obtained when working on traces at *abs*1 and *abs*2 (respectively).

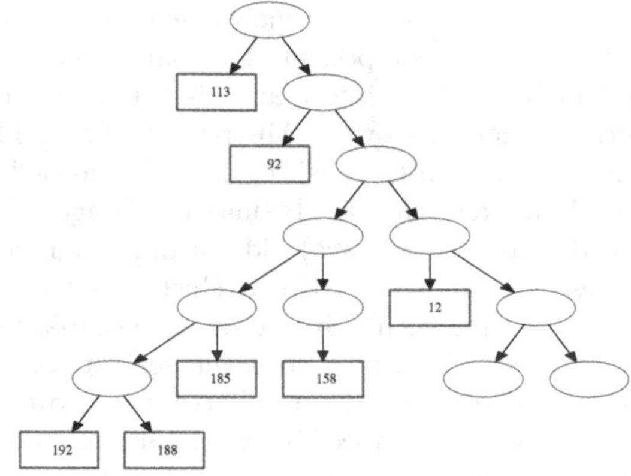

Fig. 3. Cluster tree obtained on ground traces, on the example SU

Figure 6 shows a comparison of the average homogeneity values, computed by level in the cluster trees, on ground traces vs. traces at *abs*2 and *abs*1. As it can be observed, homogeneity on abstracted traces was always higher than the one calculated on ground traces. Moreover, homogeneity at *abs*1 was never lower than the one calculated at *abs*2.

It is also interesting to study the management of outliers, i.e., in our application domain, traces that correspond to the treatment of atypical patients. This traces record rather uncommon actions, and/or present uncommon temporal constraints among their actions. For instance, trace 113 is very peculiar: it describes the management of a patient suffering from several inter-current complications (diabetes, hypertension), who required many extra-tests and many specialist counseling sessions, interleaved to more standard actions.

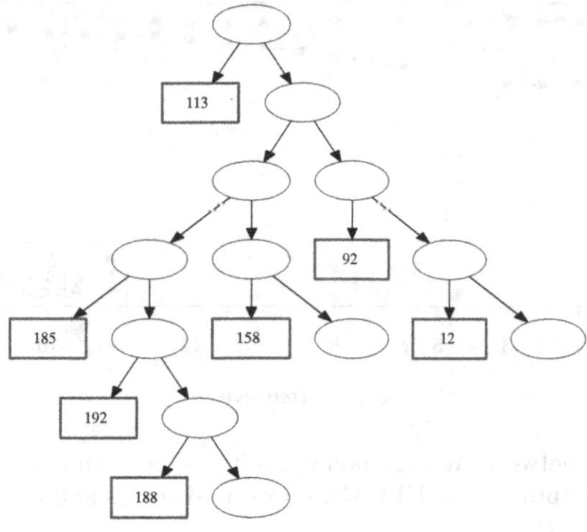

Fig. 4. Cluster tree obtained on traces at *abs*1, on the example SU

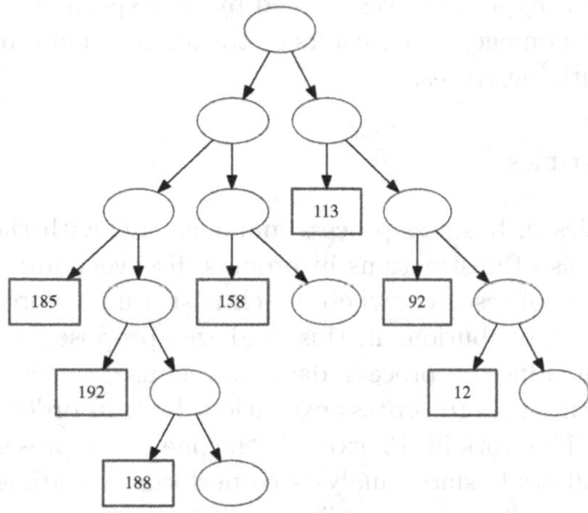

Fig. 5. Cluster tree obtained on traces at *abs*2, on the example SU

Ideally, these anomalous traces should remain isolated as a singleton cluster for many UPGMA iterations, and be merged to other nodes in the hierarchy as late as possible, i.e., close to the root (level 0).

When working on ground traces, 7 outliers of the example SU were merged very late to the hierarchy, as expected (between level 1 and level 6, see Fig. 3). In particular, trace 113 was merged only at level 1. Very interestingly, this capability of "isolating" outliers was preserved when working on abstracted traces, both at *abs*1 and at *abs*2. Indeed, as it can be seen in Figs. 4 and 5, despite some differences, all 7 outliers were early isolated, specifically between level 1 and level 7 in the trees. Trace 113 was the latest trace to be merged to the cluster trees in both abstraction levels.

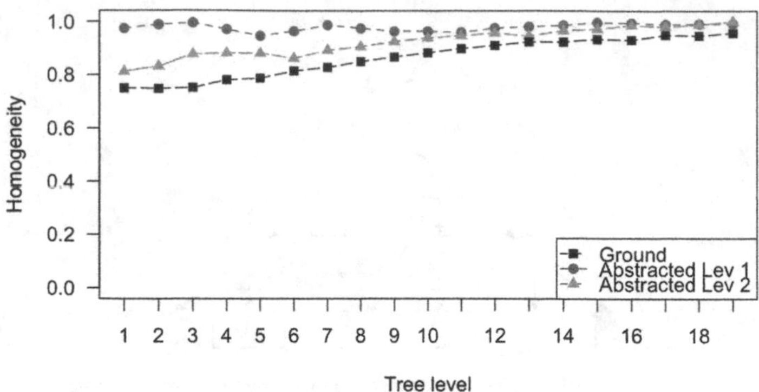

Fig. 6. Comparison between average homogeneity values, computed level by level in the two cluster trees obtained by UPGMA on ground traces and on traces at *abs*2 and *abs*1 on the example SU

In conclusion, our hypothesis was verified by the experiments, since we obtained more and more homogeneous clusters as the abstraction level increased, still clearly isolating outlying traces.

5 Related Works

The use of semantics in business process management, with the aim of operating at different levels of abstractions in process discovery and/or analysis, is a relatively young area of research, where much is still unexplored.

One of the first contributions in this field was proposed in 2002 by Casati et al. [5], who introduce a process data warehouse, where taxonomies are exploited to add semantics to process execution data, in order to provide more intelligent reports. The work in [13] extends the one in [5], presenting a complete architecture that allows business analysts to perform multidimensional analysis and classify process instances, according to flat taxonomies (i.e., taxonomies without subsumption relations between concepts).

Semantic Business Process Management is further developed in the SUPER project [24], within which several ontologies are created, such as the process mining ontology and the event ontology [23]; these ontologies define core terminologies of business process management, usable by machines for task automation. However, the authors do not present any concrete implementations of semantic process mining or analysis. Ontologies, references from elements in logs to concepts in ontologies, and ontology reasoners (able to derive, e.g., concept equivalence), are described as the three essential building blocks for semantic process mining and analysis in [7]. This paper also shows how to use these building blocks to extend ProM's LTL Checker [28] to perform semantic auditing of logs. A more recent work [15] introduces a methodology that combines domain and company-specific ontologies and databases to obtain multiple levels of abstraction for process mining and analysis. Similarly to our approach, in this paper data in databases become instances of concepts at the bottom level of the taxonomy tree structure. If consecutive tasks in the discovered model abstract as the same concepts, those tasks can be aggregated.

Most of the papers cited above, however, present theoretical frameworks, and not yet a detailed technical architecture nor a concrete implementation of all their ideas. Moreover, they do not focus on trace comparison.

As regards trace comparison, on the other hand, a few metrics have been proposed in the literature. In particular, [16] combines a contribution related to action similarity, and a contribution related to delays between actions. As regards the temporal component, it relies on an interval distance definition which is quite similar to ours. Differently from what we do, however, the work in [16] always starts the comparison from the last two action in the traces: no search for the optimal action alignment is performed. Moreover, it stops the calculation if the distance between two actions/intervals exceeds a given threshold, while we always calculate the overall distance. The distance function in [16] does not exploit action duration, and does not rely on semantic information about actions, as we do. Finally, it does not deal with different types of qualitative temporal constraints, since it cannot manage (partially) overlapping actions.

Another interesting contribution is [6], which addresses the problem of defining a similarity measure able to treat temporal information, and is specifically designed for clinical workflow traces. Interestingly, the authors consider qualitative temporal constraints between matched pairs of actions, resorting to the A-neighbors graph proposed by Freska [12], as we do. However, in [6] the alignment problem is strongly simplified, as they only match actions with the same name. In this sense, our approach is also much more semantically oriented.

The issue of trace abstraction is not considered in these works.

In conclusion, to the best of our knowledge, our metric represents one of the most complete contributions to properly account for both non temporal and temporal information, and to perform a semantic comparison between ground as well as abstracted trace actions.

6 Conclusions

In this paper we have introduced a framework for abstracting process traces at different levels of detail, on the basis of trace action semantics. Specifically, we have presented an application of the framework to the domain of stroke patient management, where trace actions can be abstracted on the basis of their medical goals/sub-goals, at the generalization level indicated by the user.

Once the traces have been abstracted, they can be compared, by resorting to a metric we have properly extended, to take into account penalties collected during the abstraction phase.

Our experiments have shown that cluster homogeneity, when operating on abstracted traces, reaches higher values; at the same time, outliers (i.e., anomalies and incorrect behaviors) are still clearly visible in abstracted traces as well. Abstracted trace comparison can be indeed adopted to cluster the existing event log, thus allowing the user to concentrate on a subset of traces, which are particularly similar, and to reduce the computational time of subsequent analysis steps, that can work on a smaller dataset. In fact, clustering can be employed as a pre-processing step for other process management tasks, including process mining [3]. Moreover, trace comparison can be adopted to retrieve similar traces in operational support, along the lines we described in [4]. Given the more robust results obtained when comparing abstracted traces with respect to ground ones, we believe that operational support would benefit of the abstraction technique. In the future, we plan to test this statement by adopting our framework in operational support as well as in different process management tasks, with particular attention to medical applications.

From a methodological viewpoint, we plan to extend our approach in different directions. First, we will consider different knowledge structures, such as ontologies, or multiple taxonomies, able to provide abstraction information from different viewpoints (e.g., not only the viewpoint of activity goals - as it happens in the single taxonomy we are currently adopting - but also the one of roles and responsibilities of the involved actors, when available). Second, we will introduce a rule-based approach to initiate abstraction. Proper rules, having as an antecedent the execution of some action registered earlier in the log, will fire, to initiate the abstraction step. This will allow us to control the abstraction process on the basis of the context, i.e., of the already executed actions. Temporal constraints (e.g., the delay since the completion of the already executed action) will also be taken into account in these rules. Finally, we also plan to introduce different definitions for the penalties to be considered when comparing abstracted traces. In particular, it would be useful to consider not only the length of the interleaved actions, but also their type.

We believe that such improvements will make our framework even more flexible and useful in practice.

References

1. Aamodt, A., Plaza, E.: Case-based reasoning: foundational issues, methodological variations and systems approaches. AI Commun. **7**, 39–59 (1994)
2. Allen, J.F.: Towards a general theory of action and time. Artif. Intell. **23**, 123–154 (1984)
3. Bose, R.P.J.C., Van der Aalst, W.: Context aware trace clustering: Towards improving process mining results. In: Proceedings of the SIAM International Conference on Data Mining, pp. 401–412. Springer (2009)
4. Bottrighi, A., Canensi, L., Leonardi, G., Montani, S., Terenziani, P.: Trace retrieval for business process operational support. Expert Syst. Appl. **55**, 212–221 (2016)
5. Casati, F., Shan, M.-C.: Semantic analysis of business process executions. In: Jensen, C.S., Šaltenis, S., Jeffery, K.G., Pokorny, J., Bertino, E., Böhn, K., Jarke, M. (eds.) EDBT 2002. LNCS, vol. 2287, pp. 287–296. Springer, Heidelberg (2002). doi:10.1007/3-540-45876-X_19
6. Combi, C., Gozzi, M., Oliboni, B., Juarez, J.M., Marin, R.: Temporal similarity measures for querying clinical workflows. Artif. Intell. Med. **46**, 37–54 (2009)

7. de Medeiros, A.K.A., van der Aalst, W.M.P., Pedrinaci, C.: Semantic process mining tools: Core building blocks. In: Golden, W., Acton, T., Conboy, K., van der Heijden, H., Tuunainen, V.K. (eds.) 2008 16th European Conference on Information Systems, ECIS 2008, Galway, Ireland, pp. 1953–1964 (2008)
8. Van der Aalst, W.: Process Mining: Discovery, Conformance and Enhancement of Business Processes. Springer, Heidelberg (2011)
9. Van der Aalst, W., van Dongen, B., Herbst, J., Maruster, L., Schimm, G., Weijters, A.: Workflow mining: A survey of issues and approaches. Data Knowl. Eng. **47**, 237–267 (2003)
10. Duda, R.O., Hart, P.E., Stork, D.G.: Pattern Classication. Wiley-Interscience, New York (2001)
11. Francis, P., Leon, D., Minch, M., Podgurski, A.: Tree-based methods for classifying software failures. In: International Symposium on Software Reliability Engineering, pp. 451–462. IEEE Computer Society (2004)
12. Freska, C.: Temporal reasoning based on semi-intervals. Artif. Intell. **54**, 199–227 (1992)
13. Grigori, D., Casati, F., Castellanos, M., Dayal, U., Sayal, M., Shan, M.C.: Business process intelligence. Comput. Ind. **53**(3), 321–343 (2004)
14. IEEE Taskforce on Process Mining: Process Mining Manifesto. http://www.win.tue.nl/ieeetfpm. Accessed 4 Nov 2013
15. Jareevongpiboon, W., Janecek, P.: Ontological approach to enhance results of business process mining and analysis. Bus. Proc. Manag. J. **19**(3), 459–476 (2013)
16. Kapetanakis, S., Petridis, M., Knight, B., Ma, J., Bacon, L.: A case based reasoning approach for the monitoring of business workflows. In: Bichindaritz, I., Montani, S. (eds.) ICCBR 2010. LNCS (LNAI), vol. 6176, pp. 390–405. Springer, Heidelberg (2010). doi:10.1007/978-3-642-14274-1_29
17. Lanz, A., Weber, B., Reichert, M.: Workflow time patterns for process-aware information systems. In: Bider, I., Halpin, T., Krogstie, J., Nurcan, S., Proper, E., Schmidt, R., Ukor, R. (eds.) BPMDS/EMMSAD -2010. LNBIP, vol. 50, pp. 94–107. Springer, Heidelberg (2010). doi:10.1007/978-3-642-13051-9_9
18. Levenshtein, A.: Binary codes capable of correcting deletions, insertions and reversals. Sov. Phys. Dokl. **10**, 707–710 (1966)
19. Montani, S., Leonardi, G.: Retrieval and clustering for business process monitoring: results and improvements. In: Agudo, B.D., Watson, I. (eds.) ICCBR 2012. LNCS (LNAI), vol. 7466, pp. 269–283. Springer, Heidelberg (2012). doi:10.1007/978-3-642-32986-9_21
20. Montani, S., Leonardi, G.: Retrieval and clustering for supporting business process adjustment and analysis. Inf. Syst. **40**, 128–141 (2014)
21. Montani, S., Leonardi, G., Striani, M., Quaglini, S., Cavallini, A.: Multi-level abstraction for trace comparison and process discovery. Expert Syst. Appl. **81**, 398–409 (2017). doi:10.1016/j.eswa.2017.03.063
22. Palmer, M., Wu, Z.: Verb semantics for English-Chinese translation. Mach. Transl. **10**, 59–92 (1995)
23. Pedrinaci, C., Domingue, J.: Towards an ontology for process monitoring and mining. In: Hepp, M., Hinkelmann, K., Karagiannis, D., Klein, R., Stojanovic, N. (eds.) Proceedings of the Workshop on Semantic Business Process and Product Lifecycle Management SBPM 2007, held in Conjunction with the 3rd European Semantic Web Conference (ESWC 2007), Innsbruck, Austria, June 7, 2007, vol. 251 of CEUR Workshop Proceedings (2007)

24. Pedrinaci, C., Domingue, J., Brelage, C., van Lessen, T., Karastoyanova, D., Leymann, F.: Semantic business process management: Scaling up the management of business processes. In: Proceedings of the 2th IEEE International Conference on Semantic Computing (ICSC 2008), August 4–7, 2008, Santa Clara, California, USA, pp. 546–553. IEEE Computer Society (2008)
25. Reichert, M., Dadam, P.: Adeptflex-supporting dynamic changes of workflows without losing control. J. Intell. Inf. Syst. **10**, 93–129 (1998)
26. Sharan, R., Shamir, R.: CLICK: A clustering algorithm for gene expression analysis. In: Proceedings of the International Conference on Intelligent Systems for Molecular Biology, pp. 260–268 (2000)
27. Sokal, R., Michener, C.: A statistical method for evaluating systematic relationships. Univ. Kansas Sci. Bull. **38**, 1409–1438 (1958)
28. Aalst, W.M.P., Beer, H.T., Dongen, B.F.: Process mining and verification of properties: An approach based on temporal logic. In: Meersman, R., Tari, Z. (eds.) OTM 2005. LNCS, vol. 3760, pp. 130–147. Springer, Heidelberg (2005). doi:10.1007/11575771_11
29. Weber, B., Wild, W.: Towards the agile management of business processes. In: Althoff, K.-D., Dengel, A., Bergmann, R., Nick, M., Roth-Berghofer, T. (eds.) WM 2005. LNCS (LNAI), vol. 3782, pp. 409–419. Springer, Heidelberg (2005). doi:10.1007/11590019_48
30. Yip, A.M., Chan, T.F., Mathew, T.P.: A Scale Dependent Model for Clustering by Optimization of Homogeneity and Separation, CAM Technical Report 03–37. Department of Mathematics, University of California, Los Angeles (2003)
31. Yujian, L., Bo, L.: A normalized levenshtein distance metric. IEEE Trans. Pattern Anal. Mach. Intell. **29**, 1085–1091 (2007)

On the Pros and Cons of Explanation-Based Ranking

Khalil Muhammad(✉), Aonghus Lawlor, and Barry Smyth

Insight Centre for Data Analytics, University College Dublin, Belfield, Dublin 4, Ireland
{khalil.muhammad,aonghus.lawlor,barry.smyth}@insight-centre.org

Abstract. In our increasingly algorithmic world, it is becoming more important, even compulsory, to support automated decisions with authentic and meaningful explanations. We extend recent work on the use of explanations by recommender systems. We review how compelling explanations can be created from the opinions mined from user-generated reviews by identifying the *pros* and *cons* of items and how these explanations can be used for recommendation ranking. The main contribution of this work is to look at the relative importance of *pros* and *cons* during the ranking process. In particular, we find that the relative importance of *pros* and *cons* changes from domain to domain. In some domains *pros* dominate, in other domains, *cons* play a more important role. And in yet other domains there is a more equitable relationship between *pros* and *cons*. We demonstrate our findings on 3 large-scale, real-world datasets and describe how to take advantage of these relative differences between *pros* and *cons* for improved recommendation performance.

1 Introduction

As we come to rely more and more on algorithms and AI in our everyday lives, it is becoming increasingly important for such systems to explain the reasoning behind their decisions, advice, and recommendations. On the one hand explanations can help make suggestions more persuasive, but on the other hand, they can play an important role in conveying the reasons underpinning these suggestions, and thus aid system transparency. Either way, explanations help to improve user acceptance and serve to build trust between a system and its users. Not surprisingly then, there has been considerable recent research on the topic of generating explanations to accompany reasoning outcomes as an important part of building trust with end-users. Indeed there is good reason to believe that such explanations will become compulsory, as institutions such as the EU roll out regulations that contain a *right to explanation* for its citizens [1]. While such regulations will introduce significant challenges for an increasingly data-driven, algorithmic world, they also motivate novel and interesting research questions and introduce new opportunities for AI research; see for example [2–5].

Of course the important role that explanations can play in reasoning systems is not new for the case-based reasoning community; see [6–17], for example.

D.W. Aha and J. Lieber (Eds.): ICCBR 2017, LNAI 10339, pp. 227–241, 2017.
DOI: 10.1007/978-3-319-61030-6_16

More recently, the field of recommender systems has attracted considerable interest in the use of explanations to support recommendations, primarily as a way to justify suggestions to users [18–22].

In this paper, we build on recent work at the intersection between recommender systems, case-based reasoning and opinion mining by harnessing user-generated reviews as a novel way to generate rich product cases and user profiles to use in recommendation [23–25]. In particular we extend the work of [26–28], which describes a way to generate explanations from such product cases by highlighting the *pros* and *cons* of a recommended product-case, as well as how it relates to other recommendations and shows how to use these explanations during recommendation ranking; thus, tightly coupling the recommendation and explanation tasks. This provides a starting point for the present work where we turn our attention to the positive and negative features (*pros* and *cons*) of explanations with a view to understanding how these features interact during recommendation and explanation. The main contribution of this work is to demonstrate how the relative importance of *pros* and *cons* can vary across different recommendation domains and how this impacts recommendation performance. We further demonstrate how to adapt explanation-based ranking in order to factor such *pro/con* differences during the ranking process and so improve recommendation performance. This work benefits the CBR community in that it provides an approach for mining product cases from reviews; and these cases can then be used to produce explanations that can justify and rank recommendations.

In the next section, we will briefly review related work, focusing on the recent interest in so-called *Explainable AI* and the origins of explanations in CBR and recommender systems. Next, we will briefly review the *opinionated recommendation and explanation* approach which has been detailed in [27,29] previously; this will provide a starting point for this research. We will describe a new approach to measuring the strength of explanations, for the purpose of recommendation ranking, in which the relative importance of *pros* and *cons* can be varied. We provide a detailed evaluation using 3 large-scale, real-world datasets to demonstrate the benefits of explanation-based ranking across multiple domains, compared to a more conventional similarity-based, case-based recommendation approach. Moreover, we analyse how the importance of *pros* and *cons* varies across domains and how this influences recommendation performance.

2 Related Work

The need for, and recent interest in, *Explainable AI* has gathered pace quickly as organisations introduce initiatives to encourage more research into the generation of explainable models, which are capable of preserving predictive performance while helping end-users to understand, trust, and appreciate the reasoning behind predictions and suggestions; see for example, DARPA *Explainable AI (XAI)* programme[1]. One would be forgiven for believing that this is a new

[1] http://www.darpa.mil/program/explainable-artificial-intelligence.

interest for the wider AI community but in fact AI research has a long history in explanations and explanation-based reasoning and learning [30–36]. For instance Explanation-based Learning (EBL) systems explicity seek to *explain* what they learn from training examples for the purpose of identifying relevant training aspects during learning. Explanations are translated into a form that a problem-solving program can understand and generalised for use in other problem settings. For example, the PRODIGY system used problem-solving traces to construct explanations in order to learn control rules to help guide problem-solving search; see [37].

Likewise early work in the case-based reasoning community has long been influenced by explanations [6–17]. For example, the work of [7] describes SWALE, a project to study creative explanation of anomalous events. SWALE is a story understanding program that detects anomalous events and uses CBR to explain these anomalies by storing and reusing prior explanation cases or *explanation patterns*. Elsewhere, the work of [8] proposed an approach to *explanation-driven case-based reasoning* to produce context dependent explanations to support the main tasks within the standard CBR model; in essence explanations played different roles during retrieval, reuse, and learning.

In the field of recommender systems, work on explanations has mainly focused on different styles of explanation interfaces – how explanations should be presented [18,19] and how they are perceived by users [20] – with a view to improving transparency, persuasiveness, and trust. More recently, [21] presented a framework that uses information from the Linked Open Data (LOD) cloud to generate personalised natural language explanations. Likewise, [22] also demonstrated a process that combines crowd-sourcing and computation to explain movie recommendations using natural language explanations. While these content-based approaches use semi-structured information, it can more challenging to explain recommendations from Matrix Factorisation algorithms, which are commonplace. The latent factors created during the recommendation process tend to be less interpretable by the user. For example, the work of [38] addresses this by describing ways to explain latent factors with topics that have been extracted from content descriptions.

More relevant to this work is research by [27,28] which has argued for a more intimate connection between recommendation and explanation. Specifically, it is proposed that explanations can and should play a role during the recommendation process itself, only then can we truly say that explanations are 'authentic' because then they are driving the recommendation and ranking process. In this paper, we extend these *explanation-based recommendation* ideas by describing a number of different variations of explanation-based recommendation and comparing their relative performance in 3 different recommendation domains.

3 Opinionated Recommendation

The starting point for this work is a recommender system that harnesses user-generated reviews as its core recommendation knowledge, based on the work of

[24,25]. In summary, item descriptions and user profiles are generated by mining features and sentiment from user reviews as summarised in the following sections.

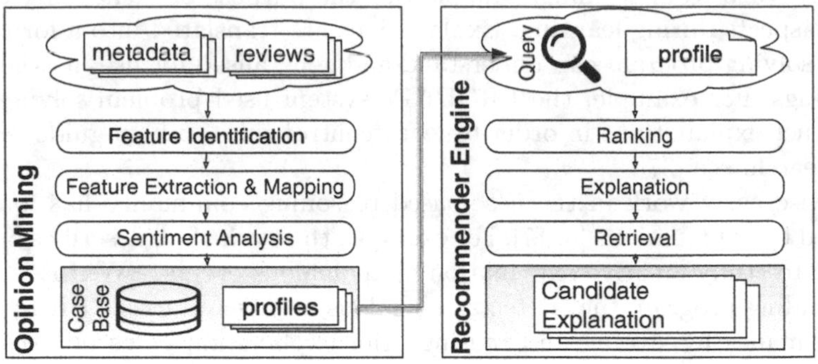

Fig. 1. An architecture of the opinion mining and explanation-based ranking system.

3.1 Generating Item Cases and User Profiles

An item or product (i_i) is associated with reviews $reviews(i_i) = \{r_1, \ldots, r_n\}$ and the opinion mining process extracts a set of features, $F = \{f_1, \ldots, f_m\}$, from these reviews. It does this by looking for frequent patterns of sentiment-rich words and phrases such as *"a beautiful view"* or *"terrible service"* using techniques described by [23–25].

Briefly, each *item-case* is composed of the features extracted from the item's reviews; see (Eq. 1). And each feature, f_j (e.g. "location" or "customer service") is associated with an *importance* score and a *sentiment* score, as per Eqs. 2 and 3. Briefly, importance score of f_j, $imp(f_j, i_i)$, is the relative number of times that f_j is mentioned in the reviews of item i_i. The sentiment score of f_j, $s(f_j, i_i)$, is the degree to which f_j is mentioned positively or negatively in $reviews(i_i)$. Note, $pos(f_j, i_i)$ and $neg(f_j, i_i)$ denote the number of mentions of f_j labelled as positive or negative during the sentiment analysis phase.

$$item(i_i) = \{(f_j, s(f_j, i_i), imp(f_j, i_i)) : f_j \in reviews(i_i)\} \tag{1}$$

$$imp(f_j, i_i) = \frac{count(f_j, i_i)}{\sum_{f' \in reviews(i_i)} count(f', i_i)} \tag{2}$$

$$s(f_j, i_i) = \frac{pos(f_j, i_i)}{pos(f_j, i_i) + neg(f_j, i_i)} \tag{3}$$

For a target user u_T, a user profile can be generated in a similar manner, by extracting features and importance information from u_T's reviews (reviews the user has authored or perhaps rated positively) as in Eq. 4.

$$user(u_T) = \{(f_j, imp(f_j, u_T)) : f_j \in reviews(u_T)\} \tag{4}$$

4 From Opinions to Explanations

Our aim is to describe an approach for generating explanations for each item, i_i, in a set of recommendations $I = \{i_1...i_k\}$ generated for some user u_T, as summarised in see Fig. 1.

4.1 Generating Explanations

Each item/product selected for recommendation is associated with an explanation which is composed of a set of features, their sentiment scores, and information about how these features relate to other items selected for recommendation. The starting point for this is a *basic explanation structure*, and in Fig. 2 we show an example of this structure for a BeerAdvocate beer. The explanation features have been divided into *pros* and *cons* based on their average sentiment scores. In this example, the *pros* reflect aspects of the beer (appearance, bottle, palate) that users have generally liked, and as such are reasons to choose the beer. On the other hand, *cons* may be reasons to reject the beer, because they are things people have reviewed poorly, such as the aroma, price, and taste in this instance.

	Feature	Importance	Sentiment	better/worse
PROS	Appearance	0.11	0.70	89%
	Bottle	0.10	0.73	78%
	Palate	0.05	0.74	78%
CONS	Aroma	0.18	0.62	56%
	Price	0.01	0.50	22%
	Taste	0.54	0.70	22%
	:	:	:	:

Fig. 2. An example of an explanation structure for a BeerAdvocate beer showing *pros* and *cons* that matter to the user along with associated sentiment, and *better/worse* than scores.

The so-called *better* and *worse* scores in Fig. 2 reflect how the sentiment of these *pros* and *cons* compare to the same features in other items that have been selected for recommendation; see Eqs. 5 and 6. For instance, the appearance of this beer has been more positively reviewed than 89% of the other recommendations but the aroma has been reviewed more negatively than 56% other recommendations.

$$better(f_j, i_i, I') = \frac{\sum_{h_a \in I'} 1[s(f_j, i_i) > s(f_j, h_a)]}{|I'|} \tag{5}$$

$$worse(f_j, i_i, I') = \frac{\sum_{h_a \in I'} 1[s(f_j, i_i) \leq s(f_j, h_a)]}{|I'|} \tag{6}$$

Generally speaking, not every *pro* or *con* in an explanation will make for a *compelling* reason to choose or reject the item in question and, in practice, explanations will be filtered to exclude *pros*/*cons* with better/worse scores that are less than 50%. In other words, the remaining *pros*/*cons* will all be better/worse than a majority of recommendations.

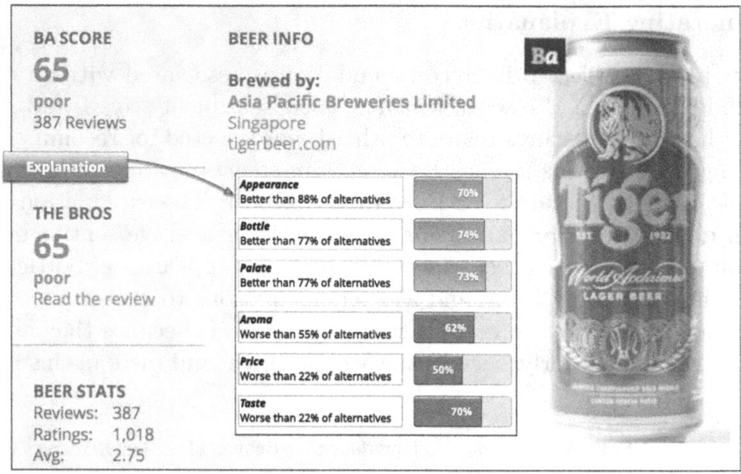

Fig. 3. An example explanation for a BeerAdvocate beer showing *pros* and *cons* that matter to the target user along with sentiment indicators (horizontal bars) and information about how this item fares with respect to alternatives.

The resulting explanation structures are then used as the basis for the explanation that is presented to the user. An example, based on the explanation structure shown in Fig. 2, is depicted in Fig. 3 within the standard BeerAdvocate interface.

4.2 Explanation-Based Ranking

The earlier work of [27] proposed a way to estimate the *strength* of an explanation based on the difference between the better scores of *pros* and the worse scores of *cons*; see Eq. 7. They went on to show that this strength metric could then be used to rank recommendations so that items associated with stronger or more compelling explanations would be ranked ahead of items that have weaker, less compelling explanations, in which *cons* dominate *pros*.

$$Strength(u_T, i_i, I') = \sum_{f \in Pros(u_T, i_i, I')} better(f, i_i, I') - \sum_{f \in Cons(u_T, i_i, I')} worse(f, i_i, I') \quad (7)$$

A key benefit of this approach is that it ties together the recommendation and explanation process. It means that the explanations shown to the user provide a more genuine account of why a particular product was suggested because the strength of the explanation was responsible for the ranking. This is in contrast to more conventional approaches to recommendation and explanation in which the recommendation and explanation process are largely independent, with explanations providing a justification for a recommendation that may have little to do with how and why it was actually suggested.

An important limitation of this approach is that the strength metric we are using gives equal weight to *pros* and *cons*. The central question in this work is whether this is appropriate, or whether *pros* and *cons* should be weighted differently to improve recommendation performance. Furthermore, might such weighting decisions depend on the properties of a particular recommendation domain? To test this, we evaluated a simple weighted form of this strength metric as shown in Eq. 8. Now the relative importance of *pros* and *cons* in the calculation of explanation strength can be adjusted by changing α. When $\alpha = 0.5$ we have the *uniform* strength metric used by [27] but for values of α greater than 0.5 or less than 0.5 we can emphasise *pros* or *cons*, respectively. By changing α in this way we can test its effect on recommendation performance to evaluate the relative importance of *pros* and *cons* in different domains.

$$Strength(u_T, i_i, I') = \sum_{f \in Pros(u_T, i_i, I')} \alpha better(f, i_i, I') - \sum_{f \in Cons(u_T, i_i, I')} (1 - \alpha) worse(f, i_i, I') \quad (8)$$

5 Evaluation

The primary objective of this evaluation is to evaluate: (a) the performance benefits of explanation-based ranking across different domains; (b) whether the relative importance of *pros* and *cons* matters in explanation-based ranking and if it does, whether it varies across domains; and (c) if the relative importance of *pros* and *cons* varies from domain to domain then why?

5.1 Datasets

In this evaluation we use three real-world datasets: (1) beer reviews from BeerAdvocate (BA) [39]; (2) restaurant reviews from Yelp (YP); and (3) hotel reviews TripAdvisor (TA). The TA dataset was collected during the period June to August 2013, whereas the YP dataset was provided as the basis for the Yelp dataset challenge[2], and the BA dataset was provided by BeerAdvocate on request.

Summary details of these datasets are presented in Table 1. All 3 domains involve thousands of items and users and tens of thousands of reviews. TA has many reviews per item but fewer reviews per user, whereas BA has lots of reviews per user but fewer reviews per item, and YP has few reviews per user or item.

[2] www.yelp.com/dataset_challenge.

TA reviews and items tend to be longer and feature rich – the average TA item contains 19 features – whereas the BA and YP items have only 6 and 8 features per item, respectively.

Table 1. Summary descriptions of the three datasets used in this evaluation.

				Average reviews		
Dataset	Reviews	Items	Users	Items	Users	# Features
BeerAdvocate (BA)	131,418	17,856	5,710	8 ± 21.6	23 ± 83.6	6
TripAdvisor (TA)	43,528	1,982	10,000	22 ± 24.1	4 ± 1.93	19
Yelp (YP)	23,109	8,048	10,000	3 ± 3.9	2.3 ± 4.1	8

5.2 Methodology

We use a standard leave-one-out evaluation using user profiles as a source of user-item test-pairs, (u_T, i_R), for all reviewed items in u_T's profile. For each (u_T, i_R) pair, our datasets also contain the 10 best recommendations $(i_1, ..., i_{10})$ suggested alongside i_R by TA, BA, or YP. These items serve as recommendations with a given i_R as a test query. We will use explanation-based ranking to re-rank these recommendations based on the strength of their corresponding explanations, varying α to test the relative importance of *pros* and *cons* in each domain. Each of these recommended items has an overall rating score, which is currently used by TA, BA, and YP for ranking and we will use these ratings-based rankings as our *ideal, ground-truth* for the purpose of this evaluation.

Our main objective is to compare the quality of the default, unweighted explanation-based ranking ($\alpha = 0.5$), to the rankings produced for different values of α in order to emphasise *pros* or *cons* as discussed. In addition, we will implement a further baseline in order to evaluate the performance of explanation-based ranking more generally across the 3 test domains; this will be the first time that explanation-based ranking has been evaluated outside of the TA domain. This additional baseline uses cosine similarity between the features in the query item and those in a recommendation candidate with the importance scores of features (Eq. 2) as feature values.

5.3 Explanation-Based vs. Similarity-Based Ranking

To evaluate ranking quality we compare the rankings produced for each test query, i_R, using explanation-based and similarity-based approaches, to the ideal rating-based rankings used by TA, BA, and YP. We use the normalised, discounted cumulative gain $nDCG$, for the test rankings at each rank position, k, as our evaluation metric and the results are presented in Fig. 4(a–c).

In each of the test domains, we can see that explanation-based ranking outperforms similarity-based ranking for all values of k. For TA and YP domains, this performance improvement is especially significant. For example, for Yelp at

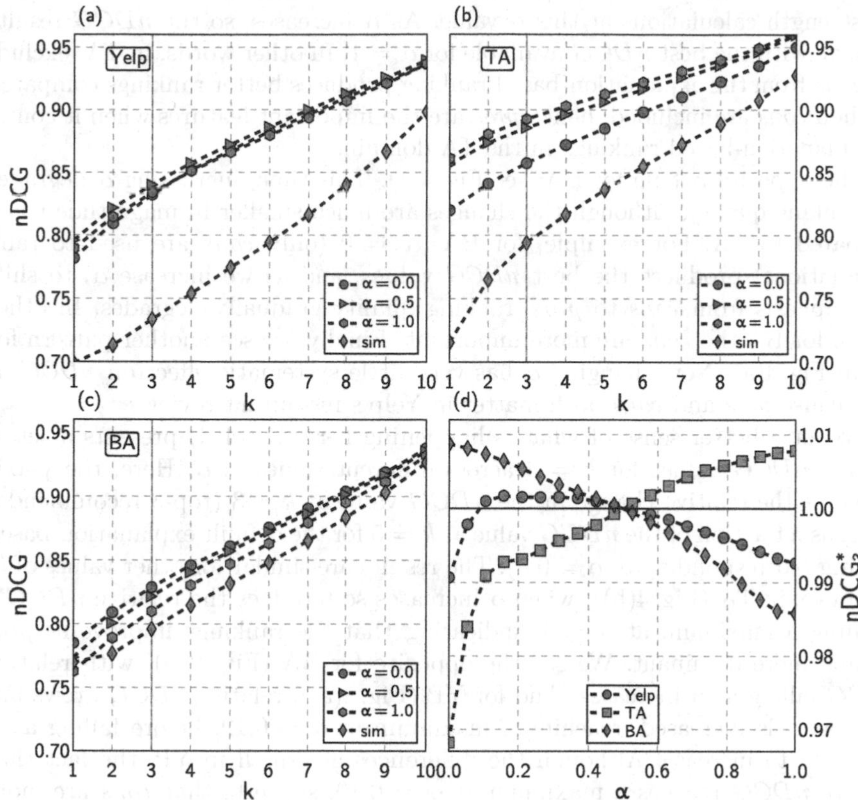

Fig. 4. The nDCG results at different rank positions (k) for explanation-based and similarity-based ranking for YP, TA and BA (a–c). In (d) the relative $nDCG^{*}$ for YP, TA and BA at different levels of α. The relative $nDCG_3^{*}$ measures the relative change in $nDCG$ for a given α compared to the $nDCG$ for $alpha = 0.5$, at $k = 3$.

$k = 3$ the $nDCG$ of the default explanation-based ranking ($\alpha = 0.5$) is approximately 0.84, compared to 0.73 for similarity-based ranking. In other words, the $nDCG$ for explanation-based ranking at $k = 3$ is over 15% better than a similarity-based ranking; the corresponding improvement for TA is over 11% at $k = 3$, but only about 3% for BA. Thus, explanation-based ranking produces rankings that are closer to the ideal, ratings-based ranking.

5.4 On the Importance of *pros* and *cons*

In Fig. 4(a–c) we can also see the effect of changing α; for reasons of clarity we only show results for $\alpha = 0$, 0.5, and 1. Interestingly, each test domain responds differently to changes in α. For example, the largest response is evident in TA (Fig. 4(b)) in which the lowest $nDCG$ values are obtained for $\alpha = 0$, where *pros* have no impact on explanation ranking since only the *cons* participate in the strength calculations at this α value. As α increases, so the $nDCG$ results improve with the best $nDCG$ available for $\alpha = 1$. In other words, in TA, excluding *cons* from the explanation-based ranking produces better rankings compared to when *cons* are included; hence *pros* are the important features when it comes to explanation-based rankings in the TA domain.

The opposite is true for BA; see Fig. 4(c). This time, increasing α degrades the ranking quality, although the changes are much smaller in magnitude when compared to TA. For example, for BA, $\alpha = 0$ (only *cons* are used to rank explanations) produces the best $nDCG$ values, and as we increase α, to shift the emphasis from *cons* to *pros*, ranking quality gradually degrades. In other words, for BA the *cons* are more important. Finally, we see another pattern for YP in Fig. 4(a). Now changing α has very little systematic effect on $nDCG$ at all because *pros* and *cons* both matter in Yelp's restaurant reviews.

To get a better sense of what is happening here, Fig. 4(d) presents a set of *relative* $nDCG$ values for $k = 3$ across different values of α. Here, the y-axis measures the relative change in the $nDCG$ value at $k = 3$ (top-3 recommendations) as a fraction of the $nDCG$ value at $k = 5$ for the default explanation-based ranking (corresponding to $\alpha = 0.5$). The results are similar for other values of k. As above, in TA (Fig. 4(b)), when α increases so too does the relative $nDCG^*$, reaching a maximum at $\alpha = 1$, indicating that the rankings improve as *pros* become more dominant. We see the opposite for BA (Fig. 4(c)), with relative $nDCG$ falling as α increases. And for YP (Fig. 4(a)), relative $nDCG$ grows initially as α is increased, reaching a maximum at $\alpha = 0.19$, before falling as α continues to increase. Although the differences are small in YP, the fact that relative $nDCG$ reaches a maximum at $\alpha = 0.19$, suggests that *cons* are more important than *pros* in YP.

5.5 On the Structure of Explanations

So far we have shown how explanation-based rankings are closer to the ideal rankings than the similarity-based rankings in all 3 test domains. Further, adjusting the relative importance of *pros* and *cons* in these explanation rankings can have an additional effect on overall ranking quality, but the nature and scale of this effect is domain dependent; some domains benefit from a greater emphasis on *pros*, others on *cons*, and yet others benefit from a more equitable weighting between *pros* and *cons*. What is it about a domain that makes *pros* or *cons* more or less important?

To begin to shed light on an answer to this question we will look at the distribution of *pros* and *cons*, and their better and worse scores, across recommendation sessions. For example, Fig. 5(a–c) shows the average number of *pros* and *cons* (and their average better/worse scores) for the explanations that are associated with recommendations at each rank position for the 3 test domains. As we should expect, in each domain, recommendations at the top of the rankings tend to be associated with explanations that are composed of more *pros* than *cons*, whereas the lower ranked recommendations come with explanations that are more dominated by *cons*.

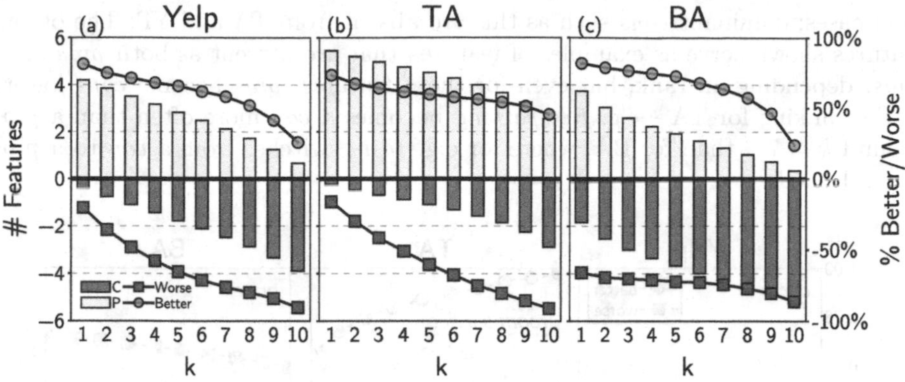

Fig. 5. The number of *pros* and *cons* and average *better/worse* scores for the explanations associated with recommendations at different rank positions, for YP, TA, and BA.

Now we can start to get a sense of some of the differences between the 3 domains in terms of the relative importance of these *pros* and *cons*. For instance, TA stands out because of the number of *pros* and their relative strength (*better* scores) across k; see Fig. 5(a). Even at the bottom of the ranking, TA explanations continue to have *pros* present in their explanations, indeed they have as many, if not more, *pros* than *cons*. And while the *better* scores of these *pros* does decrease as k increases, even at the bottom of the ranking the *better* scores remain relatively high (≈50%). Thus, in TripAdvisor item-cases tend to have many more *pros* than *cons*, but also these cases tend to appear in recommendation sets where *pros* tend to be better reviewed than most of the other recommendations in the set (they have high *better* scores). And while *cons* feature more frequently in lower-ranked items they never really dominate over *pros*. In this case, *pros* provide a stronger ranking signal than *cons*.

Contrast this with the nature of explanations in the BA rankings; see Fig. 5(c). Now *cons* are seen to play a more dominant role. They are frequent across the ranks k, even appearing in explanations at the top of the rankings. Indeed, *cons* start to dominate as early in the ranking as $k = 3$, and their *worse* scores are extremely high (>80%) throughout the ranking range. *Pros* quickly become far less frequent in the BA rankings, and their *better* scores reduce steadily for increasing k. Thus in BeerAdvocate, *cons* play a much more significant role. Even high-ranking recommendations come with the 'health warning' of many strong *cons*, and as a result, *cons* provide a stronger ranking signal than *pros*.

We can get another sense of this from Fig. 6, which shows the *better/worse* scores for two typical features from each domain, based on rank position. For instance the *room service* feature in TA is almost always positively reviewed, regardless of rank, and rarely negative reviewed and represents an extreme example of a dominant *pro*. Features like this simply don't exist in BA or YP. In contrast, in BA *price* is a feature that is usually a *con*, even in top-ranking item cases; dominant *cons* such as this are absent from TA and YP. The other features shown serve as examples of features that are present as both *pros* and *cons*, depending on rank, but even for these features, *pros* become *cons* lower in the ranking for TA – e.g. *free parking* becomes a *con* more often than a *pro* around $k = 7$ – than for BA, where for e.g. *taste* switches from a *con* to a *pro* around $k = 5$.

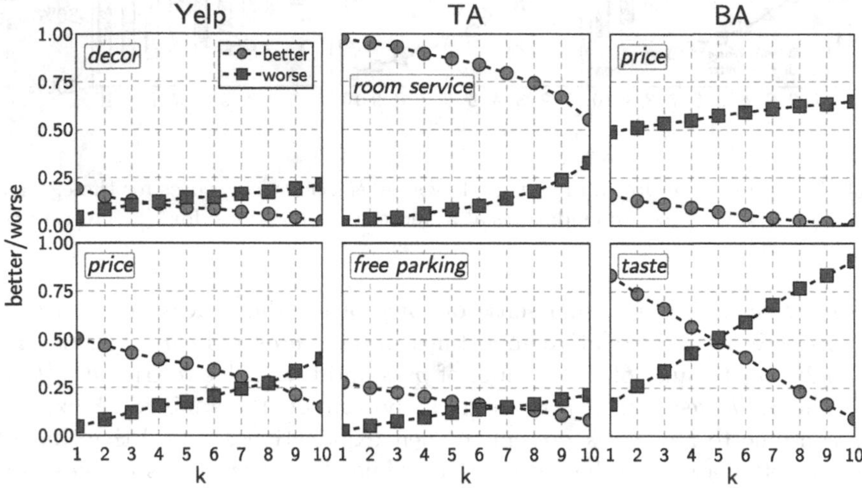

Fig. 6. Example, features from TA, YP, and BA and how their better and worse scores vary by rank position.

To sum up, TA and BA appear to represent two extremes on the continuum of explanation-based ranking, the former's explanations dominated by strong *pros*, the latter by strong *cons*. Yelp's explanations occupy a middle-ground. Strong *pros* dominate the high-ranking explanations and strong *cons* dominate the tail of the recommendation lists, but *pros* and *cons* compete on an equal footing within the middle of the rankings and both are required to produce a strong ranking signal.

6 Conclusions

We set out to demonstrate the efficacy of explanation-based ranking across multiple domains, and the above results show how this approach to recommendation ranking produces rankings that are closer to the ideal – in this case the ratings-based ranking used by TA, YP, and BA – than a conventional similarity-based approach. This means that our approach to explanation not only generates rich and meaningful explanations, but these explanations provide an authentic insight into the reasoning behind a set of suggestions because the recommendations themselves have been used to create the ranking. This intimate connection between explanation and recommendation is particularly advantageous from an explainable AI perspective and meets the authenticity test that is typically absent from other ad-hoc approaches to recommendations. The temptation, all too often, is to 'fake it' when it comes to explanations, at least in the sense that they rarely convey the genuine reasons for a decision or recommendation; this will inevitably erode trust between users and systems.

A key question for this paper was whether *pros* or *cons* are more or less important in different domains. By manipulating the relative weighting of *pros* and *cons*, during explanation-based ranking, we were able to show this in action. In TripAdvisor, *pros* played a dominant role and emphasising *pros* over *cons* resulted in even better recommendation rankings. The reverse was true in BeerAdvocate, emphasising *pros* had a deleterious effect on ranking quality, but emphasising *cons* boosted ranking quality. While the size of these effects was modest they nonetheless provided a way to further improve the ranking quality of explanation-based ranking at least in such pro-dominant or con-dominant domains. Improved rankings could not be demonstrated for Yelp, however, and manipulating the relative importance of *pros* or *cons* made little meaningful difference in ranking quality.

An interesting consideration here is that, while this suggests a way to further fine-tune explanation-based ranking, by identifying a setting for α that optimises ranking quality, the implications for the explanations that are presented to the end-user remain unclear and remain as a matter for future work. For example, given that the best rankings for TA arose from $\alpha = 1$, effectively eliminating *cons* and their worse scores from the ranking signal, does this mean that *cons* should not be presented to the end user in the final explanation? Likewise, should *pros* be dropped from explanations in BA? This is certainly a logical conclusion, particularly if we wish to maintain explanation authenticity, even though it feels unnatural to eliminate *pros* or *cons* in their entirety. It suggests a more fine-grained analysis of *pros* and *cons* is required in order to better understand whether the observed effects are limited to certain features of item cases in particular recommendation contexts.

Acknowledgement. This research is supported by Science Foundation Ireland through the Insight Centre for Data Analytics under grant number SFI/12/RC/2289.

References

1. Goodman, B., Flaxman, S.: EU regulations on algorithmic decision-making and a right to explanation. In: ICML Workshop on Human Interpretability in Machine Learning (2016)
2. Ribeiro, M.T., Singh, S., Guestrin, C.: "Why should I trust you?": explaining the predictions of any classifier. In: SIGKDD-2016, pp. 1135–1144 (2016)
3. Jordan, M., Mitchell, T.: Machine learning: trends, perspectives, and prospects. Science **349**(6245), 255–260 (2015)
4. Chakraborti, T., Sreedharan, S., Zhang, Y., Kambhampati, S.: Explanation generation as model reconciliation in multi-model planning. arXiv preprint arXiv:1701.08317 (2017)

5. Hendricks, L.A., Akata, Z., Rohrbach, M., Donahue, J., Schiele, B., Darrell, T.: Generating visual explanations. In: Leibe, B., Matas, J., Sebe, N., Welling, M. (eds.) ECCV 2016. LNCS, vol. 9908, pp. 3–19. Springer, Cham (2016). doi:10.1007/978-3-319-46493-0_1
6. Redmond, M.: Combining case-based reasoning, explanation-based learning, and learning form instruction. In: Proceedings of the Sixth International Workshop on Machine Learning (ML 1989), Cornell University, Ithaca, New York, USA, 26–27 June 1989, pp. 20–22 (1989)
7. Leake, D.B.: An indexing vocabulary for case-based explanation. In: Proceedings of the 9th National Conference on Artificial Intelligence, Anaheim, CA, USA, 14–19 July 1991, vol. 1, pp. 10–15 (1991)
8. Aamodt, A.: Explanation-driven case-based reasoning. In: EWCBR-1993, pp. 274–288 (1993)
9. Schank, R.C., Kass, A., Riesbeck, C.: Inside Case-based Explanation. Artificial Intelligence Series. Lawrence Erlbaum, Hillsdale (1994)
10. Cunningham, P., Doyle, D., Loughrey, J.: An evaluation of the usefulness of case-based explanation. In: Ashley, K.D., Bridge, D.G. (eds.) ICCBR 2003. LNCS, vol. 2689, pp. 122–130. Springer, Heidelberg (2003). doi:10.1007/3-540-45006-8_12
11. Roth-Berghofer, T.R.: Explanations and case-based reasoning: foundational issues. In: Funk, P., González Calero, P.A. (eds.) ECCBR 2004. LNCS, vol. 3155, pp. 389–403. Springer, Heidelberg (2004). doi:10.1007/978-3-540-28631-8_29
12. Roth-Berghofer, T.R., Cassens, J.: Mapping goals and kinds of explanations to the knowledge containers of case-based reasoning systems. In: Muñoz-Ávila, H., Ricci, F. (eds.) ICCBR 2005. LNCS, vol. 3620, pp. 451–464. Springer, Heidelberg (2005). doi:10.1007/11536406_35
13. Sørmo, F., Cassens, J., Aamodt, A.: Explanation in case-based reasoning-perspectives and goals. Artif. Intell. Rev. **24**(2), 109–143 (2005)
14. Nugent, C., Doyle, D., Cunningham, P.: Gaining insight through case-based explanation. J. Intell. Inf. Syst. **32**(3), 267–295 (2009)
15. Leake, D.B., McSherry, D.: Introduction to the special issue on explanation in case-based reasoning. Artif. Intell. Rev. **24**(2), 103–108 (2005)
16. Bergmann, R., Pews, G., Wilke, W.: Explanation-based similarity: a unifying approach for integrating domain knowledge into case-based reasoning for diagnosis and planning tasks. In: EWCBR-1993, pp. 182–196 (1993)
17. Wang, L., Sawaragi, T., Tian, Y., Horiguchi, Y.: Integrating case based reasoning and explanation based learning in an apprentice agent. In: ICAART-2010, pp. 667–670 (2010)
18. Herlocker, J.L., Konstan, J.A., Riedl, J.: Explaining collaborative filtering recommendations. In: Proceedings of the 2000 ACM Conference on Computer Supported Cooperative Work, Philadelphia, USA, pp. 241–250, December 2000
19. Gedikli, F., Jannach, D., Ge, M.: How should I explain? A comparison of different explanation types for recommender systems. Int. J. Hum Comput Stud. **72**(4), 367–382 (2014)
20. Tintarev, N., Masthoff, J.: The effectiveness of personalized movie explanations: an experiment using commercial meta-data. In: Proceedings of the 5th International Conference on Adaptive Hypermedia and Adaptive Web-Based Systems, Hannover, Germany, vol. 5149, pp. 204–213, July 2008
21. Musto, C., Narducci, F., Lops, P., De Gemmis, M., Semeraro, G.: ExpLOD: a framework for explaining recommendations based on the linked open data cloud. In: RecSys-2016, pp. 151–154. ACM (2016)

22. Chang, S., Harper, F.M., Terveen, L.: Crowd-based personalized natural language explanations for recommendations. In: RecSys-2016, pp. 175–182. ACM (2016)
23. Dong, R., Schaal, M., O'Mahony, M.P., McCarthy, K., Smyth, B.: Mining features and sentiment from review experiences. In: Delany, S.J., Ontañón, S. (eds.) ICCBR 2013. LNCS, vol. 7969, pp. 59–73. Springer, Heidelberg (2013). doi:10.1007/978-3-642-39056-2_5
24. Dong, R., Schaal, M., O'Mahony, M.P., Smyth, B.: Topic extraction from online reviews for classification and recommendation. In: IJCAI-2013, pp. 1310–1316 (2013)
25. Dong, R., O'Mahony, M.P., Smyth, B.: Further experiments in opinionated product recommendation. In: Lamontagne, L., Plaza, E. (eds.) ICCBR 2014. LNCS, vol. 8765, pp. 110–124. Springer, Cham (2014). doi:10.1007/978-3-319-11209-1_9
26. Muhammad, K., Lawlor, A., Rafter, R., Smyth, B.: Generating personalised and opinionated review summaries. In: UMAP-2015 (2015)
27. Muhammad, K., Lawlor, A., Rafter, R., Smyth, B.: Great explanations: opinionated explanations for recommendations. In: Hüllermeier, E., Minor, M. (eds.) ICCBR 2015. LNCS, vol. 9343, pp. 244–258. Springer, Cham (2015). doi:10.1007/978-3-319-24586-7_17
28. Muhammad, K., Lawlor, A., Smyth, A.: A live-user study of opinionated explanations for recommender systems. In: IUI-2016, Sonoma, USA, pp. 256–260. ACM (2016)
29. Lawlor, A., Muhammad, K., Rafter, R., Smyth, B.: Opinionated explanations for recommendation systems. In: Bramer, M., Petridis, M. (eds.) Research and Development in Intelligent Systems XXXII, pp. 331–344. Springer, Cham (2015). doi:10.1007/978-3-319-25032-8_25
30. Mitchell, T.M., Keller, R.M., Kedar-Cabelli, S.T.: Explanation-based generalization: a unifying view. Mach. Learn. **1**(1), 47–80 (1986)
31. Almonayyes, A.: Improving problem understanding by combining explanation-based learning and case-based reasoning: a case study in the domain of international conflicts. Ph.D. thesis, University of Sussex, UK (1994)
32. DeJong, G., Mooney, R.J.: Explanation-based learning: an alternative view. Mach. Learn. **1**(2), 145–176 (1986)
33. Pazzani, M.J.: Explanation-based learning for knowledge-based systems. Int. J. Man Mach. Stud. **26**(4), 413–433 (1987)
34. Pazzani, M.J.: Explanation-based learning with week domain theories. In: Sixth International Workshop on Machine Learning, pp. 72–74 (1989)
35. Bhatnagar, N.: Learning by incomplete explanation-based learning. In: Proceedings of the Ninth International Workshop on Machine Learning, pp. 37–42 (1992)
36. Sun, Q., Wang, L., DeJong, G.: Explanation-based learning for image understanding. In: IAAI-2006, pp. 1679–1682 (2006)
37. Knoblock, C.A., Minton, S., Etzioni, O.: Integrating abstraction and explanation-based learning in PRODIGY. In: AAAI-1991, pp. 541–546 (1991)
38. Rossetti, M., Stella, F., Zanker, M.: Towards explaining latent factors with topic models in collaborative recommender systems. In: DEXA 2013, pp. 162–167 (2013)
39. McAuley, J., Leskovec, J., Jurafsky, D.: Learning attitudes and attributes from multi-aspect reviews. In: ICDM-2012, pp. 1020–1025. IEEE (2012)

A User Controlled System for the Generation of Melodies Applying Case Based Reasoning

María Navarro-Cáceres[✉], Sara Rodríguez, Diego Milla, Belén Pérez-Lancho, and Juan Manuel Corchado

BISITE Research Group, University of Salamanca, Calle Espejo, S/N, 37007 Salamanca, Spain
maria90@usal.es

Abstract. The automatic generation of music is an emergent field of research that has attracted a wide number of investigators. Many systems allow a collaboration between human and machine to generate valuable music. Among the different approaches developed in the state of the art, the present research is focused on an intelligent system that generates melodies through a mechanical device guided by the user. The system is able to learn from previous compositions created by the users to improve future results. A Case-Based Reasoning architecture was developed with a Markov model to obtain the probabilities of a given note following the last note incorporated in the melody. This probability also depends on the mechanical device connected to the system that can be used at any moment to control the pitches and the duration of the musical notes. As a result of the collaboration between machine and user, we obtain a melody that will be rated and, according to the rating, incorporated into the memory of the system for future use. Several experiments were developed to analyze the quality of the system and the melodies created. The results of the experiments reveal that the proposed system is able to generate music adapted and controlled by the users.

Keywords: Case-based reasoning · Music generation · Markov models

1 Introduction

The different computational advances in the field of Artificial Intelligence that have occurred during the last years have attracted the attention of researchers from multiple backgrounds and motivations, thus creating innovative fields that unite seemingly disparate concepts such as Artificial Intelligence and Art. It is from these two disciplines that the area of Artificial Creativity was born, which can be loosely defined as the partial or completely automated computational analysis or synthesis of works of art [6]. Given the particular nature of this field of research, which brings together two sectors with very different (sometimes even opposing) methods and objectives, the state of the art is very diverse and difficult to compare. In the area of music generation, there are some interesting proposals about music generation and bio-inspired algorithms [20,21] or Markov Models [17,18].

D.W. Aha and J. Lieber (Eds.): ICCBR 2017, LNAI 10339, pp. 242–256, 2017.
DOI: 10.1007/978-3-319-61030-6_17

Among the works in music generation, some emerging devices have been developed to allow different users to control artistic creation with a certain level of abstraction. This is the case, for example, of MotionComposer [3], a research in which a device is used to compose music from the movement of people with disabilities. [2] propose a new interface to generate music by using touchscreens. The Hand Composer [14] is a novel and interactive framework that enables musicians to generate/compose through hand motion and gestures.

Additionally, some research works have analyzed the user-guided generation to learn from it. Flow Machines [18] is a project that proposes new ways to create music collaborating with human musicians, which can lead the creative process. VirtualBand [15] generates jazz compositions following the performance of a melodic instrument in real time. GimmeDaBlues [7] automatically generates the bass and drums parts while the user plays keyboard and/or solo instruments, responding to the user's activity.

This work proposes an intelligent system to generate melodies based on user guidelines and previous melodies generated. The system can be adapted to various musical styles and to different users, based on their preferences. Additionally, the user can guide the generation of melodies by moving a connected mechanical device. The selection of the sequence of notes is stored according to the user's satisfaction and retrieved in a future execution to improve the user experience.

Some works have proposed the integration of a Case Based Reasoning (CBR) architecture. Of special interest is the one developed by [9], which proposed a CBR to transform the tempo of a musical composition. The successful results of this paper encouraged us to integrate a CBR into our system to learn from previous melodies generated by the user. The CBR architecture automatically retrieves files encoded in MIDI [12] which follow different styles, such as classical or jazz music and are stored in a memory of the previous melodies. It then trains a Markov Model, which is a statistical tool that allows generating music influenced by diverse styles in order to extract the most likely sequence of notes. These probabilities are also influenced by the users' preferences expressed with the movements of the mechanical device incorporated in the system. Once the melody is generated, the user can rate the final result and the score given is then applied as feedback for the system to improve future results. The work done in [9] work also presents a CBR architecture, but for a different motivation in which a mechanical device is not involved.

This article details the entire research process carried out. The next section introduces some recent research related to music and learning. Section 3 describes the overall methodology followed. Section 4 details the experiments and provides a preliminary discussion of the results. Finally, Sect. 5 collects the main conclusions and suggests some possible future studies.

2 Related Work

Music composition traditionally performs a number of tasks such as defining melody and rhythm or harmonization to generate a worthy artistic work.

Today, all these tasks can be partially automated by computers [8]. In the case of relatively small degrees of automation, specific programming languages and graphical tools have emerged to assist the composer by automating specific tasks, or by providing a basis upon which to be inspired, such as the Csound language developed by MIT researchers [5] or the visual language MAX/MSP [19]. The project ReacTable [11] explores the possibility of generating live music composing in real time using a table and certain pieces as interface.

In terms of algorithmic composition with higher degrees of automation, many different works with different methodologies are proposed. It is worth mentioning the NIME (New Interfaces for Musical Expression) conference which, despite being a broad spectrum event, brings together a large amount of research in this sector. Closer to the present line of this study could be the research presented in [13] in which 3 melodies are generated simultaneously, controlled by 2 hamsters each. However, unlike the proposed methodology, the work cited is based on predefined rules and is not trained with a set of data. The Continuator [16] composes music from a brief introduction that the musician makes with a musical instrument to provide some guidelines. Conchord [4] harmonizes a melody in real-time, following the user indications through a keyboard.

It is also worth mentioning Pachet, who has worked intensively on the automatic generation of melodies and in 2011 published a study on the possibility of adapting a Markov model to certain predefined constraints [17]. Given these circumstances, a simple and innovative approach was sought based on a previously used model for the automatic generation of melodies: the Markov models.

3 The CBR Architecture

Any entity built with a CBR architecture follows the four stages of the CBR cycle, namely, Retrieve, Reuse, Review and Retain [1]. In our case, the system is encoded to plan actions according to the information retrieved from the context. The system is provided with a memory to store different solutions or melodies according to past experiences. Initially, the user should enter any credentials so that the system can load the corresponding profile with the past experiences the user might have had. The user can then select the type of composition to make, according to the style (jazz, classical, pop), composer (Bach, Mozart, etc.) and optional tonality, meaning major or minor tonality. The system then selects the melodies or solutions previously generated (Retrieval Step) to train a Markov Model with these solutions (Reuse Stage). If the system memory does not have enough solutions to train the Markov Model, it searches for some external files that satisfy the user preferences and add them to the training set. The probabilities extracted with the Markov Model with regard to musical pitch and duration can be controlled partially by the user through the mechanical device. The whole adaptation process produces a final melody that should be evaluated in terms of musical quality and user satisfaction (Review Stage). Finally, the system stores the case if it is deemed useful based on the user's ratings. Figure 1 illustrates the main process.

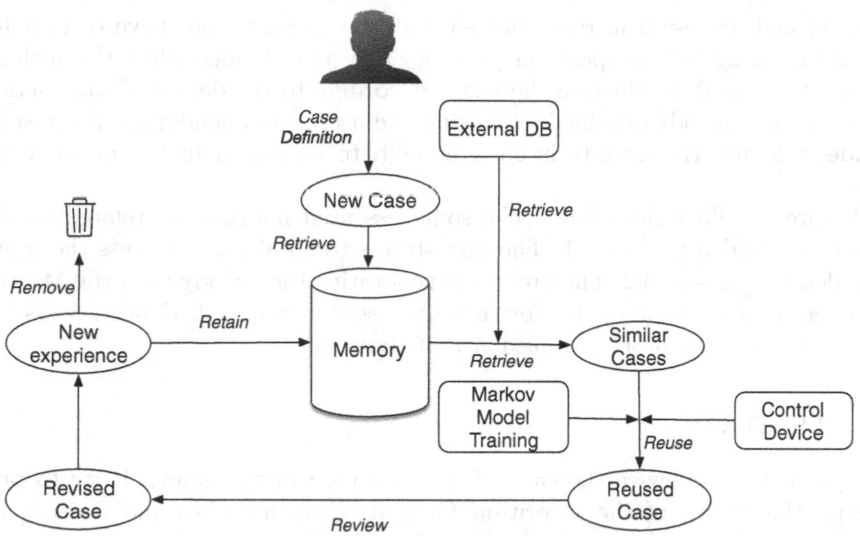

Fig. 1. Schema with the CBR process. The case represents a formalization of a melody with the initial requirements submitted by the user, and the final review obtained.

The entire process is based on the concept of case, which can be defined as a 3-tuple $C = <P, S, R>$. P is the problem domain that represents the information of the composition we want to create, which includes the style, the tonality and the author, all encoded as labels. S is a set of notes that form the melody, formalized as a tuple $<N, D>$ where N is the musical note considered and D corresponds to its duration. Finally, R represents the subjective rating of the user about the musical melody obtained. R is calculated by collecting the user opinion after the melody is constructed.

The proposal is divided into several stages which go from the extraction of the data to the final analysis to the results. The scheme of Fig. 1 gives a general idea of the different stages that are carried out for the generation of the melody directed by the device. Initially, the user submits a description of the case, which consists of formal parameters, to generate the melody. These parameters or labels are related to the musical style, composer and tonality. The system then retrieves a list of cases (previous melodies composed) with similar labels in its memory. If the system cannot find any case or the list is not wide enough, an external module is implemented to search for new cases in the web that comply with the initial requirements (style, author and tonality). With these new cases, a statistical model based on Markov models is trained to calculate the probability of a note being selected depending on the previous notes that appear in the melody. This training can take a few seconds. Once the Markov model is ready, the system starts composing a new melody following two constraints: the probabilities calculated by the statistical model and the indications of the user through the mechanical device connected. The indications of the device are received and processed in real time, so that the user does not have to wait for any delay along the composition process. The user decides when the melody is completed, and should rate the result according to the degree of satisfaction with the final melody obtained. Following the user recommendation, the system decides whether the melody is good enough to be stored in the memory for future use.

Before detailing the CBR cycle, some essential information related to the device is included in Sect. 3.1. The first step is to obtain and process the input data detailed in Sect. 3.2. The process of generating the melody from the Markov model and the position of the device is discussed in Sect. 3.3. Finally, Sects. 3.4 and 3.5 explain the process of feedback of the results.

3.1 The Device

The joystick-type device shown in Fig. 2 was used in this study. Prior to presenting the details of the algorithm for generating melodies and control, an explanation of the device will be provided.

Fig. 2. Illustration of the mechanical device.

The main researcher behind this device is Professor Wataru Hashimoto of the ILO (Osaka Institute of Technology). Technically, the device consists of 2 motors anchored to an aluminum frame and interconnected forming a pantograph with 2 degrees of freedom. One of the advantages of this configuration is that the motors can exert a force of feedback that can be useful for the user in diverse applications.

This device consists of an articulated arm with 4 rigid segments and 3 joints, connected at both ends to two motors anchored in an aluminum frame. He user can manipulate the position of the central articulation to place it in any interior point (p_x, p_y) of the rectangle that delimits the frame. The arm translates the position of that point at the angles $(\theta_1,\ \theta_2)$ of the motors, according to the trigonometric relationships drawn from the diagram in Fig. 3. These values encoded by the motors are sent to the main system through a transducer card.

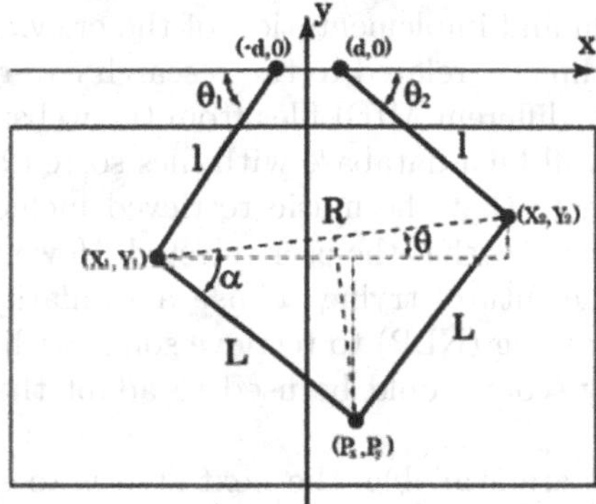

Fig. 3. Representation of the coordinates and angles existing in the device.

One of the advantages of using this device instead of a simple machine such as a mouse or a keyboard is that the motors can exert a force of feedback that can be useful for a future study in which the machine could assist the users in their movements, creating a resistance force.

3.2 Searching for Previous Solutions (Retrieve Stage)

Since the goal is to generate and control a melody by varying both pitch and rhythm, input data representing these two characteristics are required. Currently, there is no standard format that only deals with these two variables which makes it necessary to extract data from other types of files. The decision was made to use MIDI files (Musical Instrument Digital Interface, [12]) due to its availability throughout the Net, and its structure, which allows easy access to notes and durations. The files do not contain the sounds. Instead they include instructions that permit the reconstruction of the song by using a sequencer and a synthesizer that work with the MIDI specifications. Thus, the files are quite light files since they allow encoding a complete song in a few hundred lines.

In the retrieval stage, all the previous cases, meaning 3-tuples $C = <P, S, R>$ stored from previous experiences, are grabbed and the P component is compared with the input description. This $P = <L, T>$ is a list of labels L_i with the style and author features and/or the main tonality of the composition T, which can be optional. The system searches for those solutions that are associated with such labels and composed in the same tonality (major or minor). The solutions are presented in MIDI files, from which the data will be extracted for the reuse stage.

In some circumstances, a low number of solutions could be retrieved. This might happen at initial stages of the system, when the user requires new compositions according to new musical styles or when new users are running the system. For such cases, a large amount of external files should be collected from different sources. To do so, a simple crawler was developed, allowing the user to download all MIDI files that meet certain conditions according to the labels L_i.

Details on the operation and implementation of the crawler are not included in this article as it is not directly related to the research covered in this document. The crawler can retrieve different MIDI files from the websites that are available in the Net. The result will be a database with files sorted by labels.

It is important to note that the music retrieved includes those files whose labels coincide with the labels that the user selected. However, we could improve this methodology in the future, trying to use a similarity measure based on Natural Language Processing (NLP) to retrieve songs with similar descriptions. In fact, those similarity scores could be used to adapt the probabilities in the Markov model.

Once the MIDI files are available, the next step is to extract the necessary information for the project: notes or pitches N_i and their duration D_i. According to MIDI specifications [12], each MIDI file represents a sequence (usually corresponding to a song or composition), which in turn consists of one or more tracks. These tracks are characterized by a sequence of MIDI events, meaning MIDI messages associated with a particular time. The different tracks of the sequences are played simultaneously, so generally each track is used for one instrument. Given these MIDI format specifications, only a few steps are needed to extract the required data. First, the MIDI files are iterated to obtain the corresponding sequence. Then, they are iterated along the tracks, and search is made for different events for each track. As both the notes and their durations are needed for the present work, we are interested in events of type NOTE_ON and NOTE_OFF [10]. As the name implies, these events are used to determine the start and end of a particular note, specified in the first byte of the MIDI message. The second byte of the message indicates the strength of the note (called "speed" in MIDI messages). This byte is ignored in this project except in the case of a message of type NOTE_ON with speed 0, which is the equivalent of a Message type NOTE_OFF [10]. It also takes into account events of type PROGRAM_CHANGE, which indicate a change of instrument. It could be interesting for a future analysis, to group the melodies according to the instrument for which the user wants to compose.

Once the melodies are extracted from the MIDI files, they are used as the training set for the Markov Model. The next section details the process to obtain the new melody that the user has to validate.

3.3 Generating the New Melody (Reuse Stage)

For the intelligent and controlled generation of melodies, a hybrid approach is proposed in which the algorithm that generates the melody takes into account both the Markov model and the position of the device. Figure 4 below summarizes the different steps related to the control algorithm and the generation of melodies detailed throughout this subsection.

The first step in this stage is to obtain a model that represents the data obtained and serves for the generation of melodies. Markov models are a widely used tool for modeling the temporal properties of various phenomena, from the structure of a text to economic fluctuations. Since these models are relatively

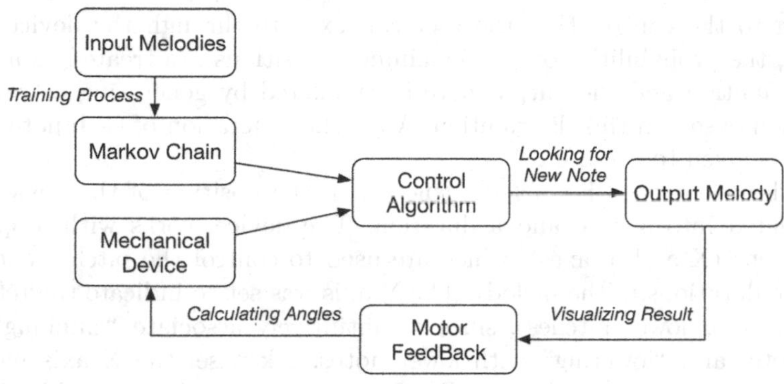

Fig. 4. Schema with the adaptation process to generate a new melody.

easy to generate, they are also used for content generation applications, such as text or music generation. Thus, this type of model was chosen for the study, although its results are adapted to the interactive control of the device. Briefly, a Markov Model represents a special type of stochastic process in which the probability of an event occurring (in this case a note or silence with a given duration) depends only on the n previous events. This feature of "limited memory" is what is known as Markov property. In our case, n was empirically set to 4.

For the purposes of this work, and with specific regard to the control device to express the duration and the pitch of the notes in the melody, the training set is compressed in three octaves. For this process, an algorithm was developed to determine which the three consecutive octaves regroup the largest number of notes of the input data and will be considered the reference octaves for the corresponding data. The notes below this reference are moved to the lower reference octave, and those above the upper reference octave are also moved to the upper octave. This allows all notes to be placed within the reference octaves.

Additionally, the durations are normalized to 8 possible durations: sixteenth note, eighth note, eighth note with dot, black, dotted black, white, dotted white and round (a dot is equivalent to multiplying the length of the note by 1.5). In this way the 8 possible durations expressed in the sixteenth notes are 1, 2, 3, 4, 6, 8, 12 and 16. The Markov model is then constructed through a simple training process. It is iterated between all the sequences of notes with the duration extracted for the selected group of instruments, constantly updating the state of the Markov model as it is iterating the algorithm and progressively adding each transition to its corresponding state. Once the loop is finished, it is iterated again from the beginning of the data to ensure the stability of the Markov model (i.e., there is at least one transition for each state), but this time the sequence of the final note is considered a state.

Although the Markov model determines the possible transitions for each state and the initial probabilities of each one of them, the selection of the notes are always influenced by the position of the mechanical device. Thus, the probabilities of each note being selected as the next note in the melody are adjusted according to the control that the user can execute through the device. After this step, the probabilities of the remaining transitions are treated as a statistical distribution and the output note is calculated by generating a "random" observation based on this distribution. After the generation of each note, visual feedback is given to the user.

In order to control the melody generation, the position of the device must be translated into a note and a duration. The device works with a space of 2 dimensions (X and Y axes), which are used to control the pitches and their respective durations in the melody. The Y axis was set to indicate the reference note (higher or lower pitches), since we intuitively associate "climbing" with higher notes and "lowering" with lower notes. Likewise, the X axis indicates a reference duration. In order for all reference points to be accessible through the use of the device, the coordinate space of the device has to be reduced by transforming it into a rectangular space. Once the coordinates to be used are delimited, and for a greater generalization of the control algorithm, both axes are given minimum and maximum values. On the X axis, 1 represents the shortest note (sixteenth note) and 16 represents the longest note (round). On the Y axis, 0 represents silence; 3 reference octaves are represented, from the lowest to the highest. These delimitations of the space of reference facilitate the training process of the model.

Another feature added is the possibility of shortening notes that are considered too long from the users's point of view. The logic behind this is very simple: if a long note has been generated and the device is shifted to the left on the X-axis (towards 0, that is to shorter notes), the current note is "cut" and the duration of the next one is calculated.

It is important to note that the user can partially control the melody. That means that the position the user moves the device does not correspond to a particular pitch, but to a "reference" note or setpoint and duration t_r. Therefore, it is possible that the same position can give different notes (although similar en duration and pitch) at different moments, according to the Markov Model that limits which notes are eligible for the melody.

Consequently, after translating the position of the device into a note and a reference duration, we aim to modify the probabilities of the different possible transitions of the Markov model according to these values, where a transition means a new note added to the melody. This modification is made in two complementary ways: on the one hand, the probability of the transitions closest to the reference value t_r is increased, and on the other, all transitions that are too far from this value are avoided.

The probability of the transitions close to the control is increased as a function of the parameter k, which determines the weight of the reference when calculating the probability of each transition t_i according to the following formula:

$$P(t) = k \cdot P_D(t_i, t_r) + (1 - k) * P_M(t_i) \quad (1)$$

where $P_D(t)$ means the probability of the transition following the device, and $P_M(t)$ represents the probability of a transition following the Markov model. k is empirically set to 0.55.

To begin, we selected the most probable transitions t_i that the Markov Model provided. $P_D(t)$ is then calculated as a function of the distance between each transition t_i and the reference transition t_r. As each transition is characterized by a note and a duration, there are two distances for each transition t_i: the distance of the note and the distance of the duration. The probability has been defined as the inverse of the normalized Euclidean distance between the reference pitch and the pitch d_p, and the reference duration and the duration of the transition d_d.

$$P_D(t_i, t_r) = \begin{cases} 1 - ||d_p(t_i, t_r) + d_d(t_i, t_r)|| : d_d < d_{dmax}, d_p < d_{pmax} \\ 0 \qquad\qquad\qquad\qquad\qquad\qquad\quad : otherwise \end{cases} \tag{2}$$

Equation 2 shows that the probability of a transition t_i to be selected is inversely proportional to the distance between t_i and the reference transition t_r. The distance d_p is calculated applying the Euclidean distance of the coordinate x of t_r and the x-coordinate that would correspond to the transition t_i when represented in the space of the mechanical device. The distance d_d also represents the Euclidean distance of the x coordinate of t_r and t_i.

It is also possible that the Markov Models might provide some transitions that must be discarded because they are too far from the reference transition t_r. This limitation of the output range is made by two adjustable parameters that determine the maximum distance of the notes MAX_{DP} and the duration of the output transitions with respect to the reference values MAX_{DD}. When the duration or the pitch distance between t_i and t_r are above these thresholds, the probability P_D is automatically set to 0, to avoid transitions too far from the device control. MAX_{DD} and MAX_{DP} values are adjustable, and for the purposes of our work have been empirically set to 4.3 and 6.2, respectively.

In some situations, the system cannot find any transition that is within the control range (i.e. all possible transitions are too far from the control point). For such cases, the transition of the Markov model with the minimum distance between t_i and the t_r is selected as the next note in the melody. The next transition is then calculated by applying the probabilities of the Markov model and the position of the device.

The process finishes when the user decides to do so. This set of notes is then re-played for the user evaluation which is fully explained in the next section.

3.4 Validating the Final Melody (Review Stage)

The melody obtained from the machine and the user collaboration needs to be evaluated in order to measure the usefulness of the creation for the present user. Such evaluations can be done by following multiple automatic methods. However, in our proposal both the nature of the results, a creative product that depends entirely on user performance, and the validation through the opinion of the user are essential. However, it is important to note that such musical evaluations always depend on personal preferences, mood, musical training or social culture. A melody valid for one user may not be pleasant for another. Therefore, the profiles are created so that the machine can adapt to the preferences of different users.

The musical result is presented to the user, who can play the melody several times. They are asked for their degree of satisfaction with the final result, meaning the degree to which the melody has been adapted to their requirements manifested with the mechanical device. A categorical scale is set from 0 to 10, where 0 means totally unsatisfied and 10 means totally satisfied with the result. The rate is then used to decide whether the melody should be stored in the memory for a future use.

3.5 Storing a New Case (Retain Stage)

All the melodies generated have assigned a global rate. The system should decide whether the case is stored for a future use. To do so, it dynamically establishes a threshold which depends on the previously stored cases. If we are storing cases with low values, a case with a global score of 6 can be very useful. On the contrary, for cases with high ratings, a case evaluated as 6 are not so interesting. Thus, the threshold is established by calculating the mean for all the global rates for the compositions when there are more than 10 songs stored. Initially, the threshold will be established in 4 when there are not enough data stored in the memory.

4 Results and Discussion

The evaluation for this system is twofold. To begin, we aim to demonstrate that the system can be adapted to the preferences of multiple users as they are running the system. To do so, we show the evolution of the overall rates of the different melodies when the number of executions are gradually increasing, and compare it with the same system but without the CBR architecture implemented. Secondly, we aim to determine the usefulness of the system to generate melodies adapted to users preferences. Therefore, we prepared a poll that collects the overall opinion of the people who have used the system.

The system was performed as an application installed in a PC, and connected to the mechanical device. Initially, the user can register in the system. Once they have logged in, they can select if they want to generate a melody based on a general musical style or a specific author. For this preliminary test, they can choose among "barroque", "classical", "romanticism", "jazz" and "pop-rock" styles. They also can select a specific composer among "Vivaldi", "Bach", "Haydn", "Mozart", "Beethoven", "Schubert", "Albéniz", "Louis Armstrong" and "The Beatles". In a future study we plan to extend the list to include different composers and styles, slightly modifying the crawler to collect any kind of music and classify the files. As an option, for each composer or style chosen, they can also select the preferred tonality to generate the melody.

The system then starts to create music. The user can guide this melody generation by moving the mechanical device ahead if they want a faster melody, backwards if they are looking for a slower composition, up for higher pitches or down for lower pitches. Each transition is registered in the system to modify the direction of the melody, and all the movements are reflected in the screen in real time (Fig. 5). Users can finish the melody whenever they want, clicking the "Stop" button of the application. At this moment, the system shows a screen with the melody line to re-play or download the MIDI file. In the same screen, a brief question is also presented to ask users whether they are satisfied with the melody composed. Once the user has answered the question, he or she can start the process again for a new melody.

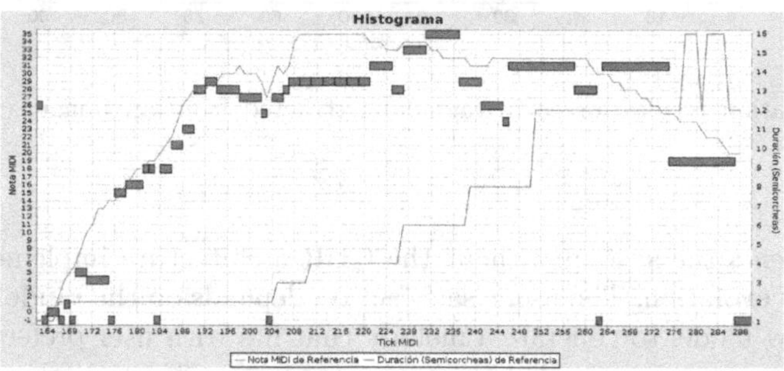

Fig. 5. Captures the screen while the user is composing a melodic line

For all the results, a total of 21 users were selected to create their profile and test the system during 16 days. About 100 melodies per user were generated with the system, applying our CBR architecture to learn the preferences of the users. We began by analyzing whether in general, the melodies generated in the final days were better evaluated than those melodies created in the initial stages of the test. Although the users are able to improve their skills in using the mechanical device, as it is quite easy to use accurately, we assume the main reason for improvement in the evaluation score is that the system is adapted to individual user preferences, and better trained according to the previous melodies generated. To demonstrate this hypothesis, we calculated the mean for all the evaluations given every 5 iterations of the system.

Additionally, we made a comparative study between the results obtained from the system with and without the CBR architecture to check if the system is adapted to user preferences. In this case, the users also compose melodies, but their feedback is not considered for future use, the new melodies are only based on the training set that the crawler provides. We also calculated the mean for all the evaluations for every 5 iterations of the system, and plotted them in the same graph. Figure 6 shows the final results.

The horizontal axis of the plot represents the number of melodies (iterations) generated before collecting the evaluation, while the vertical axis represents the global rate that the melodies have obtained. The graph in blue, which represents the system with CBR architecture, shows a general increase as the number of iterations are rising. This tendency demonstrates that the system is able to learn from the user and reflects their preferences in future melodies. The graph in red,

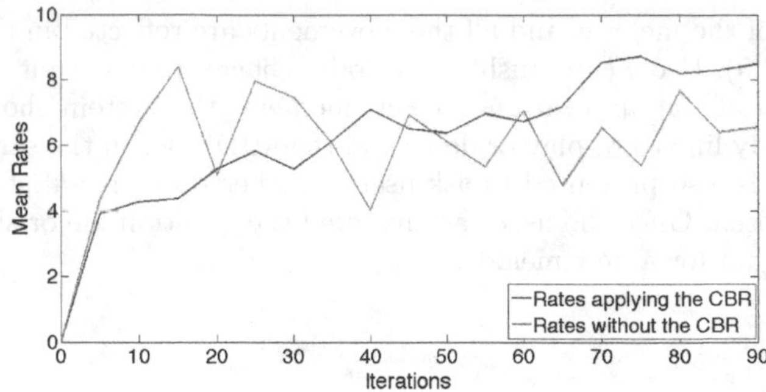

Fig. 6. Evaluation results for each fragment extracted from an image. (Color figure online)

which represents the system without the CBR architecture implemented, does not show any evolution. The users' satisfaction depends on the random variables of the Markov model to generate a melody that fits with user preferences. Nevertheless, we must point out that the scores are very subjective, thus the results might differ from user to user.

The second part of the test consist of validating the usefulness of the system. To this end, the users were asked for their experience with the system after 16 days of use. They answered a questionnaire about whether the system is easy to use, the interface conforms to the real movements of the device, the device is a helpful tool to control the melody, or the system adapts correctly to their preferences after a few times of use, as well as the overall score for the system and possible suggestions. All the questions could be rated from 1 ("Completely disagree") to 5 ("Completely agree"). Table 1 shows the mean scores for all these questions.

Table 1. Shows the final statistics when the users finished testing the system.

	Easy to use	Interface	Control quality	Adaptation degree	Overall ratings
Ratings	4.22 ± 0.86	3.23 ± 1.53	3.88 ± 1.27	4.09 ± 0.94	4.01 ± 1.09

The Table shows that the general satisfaction degree is quite high, with a mean of 4.01. The users consider the system to be very easy to use even for people without any musical training (4.22), although the interface could be improved (3.23). Some users have suggested the addition of a complete score with notes and rhythms instead of a melodic line in the interface. However, we can conclude the system can make melodic compositions that adapt to user preferences. Among the suggestions that could be added for a future study, we would particularly note the freedom to choose the author and the style, and a system update to allow the use of different mechanical devices such as joysticks.

5 Conclusions

This proposal has presented an intelligent system to compose melodies using a mechanical device to control the duration and the pitch of the generated notes. The melodies adapt to user preferences gy applying a CBR architecture that learns from previous solutions provided by the system. As a first step, the proposed approach retrieves a set of MIDI files from which the notes are extracted with their respective durations. A Markov model is then trained with the data of the desired group, and the transition probabilities of this model are modified according to the control device to generate a melody that respects these "controlled" probabilities.

The results of the different experiments carried out emphasize the importance of user preferences in the melody generation. The system implemented with the CBR architecture shows an overall tendency to improve their results as the number of melodies for a specific use increases. That means the CBR can easily capture movements with the mechanical device and the user profile. Likewise, the results also show an optimistic view of the users, who consider the system to be easy to use and helpful to generate music adapted to their personal preferences.

However, when training only with two dimensions, the notes and their durations, the Markov model does not capture other important characteristics of the melodies such as musical phrases or harmony. We can study the possibility of introducing this information in the system to generate more phrasal music or to add harmonization to the melody. It would also be interesting to explore the possibilities offered by a relative control (perhaps based on fuzzy logic) that is not limited to 3 octaves and which avoids the need to normalize the input data.

Acknowledgments. This work was supported by the Spanish Ministry of Economy and FEDER funds. Project. SURF: Intelligent System for integrated and sustainable management of urban fleets TIN2015-65515-C4-3-R. And by the Spanish Government through the FPU program of the Ministry of Education and Culture.

References

1. Bajo, J., Corchado, J.M., Rodríguez, S.: Intelligent guidance and suggestions using case-based planning. In: Weber, R.O., Richter, M.M. (eds.) International Conference on Case-Based Reasoning. LNCS, pp. 389–403. Springer, Heidelberg (2007). doi:10.1007/978-3-540-74141-1_27
2. Berdahl, E., Holmes, D., Sheeld, E.: Wireless vibrotactile tokens for audio-haptic interaction with touchscreen interfaces. In: Proceedings of the International Conference on New Interfaces for Musical Expression, vol. 16, pp. 5–6. Queensland Conservatorium Griffth University, Brisbane, Australia (2016). ISSN 2220-4806
3. Bergsland, A., Wechsler, R.: Composing interactive dance pieces for the motioncomposer, a device for persons with disabilities. In: Proceedings of the International Conference on New Interfaces for Musical Expression, pp. 20–23 (2015)
4. Bernardes, G., Cocharro, D., Guedes, C., Davies, M.E.: Conchord: an application for generating musical harmony by navigating in a perceptually motivated tonal interval space. In: Proceedings of the 11th International Symposium on Computer Music Modeling and Retrieval (CMMR), pp. 71–86 (2015)

5. Boulanger, R.C.: The Csound Book: Perspectives in Software Synthesis, Sound Design, Signal Processing, and Programming. MIT Press, Cambridge (2000)
6. Dartnall, T.: Artificial Intelligence and Creativity: An Interdisciplinary Approach, vol. 17. Springer Science & Business Media, Berlin (2013)
7. Dias, R., Marques, T., Sioros, G., Guedes, C.: GimmeDaBlues: an intelligent Jazz/Blues player and comping generator for ios devices. In: Proceedings of the Conference on Computer Music and Music Retrieval (CMMR 2012), London (2012)
8. Fernández, J.D., Vico, F.: Ai methods in algorithmic composition: a comprehensive survey. J. Artif. Intell. Res. **48**, 513–582 (2013)
9. Grachten, M., Arcos, J.L., de Mántaras, R.L.: A case based approach to expressivity-aware tempo transformation. Mach. Learn. **65**(2–3), 411–437 (2006)
10. Huber, D.M.: The MIDI Manual: A Practical Guide to MIDI in the Project Studio. Taylor & Francis, Abingdon (2007)
11. Jordà, S., Geiger, G., Alonso, M., Kaltenbrunner, M.: The reactable: exploring the synergy between live music performance and tabletop tangible interfaces. In: Proceedings of the 1st International Conference on Tangible and Embedded Interaction, pp. 139–146. ACM (2007)
12. Jungleib, S.: General Midi. AR Editions, Inc., Middleton (1996)
13. Lorenzo Jr., L.M.I.: Intelligent MIDI sequencing with hamster control. Ph.D. thesis, Cornell University (2003)
14. Mandanici, M., Canazza, S.: The "hand composer": gesture-driven music composition machines. In: Proceedings of the 13th International Conference on Intelligent Autonomous Systems, pp. 15–19 (2014)
15. Moreira, J., Roy, P., Pachet, F.: Virtualband: interacting with stylistically consistent agents. In: ISMIR, pp. 341–346 (2013)
16. Pachet, F.: The continuator: musical interaction with style. J. New Music Res. **32**(3), 333–341 (2003)
17. Pachet, F., Roy, P.: Markov constraints: steerable generation of Markov sequences. Constraints **16**(2), 148–172 (2011)
18. Papadopoulos, A., Roy, P., Pachet, F.: Assisted lead sheet composition using flowcomposer. In: Rueher, M. (ed.) CP 2016. LNCS, pp. 769–785. Springer, Cham (2016). doi:10.1007/978-3-319-44953-1_48
19. Puckette, M.: Max at seventeen. Comput. Music J. **26**(4), 31–43 (2002)
20. Serrà, J., Arcos, J.L.: Particle swarm optimization for time series motif discovery. CoRR abs/1501.07399 (2015)
21. Serrà, J., Matic, A., Arcos, J.L., Karatzoglou, A.: A genetic algorithm to discover flexible motifs with support. In: 2016 IEEE 16th International Conference on Data Mining Workshops (ICDMW), pp. 1153–1158. IEEE (2016)

Towards a Case-Based Reasoning Approach to Dynamic Adaptation for Large-Scale Distributed Systems

Sorana Tania Nemeş(✉) and Andreea Buga

Christian Doppler Laboratory for Client-Centric Cloud Computing,
Johannes Kepler University of Linz, Software Park 35, 4232 Hagenberg, Austria
{t.nemes,andreea.buga}@cdcc.faw.jku.at

Abstract. The ever growing demands from the software area have led to the development of large-scale distributed systems which bring together a wide pool of services and resources. Their composition and deployment come in different solutions tailored to users requests based on business models, functionality, quality of service, cost, and value. Bridging different parts into one software solution is brittle due to issues like heterogeneity, complexity, lack of transparency, network and communication failures, and misbehavior. The current paper proposes a decision-based solution for the dynamic adaptation part of a middleware which addresses the aforementioned problems for large-scale distributed systems. The envisioned architecture is built on case-based reasoning principles and stands at the base of the adaptation processes that are imperative for ensuring the delivery of high-quality software. The solution is further extended through ground models with a focus on reliability, availability of components, and failure tolerance in terms of abstract state machines. The novelty of the approach resides in making use of formal modeling for one of the emerging problems and introducing an adequate prototype, on top of which one can apply reasoning and verification methods.

Keywords: Case-based reasoning · Formal modeling · Abstract state machines · Large-scale distributed systems · Adaptation

1 Introduction

Large-scale distributed systems (LDS) have appeared as a solution to the continuously expanding computing and storage demands. Their evolution has been favored by the advances in the area of service-oriented architecture (SOA) and the development of high-speed communication networks. Services offered through such architectures bring an increased value to the end client, but there are still many open questions posed by issues like heterogeneity, network failures, and random behavior of components, as identified by [15,33,34].

One of the biggest current trends in the IT community represented by cloud computing searches for efficient methods of expanding. The direction is set

D.W. Aha and J. Lieber (Eds.): ICCBR 2017, LNAI 10339, pp. 257–271, 2017.
DOI: 10.1007/978-3-319-61030-6_18

towards cloud federations which aim to improve the quality of service (QoS) by better resource usage, by costs reduction and by failure rate diminution. The Cloud federation approach matches the description of an LDS and serves as a good case study for our purposes. The processes and resources are mediated and coordinated normally with the aid of a middleware. In this case we talk about cloud-enabled LDS (CELDS).

The solution envisioned in our project proposes distributed middleware components containing different units, each in charge of a specific task, as for instance: service integration, process optimization, communication handling. In addition to these units, there is an abstract machine containing formal specifications for different levels: execution, monitoring and adaptation. The adoption of CELDS demands a deep understanding of the underlying infrastructure and its running mechanisms. Delivery of reliable services requires a continuous evaluation of the system state and adaptation in case of abnormal execution. Diagnosis is strongly correlated with the high-level interpretation of collected data. Moreover, different cloud providers tie their clients to their own services, making it extremely hard to create extensible solutions. The abstract machine supports formal specifications which can serve as starting points for establishing communication and interface models for inter-clouds.

The current paper presents a formal approach for modeling the decision process for the adaptation components inside a CELDS. Our project promotes a service-oriented approach to heterogeneous, distributed computing that enables on-the-fly run-time adaptation of the running system based on the replacement of sets of employed services by alternative solutions. For this we develop an advanced architecture and an execution model by envisioning and adapting a wide spectrum of adaptation means such as re-allocation, service replacement, change of process plan, etc. The approach we follow is an extension on top of cloud proprietary adaptation components. Model description is done using abstract state machines (ASMs) [9], which do not constrain the user to explicit implementation frameworks or programming languages and which are easily refined into more comprehensive specifications. We choose ASM method, because it is suitable for designing concurrent systems by expressing action parallelism and allowing modeling a component in separation with the others.

The remainder of the paper is structured as follows. Section 2 provides an overview of the system, followed by the structure of the proposed adaptation framework and its underlying decision support approach in Sect. 3. The formal specification of the adaptation framework is detailed in Sect. 4. Related work is discussed in Sect. 5, after which conclusions are drawn in Sect. 6.

2 System Overview

LDS aim to incorporate a large number of software services available through different delivery models (web services, cloud computing applications). The system specification needs to capture the communication, interaction and dependence among different components with the purpose of fulfilling users' requests. The focus of our work is directed towards CELDS which are the future solution for delivering high quality services to a wide spectrum of areas.

A clear image of the whole cloud organization requires collecting data from all the components and transforming it into domain-specific knowledge. However, before deciding the system state, it is important to build a robust infrastructure for communication and deployment of services. The solution we envision is part of a CELDS architecture model concerned with tackling heterogeneity of numerous constituent parts. Inner processes in charge with coordination and administration are included in the middleware specification. The architecture of the system is expressed as an abstract machine model containing different modules replicated at node level [10].

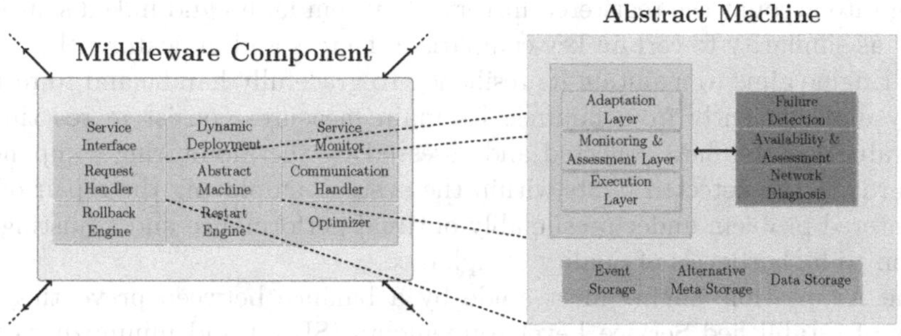

Fig. 1. Structure of the middleware and its internal abstract machine

As illustrated in Fig. 1, the middleware component is in charge with different operations needed for ensuring the composition of services, their communication, and optimal delivery to the end-client. The core makes use of ASMs for expressing the specification of the other components. Among all the aspects, we focus our attention on the monitoring and adaptation layers. As we rely on formal specifications, we neglect implementation details of individual services.

The monitoring and adaptation layers are cooperating for the delivery of expected QoS, each fulfilling clearly defined tasks, but closely collaborating. Monitors are responsible for collection of data, aggregating it into meaningful information and communicating observations about abnormal executions to the adaptation framework. The latter deals with recovering from anomalous situations, logging them, and finding the best remedy to restore the CELDS to normal running mode. The main issues tackled by the system are failure detection at both process and network level, and availability. Flaws and the solution chosen for their resolution are stored in a case-based repository which is continuously improved by the adaptation component.

With the focus on CELDS the project addresses service-oriented systems “in the large” and cloud interoperability, considering resilience, fault tolerance and performance. With respect to resilience the project targets system architectures that guarantee that a CELDS keeps running and producing desired results, even if some services become unavailable, change or break down. With respect to fault tolerance the project targets assessment methods that permit the detection of failure situations and adaptive repair mechanisms. With respect to performance the project targets likewise the detection of performance bottlenecks and subsequent correction. Therefore for CELDS, adaptability is a valuable and an almost inevitable process. This is mainly because cloud environments are not static, they are subject to continuous change: they evolve, and the parties must respectively adapt to new contexts varying from network traffic fluctuations to unavailability of different system components.

3 Organization of the Adaptation Framework

Taking into account the registered history of system faults and linked solutions, as well as similarity to certain key comparison factors and indicators, the Adaptation Engine aims to maintain its resiliency to gracefully handle and to recover quickly and efficiently from failures. Its main measures consist in reacting to and evaluating the data collected and assessed by the monitoring components in regards to the detected faults within the system, employing the repair of the encountered problem under presumably optimal performance and adjusting the solution to higher levels of quality compliance.

The Adaptation Engine must abide by a balance between preventing violations of established Service Level Agreements (SLAs) and minimizing costs, risks and resource consumption, while employing the recovery method that triggers the repair action associated to a particular problematic event. Maintaining such a balance will provide a useful basis for maximizing and maintaining high quality levels with respect to internal and external quality characteristics like reliability, efficiency and in use quality characteristics like effectiveness, safety and satisfaction [1].

As an inner component of the abstract machine included in the middleware described in Sect. 3, the Adaptation Engine is comprised of two major parts: the decision phase defined by solution exploration, identification and maintenance, and solution management and enactment phase, each with well delimited responsibilities and areas of inference and control. The current paper focuses on the decision phase of the adaptation process, emphasis on the actual adaptation being presented in a different article.

3.1 Case-Based System Development Procedure

At any point in time, the Adaptation Engine should be able to react to the input measurements and notifications from the monitoring component and output an adaptation solution that optimally avoids complete system failure and remedies the reported problems. By exploiting existing information resources like components' properties and states (average of QoS failures, cloud's violations responsiveness etc.) and by sharing monitoring knowledge, the Adaptation Engine builds and continuously enhances a catalog of cases that consist of linked adaptation actions and deterministic behaviors and events between them. In order to organize the cases in a manageable manner that supports efficient search and retrieval, the flat/linear model [32] (all cases are organized in the same level) is used, where the model is characterized by maximum accuracy, easy maintenance, and easy but rather slow retention in large case repositories.

The case repository is maintained and handled by the Case Manager which oversees the activities typical for a Case-Based Reasoning (CBR) cycle [3]. The CBR cycle is defined by identifying, applying, adapting and storing past registered solutions/experiences to similar problems by heavily relying on the quality and amount of the collected data, the background knowledge and the pattern discovery mechanism that determines the similarity between two problems/cases [5]. The development procedure adapted from [12] includes several steps, our focus at the moment being on the first three steps:

- **Knowledge acquisition.**
- **Case representation** which is described in detail in Sect. 3.2.
- **Verification and validation** in order to demonstrate the correctness, completeness and consistency of the chosen solution [4]. This is covered in detail in Sect. 4.
- **System implementation** which includes case repository and retrieval, and indexing process implementation.

3.2 Implementation of Case-Based Reasoning

A case C_r represents a formatted instance of a problem P_r linked to a recorded solving experience S_r. Table 1 singles out the role of each constituent part of the case in relation to different viewpoints of the adaptation process.

Table 1. The problem part versus the solution part of a case

Viewpoint	Problem part (P_r)	Solution part (S_r)
Contents	Problem	Solution
Objective	Selection	Adaptation
Access	Frequent	Not frequent
Representation	Feature-value tuples	Divergent
Data length	Fixed	Variable

Each case is defined as a universally unique identifier *uuid*, a collection of description features subject to a common pattern recognition mechanism and a finite set of repair actions also known as adaptation schema:

$$C_r = \{uuid, \{f_1, f_2, \dots, f_n\}, \{a_1, a_2, \dots, a_m\}\} \tag{1}$$

Such repair actions can be the replacement of a component service by an equivalent one or the change of location for a service, up to the replacement of larger parts of the CELDS, i.e. a set of running services, by a completely different, alternative solution. A description feature f_i is a predicate of the form:

$$f_i = \{param\_name_i, rel, param\_goal_i, param\_value_i\}$$
$$where\ rel \in \{<, \leq, =, \geq, >, |\},\ 1 \leq i \leq n \quad (2)$$

where $param\_name_i$ represents the parameter name, $param\_goal_i$ the parameter goal, $param\_value_i$ the provided value and rel the appropriate relation operator between the parameter and its goal. An example of such a feature can be seen in Eq. 3:

$$f_1 = \{\text{"Outgoing bandwidth"},\ \geq,\ 33.0,\ 30.0\} \quad (3)$$

In order to ensure a better case organization that deals with repository overflow without the loss of generality for the adaptation cases, the parameter value is represented either by a fixed value, a set of values or a numeric interval - this will be further detailed in the retain phase of an adaptation case. A feature without a goal filled in represents a measurement feature intended to convey a broader system state overview and to strengthen the querying process accuracy. An example of such a feature is showcased in Eq. 4:

$$f_2 = \{\text{"Availability"},\ |,\ null,\ 99.8\} \quad (4)$$

Typical cloud computing specific attributes that would constitute a case feature would include, to name a few:

- **Response time:** the duration of time between sending a request to a service and receiving a response [2].
- **Throughput:** the number of requests a service can handle in a certain amount of time [2].
- **Price:** it is a unit price per hour for usage of the cloud service [6].
- **Availability:** defines the amount of time the system is operational and accessible when required for use. In cases of downtime service providers generally pay penalties in different forms for consumers [27].
- **Message reliability:** services typically communicate with each other or with consumers through messages. These are dependent on the network performance. This means that if the connection channel is not reliable, then message delivery assurance is necessary.
- **Portability:** as cloud computing services are accessed over the internet through interfaces, service consumers need to be sure that the services will be working on different devices or on different platforms.
- **Region:** systems need to comply with legislation of the county/territory they are hosted in. Services should provide their locations to reflect the legal obligations the consumer would have if he used the service [6].

Based on the above-mentioned attributes, an example list of knowledge features constituent for a problematic case reported by the monitoring component is the following:

$$P_1 = \{\{\text{“Outgoing bandwidth”},\ \geq,\ 33.0,\ 30.0\}, \{\text{“Availability”}, |, \mathit{null}, 99.8\}, \{\text{“Physical machines”}, |, \mathit{null}, 18\}, \{\text{“Price”}, |, \mathit{null}, 50\}, \{\text{“Incoming bandwidth”},\ >,\ 20,\ 15\}\} \quad (5)$$

Once an error is reported by the monitoring component, the requested case *ReqC* is constructed based on the aggregated data provided by the monitoring component. The case deemed most similar to the specified problem is retrieved from the repository through application of domain-specific similarity measures between the new requested case *ReqC* and the existing ones *ExistC*:

$$i : \mathit{Sim}(\mathit{ReqC} : f_i, \mathit{ExistC} : f_i), 1 \leq i \leq \mathrm{m} \quad (6)$$

for each of the specified features f_i in the case base and a nearest neighbor assessment, based on a strengthened or weakened weighted (w) sum of all the features, initial and best matches are identified:

$$x = \frac{\sum_{i=1}^{m} w_i \times \mathit{Sim}(f_i^{\mathrm{ReqC}}, f_i^{\mathrm{ExistC}})}{\sum_{i=1}^{m} w_i} \quad (7)$$

Table 2 exemplifies some of the case features and assigned weights that reflect their relative significance to the case.

Table 2. Assigned weights for the case's features

Feature	Weight	Feature	Weight
CPU power	3.0	Storage	3.0
Throughput	4.0	Response time	4.0
Memory	3.0	Number of physical machines	3.0
Region	4.0	Outgoing bandwidth	3.0
Security	3.0	Maximum number of users	5.0
Availability	4.0	Price	6.0

The retrieved case's configuration is loaded and through simple operations of copy and adapt, the retrieved case is mapped to the current situation by abstracting away the differences $D_{ReqC,SolC}$ between the target problem P_{ReqC} and the retrieved case solution part S_{SolC}:

$$\mathit{ReqC} = P_{\mathrm{ReqC}} + D_{\mathrm{ReqC,SolC}} + S_{\mathrm{SolC}} \quad (8)$$

Reusing this case implies passing the configured solution to the Adaptation Manager where the defining adaptation actions are executed according to the action workflow schema describing them. For the case mentioned in Eq. 5, based on the portfolio of cases stored in the repository, a potential adaptation action would be to “increase the outgoing bandwidth by 5%”.

$$\mathit{ReqC} = \{\mathit{null},\ P_1, \{\text{“\textit{increase the outgoing bandwidth by} 5\%”}\}\} \quad (9)$$

Once the solution is carried out according to its specification, the monitoring component is requested to apply the workflow analysis and performance, accuracy and output evaluation to specific threshold values. In case of continuous failures (problem persists or is partially satisfactory) the Case Manager needs then to revise the case and optimize the definition of aggregated features by applying a set of rules to the feature values of the current case, making it a better fit with the new case requirements. And thus the cycle is repeated for the newly refined case. If there is no possibility to improve the unsatisfactory case, then the same limitation of no automated adaptation option is reached, as when retrieving an initial case from the repository.

Coming back to the case in Eq. 9, increasing the outgoing bandwidth by 5% does not solve the problem as the bandwidth would be at 31.5 instead of the targeted value of 33. The revised case's solution would need to increase the outgoing bandwidth by a minimum of 10%. So the minimum viable and satisfactory solution includes an increase by 10%:

$$ReqC = \{null,\ P_1, \{\text{``}increase\ the\ outgoing\ bandwidth\ by\ 10\%\text{''}\}\} \tag{10}$$

If the registered results are positive after applying the case, then the Case Manager is notified to classify the case as a valid solution, index it and retain the new problem-solving experience into the case repository for future problem solving:

$$ReqC = \{eaa61774 - 2aa7 - 11e7 - 93ae - 92361f002671,\ P_1,\\ \{\text{``}increase\ the\ outgoing\ bandwidth\ by\ 10\%\text{''}\}\} \tag{11}$$

One of the risks with case repositories specialized for exact parameter value is that the uncontrolled date growth causes performance loss as retrieval and identification of correct and incorrect cases degrades significantly. A countermeasure to that is to consider the cases' parameter value a set or an interval (as mentioned above) that would deem a stored case similarly suitable to a wider spectrum of input cases. Coverage and reachability are well known criteria for assessing and improving case base quality [31]. In other words, the retaining operations include review options as to generalize the retrieved case by exposing a larger value interval that includes the updated case as well (if the differences between the two cases are limited to the parameter values) or index and persist it as new valid case.

Leake and Wilson [20] examine and underline the importance of a strategic approach for addition and deletion of cases by balancing the performance and coverage criteria of such cases. A competence-based maintenance approach is suggested also by Smyth [30]. A future extension in this direction would be to consider the performance of a given case as a constituent parameter in the case's configuration and a determining factor in the similarity assessment process between cases.

In order to better understand the intrinsic problems that the decision process can face, attention is focused also on building ground models in terms of

ASMs. The designed ground model, subject to subsequent high level stepwise refinements, guarantees that requirements can be validated and properties as safety and liveness can be verified, already at the early stages of the system development.

4 Formal Specification

4.1 Abstract State Machine Theory

Model-driven engineering facilitates the collaboration of stakeholders for defining specific concepts and entities. Models evolve from natural-language requirements, use cases or user stories to formal specifications standing at the base of the software development process. Focus is oriented towards capturing correctly the functional and non-functional requirements of the system. Spotting errors later in the development process leads to higher costs for software projects. By model-driven techniques, properties can be validated and verified before development through simulation and model checking. Thus, costs incurred by wrong design and development are reduced.

ASMs rely on the concept of evolving algebras proposed by Gurevhich in [16]. Their proposal was motivated by their power to improve Turing machines with semantic capabilities. The ASM method allows a straightforward transition from natural-language requirements to ground model and control state diagrams, which can be easier formalized. An ASM machine M is represented by a tuple $M = (\Sigma, S_0, R, R_0)$, where Σ is the signature (the set of all functions), S_0 is the set of initial states of Σ, R is the set of rule declarations, R_0 is the main rule of the machine.

ASMs derive from the notion of Finite State Machine (FSM) to which they add synchronous parallelism capabilities. Hence, in an operation step several locations can be updated at the same time). They also enhance the *in* and *out* states with the possibility to express data structures.

The specification of an ASM consists of a finite set of *transition rules* of the type: **if** *Condition* **then** *Updates* [9], where an *Update* consists of a finite set of assignment $f(t_1, \ldots, t_n) := t$. As ASMs allow synchronous parallelism execution, two machines might try to change a location with two different values, triggering an inconsistency. In this case the execution throws an error.

Rules consist of different control structures that reflect parallelism (`par`), sequentiality (`seq`), causality (`if...then`) and inclusion to different domains (`in`). With the `forall` expression, a machine can enforce concurrent execution of a rule R for every element x that satisfies a condition φ: `forall` x `with` φ `do` R. Non-determinism is expressed through the `choose` rule: `choose` x `with` φ `do` R.

Definition 1. *A* control state ASM *is an ASM following the structure of the rules defined by: any control state i verifies at most one true guard,* $cond_k$*, triggering, thus,* $rule_k$ *and moving from state i to state* s_k*. In case no guard is fulfilled, the machine does not perform any action.*

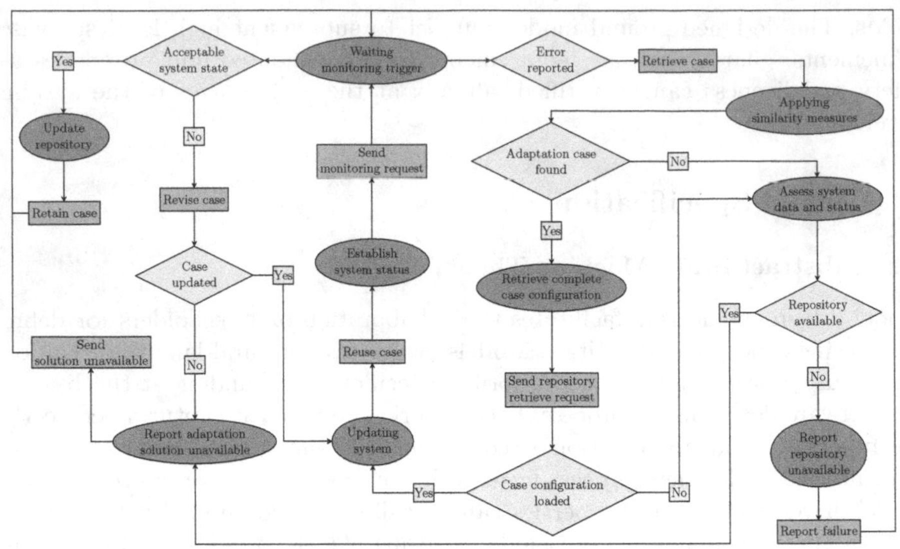

Fig. 2. Control state ASM ground model of the Case Manager

Functions in ASMs are classified according to whom the permissions on different operations belong. Static functions refer specifically to constants, while dynamic functions can be updated during execution. *Controlled* functions are written only by the machine, while *monitored* ones are written by the environment and read by the machine. Both the machine and its environment can update *shared* functions.

4.2 Control State ASMs

Based on the overall specification of the adaptation framework mentioned in Sect. 3.2, we define the specific states and transitions of the adaptation processes, with emphasis on the decision phase. The model contains CaseManager ASM agents, each of which carries out its own execution. The ground model illustrated in Fig. 2 emphasizes the whole decision system rather than the individual components and represents the starting point for further refinements and specifications. The CaseManager can pass through several states by various rules and guards.

At initialization, the CaseManager is in the *Waiting monitor trigger* state. This initial state is reached again either when the monitors are requested to evaluate the overall system state after the underlying system update or when a particular event has been successfully handled and new reported scenarios need to be tackled through new or existing adaptation cases. Every CaseManager agent exposes all CBR specific phases, which in turn each determines a state change (e.g.: once a case is retrieved and reused, the CaseManager is in the *Updating system* state; when the new case needs to be retained in the repository, the CaseManager is in the state *Updating repository*; once the case is reused, the CaseManager is in state *Establish system data* and the monitoring component is signaled on assessing the system status as a result of the latest update).

As a first step towards building the knowledge base or case repository, the case repository is filled with some meaningful initial cases, representative for the actions that can be triggered. We need to make sure that the repository is not filled in with superfluous cases, as although we are talking about a generative system that must account for all possible problems, the Case Manager needs to be able to handle the types of problems that do occur in practice [19].

Given this context, there might not be an existing case to match the newly reported problem; thus the CaseManager is in state *Report adaptation solution unavailable* and the system is notified on the lack of possibility of automatic adaptation for this particular system failure. Another scenario for not retrieving an adaptation case is of course the unavailability of the case repository. The CaseManager's state then changes to *Report repository unavailable* and thew problem is handled and reported as a system failure by the monitoring component. Thus when a case is not retrieved, the system status and data are assessed and handled accordingly.

4.3 AsmetaL Specification

ASMETA[1] consists of a toolset for simulating, validating and model-checking ASM models. The specification is written in the AsmetaL language, which is able to capture specific ASM control structures and functions. The first part of the specification contains signatures of the domains and of the functions. *Dynamic* domains can be extended with new instances. The *CaseManager* domain is part of the Agent universe and it behaves as an ASM machine, having its own states and transitions. The *Notification* domain is left abstract. The states of an CaseManager are part of an enumeration domain. In a future refinement we want to add the *Adaptation Engine* agent that handles the deployment of the CaseManagers in the system.

```
dynamic domain CaseManager subsetof Agent
dynamic abstract domain Notification
dynamic abstract domain Case
enum domain CaseManager_States = {WAITING_MONITOR_TRIGGER |
      APPLYING_SIMILARITY_MEASURES | RETRIEVE_COMPLETE_CASE_CONFIGURATION |
      ASSESS_SYSTEM_DATA_AND_STATUS | REPORT_REPOSITORY_UNAVAILABLE |
      UPDATING_SYSTEM | ESTABLISH_SYSTEM_STATUS |
      REPORT_ADAPTATION_SOLUTION_UNAVAILABLE | UPDATE_REPOSITORY }

controlled manager_state: CaseManager -> CaseManager_States
controlled case_repository: CaseManager -> Seq(Case)
controlled monitor_notification: CaseManager -> Seq (Notification)
monitored update_successful: CaseManager -> Boolean
```

Listing 1.1. Signatures of domains and functions

CaseManager state is expressed as a controlled function which is updated by the CaseManager agent itself. The cases assigned to a specific adaptation solution are expressed as a sequence. Each CaseManager contains a sequence of

[1] http://asmeta.sourceforge.net/.

Notification requests it sends to the monitor components to validate the post-update state of the system. A brief extract of the signature of domains and functions related to the CaseManager agent are captured in Listing 1.1. All the syntax and control structures presented in Sect. 4.2 are expressed with AsmetaL and are subject to further simulation and validation. The following code excerpt in Listing 1.2 represents the rule for retaining a case in the repository.

```
rule r_RetainCase($c in Controller) =
  seq
    while (manager_state($c) = WAITING_MONITOR_TRIGGER and not(notification_response_arrived($c)) do
      wait
    if (update_successful($c))
      seq
        manager_state($c) := UPDATING_REPOSITORY
        extend Case with $case do
          seq
            case_repository($c) := append(case_repository($c), $case)
            manager_state($c) := WAITING_MONITOR_TRIGGER
          endseq
      endseq
    else
      r_ReviseCase[$c]
    endif
endseq
```

Listing 1.2. Retain case rule

The rule validates the status of the manager and the monitor's response to the sent notification for system status analysis. If the update as part of the adaptation solution was successful, then the case is added to the already collected cases. Otherwise, the rule to revise the case in question is triggered.

More than one system failure can be reported in a short time frame. Therefore, the handling part is done in a sequential mode because, although a case is locked while it's associated solution is executed, a parallel execution of simultaneous adaptations may try to update system parts or components with different values at the same time. The inconsistency error was detected at simulation time with the aid of the AsmetaS tool. We leave as a future work the elaboration of transaction specific operations, which would permit triggering simultaneously multiple adaptions within the system. This could be supported by annotating the case with extensive knowledge on the area of inference in the system of each case, which would later on be considered in the retrieval phase of the process.

Another aspect to consider as future refinement of the ground model is to take into account the level of compliance or possible valid states of the system in relation to the cost of adaptation. This became apparent also through simulations with AsmetaS tool. One adaptation can bring the system to a state that is compliant to a certain percentage (not 100%) to the users' needs but is acceptable given its cost and the cost of executing a revised case for the needed system adaptation. The cost in question can represent time, price, resources etc.

5 Related Work

Most of the related work on knowledge management and assessment in self-adaptable Clouds covers SLA management and, in some cases, preventive SLA management. Some papers like [14] only describe the monitoring of SLAs while other papers like [13] do describe in depth the process of how to fulfill an SLA, which very often are limited to just one Service Level Objective (SLO) like CPU usage or response time.

A rule based approach for dynamically dealing with SLAs in combination with ContractLog is described by Paschke and Bichler [28]. The Conversion Factory introduced by Hasselmeyer et al. [17] creates at design level Operational Level Agreements (OLA's) by combining the SLA, the Business Level Objectives (BLO) and the system status. The Reservoir model [29] is a framework for cloud computing which underlines the importance for resource adjustment to established SLAs, but does not indicate a way to do that.

Successful CBR-based approaches to process or product design have been developed in [18,22–26]. The main contribution of this paper consists in formally modeling, validating and verifying properties of the Case Manager, which increase the safety of the LDS. Ensuring from design time that the adaptation components behave correctly and react to the identified problems of the system enhances the reliability of the whole system.

Formal modeling, more specific the ASM technique, contributed to the description of the job management and service execution in [7], work that was further extended by [8]. ASMs have been also proposed for realization of web service composition. In [21], authors introduced the notion of Abstract State Services and showed an use case for a cloud service for flight booking. Service composition and orchestration in terms of ASMs has been researched by [11].

6 Conclusions

This article pertains to formally capturing, with the required generality, the conceptual and behavioral range of possible dynamic adaptation changes and corresponding evolution patterns of LDS, while achieving the right trade-off between the functional and non-functional adaptation requirements and the adaptation cost itself. We defined the specific rules and states of the adaptation model through the AsmetaL language part of the Asmeta toolset that allows model simulations.

In the future steps of our work we aim to improve the formal model in order to capture finer-level details. We plan to achieve separation of concerns by employing ASM modules for different constituent actors of the adaptation framework. A parallel development will be to ensure the correctness of the detailed solution by undergoing verification through the AsmetaSMV tool.

References

1. Iso 9126:2001 software engineering - product quality (2001)
2. OASIS Committee Specification 01: Advancing open standards for the information society (oasis), 22 September July 2011
3. Aamodt, A., Plaza, E.: Case-based reasoning: foundational issues, methodological variations, and system approaches. AI Commun. **7**(1), 39–59 (1994)
4. Adrion, W.R., Branstad, M.A., Cherniavsky, J.C.: Validation, verification, and testing of computer software. ACM Comput. Surv. **14**(2), 159–192 (1982)

5. Althoff, K.-D.: Case-based reasoning. Handb. Softw. Eng. Knowl. Eng. **1**, 549–587 (2001)
6. Becha, H., Amyot, D.: Non-functional properties in service oriented architecture - a consumer's perspective. JSW **7**(3), 575–587 (2012)
7. Bianchi, A., Manelli, L., Pizzutilo, S.: A distributed abstract state machine for grid systems: a preliminary study. In: Topping, B.H.V., Iványi, P. (eds.) Proceedings of the 2nd International Conference on Parallel, Distributed, Grid and Cloud Computing for Engineering. Civil-Comp Press, April 2011
8. Bianchi, A., Manelli, L., Pizzutilo, S.: An ASM-based model for grid job management. Informatica (Slovenia) **37**(3), 295–306 (2013)
9. Börger, E., Stark, R.F.: Abstract State Machines: A Method for High-Level System Design and Analysis. Springer-Verlag New York Inc., Secaucus (2003)
10. Bósa, K., Holom, R.-M., Vleju, M.B.: A formal model of client-cloud interaction. In: Thalheim, B., Schewe, K.-D., Prinz, A., Buchberger, B. (eds.) Correct Software in Web Applications and Web Services, pp. 83–144. Springer, Heidelberg (2015). doi:10.1007/978-3-319-17112-8_4
11. Brugali, D., Gherardi, L., Riccobene, E., Scandurra, P.: Coordinated execution of heterogeneous service-oriented components by abstract state machines. In: Arbab, F., Ölveczky, P.C. (eds.) FACS 2011. LNCS, vol. 7253, pp. 331–349. Springer, Heidelberg (2012). doi:10.1007/978-3-642-35743-5_20
12. Chan, C.W., Chen, L.-L., Geng, L.: Knowledge engineering for an intelligent case-based system for help desk operations. Expert Syst. Appl. **18**(2), 125–132 (2000)
13. Chen, Y., Iyer, S., Liu, X., Milojicic, D., Sahai, A.: Translating service level objectives to lower level policies for multi-tier services. Clust. Comput. **11**(3), 299–311 (2008)
14. Fakhfakh, K., Chaari, T., Tazi, S., Drira, K., Jmaiel, M.: A comprehensive ontology-based approach for SLA obligations monitoring. In: 2008 the 2nd International Conference on Advanced Engineering Computing and Applications in Sciences, pp. 217–222, September 2008
15. Grozev, N., Buyya, R.: Inter-cloud architectures and application brokering: taxonomy and survey. Softw. Pract. Exper. **44**(3), 369–390 (2014)
16. Gurevich, Y.: Evolving algebras 1993: Lipari guide. In: Börger, E. (ed.) Specification and Validation Methods, pp. 9–36. Oxford University Press Inc., New York (1995)
17. Hasselmeyer, P., Koller, B., Schubert, L., Wieder, P.: Towards SLA-supported resource management. In: Gerndt, M., Kranzlmüller, D. (eds.) HPCC 2006. LNCS, vol. 4208, pp. 743–752. Springer, Heidelberg (2006). doi:10.1007/11847366_77
18. Henninger, S., Baumgarten, K.: A case-based approach to tailoring software processes. In: Aha, D.W., Watson, I. (eds.) ICCBR 2001. LNCS, vol. 2080, pp. 249–262. Springer, Heidelberg (2001). doi:10.1007/3-540-44593-5_18
19. Leake, D.B.: Case-Based Reasoning: Experiences Lessons and Future Directions, 1st edn. MIT Press, Cambridge (1996)
20. Leake, D.B., Wilson, D.C.: Remembering why to remember: performance-guided case-base maintenance. In: Blanzieri, E., Portinale, L. (eds.) EWCBR 2000. LNCS, vol. 1898, pp. 161–172. Springer, Heidelberg (2000). doi:10.1007/3-540-44527-7_15
21. Ma, H., Schewe, K.D., Wang, Q.: An abstract model for service provision, search and composition. In: 2009 IEEE Asia-Pacific Services Computing Conference (APSCC), pp. 95–102, December 2009
22. Madhusudan, T., Zhao, J.L., Marshall, B.: A case-based reasoning framework for workflow model management. Data Knowl. Eng. **50**(1), 87–115 (2004)

23. Maher, M.L., de Silva Garza, A.G.: Case-based reasoning in design. IEEE Expert **12**(2), 34–41 (1997)
24. Mukkamalla, S., Muñoz-Avila, H.: Case acquisition in a project planning environment. In: Craw, S., Preece, A. (eds.) ECCBR 2002. LNCS, vol. 2416, pp. 264–277. Springer, Heidelberg (2002). doi:10.1007/3-540-46119-1_20
25. Müller, G., Bergmann, R.: Workflow streams: a means for compositional adaptation in process-oriented CBR. In: Lamontagne, L., Plaza, E. (eds.) ICCBR 2014. LNCS, vol. 8765, pp. 315–329. Springer, Cham (2014). doi:10.1007/978-3-319-11209-1_23
26. Munoz-Avila, H., Weberskirch, F.: Planning for manufacturing workpieces by storing, indexing and replaying planning decisions (1996)
27. O'Brien, L., Merson, P., Bass, L.: Quality attributes for service-oriented architectures. In: Proceedings of the International Workshop on Systems Development in SOA Environments, SDSOA 2007, p. 3. IEEE Computer Society, Washington, DC (2007)
28. Paschke, A., Bichler, M.: Knowledge representation concepts for automated SLA management. Decis. Support Syst. **46**(1), 187–205 (2008)
29. Rochwerger, B., Breitgand, D., Levy, E., Galis, A., Nagin, K., Llorente, I.M., Montero, R., Wolfsthal, Y., Elmroth, E., Caceres, J., Ben-Yehuda, M., Emmerich, W., Galan, F.: The reservoir model and architecture for open federated cloud computing. IBM J. Res. Dev. **53**(4), 4:1–4:11 (2009)
30. Smyth, B.: Case-base maintenance. In: Pasqual del Pobil, A., Mira, J., Ali, M. (eds.) IEA/AIE 1998. LNCS, vol. 1416, pp. 507–516. Springer, Heidelberg (1998). doi:10.1007/3-540-64574-8_436
31. Smyth, B., Keane, M.T.: Remembering to forget: a competence-preserving case deletion policy for case-based reasoning systems. In: Proceedings of the 14th International Joint Conference on Artificial Intelligence. IJCAI 1995, vol. 1, pp. 377–382. Morgan Kaufmann Publishers Inc., San Francisco (1995)
32. Soltani, S.: Case-based reasoning for diagnosis and planning. Technical report, Queens University, Kingston (2013)
33. Toosi, A.N., Calheiros, R.N., Buyya, R.: Interconnected cloud computing environments: challenges, taxonomy, and survey. ACM Comput. Surv. **47**(1), 7:1–7:47 (2014)
34. Villegas, D., Bobroff, N., Rodero, I., Delgado, J., Liu, Y., Devarakonda, A., Fong, L., Sadjadi, S.M., Parashar, M.: Cloud federation in a layered service model. J. Comput. Syst. Sci. **78**(5), 1330–1344 (2012)

Evolutionary Inspired Adaptation of Exercise Plans for Increasing Solution Variety

Tale Prestmo[1], Kerstin Bach[1(✉)], Agnar Aamodt[1], and Paul Jarle Mork[2]

[1] Department of Computer Science,
Norwegian University of Science and Technology, Trondheim, Norway
[2] Department of Public Health and Nursing,
Norwegian University of Science and Technology, Trondheim, Norway
http://www.idi.ntnu.no, http://www.ntnu.no/ism

Abstract. An initial case base population naturally lacks diversity of solutions. In order to overcome this cold-start problem, we present how genetic algorithms (GA) can be applied. The work presented in this paper is part of the selfBACK EU project and describes a case-based recommendation system that creates exercise plans for patients with non-specific low back pain (LBP). In selfBACK Case-Based Reasoning (CBR) is used as its main methodology for generating patient-specific advice for managing non-specific LBP. The sub-module of selfBACK presented in this work focuses on the adaptation process of exercise plans: A GA inspired method is created to increase the variation of personalized exercise plans, which today are crafted by medical professionals. Experiments are conducted using real patients' characteristics with expert-crafted solutions and automatically generated solutions. In the evaluation we compare the quality of the GA-generated solutions to null-adaptation solutions.

Keywords: Case-Based Reasoning · Similarity assessment · Adaptation · Genetic algorithm · Cold start problem

1 Introduction

Up to 80% in the adult population of Norway will experience low back pain during their lifetime, and a study showed that 50% of them had experienced pain during the last 12 months [18]. About 85% of these will experience non-specific low back pain, i.e., pain without a known pathomechanism [6]. As an example, back pain is the largest single cause of sickness leave in Norway, and it costs about 2% of the gross domestic product. Even though the amount of research in the area has increased, as well as the access to treatment and less physically demanding work, the costs have significantly increased over the last 30 years. General physical activity along with specific strength and stretching exercises constitute the core components in the prevention and management of non-specific low back pain.

CBR has been used in the domain of health science for a long time, because its method of using past experiences to solve a new problem lies very close to

D.W. Aha and J. Lieber (Eds.): ICCBR 2017, LNAI 10339, pp. 272–286, 2017.
DOI: 10.1007/978-3-319-61030-6_19

how clinical medicine is performed by specialists today. It is also a field where one often have the advantage of already having a collection of past cases to use when reviewing a new problem. The use of CBR in health sciences has proven to be so popular that over the past 10 years it has become a specialized sub-area within CBR research and application. There exist CBR-systems that are used commercially in the field of medicine, but it has still not become as successful here, in terms of successfully deployed applications, as in many other domains [5,9].

The SELFBACK project aims at creating a self-management tool for patients with non-specific low back pain, which will support them to self-manage their pain by obtaining personalized advice and continuous follow-up. After an initial screening of the patient using questionnaires, the patient gets access to a wearable and a smart phone app that is the interface to the decision support system. The wearable will be used to track activities and obtain objective measurements while the smart phone app displays feedback, shows progress in achieving the patient's goals, and obtains regular follow-up on pain, function and self-efficacy development. This includes for example whether the pain level decreases, the functionality increases and coping with pain improves. Figure 1 gives an overview of the architecture. A more thorough description of the CBR approach in SELFBACK is given in [2]. This work focuses on how an adaptation phase can further improve the creation of exercise plans.

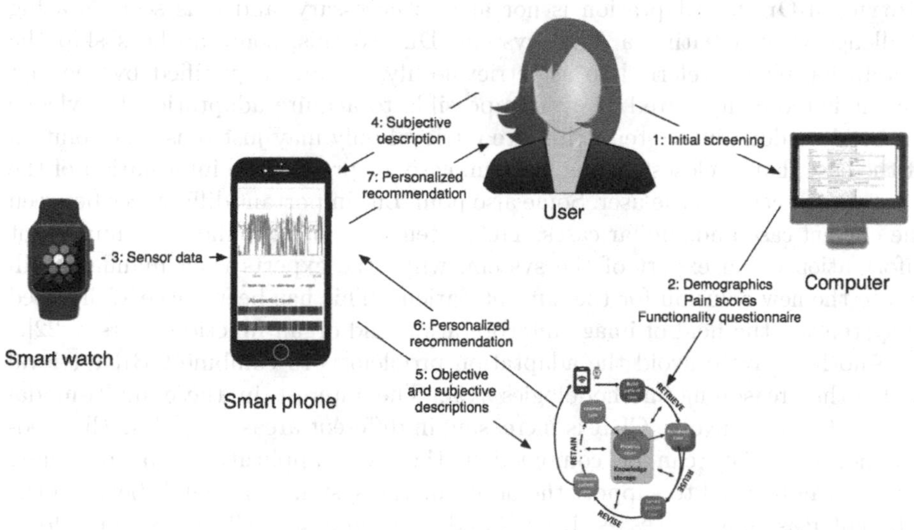

Fig. 1. The overall SELFBACK architecture

1.1 Background

The adaptation part of CBR is one of the most challenging issues for CBR systems in general as well as in the health sciences, where it has traditionally been carried out manually by experts of the domain. In recent years, however, the problem has been more focused. Several systems explore different approaches to automatic and semiautomatic adaptation strategies [4]. It has also been argued that the adding of adaptation is what makes the CBR system an artificial intelligence method, and that without it can be seen as a simple pattern matcher [12]. The challenge with the adaptation phase is that it is hard to find a general strategy for case adaptation, and therefore the adaptation techniques generated are often domain specific. Adaptation is a challenge not only in the medical domain, but it is usually more complex here because cases often consist of a large number of features [22]. The reason for doing adaptation is because usually you can't reuse solutions of cases directly when you have a new case [8].

One of the reasons for the focus on adaptation in the work reported here is to deal with the cold-start problem in the beginning of the deployment of a CBR system. The cold-start problem describes the situation where the amount of cases is too low to create a good solution to the new problem [14]. Or, alternatively, if you want to introduce some variations of the solution to make a system more personalized or adaptive.

Retrieval-Only. Adaptation is not always necessary, and it is seen as a big challenge when creating a CBR system. Due to this, some authors skip the adaptation phase, referred to as retrieval-only. It can be justified by the fact that it is too complicated or even impossible to acquire adaptation knowledge in the given domain. Systems that are retrieval-only may just reuse the solution of the case that is closest to the problem case, or present the information of the most similar cases to the user. Some also point out important differences between the current case and similar cases. The system may present the most important information to an expert of the system, while the experts then manually will create the new solution for the current patient. This has been successfully used in systems in the field of image interpretation and organ function courses [22].

Another way to avoid the adaptation problem is to combine CBR retrieval with other reasoning methodologies [19]. The interest in these multi-modal approaches that involve CBR is increasing in different areas, including the medical domain. They can be combined in the same application, one reasoning process can be used to support the other, or the system can switch between the different reasoning processes. Rule Based Reasoning as well as reasoning form extended probabilistic and multi-relational models may be combined with CBR. A straight-forward combination is that rules and cases cooperate such that rules deal with reasonably standard or typical problems, while CBR faces exceptions, but they can be integrated in other ways [22]. Another example is to use rules or other generalized models an explanatory support to the case process [16].

Genetic Algorithms. Genetic Algorithms (GA) are adaptive heuristic search algorithms that are based on the natural process of evolution, known as the survival of the fittest. Systems that use GA are modeled loosely on the principles of evolution via natural selection through variation-inducing operators such as mutation and crossover. To have success you have to have a meaningful fitness evaluation and an effective GA representation. One reason to use this method is that it is capable of discovering good solutions in search spaces that are large, complex or poorly understood, where the domain knowledge is limited or the expert knowledge is difficult to encode in rules or other models. The use of GA may not find the optimal solution, but it usually comes up with a partially optimal solution [13].

1.2 Related Work

GAs have already been combined with CBR to optimise case retrieval, clean up case memory and create new and unique cases. They have also been used in the adaptation step, to achieve an adaptation technique that is not domain specific [12]. One of the most well known approaches for applying evolutionary algorithms to case adaptation is [11], in which the incremental evolution of solution candidates creates novel solutions. While this approach is general and knowledge independent, the work we are presenting in this paper includes domain knowledge from the case representation for guiding the evolution process.

Case-based reasoning is used in several health systems today, within a lot of different areas such as clinical diagnosis and treatment in psychiatry [21]. It has become a recognized and well-established method for the health sciences, and since the domain of health sciences is offering a variety of complex tasks which are hard to solve with other methods and approaches, it drives the CBR research forward. Since CBR is a reasoning process that works similarly to the reasoning of a clinician, with the use of previous experiences to solve the same or similar cases, it has become medically accepted and is also getting increased attention from the medical domain [4]. There are several advantages of using CBR in the medical domain, one is that with the use of CBR it is possible to find solutions to problems even though the complete understanding of the domain is not captured, or if the domain is very complex. The reuse of earlier solutions saves time since it is not necessary to solve every problem from scratch, and it allows learning from mistakes. The fact that cases hold a lot of information makes it usable for a number of different problem-solving purposes, compared to rules that can only be used for the purpose they were designed for [21].

Looking into previous work, we will now focus on relating our approach to existing CBR applications in the medical field and later on discuss how genetic algorithms come into play.

CASEY [17] is one of the earliest medical decision support systems that applied CBR, and it deals with heart failure diagnosis. It first retrieves similar cases, then looks at the differences between the current case and the similar case. If the differences are not too important it transfers the diagnosis of the similar case to the current one, and if the differences are too large it attempts to explain and modify the diagnosis. It falls back on a probabilistic network type of domain model if this does not work, or if no similar case can be retrieved.

Protos [3] is another well-known early medical CBR system. It addressed the problem of concept learning and classification in weak-theory domains, such as medicine. It combined cases with a multi-relational network model used to explain case matching if features were syntactically different bu semantically related. Its domain was hearing disorders, and in the final testing it performed very well compared to clinical audiologists.

Another system that uses CBR deals with anterior cruciate ligament (ACL) injury [23], and it combines fuzzy logic with CBR. The system is not intended to interact directly with the user, but with experts such as sport trainers, coaches, and clinicians for multiple purposes in context of the ACL injury such as monitoring progress after an injury and predicting performance. It uses body-mounted wireless sensors to retrieve the input data for the case, while the solution part consists of recovery classification, treatment at different stages, as well as performance evaluation and prognosis. All the information is stored in the knowledge base with a profile of the patient and information about the recovery sessions.

One of the top fatal diseases in the world is cancer, and as part of their cancer treatment patients get diets to reduce the side-effects of the treatment, as well as making sure they get sufficient nutrition to boost the recovery cycle. This personalized diet recommendation system for cancer patients [13] makes use of the data mining techniques of CBR, and combines them with rule-based reasoning and a genetic algorithm. The CBR part of the system retrieves a set of diet plans from the case base, while the rule-based reasoning is used on this set to do further filtering of irrelevant cases. Then the genetic algorithm is used for the adaptation phase to make sure each diet menu is customized according to the patient's personal health condition. The solution part of this system consists of a menu recommendation that suggest dishes for the patient, as well as a list of specific nutritional values to be taken daily.

Radiotherapy treatment tries to destroy tumor cells with radiation, and radiotherapy treatment planning tries to make sure the radiation dose is sufficient to destroy the cells without damaging healthy organs in the tumour-surrounding area. The normal process of creating a solution to this problem can take everything from 2–3 h to a few days, which makes it time-consuming, and it includes a group of experts in the area that you are dependent on. The Radiotherapy treatment planning CBR-system [21] was created to attempt to make the process faster and without the need to have several experts involved. The case base in the created system consists of cases made out of brain cancer patient descriptions as well as the plan used for the treatment. The treatment, i.e. the solution part, consists of the number of beams applied to the tumour and the angles of those beams. The system creates a new solution for the patient based on earlier patient cases and their treatment plans.

2 Case Representation

The case representation is based on the SELFBACK questionnaire, as this creates the basis for the data used in the experiments. The questionnaire describes the characteristics of a patient with non-specific low back pain. It covers areas such as the pain level, their quality of life despite the pain, functionality, coping capabilities and their physical activity level.

From the overall characteristics three different types of advice will be generated to support self-management:

- Goals for physical activity: number of steps/day, maximum of inactive periods during hours the patient is awake
- Education: Tailored list of educational exercises that support and reassure the patient in his/her self-management.
- Exercise: A customized list of exercises that combine clinical guidelines for low back pain with past cases into action items.

In the following, we will focus on the generation of exercise lists based on given cases. Therefore our case representation consists of two different concepts, the patient characteristics and the list of exercises at a given time. The patient characteristics are taken as problem description and the exercise are describing the solution part. These two different concepts are explained in further detail in Sects. 2.1 and 2.2.

2.1 Patient Concept

The patient concept consists of 44 attributes that describe different aspects of the patient's health. These attributes can be divided into different groups of information collected by (1) the SELFBACK questionnaire, (2) a physical activity detecting wristband worn by the patient, and (3) an interaction module in the SELFBACK app. The attributes collected by the questionnaire are a combination of important prognostic factors and outcome measures. Pain self-efficacy and beliefs about back pain have been shown to have great impact on the future course of low back pain [15]. Likewise, baseline pain and pain-related disability have strong influence on the course of low back pain but these attributes are also important outcome measures [7]. Quality of life at baseline may also influence the course of low back pain but this is mainly considered an important outcome measure [10]. An example of a patient concept can be seen in Fig. 2. The patient data in this case is made up, as data from real patients are confidential.

Demographics. With a new patient it is necessary to know some simple demographics, such as height, weight, age and gender. These are the basis for each patient, and are all quite easy attributes to measure. All of these attributes may influence the solution, as all attributes can be an indication of how well a patient is able to perform and follow-up on a particular exercise plan. Young people are usually stronger and more fit than older people, men are in general stronger than women, and younger people are usually able to carry out more intense physical activity or exercises than older people. Obese people may need to focus on other exercises than normal weight people.

Quality of life. The impact of low back pain on quality of life is another important measure of the severity and consequences of the back pain. As an additional measure, the patient also provides a score in his/her own health from 0 (worst) to 100 (best).

Pain self-efficacy and beliefs about back pain. Scoring of pain self-efficacy indicates if the patient is confident that he/she can do various activities regardless of the pain and is therefore an important measure of how the patient copes with the pain. A related measure is fear-avoidance beliefs, i.e., to what extent the patient believes that physical activity will be harmful and exacerbate the back pain.

Physical activity and exercise. Information about general physical activity is assessed by the SELFBACK questionnaire and the physical activity detecting wristband. The attributes assessed by the questionnaire include work characteristics (i.e., physical work demands), physical activity limitations in everyday activities (work and/or leisure) due to back pain, and level of leisure time physical activity. Physical activity information that can be derived from the wristband data includes several attributes, such as step count (including intensity [i.e., step frequency] during walking/running), and distribution of active and inactive periods during wake time. The interaction module in the SELFBACK app will ask the patient about accomplishment and adherence to the exercises prescribed in the self-management plan as well as a rating of whether the patient perceived the prescribed exercises as useful and enjoyable. All these attributes will say something about how active the patient is and the coping behavior related to his/her low back pain.

Pain and pain-related disability. Information about various aspects and characteristics of pain is relevant for the case, both because it can track progress and it provides an indication on how severe the case is. History of low back pain provides information about whether the patient has experienced similar problems before, if it is a recurrent problem, or if it is a long-lasting ('chronic') problem. Number of pain sites reported by the patient is important to assess musculoskeletal co-morbidity while the scoring of pain-related disability provide information about how the back pain influence function.

Exercise list. To connect the two concepts, the patient has an attribute that is a list of all the exercises the patient has in his solution part. This consists of cases on the form of the exercise concept that is further described below.

2.2 Exercise Concept

The exercise concept consists of four different attributes. An example of how the exercise concept looks like can be found in Table 1 in the results section.

Description. The descriptive name and type of the exercise. The type can be a strength, flexibility or pain-relief exercise. All patients are encouraged to perform strength and/or flexibility exercises each week, unless they are unable because of strong pain. In general, strength and flexibility exercises

Feature	Example
Gender	Male
Age	45
Height	1.89m
Weight	82
BMI	23
Quality of life (EQ5D)	90
Disability (RMDQ)	9
Pain (NPRS)	8
Work type	Mostly sitting
Self-Efficacy (PSFS) activity	Prolonged standing
Self-Efficacy (PSFS) Σ	8
FABQ Physical Activity	2
Pain medication	none
Pain history	none

Fig. 2. Patient example

are not recommended in the acute stage of a low back pain episode. By performing exercises regularly the patient will increase strength and improve flexibility, which over time will prevent relapse. In the acute stage or in case of a relapse, pain-relief exercises can be recommended to help the patient to relax and reduce the most intense pain. These exercises will mainly help to relieve acute pain but will have limited relevance when the patient is pain-free.

Level. An exercise can have different levels. The strength exercises used in this project have up to six different levels, where each level is a new variation of an exercise that is more demanding than the former. The patient changes levels as he/she progresses, i.e., first by increasing the number of repetitions within a level before moving to the next level.

Repetitions. Each exercise is performed in sets, with a given number of repetitions for each set. There are four levels of repetitions, 8, 10, 12 and 15 repetitions respectively. When the patient is able to perform 12–15 repetitions per exercise the patient moves up a level in the exercise.

Set. The set indicates the number of times the patient should perform the given repetitions for the exercise.

3 Experiments

In the experiments two different approaches are used, a no-adaptation and a genetic algorithm, to see how they compare in regard to solution variety as well as solution quality. Our hypothesis is that the GA inspired approach will produce better solution variety, but it will also have to produce solutions of good quality to be useful.

3.1 Case-Set

The cases used for testing the algorithm are gathered from the SELFBACK project, and consist of data from real patients who experience low back pain. A total of nine cases were created with an associated solution crafted by medical professionals.

3.2 No-Adaptation

The first approach, and also one of the most used approaches, is the no-adaptation approach. This approach did not require any design choices as this solution was built out of the box from the myCBR workbench REST API[1]. This approach is dependent on a comprehensive case base with a high case variation to be able to provide a good solution for all the different patients, as this does not evolve over time. The number of solutions will always be equal to or less than the number of cases you have, and this does not give enough room for patients having different needs and different baselines. In addition, this solution does not allow the patient to increase his or her level, nor the number of exercises or the frequency of the exercises.

3.3 Genetic Algorithm

A genetic algorithm was incorporated in the CBR cycle to perform an adaptation on the cases. The idea behind the genetic algorithm is to retrieve the two most fit cases, and combine them to create a new case. This approach is based on how nature evolves, and the assumption behind this approach is that the combination of the two best cases will give a satisfactory solution.

General Algorithm Structure. A genetic algorithm consist of different parts. It has a fitness function, i.e. a function that helps you describe how good a given specimen is. It also has a crossover function, which creates a new solution based on the two fittest individuals. The algorithm is programmed to stop at a termination condition, where the new solution satisfies the given condition. In the genetic algorithm you also have a probability of a mutation to happen. This changes one of the attributes in the solution at random, to possibly create better solutions, and avoid getting stuck in local maxims.

Adapted Algorithm. The general structure of a genetic algorithm was adapted to fit the domain. The fitness function in the adapted algorithm is based on the similarity scores between the cases, and the two fittest individuals from the population are chosen by retrieving the two most similar cases to the new problem description from the case base. From these two cases we retrieve their solution, the exercise list, and we create a chromosome of the solutions that is used by the genetic algorithm. The chromosome is built up such that all exercises for the same muscle group are placed inside the same gene, as each gene represents a specific trait of an individual. An example of how this mapping is done can be seen in Fig. 3.

[1] https://github.com/kerstinbach/mycbr-rest-example.

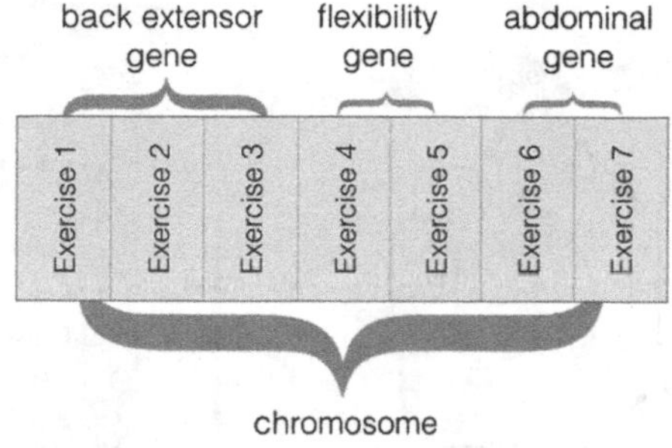

Fig. 3. The exercise list mapped to a GA chromosome

In Fig. 4 you see the description of the 4R cycle in this work with an adaptation example. Based on the patient description the two best matching cases are retrieved (C1 and C2). The two chromosomes representing the solution parts (S(C1) and S(C2)) are then sent to the crossover function. Here a new individual is created of the parent chromosomes, and it is done with a uniform crossover [24]. The mixing ratio is set to 0.5, since the solution is desired to have a close to equal mix of the parents' genes. The adapted algorithm finishes after one crossover at this moment as there exists no good measures to describe how well a patient will progress before they have executed the exercise plan. Measurements on progress will be added at later iterations, but in this work we only address the initial creation of plans.

The exercises also have a probability of 1.5% to have a mutation. These mutations are given some restrictions, such as that the type of exercise will be kept, but the level and the number of repetitions may alter by one level. The reason for such restrictions to the mutation is so that the algorithm should produce a solution that is feasible for the patient to fulfill and therefore not demotivating for the patient. Further the suggested solution should not be too easy and ensure the optimal progress for the patient.

3.4 Experimental Setup

Experiments were conducted in such a way that the solutions to the new problems created by the different approaches were compared to each other. Every unique solution was counted in order to check the increase in solution variety, and then the solution quality was checked. To define how good a solution is it was compared to the solution created by a medical professional. To be able to do this, the respective case to be tested was removed from the case base. The problem part of the case was fed as input to the system, and a new solution to the problem was generated. This solution, for both the no-adaptation and the GA systems, was in turn compared to the one that the medical expert had crafted. The comparison of the two solutions was done by using similarity measures to check how close the generated result was to the original solution. The no-adaptation method always creates the same solution, while the GA will return

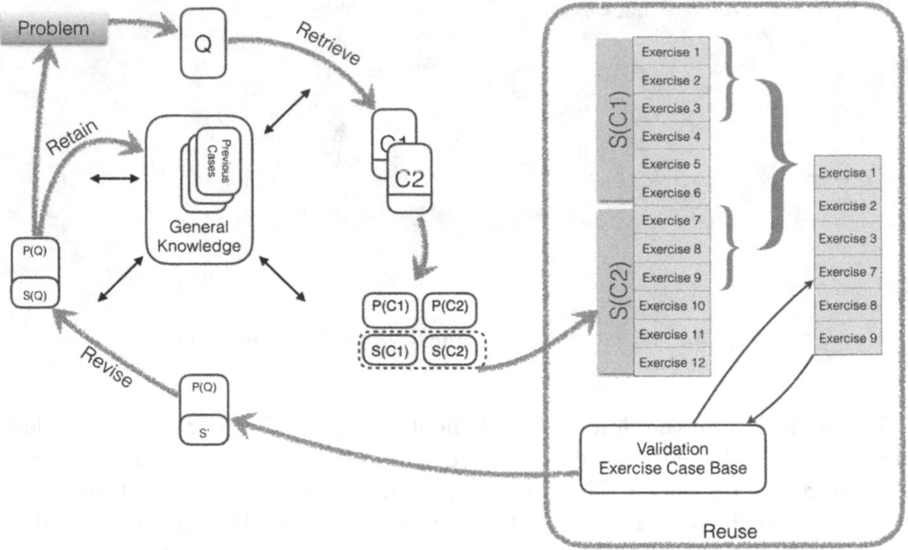

Fig. 4. Overview of the 4R cycle (based on [1]) including an adaption example of a result from a crossover between to chromosomes

solutions that differ from each other. As a result of the fact that the GA will provide results with varied scores this approach was tested a total of five times to see how well it performed in terms of best case, worst case and average case.

3.5 Results

Regarding the cold-start problem the number of solutions in the case base will improve with the GA-approach, as hypothesized. In Fig. 5(a) the evolution of different solutions in the case base is presented. The number of different solutions of the no-adaptation method is, also as expected, staying constant, while for the GA the variation increases. It is expected that the GA-graph will converge with more iterations as the number of exercises to choose from is a finite number, as well as the specifications of level, repetitions and set. It still verifies that the GA-approach creates a greater solution variety when starting with a small case base.

While the GA clearly is a better solution for increasing solution variety, it only adds value to the user if the created solutions have an appropriate quality. To assure that the solutions are satisfactory they are compared to the ones created by the no-adaptation approach and the expert solutions. A textual example of the difference in results between the created solutions can be seen in Table 1. Here none of the solutions created consists of exactly the same exercises. Some are the same on all three solutions, other exercises differ between the three solutions. The flexibility exercises recommended are for instance the same three variation of exercises in all three cases, while in the "strong in mid-position"

Table 1. Textual representation of different solutions: each solution has a level and most of them have the number of repetitions specified

Exercise description	Expert solution	GA solution	Retrieval only solution
Strength exercises for the back extensors	*Level:* 4 *Repetitions:* 10 *Set:* 2 *Level:* 5 *Repetitions:* 10 *Set:* 2	*Level:* 6 *Repetitions:* 10 *Set:* 3	*Level:* 6 *Repetitions:* 10 *Set:* 3
Strength exercises for the gluts and back extensors	*Level:* 1 *Repetitions:* 10 *Set:* 3	*Level:* 1 *Repetitions:* 10 *Set:* 3	*Level:* 2 *Repetitions:* 10 *Set:* 3
Strength exercises for the abdominal muscles	*Level:* 4 *Repetitions:* 10 *Set:* 3	*Level:* 4 *Repetitions:* 10 *Set:* 3	*Level:* 4 *Repetitions:* 10 *Set:* 3
Strength exercises for the oblique abdominal and rotators muscle in the back	*Level:* 3 *Repetitions:* 10 *Set:* 3	*Level:* 3 *Repetitions:* 10 *Set:* 3	*Level:* 3 *Repetitions:* 10 *Set:* 3
Strength "Strong in mid-position"	*Level:* 1 *Repetitions:* 10 *Set:* 3	*Level:* 1 *Repetitions:* 10 *Set:* 3	*Level:* 2 *Repetitions:* 8 *Set:* 3
Flexibility	*Level:* 2 *Level:* 3 *Level:* 4	*Level:* 2 *Level:* 3 *Level:* 4	*Level:* 2 *Level:* 3 *Level:* 4

exercise we see that the solution from the expert and the GA match and that the retrieval only solution has another suggestion. If we look at the first exercise-type suggested we can see that the expert solution suggests two exercises while the two other only suggest one and the same exercise, and the suggested exercise is neither one of the one's suggested by the expert.

The different suggested plans from both the no-adaptation and GA methods were scored against the exercise plan the medical expert created based on their similarity. Since the GA creates different solutions the results show how they scored in the best case, the worst case and the average case out of the five runs. The three different rankings are compared to the similarity scores for the no-adaptation result and can be seen in Fig. 5(b), (c) and (d). The results are sorted by the no-adaptation score as this will be similar in all figures, and therefore give a better impression of the differences between the measures. Both methods score quite well against the expert crafted solutions, which makes sense as all the solutions are built up with the same type of exercises. In the best case scenario for the GA it scores better or equal on eight out of nine cases which suggests that this method performs better than without any adaptation. The worst case scenario on the other hand gives another impression, and in this case only two cases are better on the GA approach while five actually give a worse solution with adaptation. The average case is still probably the best to look at to get a good impression on the performance of the two solutions. The average case shows that the GA performs better in five cases and worse in four. This makes the GA approach seem only somewhat better than without any adaptation, but it has another interesting trait to it. If you compare the similarity measures it shows that the solutions that scores higher with the GA have a larger benefit

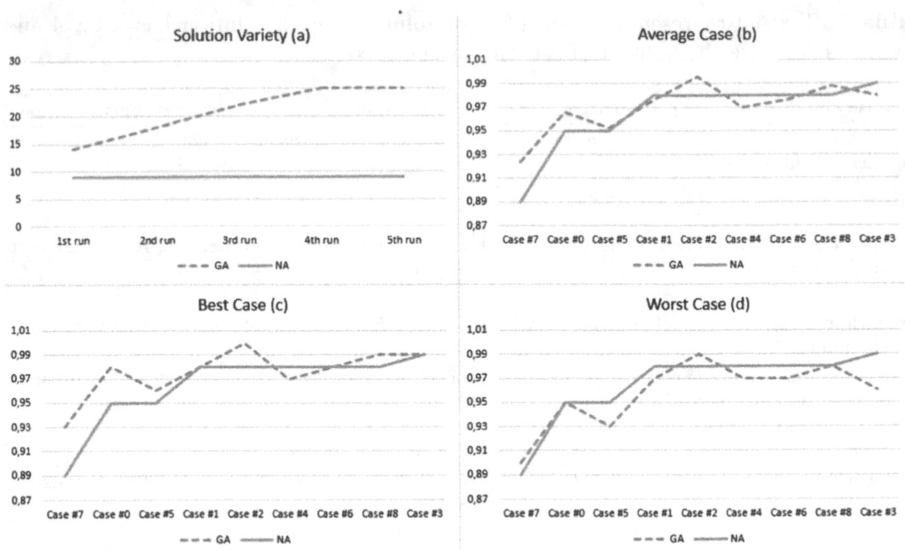

Fig. 5. The results from the experiments. Figure (a) shows the change in solution variety after testing nine different cases in five runs. The y-axis is the number of solutions created after each run. Figure (b), (c) and (d) show the quality of the exercise plans created in average case, best case and worst case respectively. Here the y-axis is the similarity score with an expert crafted solution.

compared to no-adaptation, while the solutions that perform worse are quite close in scores. On average the solutions with the GA in fact score 4.8% better, which shows that in general the gain is larger with the use of this method.

4 Conclusion and Future Work

In this paper we have presented how to apply genetic algorithms for adapting cases in order to increase the solution variety, which might be necessary when deploying a new CBR system.

The results from the experiments show that the solutions created by the genetic algorithm copes better with the cold start problem since it creates a variation of solutions that are of good quality. With information obtained during the follow-up periods within the SELFBACK project, we will gather more information on user preferences and outcomes in terms of pain and function. This information will then allow us to create a better fitness function to further improve the results.

Within SELFBACK this approach can be used for recommending behavioral change or educational sessions. More generally, the approach could fit applications where some degree of creativity is possible with user feedback available. This could, for example, be exercises for other rehabilitation programs, product recommendations, or meal planning.

In our further research, additional adaptation processes will be explored. First we would like to include adaptation rules based on clinical guidelines in order to see how they compare with the genetic algorithm. As part of our CBR research more generally, we have a focus on combining CBR with general domain models beyond rules, most recently by incorporating graphical models in the form of Bayesian networks [20]. This is a line of research that will extend our work on case adaption as well as other CBR processes within the selfBack architecture. As a further study we also plan to extend the method presented here to become not only GA-inspired, but more GA-like, as mentioned in Sect. 3.3. In order to incorporate direct feedback from patients, we plan to provide them with a web-application where they can rate the generated exercise lists.

Acknowledgement. The work has been conducted as part of the SELFBACK project, which has received funding from the European Union's Horizon 2020 research and innovation programmer under grant agreement No. 689043.

References

1. Aamodt, A., Plaza, E.: Case-based reasoning: foundational issues, methodological variations, and system approaches. Artif. Intell. Commun. **7**(1), 39–59 (1994)
2. Bach, K., Szczepanski, T., Aamodt, A., Gundersen, O.E., Mork, P.J.: Case representation and similarity assessment in the selfBACK decision support system. ICCBR 2116 (accepted for publication)
3. Bareiss, R.: Exemplar-based knowledge acquisition (1989)
4. Begum, S., Ahmed, M.U., Xiong, N., Folke, M.: Case based reasoning systems in the health sciences a survey of recent trends and developments. IEEE Trans. Syst. Man Cybern. Part C (Appl. Rev.) **41**, 421–434 (2010)
5. Bichindaritz, I.: Case-based reasoning in the health sciences: why it matters for the health sciences and for CBR. In: Althoff, K.D., Bergmann, R., Minor, M., Hanft, A. (eds.) Advances in Case-Based Reasoning. LNCS, vol. 5239, pp. 1–17. Springer, Heidelberg (2008). doi:10.1007/978-3-540-85502-6_1
6. Brox, J.: Ryggsmerter. In: Aktivitetshåndboken - Fysisk aktivitet i forebygging og behandling, pp. 537–547. Helsedirektoratet (2009)
7. da Costa, LCM., Maher, C.G., McAuley, J.H., Hancock, M.J., Herbert, R.D., Refshauge, K.M., Henschke, N.: Prognosis for patients with chronic low back pain: inception cohort study (2009)
8. Chang, C., Cui, J., Wang, D., Hu, K.: Research on case adaptation techniques in case-based reasoning. In: Proceedings of 2004 International Conference on Machine Learning and Cybernetics. IEEE (2004)
9. Choudhury, N.: A survey on case-based reasoning in medicine. Int. J. Adv. Comput. Sci. Appl. **7**(8), 136–144 (2016)
10. Deyo, R.A., Battie, M., Beurskens, A., Bombardier, C., Croft, P., Koes, B., Malmivaara, A., Roland, M., Von Korff, M., Waddell, G.: Outcome measures for low back pain research. A proposal for standardized use **23**(18), 2003–2013 (1998)
11. de A, G., Maher, M.L.: An evolutionary approach to case adaptation. In: Althoff, K.-D., Bergmann, R., Branting, L.K. (eds.) ICCBR 1999. LNCS, vol. 1650, pp. 162–173. Springer, Heidelberg (1999). doi:10.1007/3-540-48508-2_12

12. Grech, A., Main, J.: Case-base injection schemes to case adaptation using genetic algorithms. In: Funk, P., González Calero, P.A. (eds.) ECCBR 2004. LNCS (LNAI), vol. 3155, pp. 198–210. Springer, Heidelberg (2004). doi:10.1007/978-3-540-28631-8_16
13. Husain, W., Wei, L.J., Cheng, S.L., Zakaria, N.: Application of data mining techniques in a personalized diet recommendation system for cancer patients. In: IEEE Colloquium on Humanities, Science and Engineering. IEEE Xplore (2011)
14. Ben Schafer, J., Dan Frankowski, J.: Collaborative filtering recommender systems (2007)
15. Fritz, J.M., George, S.Z., Delitto, A.: The role of fear-avoidance beliefs in acute low back pain: relationships with current and future disability and work status. Pain **94**(1), 7–15 (2001)
16. Kofod-Petersen, A., Cassens, J., Aamodt, A.: Explanatory capabilities in the creek knowledge-intensive case-based reasoner. In: Proceedings of the 2008 Conference on Tenth Scandinavian Conference on Artificial Intelligence: SCAI 2008, pp. 28–35. IOS Press, Amsterdam, The Netherlands (2008)
17. Koton, P.: Reasoning about evidence in causal explanations. In: Proceedings of the Seventh AAAI National Conference on Artificial Intelligence, AAAI 1988, Saint Paul, Minnesota, pp. 256–261. AAAI Press (1988). http://dl.acm.org/citation.cfm?id=2887965.2888011
18. Lærum, E., Brox, J.I., Storheim, K., Espeland, A., Haldorsen, E., Munch-Ellingsen, J., Nielsen, L., Rossvoll, I., Skouen, J.S., Stig, L., Werner, E.L.: Nasjonale kliniske retningslinjene for korsryggsmerter. Formi (2007)
19. Marling, C., Rissland, E., Aamodt, A.: Integrations with case-based reasoning. Knowl. Eng. Rev. **20**(3), 241–245 (2005)
20. Nikpour, H., Aamodt, A., Skalle, P.: Diagnosing root causes and generating graphical explanations by integrating temporal causal reasoning and CBR. In: Coman, A., Kapetanakis, S. (eds.) Workshops Proceedings for the Twenty-Fourth International Conferenceon Case-Based Reasoning (ICCBR 2016), Atlanta, Georgia, USA, 31 October–2 November 2016. CEUR Workshop Proceedings, vol. 1815, pp. 162–172. CEUR-WS.org (2016)
21. Petrovic, S., Khussainova, G., Jagannathan, R.: Knowledge-light adaptation approaches in case-based reasoning for radiotherapy treatment planning. Artif. Intell. Med. **68**, 17–28 (2016). ScienceDirect
22. Schmidt, R., Montani, S., Bellazzi, R., Portinale, L., Gierl, L.: Cased-based reasoning for medical knowledge-based systems. Int. J. Med. Inf. **64**, 355–367 (2001). ScienceDirect
23. Senanayke, S., Malik, O.A., Iskandar, P.M., Zaheer, D.: A hybrid intelligent system for recovery and performance evaluation after anterior cruciate ligament injury. In: 2012 12th International Conference on Intelligent Systems Design and Applications (ISDA). IEEE (2012)
24. Spears, V.M., Jong, K.A.D.: On the virtues of parameterized uniform crossover. In: Proceedings of the Fourth International Conference on Genetic Algorithms, pp. 230–236 (1991)

Intelligent Control System for Back Pain Therapy

Juan A. Recio-Garcia[1(✉)], Belén Díaz-Agudo[1], Jose Luis Jorro-Aragoneses[1], and Alireza Kazemi[2]

[1] Department of Software Engineering and Artificial Intelligence, Instituto de Tecnología del Conocimiento, Universidad Complutense de Madrid, Madrid, Spain
jareciog@fdi.ucm.es, {belend,jljorro}@ucm.es
[2] Institute of Physiotherapy and Sports, Guadalajara, Spain

Abstract. Back pain is a pending subject in our society despite scientific advances. The Kazemi Back System (KBS) is a therapy machine that allows the patient to correctly perform manipulation exercises to heal or relieve pain. In this paper we describe and evaluate a CBR approach to suggest an stream of configuration values for the KBS machine based on previous sessions from the same patient or other similar patients. Its challenge is to capture the expertise knowledge of physiotherapists and reuse it for future therapies. The CBR system includes two complementary reuse processes and an explanation module. Within our experimental evaluation we discuss the problem of incompleteness and noise in the data and how to solve the cold start configuration for new patients.

1 Introduction

Back problems are among the leading causes of workplace absences and, in Spain, represent an annual expenditure of between 1.7–2.1% of the gross domestic product (GDP), approximately 16,000 million euros. It comes from a variety of different causes and affects men and women of different ages. Although back pain is a universal complaint, an overall exercise routine can be an important thing for pain relief. Dr. Alireza Kazemi is a physiotherapist and expert in the treatment of back pain. After more than 10 years using manual therapies in his clinic consisting of the manipulation of different segments of the back, he invented a therapy machine (KBS, as in Kazemi Back System) that allows the patient to correctly perform the same manipulation exercises to heal or relieve pain, increase elasticity and improve quality of live. The machine has several pneumatic actuators that apply certain pressure to the back, arms and legs,

Supported by the UCM (Group 921330) and the Spanish Committee of Economy and Competitiveness (TIN2014-55006-R). The KBS machine is developed by Kazemi Back Health Inc. and funded by the Centre for the Development of Industrial Technology of the Spanish Committee of Economy and Competitiveness.

D.W. Aha and J. Lieber (Eds.): ICCBR 2017, LNAI 10339, pp. 287–301, 2017.
DOI: 10.1007/978-3-319-61030-6_20

and allow the whole body of the patient to be manipulated. This way, it substitutes the manual manipulations applied by the physiotherapist. The machine is a prototype in the final evaluation phase, and it is being evaluated with patients in Dr. Kazemi's medical clinic. The KBS machine integrates the concepts of biomechanics, strength, flexibility, neurodynamics and structural and functional readaptation, which will indirectly activate the local metabolism through movement, which stimulates cellular activation.

Every therapy session is led by the physiotherapist in charge of configuring the KBS machine with a specific set-up for the list of pneumatic pressures. Each configuration depends on different factors such as the patient's clinical data, previous sessions, personal risks, and others. Physiotherapists interview patients who fill out a set of questionnaires before the first training session and after each training session: disability, satisfaction, pain, gain in strength, and gain in mobility. The use of questionnaires facilitates clinical evaluation as they capture self-reported general information and the progress of symptoms.

Although the configuration process requires very specialized skills, the set of values for the configuration stream is similar between different sessions for the same patient (intrapatient reuse) and between sessions for similar patients (interpatient reuse). This fact makes Case-Based Reasoning (CBR) a very good approach to help the expert with the cumbersome task of manually configuring the machine from scratch in every session.

In this paper we describe a CBR approach to facilitate this process by suggesting a stream of configuration values based on previous sessions. We propose a data-driven predictive configuration system that reuses cases from patients in order to suggest the most suitable machine configuration, personalized for each individual patient. The CBR system learns from the initial manual configuration by experts and transfers their expertise into the case base to be reused later by other physiotherapists with other patients. We discuss the representation difficulties, how to deal with missing values in the tests or in the configuration stream, as well as with various challenges that arise when comparing the resulting value streams. Within our experimental evaluation we have found a clear problem of uncertainty and noise in the data because machine sensors may fail and patients usually drop out of long-term therapies. Among other well-known advantages, the CBR approach allows noisy and incomplete input data to be dealt with; facilitates cold start configuration for new patients and it offers doctors certain explanations on the proposed configuration stream.

The paper runs as follows: Sect. 2 describes the main features of the Kazemi Back System and some of the existing related work. Section 3 describes the problem formalization. Section 4 describes a solution to the problem using two approaches: transformational intrapatient and constructive interpatient reuse. Section 5 describes our case study and explains the experimental results. Finally, Sect. 6 summarizes the main results achieved and describes forthcoming work.

2 The Kazemi Back System and Related Work

There is ample literature on CBR systems applied to health sciences. Concretely, biomedicine is one of the most successful application areas as expertise comes

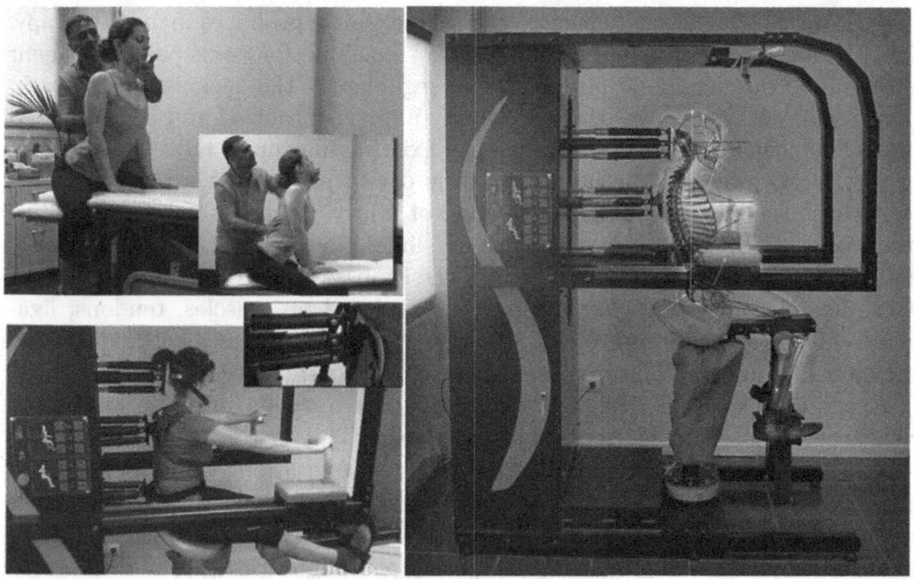

Fig. 1. The KBS machine (Patented). Manual therapy (top-left) vs. KBS therapy (bottom-left). KBS global schema on the right.

from learning by solving problems in practice [4]. One of the most related projects is the self-BACK project, which aims to develop a decision support system for patients suffering from non-specific lower back pain. The system will give users advice in the form of a self-management plan that is based on self-reported physical and psychological symptoms as well as activity stream data collected by a wristband. In [3] the authors describe various challenges when representing and comparing human physical activities using time series data within the self-BACK project.

A key component of these CBR systems is explanation capabilities [6]. This feature enhances the acceptance of the solutions proposed by the system. On the other hand, a recurrent problem is query elicitation due to the complex data format where case descriptions come from sensors or free text [2]. This problem is also shared by our system as it is very difficult to describe the back's (complex) anatomy and spine dysfunctions in a structured way. Regarding similar applications, the RHENE system [8] uses CBR to adjust the parameters of a hemodialysis machine.

The benefits of Dr. Kazemi's treatment have been demonstrated over 10 years with a large number of patients that have received this therapy, as shown in Fig. 1 (top-left). However the manual manipulation of the patient's back has many disadvantages: it requires very specialized expertise and has several risks concerning the incorrect manipulation of the patient. Therefore, this therapy cannot be widely applied to patients due to the lack of experts. As a solution to these problems Dr. Kazemi invented a machine that reproduces manual therapy. The Kazemi Back System (KBS) is a novel machine for back pain treatment consisting of the manipulation of different segments of the back as illustrated in Fig. 1 (right). KBS involves the whole body in a synchronized and concatenated way, a fact that distinguishes it from other existing machines. KBS allows movements of the spine to be induced, from the use of pressure in a predetermined position, preselected according to the kind of injury, intensity of pain and alterations in posture. This series of movements, linked in the body and consequently along the spine with the intervention of the limbs, involves different structures of the locomotor system closely related to each other (muscles, tendons, ligaments, fascia, intervertebral nerve discs, etc.), and activates local and general metabolism, joint physiology and muscle contraction-relaxation.

Every time a patient receives treatment, the machine must be configured with a specific set-up for pneumatic pressures. This is the so-called configuration stream. Experts are in charge of this configuration, which depends on different factors such as the patient's clinical data, previous sessions, etc. This configuration process requires very specialized skills, so the goal of the CBR system is to facilitate this process by suggesting a configuration based on previous sessions. The CBR system learns from the experts and is able to capture their expertise in the case base to be reused later by other physiotherapists with the same or other patients.

The choice of a lazy learning approach such as CBR is motivated by several factors. First, it eases the knowledge acquisition bottleneck, as the experts are not able to clearly formalize how the therapy is conducted. Next, the system learns as the therapy evolves. Experts have confirmed that this machine can be used for different therapies that have not yet been explored. Therefore, the A.I. behind the control software must be able to learn future patterns. Additionally, this treatment requires specialized training and the CBR system can be used as a teaching tool as it includes explanation capabilities. CBR is also appropriate with uncertainty and noise in the data because machine sensors may fail and patients usually drop out of long-term therapies.

Next we formalize the problem and the details of the CBR process.

3 Problem Formalization and Query Elicitation

The CBR system receives a description of the patient and returns a configuration of the KBS machine for a session. This is the basic structure of the cases:

$$\mathcal{CB} = \{\mathcal{C} = \langle \mathcal{D}, \mathcal{S} \rangle\} \tag{1}$$

Queries are identified by a unique patient id (*pid*) and contain basic patient information (*bd*), a clinical description (*cd*) of the patient that includes a structured representation of the back's anatomy and associated problems, and the session number (*sn*). This data is formalized as follows:

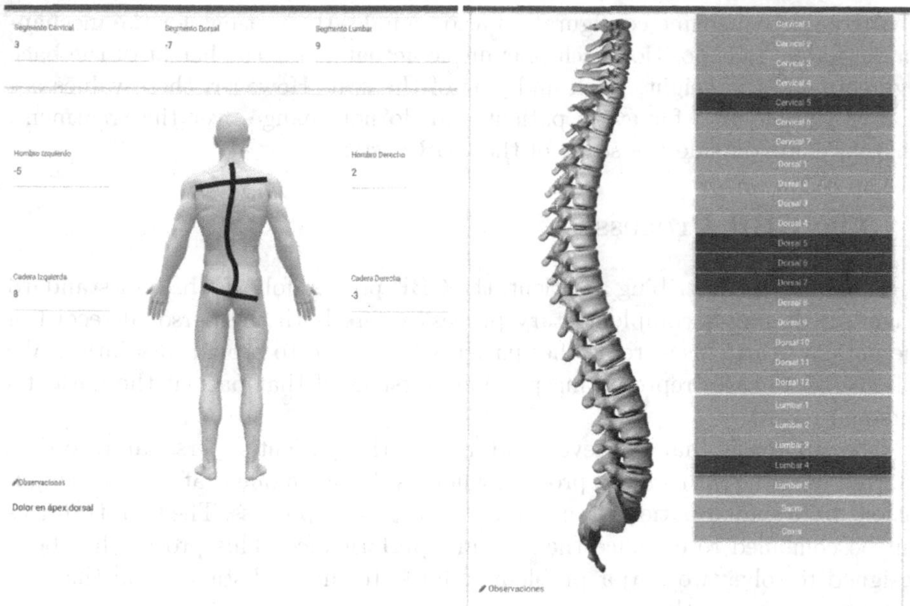

Fig. 2. Interface to obtain structured data for the back's anatomy (left) and spine problems (right).

$$\mathcal{Q} = \mathcal{D} = \langle pid, bd, cd, sn \rangle \tag{2}$$

where

$$bd = \langle gender, age, height, weight, bmi, text \rangle$$
$$cd = \langle \overline{rear}, \overline{lateral}, \overline{spine}, text \rangle$$

The basic data and the clinical description are obtained through the interfaces shown in Fig. 2. These interfaces include a *text* field where physiotherapists can include a text description of the structured data. To exploit this text data in the CBR cycle we will apply textual CBR techniques [5,10]. The clinical description includes a structured representation of the back anatomy from *rear* and *lateral* points of view as shown in Fig. 2 (left). They are encoded as vectors of numbers that represent the deviation of the different spine segments in the back. However, the *spine* element of the clinical description is a list of boolean values indicating concrete vertebra problems. It also includes a text description as shown in Fig. 2 (right).

Given a description $\mathcal{D}$ the CBR system returns a solution $\mathcal{S}$ with the most suitable configuration for the machine. A solution is composed of a list of pressures for every pneumatic actuator. There are five pneumatic actuators for the cervical, dorsal and lumbar back segments, arms and legs. Each actuator can be configured with a pressure from 0 to 3 bars with increments of 100 millibars. Therefore solutions are formalized as:

$$\mathcal{S} = \langle p_{cervical}, p_{dorsal}, p_{lumbar}, p_{arms}, p_{legs} \rangle \tag{3}$$

A complete machine configuration also includes the settings for the mechanical actuators that position each pneumatic actuator, i.e. the height of the back segments and the height, pitch and yaw of the seat. However, these values are manually configured for every patient and do not change over the treatment. Therefore, are outside the scope of the CBR system.

4 The CBR Process

Given a query Q describing a patient, the CBR process follows the four standard stages [1]. It uses a complementary process where both the personal record of the patient Q and cases from other patients are reused to provide a solution. We call the set of cases representing previous sessions of that patient the patient's personal record.

The approach that retrieves and reuses the patient's personal record is referred to as the *intrapatient* process, whereas the technique that reuses configurations from similar patients is called the *interpatient* process. The two strategies can be combined to enhance the system's performance. This process has been designed to solve two major problems related to the cold-start [7,12] that the system must deal with:

1. The personal record is in cold-start: there are few cases because the patient is at the beginning of the therapy. Thus, the intrapatient process is not suitable. Therefore the system uses similar patients (interpatient process) to provide a solution.
2. The case base is in cold-start: there are few patients in the case base. In this scenario, the intrapatient approach provides a base configuration with better results than the interpatient process, as similarity values are very low.

The CBR system also includes an explanation module that details the reasoning cycle to experts. Although a whole description of the explanation module is outside the scope of this paper, it is a crucial component of the system as experts can review the reasons that led to the solution proposed by the system before applying it to the patient. The explanation interface is shown in Fig. 5. Next we describe the whole CBR cycle that has been implemented using the jCOLIBRI software [11].

4.1 Retrieval

Having two approaches implies specific retrieval sets that we refer to RS_{intra} and RS_{inter}. The collection of the RS_{intra} cases is quite straightforward as it only contains cases from the patient's personal record. Therefore, the system retrieves the previous cases for the query patient.

$$RS_{intra} = \{C_i \in CB, \forall i(Q.pid == C_i.pid)\} \tag{4}$$

Regarding RS_{inter} the retrieval process begins with a filtering step where cases are chosen according to gender. As reported by the experts, cases cannot be reused from patients with another gender because their anatomy is too different. Therefore we apply the following pre-filtering:

$$RS'_{inter} = \{C_i \in CB, \forall i(Q.gender == C_i.gender)\} \tag{5}$$

Once the cases have been filtered we use a nearest-neighbour method to find the most similar cases. This method is not only limited to the k best neighbours but also to a minimum similarity threshold θ to assure the quality of the cases retrieved and the subsequent adaptation processes.

$$\begin{aligned} RS_{inter} &= \{\langle C_i, w_i\rangle \in RS'_{inter}, \forall i(w_i \geq \theta)\} \\ &\text{where} \\ w_i &= simil(Q, C_i) \end{aligned} \tag{6}$$

The similarity metric that compares a query description D_q and a user description D_u is defined as the weighted average of the similarities for every component $\langle bd, cd, sn\rangle$.

The basic data (*bd*) similarity is computed as the average of numerical differences. Textual CBR metrics are used to compare textual descriptions. Specifically, we apply cosine distance. For the clinical description (*cd*), vector differences are used to compare the *rear* and *lateral* elements, but the *spine* vector is compared through an XOR function that indicates whether there are problems in the same vertebra. Again, a cosine function is applied to compare the textual descriptions.

4.2 Reuse

We have designed a complementary adaptation approach where both the patient's personal record and the most similar cases from other patients are combined to maximize system performance.

Firstly, intrapatient adaptation is performed to obtain a base configuration of the machine, S_{intra}. It is a preliminary solution that is computed using a linear regression model over the pressure series found in the patient's record. The choice of a linear regression model is motivated by the evolution of the therapy applied by the physiotherapists. They increase the machine pressures as the patient improves. Therefore a linear regression method is the most suitable option to model this behaviour.

From the patient's record, which was retrieved in RS_{intra} we obtain the pressure series for every pneumatic actuator. Given an ordered list of cases according to the session number:

$$\begin{aligned} RS_{intra} &= \{C_1, \ldots, C_i, C_{i+1}, \ldots, C_n\} \\ &\text{where} \\ C_i &= \langle D_i, S_i\rangle \\ D_i.sn &= D_{i+1}.sn + 1 \\ n &= Q.sn - 1 \end{aligned} \tag{7}$$

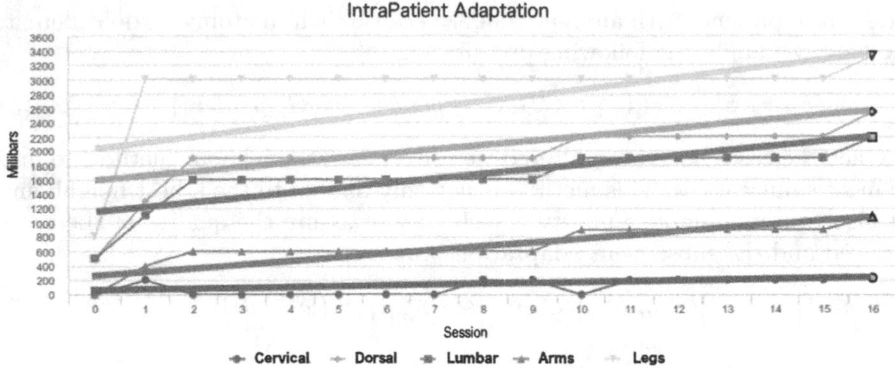

Fig. 3. Example of the intrapatient adaptation.

A paired vector of (session number, pressure) is created for every pneumatic actuator $p \in S_i$:

$$(x, y)_p = (x_1, y_1), \ldots, (x_n, y_n) \tag{8}$$

where

$$x_i = D_i.sn$$
$$y_i = S_i.p$$

To estimate the pressure of an actuator p for session $Q.sn$ given a paired vector $(x, y)_p$ that represents the previous configurations of p, we use the least squares estimation, which finds a line that minimizes the *sum of the squared errors* (SSE):

$$SSE = \sum_{j=1}^{n} r_j^2 = \sum_{j=1}^{n} (y_j - \hat{y}_j)^2 = \sum_{j=1}^{n} (y_j - \beta_0 - \beta_1 x_j)^2 \tag{9}$$

Then the regression line is:

$$y = \hat{a}\beta_0 + \hat{\beta}_1 x \tag{10}$$

And therefore, the predicted pressure can be computed as:

$$p = \hat{a}\beta_0 + \hat{\beta}_1 Q.sn \tag{11}$$

Figure 3 shows the intrapatient adaptation algorithm graphically. Lines with markers represent the pressure vectors retrieved from the personal record. Values on the horizontal axis indicate the session number. Markers with a black border at the end of the series show the estimated pressure for the session being queried ($Q.sn = 16$ in this example). This graphical representation is used to explain the process to the user, as shown in Fig. 5.

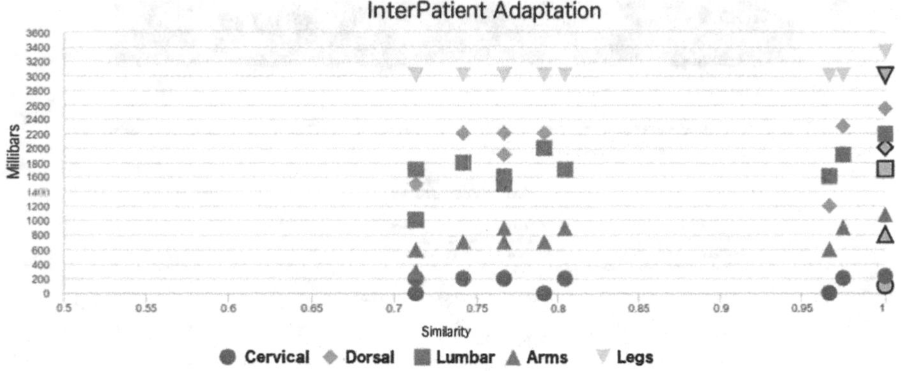

Fig. 4. Example of the interpatient process.

Next, interpatient adaptation tries to improve the base solution S_{intra} by using the nearest neighbours in $\langle C_i, w_i \rangle \in RS_{inter}$ that include similarity to the query w_i. It applies a weighted average based on those values.

Firstly, the solution provided by intrapatient adaptation, S_{intra}, is added to RS_{inter} with a similarity value of 1:

$$RS = RS_{inter} \cup \langle S_{intra}, 1 \rangle \tag{12}$$

This way the intrapatient solution has the highest weight and the remaining solutions in RS_{inter} will modify its pressure values depending on their similarity to the query. This means that this algorithm implements a transformational adaptation approach.

It is important to note that in the case of a cold-start of the personal record there is no S_{intra} and the system only reuses the solutions in RS_{inter}. In this situation the system performs a constructive adaptation process from several solutions because there is no base solution [9].

Likewise, in the case of a cold-start of the case base, RS_{inter} is empty and S_{intra} is the only solution available that will be returned for adaptation. Either way, the adaptation algorithm provides a valid solution. It applies a weighted average for every pneumatic actuator $p \in S_i$:

$$p = \frac{\sum_1^n S_i.p \cdot w_i}{\sum_1^n w_i} \tag{13}$$

where

$$C_i = \langle D_i, S_i \rangle \in RS$$
$$n = |RS|$$

The interpatient adaptation process is also explained to the physiotherapist graphically. As Fig. 4 shows it reflects similarity on the x-axis and represents every solution in RS as a set of marks over the corresponding similarity value. These marks are the pressure values for each pneumatic actuator found in the

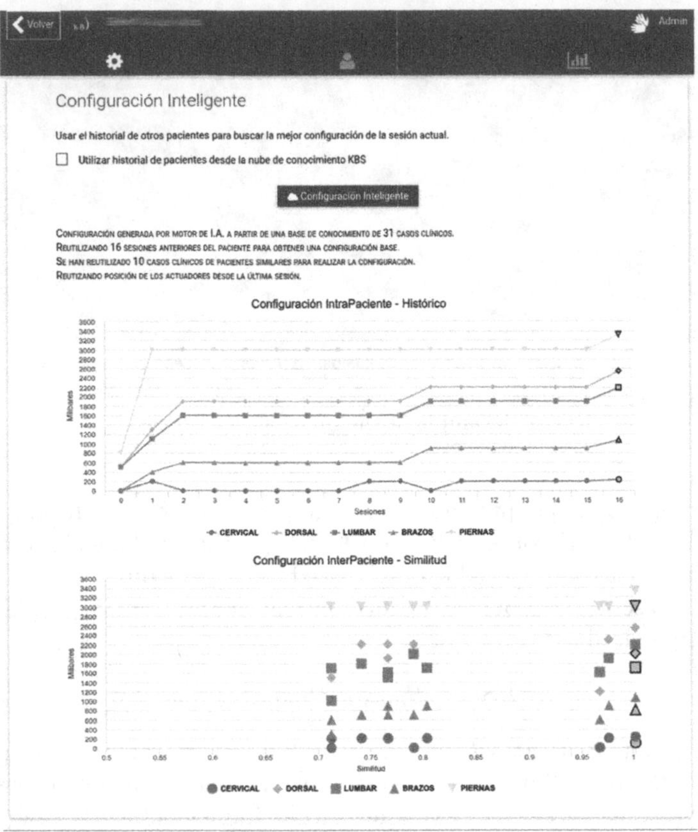

Fig. 5. Explanation interface of the KBS software

solution of that similar case. Unbordered marks over similarity value 1 represent the solution found by intrapatient adaptation. Black bordered marks are the final system's solution.

Both graphical explanation charts for the intrapatient and interpatient processes provide a sophisticated explanation system included in the CBR system. This feature is very important because physiotherapists must understand the reasons behind the proposed machine configuration. The explanation interface is shown in Fig. 5 and includes the explanation charts previously described in Figs. 3 and 4. This figure presents a clear example of the whole adaptation process. First, the intrapatient adaptation proposes a slight increase of pressure as the linear regression suggests an incremental pattern in previous configurations. However the interpatient explanation shows that two very similar cases have lower pressure values. Finally the weighted average takes those cases into consideration and leverages the final pressure values. This is illustrated over value 1 on the x-axis where every final mark (bordered) has a corresponding mark above it that represents the intrapatient solution.

These charts are displayed to the physiotherapist together with a textual description of the process that includes the number of retrieved cases, personal records reused by the intrapatient adaptation and the number of similar cases reused by the interpatient adaptation.

4.3 Revise and Retain

This kind of medical domain requires a mandatory revision stage where the physiotherapist supervises the solution proposed by the CBR system.

To perform the revision of the solution the user begins with the explanation provided by the system. There are many therapeutic or patient factors that could end up with a modification of the proposed solution. Therefore, the expert can modify the pressure values for each pneumatic actuator through the interface shown in Fig. 6. He/she can also provide a text description to explain the modifications. Moreover the revision interface allows previous configurations from past sessions to be explored. When selecting a previous configuration, the expert can explore the pressure values and the corresponding explanation of that past revision process. Once the exercise is completely configured, the revised solution is stored in the case base together with the query that launched the CBR process.

5 Evaluation

To evaluate the CBR system we used a case base of patient records obtained during the clinical evaluation of the KBS. This clinical evaluation was performed with the participation of 40 patients (14 female and 26 male) during 4 months and allowed mechanical and software problems in the machine to be detected. Consequently this case base is very noisy and incomplete as 4 months is a long period and many patients did not complete the evaluation. Moreover, several technical problems led to the corruption of several records. However, having noisy data was a precondition when developing the intelligent control software and one of the main reasons to implement a CBR system, as this kind of reasoning is able to manage noisy data. Finally, 426 patient records were available for the evaluation.

Evaluation is performed using cross-validation, specifically, a leave-one-out approach that counts the number of successful solutions. The evaluation function that compares the solution suggested by the CBR process (S') to the actual solution for the case (S) compares the difference between pressures as follows:

$$\begin{aligned} eval &: [S \times S] \rightarrow \{false, true\} \\ eval(S, S') &= (p_{cervical} - p'_{cervical}) > \delta \wedge (p_{dorsal} - p'_{dorsal}) > \delta \\ &\wedge (p_{lumbar} - p'_{lumbar}) > \delta \wedge (p_{arms} - p'_{arms}) > \delta \\ &\wedge (p_{legs} - p'_{legs}) > 2\delta \end{aligned} \tag{14}$$

The comparison of leg pressure doubles the threshold (2δ) as its values are much higher than other pressures (aprox. 2000–3000 mbars instead of 700–1000). During evaluation this threshold is set to $\delta = 100\ mbars$.

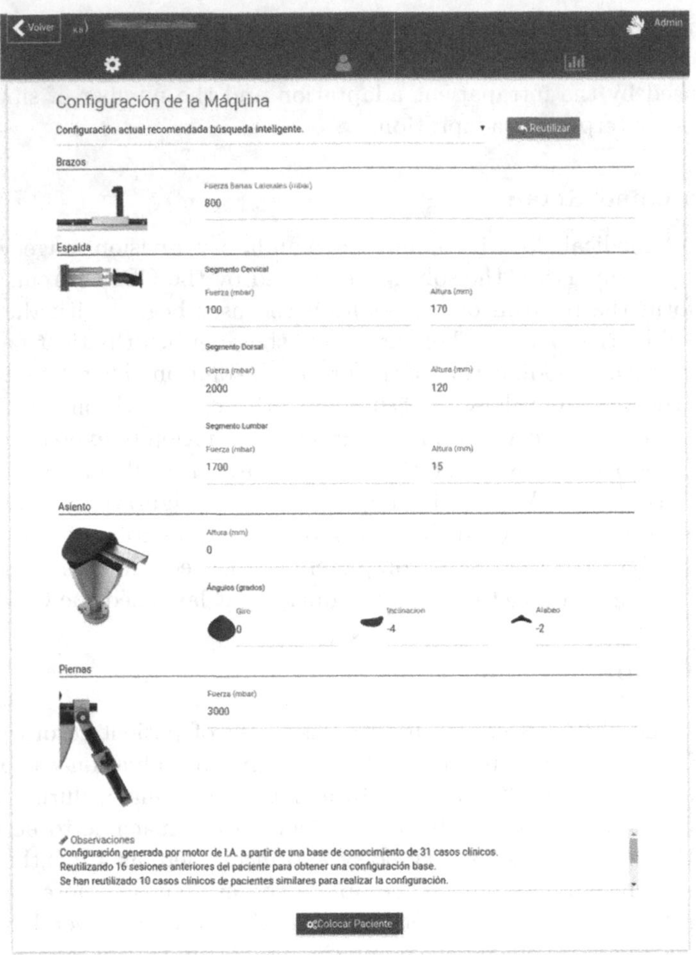

Fig. 6. Revision interface of the KBS software

The leave-one-out validation is segmented according to the session number. This way, the percentage of successful solutions for every session number is averaged independently. It allows the performance of the system to be inspected as the therapy is carried on. This segmentation is very important in order to understand the behaviour of the two adaptation approaches. During the initial sessions there are many similar cases available as almost every patient has completed these sessions but there are few previous patient's records. Therefore the intrapatient adaptation process might work poorly but the interpatient adaptation should provide good results. On the other hand, there are very few records for the final sessions, thus supplying a lot of data for the intrapatient adaptation and very few similar cases for the interpatient adaptation. This behaviour is illustrated in Fig. 7, which shows the number of patient records available for

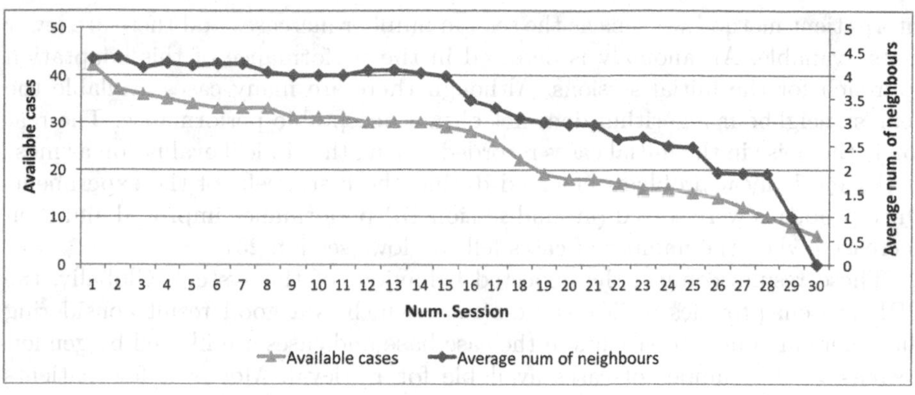

Fig. 7. Average number of available patient records (left axis) and available neighbours (right axis) for every session number.

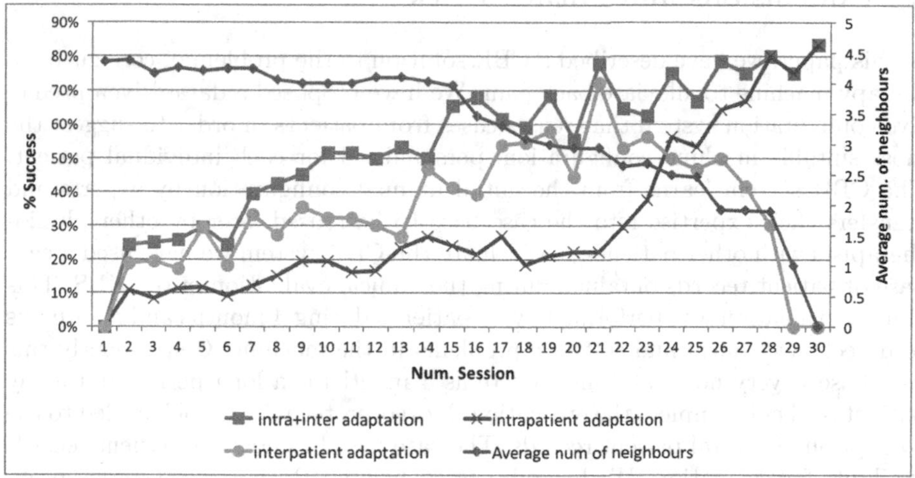

Fig. 8. Evaluation results of adaptation methods (left axis) and average available neighbours (right axis).

every session and the average number of similar neighbours retrieved by the interpatient retrieval process (Eq. 6) taking into account the θ threshold, which is configured with $\theta = 0.5$ for evaluation (no significant improvement was found when varying this value).

Several evaluations have been conducted to inspect the impact of the adaptation approaches: (1) combining intrapatient and interpatient, (2) only intrapatient and, (3) only interpatient adaptation. Results are shown in Fig. 8, which illustrates how the combination of the two adaptation approaches provides the best results. Individually, the intrapatient adaptation method improves as the number of available previous records increases. On the other hand, the interpatient method worsens as the session number increases and there are fewer cases available. An anomaly is detected in the performance of this adaptation approach for the initial sessions. Although there are many cases available the nearest neighbour algorithm does not obtain acceptable performance. The reason is the noise in the initial cases recorded during the clinical evaluation as most of the mechanical problems emerged during the first weeks of the experiment. Once problems were solved (around session 15) performance improved and then decreased when the number of cases fell too low (session 25).

These results confirm the expected behaviour of the system. Globally, the CBR system provides a 75% success rate, which is a good result considering that there are only 40 patients in the case base and cases are filtered by gender, decreasing the number of cases available for retrieval. Moreover few patients completed the therapy and there were corrupted records, which led to a very noisy case base.

6 Conclusions and Future Work

In this paper, we have described a CBR solution to the problem of configuring a therapy machine to alleviate back pain. We have proposed a data-driven predictive configuration system that reuses cases from patients in order to suggest the most suitable machine configuration, personalized for each individual patient. The CBR system learns from the initial manual configuration by experts and transfers their expertise into the case base to be reused later by other physiotherapists with other patients. To evaluate the CBR system we have used a case base of patient records obtained during the clinical evaluation of the KBS. This clinical evaluation was performed by 40 patients during 4 months and allowed us to detect mechanical and software problems in the machine. Consequently this case base is very noisy and incomplete as 4 months is a long period and many patients did not complete the evaluation. Moreover, technical problems led to the corruption of several patient records. The case base has now 426 patient records available for evaluation. We have discussed how CBR allows two major problems related to the cold-start and noisy data to be solved. We have described an *intrapatient* process that retrieves and reuses the patient's personal record, and a process that reuses configurations from similar patients, called the *interpatient* process. The evaluation results allow us to conclude that the combination of the two approaches provides the best results. Individually, the intrapatient adaptation method improves as the number of available previous records increases. On the other hand, the interpatient method worsens as the session number increases and there are fewer cases available. In the experiments we do not distinguish who the physiotherapist in charge of the case authoring and revision is. As future work, each machine will be connected to a cloud service, and the CBR system will reuse cases either from its local case base or from other case bases belonging to different experts. We will continue with the experiments and we will study the impact of the provenance of the cases in the system's performance. We also will perform experiments on the impact of the explanation capabilities on the final physiotherapist's decision.

References

1. Aamodt, A., Plaza, E.: Case-based reasoning: foundational issues, methodological variations, and system approaches. AI Commun. **7**(1), 39–59 (1994)
2. Ahmed, M.U., Begum, S., Funk, P., Xiong, N., von Schéele, B.: Case-based reasoning for diagnosis of stress using enhanced cosine and fuzzy similarity. In: Perner, P., Bichindaritz, L.S.I. (ed), 8th Industrial Conference, ICDM 2008, pp. 128–144. IBaI July 2008
3. Bach, K., Szczepanski, T., Aamodt, A., Gundersen, O.E., Mork, P.J.: Case representation and similarity assessment in the SELFBACK decision support system. In: Goel, A., Díaz-Agudo, M.B., Roth-Berghofer, T. (eds.) ICCBR 2016. LNCS, vol. 9969, pp. 32–46. Springer, Cham (2016). doi:10.1007/978-3-319-47096-2_3
4. Bichindaritz, I.: Case-based reasoning in the health sciences: why it matters for the health sciences and for CBR. In: Althoff, K.-D., Bergmann, R., Minor, M., Hanft, A. (eds.) ECCBR 2008. LNCS, vol. 5239, pp. 1–17. Springer, Heidelberg (2008). doi:10.1007/978-3-540-85502-6_1
5. Díaz-Agudo, B., Recio-García, J.A., González-Calero, P.A.: Natural language queries in CBR systems. In: 19th IEEE International Conference on Tools with Artificial Intelligence (ICTAI 2007), vol. 2, pp. 468–472. IEEE Computer Society, Patras, Greece, 29–31 October 2007
6. Doyle, D., Cunningham, P., Walsh, P.: An evaluation of the usefulness of explanation in a CBR system for decision support in bronchiolitis treatment. In: Proceedings of the Workshop on Case-Based Reasoning in the Health Sciences, Workshop Programme at the Sixth International Conference on CaseBased Reasoning, pp. 32–41 (2005)
7. Horsburgh, B., Craw, S., Massie, S., Boswell, R.: Finding the hidden gems: recommending untagged music. In: Walsh, T. (ed) Proceedings of the 22nd International Joint Conference on Artificial Intelligence, IJCAI 2011, pp. 2256–2261. IJCAI/AAAI, Barcelona, Catalonia, Spain, 16–22 July 2011
8. Montani, S., Portinale, L., Bellazzi, R., Leonardi, G.: RHENE: a case retrieval system for hemodialysis cases with dynamically monitored parameters. In: Funk, P., González Calero, P.A. (eds.) ECCBR 2004. LNCS, vol. 3155, pp. 659–672. Springer, Heidelberg (2004). doi:10.1007/978-3-540-28631-8_48
9. Plaza, E., Arcos, J.-L.: Constructive adaptation. In: Craw, S., Preece, A. (eds.) ECCBR 2002. LNCS (LNAI), vol. 2416, pp. 306–320. Springer, Heidelberg (2002). doi:10.1007/3-540-46119-1_23
10. Recio, J.A., Díaz-Agudo, B., Gómez-Martín, M.A., Wiratunga, N.: Extending jCOLIBRI for textual CBR. In: Muñoz-Ávila, H., Ricci, F. (eds.) ICCBR 2005. LNCS (LNAI), vol. 3620, pp. 421–435. Springer, Heidelberg (2005). doi:10.1007/11536406_33
11. Recio-García, J.A., González-Calero, P.A., Díaz-Agudo, B.: jcolibri2: a framework for building case-based reasoning systems. Sci. Comput. Program. **79**, 126–145 (2014)
12. Quijano-Sánchez, L., Bridge, D., Díaz-Agudo, B., Recio-García, J.A.: A case-based solution to the cold-start problem in group recommenders. In: Agudo, B.D., Watson, I. (eds.) ICCBR 2012. LNCS (LNAI), vol. 7466, pp. 342–356. Springer, Heidelberg (2012). doi:10.1007/978-3-642-32986-9_26

Dependency Modeling for Knowledge Maintenance in Distributed CBR Systems

Pascal Reuss[1,2(✉)], Christian Witzke[1], and Klaus-Dieter Althoff[1,2]

[1] Intelligent Information Systems Lab, University of Hildesheim, Hildesheim, Germany
reusspa@uni-hildesheim.de
[2] Competence Center CBR, German Center for Artificial Intelligence, Kaiserslautern, Germany

Abstract. Knowledge-intensive software systems have to be continuously maintained to avoid inconsistent or false knowledge and preserve the problem solving competence, efficiency, and effectiveness. The more knowledge a system contains, the more dependencies between the different knowledge items may exist. Especially for an overall system, where several CBR systems are used as knowledge sources, several dependencies exist between the knowledge containers of the CBR systems. The dependencies have to be considered when maintaining the CBR systems to avoid inconsistencies between the knowledge containers. This paper gives an overview and formal definition of these maintenance dependencies. In addition, a first version of an algorithm to identify these dependencies automatically is presented. Furthermore, we describe the current implementation of dependency modeling in the open source tool myCBR.

1 Introduction

Knowledge-intensive systems are using a high amount of knowledge that is not stored in a single knowledge source, but distributed over several knowledge sources. This leads to a better scalability and maintainability. Especially for systems with several case-based reasoning (CBR) systems, distributing the knowledge over several small CBR systems rather than using one large CBR system, has a great benefit on maintainability. There exist several maintenance approaches for CBR systems [8,9,16,19,20] that aim to maintain a single knowledge container in a single CBR system. However, there could be dependencies between the knowledge inside a CBR system and between the knowledge of different CBR systems. These dependencies should be considered for maintenance actions to ensure the consistency and competence of the whole knowledge-intensive system. Based on these dependencies additional maintenance actions could be identified to avoid inconsistencies or competence loss. These dependencies [13] are used by the Case Factory approach for maintaining distributed CBR systems [12]. The initial definition of the dependencies were not sufficient and we analyzed the required granularity of the dependencies and the possible knowledge on different granularity levels. In this paper we present the refined definition

D.W. Aha and J. Lieber (Eds.): ICCBR 2017, LNAI 10339, pp. 302–314, 2017.
DOI: 10.1007/978-3-319-61030-6_21

of the maintenance dependencies and the current modeling and implementation in the open source tool myCBR [5,21]. We also investigate the possibility of generating dependencies automatically to reduce the modeling effort.

In Sect. 2 we describe related research in the fields of knowledge modeling and maintenance. Section 3.1 contains a brief description of the Case Factory (CF) approach and Sect. 3.2 describes in more detail the refined dependencies required for maintenance with CFs and a first version of the dependency generation algorithm. In Sect. 3.3 we present the current modeling possibilities in myCBR. Finally, we give a summary of our paper and an outlook to future work.

2 Related Work

Knowledge modeling has been a focus of research in many communities in the last decades. Directly related to CBR, the knowledge containers [14] are a central point for knowledge modeling, which were extended with a maintenance container [10]. Our maintenance dependencies clearly belong to the maintenance knowledge of a CBR system and the Maintenance Map can be treated as an instantiation of a maintenance container. However, for our maintenance approach with explanations the knowledge from the other maintenance containers are required, too. The description of knowledge contents in CBR systems on a so-called knowledge level was first introduced by Aamodt [2]. The CBR community adapted the knowledge level view from the knowledge acquisition community to describe knowledge in CBR systems independent from the implementation. Our approach uses also the knowledge level view, but introduced different sub-levels of knowledge in the different knowledge containers. This way, we are able to build a hierarchy of knowledge for our maintenance dependencies to model abstract and detailed dependencies.

Dependency modeling between knowledge is researched in the economic domain. The focus is to model the dependencies between knowledge in firms and organizations or in business processes. Different approaches to model dependencies have been developed, for example strategic dependency diagrams [3], dependency modeling with OWL-DL, and a meta-model for dependencies [18]. In the business domain, knowledge dependencies are modeled to determine the performance of a firm or the scalability of business processes.

Software development also deals with dependency modeling. They are for example used to manage complex software applications and can identify violations of the architecture, evaluate the scalability of an application and identify hidden subsystems [17]. Also from the software development perspective comes an approach for a domain-specific dependency modeling language [1]. In our approach the knowledge levels and the defined hierarchy is a first step to a language for maintenance dependency modeling in CBR systems. An explicit language could help to identify all dependencies between the knowledge items in and between CBR systems.

3 Dependency Modeling for Knowledge Maintenance

This section describes the Case Factory maintenance approach and the required knowledge dependencies. We introduce different knowledge levels and a hierarchy of dependencies for CBR systems. In addition, we present a first algorithm for generating syntactic dependencies automatically and describe the current implementation of dependency modeling in myCBR.

3.1 Maintenance with Case Factories

The Case Factory approach is an agent-based maintenance approach for distributed CBR systems and integrated into the SEASALT architecture [4]. The SEASALT (**S**haring **E**xperience using an **A**gent-based **S**ystem **A**rchitecture **L**ayout) architecture is a domain-independent architecture for multi-agent systems to extract, analyze, share, and provide experience. The architecture consists of five components. The first component is the knowledge source component. This component contains so-called collector agents that are responsible for the extraction of knowledge from external knowledge sources. Knowledge sources could be databases, files, forums, or blogs. The second component, knowledge formalization, formalizes the extracted knowledge from the collector agents into a structural representation to be used by the third component, the knowledge provision component. This component manages the knowledge sources inside a multi-agent system (MAS) instantiated by the SEASALT architecture. Inside the knowledge provision component the so-called Knowledge Line (KL) is located. The KL contains a number of topic agents with access to internal knowledge sources like CBR systems. The basic idea is to modularize the knowledge among the topic agents and decide which knowledge is required to solve a given problem. The fourth component is the knowledge representation, that contains the underlying knowledge models for the different agents and knowledge sources. The last component is the individualized knowledge and contains the user interface for querying the system and displaying the solution [4].

The extended Case Factory approach extends the SEASALT architecture with a maintenance mechanism for CBR systems. If a topic agent has access to a CBR system, a CF is provided to maintain the CBR system. To coordinate several CFs a so-called Case Factory Organization (CFO) is provided, which consists of several agents to coordinate the overall system maintenance. A Case Factory consists of several agents that are responsible for different tasks: monitoring, evaluation, coordination, and maintenance execution. A monitoring agent will supervise the knowledge containers of a CBR system to notice changes to the knowledge like adding new cases, changing the vocabulary, or deleting cases. Monitoring agents will only notice the fact that changes have occurred and what has been changed. Evaluation agents are responsible for a qualitative evaluation of the consistency, performance, and competence of the CBR system. Which evaluation strategy is performed is up to the user. Existing approaches like utility footprint [19] or sensitivity analysis [7,22] could be applied as well as new or modified evaluation strategies. The coordination agent will collect all results from the monitoring and evaluation agents and create maintenance actions based on the collected information. In addition, the agent will use the modeled dependencies to determine additional maintenance actions that should be performed and sent them to the CFO. The dependencies and their use will be described in more detail in the next section. Maintenance execution agents are responsible for executing the confirmed maintenance actions and adapting the knowledge of the CBR system.

The Case Factory Organization is a superstructure for coordinating the maintenance activities of each Case Factory and providing the knowledge engineer with the required information to confirm or reject maintenance actions. Therefore, several agents with different tasks are part of the CFO: coordination agent, maintenance planning agent, explanation agent, and communication agent. The coordination agent gets all maintenance actions from the different CFs, derives additional maintenance actions based on the dependencies between CBR systems and passes the list of maintenance actions to the maintenance planning agent. The planning agent is responsible for creating a maintenance plan from all derived maintenance actions. Therefore, the agent checks for duplicate or conflicting maintenance actions, the order of maintenance actions and for circular maintenance actions. Based on these checks a maintenance plan is generated and passed to the explanation agent. This agent generates a human-readable explanation for each action in the maintenance plan and adds it to the corresponding maintenance action. The enhanced plan is passed to the communication agent and displayed to the knowledge engineer. After a review by the knowledge engineer, the confirmed or rejected maintenance actions are passed back to the individual CFs [11,12].

3.2 Dependencies

A central part of the Case Factory approach are the dependencies between the knowledge containers of CBR systems. These dependencies allow an overall maintenance planning with respect to connections between the individual knowledge of different CBR systems. A dependency can be defined with a source, a target, and a direction. This triple has been used by our first dependency definition and the source and target were defined as knowledge containers and can be found in Eq. (1).

$$\begin{aligned} d &= (kc_{sysS}, kc_{sysT}, t) \\ \text{where } kc &\in \{voc, sim, cb, ada\} \text{ and } sysS, sysT \in \{1 \ldots n\} \text{ and } t \in \{u, b\} \end{aligned} \tag{1}$$

The triple consists of two knowledge containers and a direction. The first knowledge container kc_{sysS} defines the left side of a dependencies, the source knowledge container. The second knowledge container kc_{sysT} defines the target knowledge container. Knowledge containers could be the vocabulary, the case base, the similarity measures or the adaptation knowledge according to Richter [15]. The indices $sysS$ and $sysT$ identifies the CBR system, the knowledge container belongs to, assuming all given CBR systems have a number between 1 and n. The last part of the triple t represents the direction of a dependency, either (u)ni-directional or (b)i-directional and determines whether the source and the target knowledge containers can be swapped or not.

Table 1. Knowledge levels, the contained knowledge, and examples

Knowledge level	Contained knowledge	Example
Knowledge level 1	CBR system	CBR system 1
Knowledge level 2	Knowledge container	Vocabulary, case base
Knowledge level 3	Specific case base	CB01, CB02
Knowledge level 4	Specific case, similarity measure, and adaptation rules	Case 123, simtax, rule23
Knowledge level 5	Attributes	Aircraft type, systems, status
Knowledge level 6	Specific values	A380, display, inoperable

But the information about the affected knowledge containers is not sufficient. For example, the information that a dependency exists between the vocabulary of CBR system A and the vocabulary of CBR system B does only allow to derive a maintenance action for changing the vocabulary. This is not enough information for specific and executable maintenance actions. Therefore, we defined six knowledge levels for CBR systems to find the required granularity of knowledge for the dependencies. The knowledge levels are shown in Table 1.

These knowledge levels are used to define a hierarchy of granularity for dependencies. The top level of the hierarchy is root level. It contains only one node, the root node. This root level could also be named as knowledge level 0. The second level represents the knowledge level 1 and contains nodes for the *CBR systems*. The third level, knowledge level 2, contains nodes for the knowledge containers of a CBR system: *vocabulary*, *case base*, *similarity measures*, and *adaptation knowledge*. The Knowledge level 3 contains nodes for the *case bases*, the other three branches have no nodes on this level. On knowledge level 4 nodes for the *cases*, *similarity measures*, and the *adaptation rules* can be found. Knowledge level 5 contains nodes for the *attributes* on all branches of the knowledge containers. The last level contains nodes for the *specific values* of each attribute. Figure 1 shows an example hierarchy with the knowledge levels.

Based on this hierarchy, dependencies with different knowledge levels could be defined. The most abstract dependencies are based on knowledge level 1 and the most detailed dependencies are based on knowledge level 6. For example, on knowledge level 1 a dependency between a CBR system A and a CBR system B could be defined. The dependency does not contain enough knowledge to derive a specific maintenance action, but the knowledge on this level is required for the knowledge levels below to differ between the more detailed dependencies. In addition, it could be used for visualization purposes of dependencies. On knowledge level 6, a dependency between two specific values could be defined. For example, there could be a dependency between the value *A380* of the attribute *aircraft type* in the knowledge container *vocabulary* of *CBR system A* and the value *A380* of the attribute *aircraft type* of the *case123* in the case base *CB01* in the knowledge container *case bases* of *CBR system A*. With this specific information, among others, a detailed maintenance action could be derived. The reworked dependency definition can be found in Eq. (2).

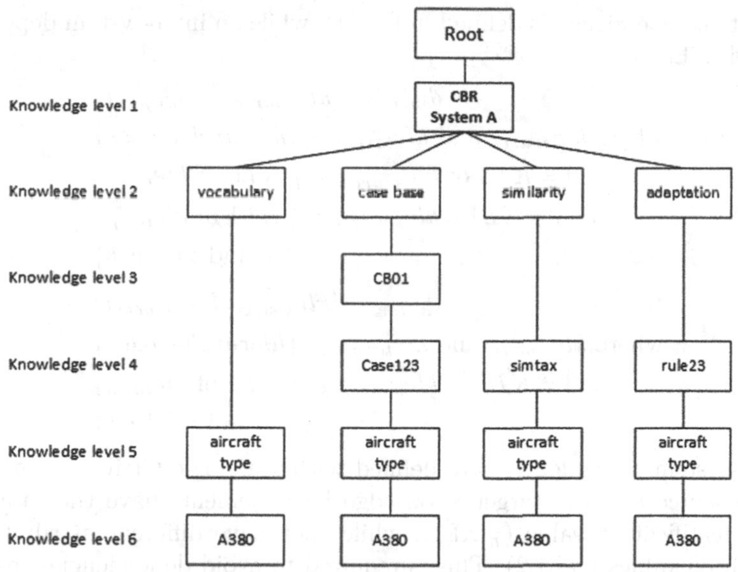

Fig. 1. Example hierarchy for granularity of dependencies

$$d = (kle_{source}, kle_{target}, t)$$
$$\text{where } kle_{source} \text{ and } kle_{target} \in \{hierarchy_n odes\} \text{ and } t \in \{u, b\} \quad (2)$$

In the previous definition, the source and target of a dependency are knowledge containers. In the current definition, the source and target are knowledge level elements (kle), that can be found in the defined hierarchy. The hierarchy is a set of nodes and edges, but a dependency references only on nodes. Therefore, $hierarchy_n odes$ is a subset of the hierarchy, that only contains the nodes. To identify an element in the hierarchy, every element gets an id code based on the knowledge level, characters, and continuous numbers. The id code consists of alphanumeric characters and starts with the number of the knowledge level. The characters for each knowledge level are combined with an underscore. The knowledge level 0 will not be considered, because it contains no knowledge. The nodes on each knowledge level will be continuously numbered. The only exception is knowledge level 2. The nodes on this level are identified with the starting character of the knowledge container. For example the id code for the *case123* node would be 1_C_1_1_0_0, the id code for the specific value *A380* in the same branch would be 1_C_1_1_1_1. A dependency between the specific value *A380* of the attribute *aircraft type* in the *vocabulary* and the same value in *case123* in *CB01* in the attribute *aircraft type* could be found in Eq. (3):

$$d = (1\_V\_0\_0\_1\_1, 1\_C\_1\_1\_1\_1, u) \quad (3)$$

The dependencies will still be differentiated between intra- and inter-system dependencies. An intra-system dependencies is defined within a CBR system, while inter-system dependencies are defined between different CBR systems. An intra-system dependency is defined in Eq. (4), while an inter-system dependency is defined in Eq. (5):

$$\begin{aligned} d_{intra} &= (kle_{source}, kle_{target}, t) \\ \text{where } kle_{source} \text{ and } kle_{target} &\in \{hierarchy_nodes\} \\ \text{and } \#KL6 \text{ of } kle_{source} &\neq \#KL6 \text{ of } kle_{target} \\ \text{and } \#KL1 of kle_{source} &= \#KL1 \text{ of } kle_{target} \\ \text{and } t &\in \{u, b\} \end{aligned} \tag{4}$$

$$\begin{aligned} d_{inter} &= (kle_{source}, kle_{target}, t) \\ \text{where } kle_{source} \text{ and } kle_{target} &\in \{hierarchy_nodes\} \\ \text{and } \#KL1 \text{ of } kle_{source} &\neq \#KL1 \text{ of } kle_{target} \\ \text{and } t &\in \{u, b\} \end{aligned} \tag{5}$$

Intra-system dependencies are defined within a single CBR system. Therefore, the source and the target knowledge level elements have the same CBR system identification value (#KL1), while they have different attribute value identification values (#KL2). This is required to avoid dependencies from specific values to themselves. This way we avoid circular processing of the same dependencies endless times. In contrast to the previous definition of intra-system dependencies, dependencies within the same knowledge container are permitted to model dependencies between different attributes of the vocabulary for example. For inter-system dependencies the CBR system identification value (#KL1) has to be different, while all other knowledge level elements could have the same identification number, for example when we have a backup copy of CBR system that contain the same knowledge.

In our first definition of dependencies [13] we introduced three trivial dependencies as intra-system dependencies. These trivial dependencies were defined between the vocabulary and the three other knowledge containers. On an abstract level this dependencies still exist, but on the knowledge levels 5 and 6, the number of dependencies cannot be defined in general, because the number of attributes and specific values depends on the knowledge modeling. Therefore, the new definition of our maintenance dependencies contain no trivial dependencies any more.

Based on the defined granularity of the dependencies, there could exist hundreds of dependencies in a CBR system. Modeling all these dependencies manually would cause a very high effort for the knowledge engineer. Therefore, the automated generation of dependencies based on a given knowledge model could reduce the effort and would allow the application of the Case Factory approach to existing CBR systems with a manageable effort. For an automated generation we have to differentiate between syntactic and semantic dependencies. A syntactic dependency is based on a syntactic compliance of values, for example the specific value *A380* of the attribute *aircraft type* in the *vocabulary* and the value *A380* in the same attribute in a given case. This dependency could be generated automatically by searching for the value *A380* in existing cases. If the value is set for the attribute in a case, a dependency is modeled. A first version of an algorithm to generate syntactic dependencies is shown in Listing 1.1.

Listing 1.1. Algorithm for generating syntactic dependencies

```
Definitions:
V_a Set of values for attribute a
C_cb Set of cases in a case base cb

v_a specific value of attribute a
c_cb specific case of case base cb
v_fct specific value in similarity measure
v_r specific value in rule

Input:
A Set of attributes in the case structure
CB Set of case bases in a CBR system
R Set of adaptation rules
S Set of similarity functions

Output:
D Set of syntactic dependencies

Algorithm:
D = \emptyset

for each (attribute a in A) {
  if (check(v_a exist in c_cb) {
     d_u = new d(v_a , v_c , u)
       if (exist(D, reverse(d_u s))) {
          d_b = new d(v_a , v_c , b)
          D = D - reverse(d_u) }
       else {
          D = D + d_b }
    }
    if (check(v_a exist in v_fct) {
       d_u = new d(v_a , v_fct , u)
       if (exist(reverse(d_u))) {
          d_b = new d(v_a , v_fct , b)
          D = D - reverse(d_u) }
       else {
          D = D + d_b }
    }
    if (check(v_a exist in v_r) {
          d_u = new d(v_a , v_r , u)
          if (exist(reverse(d_u))) {
              d_b = new d(v_a , v_r , b)
              D = D - reverse(d_u) }
          else {
              D = D + d_b }
    }}
return D
```

The algorithm iterates over all values of all attributes and compares them with the used values in the cases, the similarity measures and the adaptation rules. If a compliance is found a new dependency on the sixth knowledge level is created. The created dependency is defined a uni-directional. For every created dependency, the algorithm checks whether a reverse dependency exists, or not. If a reverse dependency is found, the newly created dependency is set to bi-directional and the reverse dependency is removed from the list. The new dependency is added to the list. This way explicit dependencies between the knowledge containers can be generated. Implicit dependencies between the knowledge containers can be found in several CBR tools and in myCBR, too. But these implicit dependencies cannot be used to generate maintenance actions and explanations. Therefore, some dependencies may exist implicitly and explicitly.

With the algorithm syntactic dependencies can be generated, but not semantic dependencies. A semantic dependency is based on user modeled connections between the knowledge items, for example a case referencing another case. A reference cannot always be identified automatically, therefore a dependency can only be generated under specific circumstances. For example, it could be checked if an attribute exists, that contains unique identifiers of cases. This attribute could be treated as reference attribute and the identifiers of the cases could be compared syntactically.

3.3 Dependency Modeling Using myCBR

We have extended the API and workbench of our tool myCBR to model and visualize dependencies. The algorithm for generation dependencies is not implemented yet. The API was extended with functions and classes to model, save, and load dependencies. Dependencies are stored in a so-called Maintenance Map. This Maintenance Map is based on the Knowledge Map [6] and stores all information about the dependencies. A Maintenance Map can be exported in RDF format. The following excerpt from an Maintenance Map shows an example modeling:

Listing 1.2. Excerpt from an example Maintenance Map

```
<rdf:Description rdf:about="'dependency1"'>
    <dep:source>1\_V\_0\_0\_1\_1</dep:source>
    <dep:target>1\_C\_1\_1\_1\_2</dep:target>
    <dep:type>1</dep:type>
    <dep:weight>1</dep:weight>
</rdf:Description>
<rdf:Description rdf:about="'dependency2"'>
    <dep:source>1\_V\_0\_0\_1\_1</dep:source>
    <dep:target>2\_V\_0\_0\_1\_2</dep:target>
    <dep:type>2</dep:type>
    <dep:weight>4</dep:weight>
</rdf:Description>
```

The example shows two dependencies, one intra-system and one inter-system dependency. For each dependency the source and the target are stored with their id code in the defined hierarchy. In addition, the direction of the dependency and the weight are stored. The direction can either be one or two, one if the dependency is uni-directional and 2 if it is bi-directional. The weight defines the importance of the dependency, the higher the weight, the more important the dependency is. A higher weight can be used to rank maintenance actions based on the according dependency. The dependencies are stored with their id code in the Maintenance Map. After reading the dependencies the id code is transformed into the more detailed knowledge level information.

The workbench was extended with a maintenance view. This view allows a user the modeling and visualization of dependencies. Figure 2 shows the maintenance view and a modeled dependency. The current implementation allows the creation of Maintenance Maps for intra- and inter-system dependencies. Intra-system dependencies are stored into a so-called local Maintenance Map, while inter-system dependencies are stored in a so-called global Maintenance Map. The differentiation is only for organizational purpose. After the creation of a Maintenance Map the associated dependencies could be modeled. A dependency can currently be modeled up to knowledge level 5. The sixth knowledge level is under development. For a local Maintenance Map the target CBR system is automatically set to be identical with the source CBR system. Dependencies with missing information cannot be saved to avoid incomplete and inconsistent dependencies.

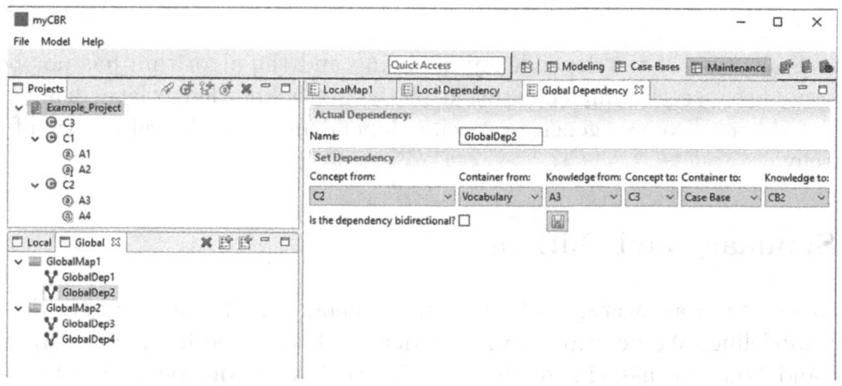

Fig. 2. Maintenance view with dependency

In addition, the workbench has a visualization component to generate an overview of the modeled dependencies. A simple list of dependencies would be adequate for few dependencies, but having dozens or hundreds of dependencies between several CBR systems, simple list would cause high effort to find and edit a specific dependency. Therefore, a graphic representation of the modeled dependencies would provide a better overview. Two different visualizations are currently implemented. Figure 3 shows a simple graph representation of the Maintenance Map. The vertices are arranged in a circle to get clear overview of the edges which represent the dependencies. Figure 4 shows a formatted graph of the Maintenance Map. The different knowledge levels are colored and associated groups of dependencies are displayed together. Vertices with no dependencies have a brighter color. Because the Maintenance Map can be edited by a user, empty dependencies can be created. An empty dependency is not treated as inconsistent and is represented with a gray vertex in the graph.

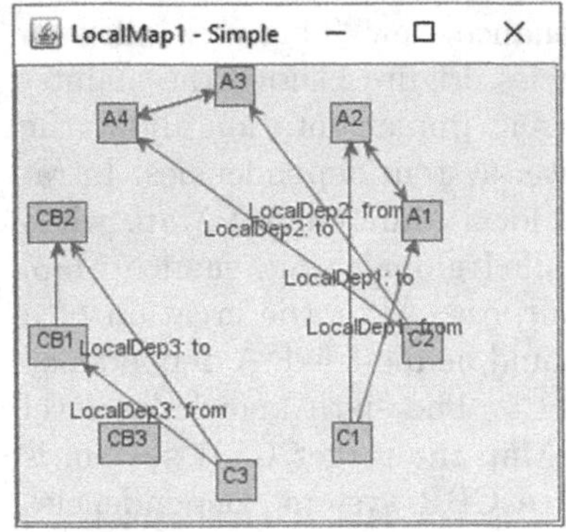

Fig. 3. Simple visualization of dependencies

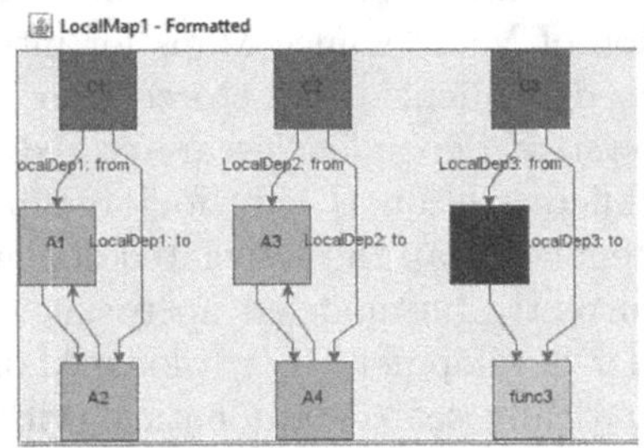

Fig. 4. Formatted visualization of maintenance map

An evaluation of the dependency modeling and the algorithm has not been done yet, because the implementation of the sixth knowledge level has to be completed before we can generate specific dependencies on the value level of the hierarchy.

4 Summary and Outlook

This paper gives an overview of the improvements for the maintenance dependency modeling. We describe the newly defined knowledge levels for CBR systems and how the associated hierarchy is used to model dependencies with different granularity. A formal definition of the improved dependencies is also given. In addition, we describe a first algorithm to generate syntactic dependencies automatically. Finally, we give an overview of the current implementation state of the dependency modeling in myCBR. Currently, we are integrating the knowledge level 6 into myCBR to be able to model all proposed granularities of maintenance dependencies. After the implementation is finished, we will use the improved dependencies in a multi-agent system with different CBR systems and associated Case Factories to evaluate the utility of the improved maintenance dependencies. In addition, we will improve the visualization of the dependencies to be able to show a scrollable visualization with different details for each knowledge level.

References

1. A Domain-Specific Language for Dependency Management in Model-Based Systems Engineering (2013)
2. Aamodt, A.: Modeling the knowledge contents of CBR systems. In: Proceedings of the Workshop Program at the Fourth International Conference on Case-Based Reasoning, pp. 32–37 (2001)
3. Al-Natour, S., Cavusoglu, H.: The strategic knowledge-based dependency diagrams: A tool for analyzing strategic knowledge dependencies for the purposes of understanding and communicating. Inf. Technol. Manage. **10**(2–3), 103–121 (2009)
4. Bach, K.: Knowledge Acquisition for Case-Based Reasoning Systems. Ph.D. thesis, University of Hildesheim , dr. Hut Verlag Muenchen 2013 (2012)
5. Bach, K., Sauer, C., Althoff, K.D., Roth-Berghofer, T.: Knowledge modeling with the open source tool myCBR. In: Nalepa, G.J., Baumeister, J., Kaczor, K. (eds.) Proceedings of the 10th Workshop on Knowledge Engineering and Software Engineering (KESE 2010). Workshop on Knowledge Engineering and Software Engineering (KESE-2014), located at 21st European Conference on Artificial Intelligence, August 19, Prague, Czech Republic. CEUR Workshop Proceedings (2014). http://ceur-ws.org/
6. Davenport, T.H., Prusak, L.: Working Knowledge: How Organizations Manage What they Know. Havard Business School Press, Boston (2000)
7. Du, J., Bormann, J.: Improved similarity measure in case-based reasoning with global sensitivity analysis: an example of construction quantity estimating. J. Comput. Civ. Eng. **28**(6), 04014020 (2012)
8. Ferrario, M.A., Smyth, B.: Distributing case-based maintenance: The collaborative maintenance approach. Comput. Intell. **17**(2), 315–330 (2001)
9. Nick, M.: Experience Maintenance Loop through Closed-Loop Feedback. Ph.D. thesis, TU, Kaiserslautern (2005)
10. Patterson, D., Anand, S., Hughes, J.: A knowledge light approach to similarity maintenance for improving case-base competence. In: ECAI Workshop Notes, pp. 65–78 (2000)
11. Reuss, P., Althoff, K.D.: Explanation-aware maintenance of distributed case-based reasoning systems. In: Workshop Proceedings of the LWA 2013, Learning, Knowledge, Adaptation, pp. 231–325 (2013)
12. Reuss, P., Althoff, K.: Maintenance of distributed case-based reasoning systems in a multi-agent system. In: Proceedings of the 16th LWA Workshops: KDML, IR and FGWM, Aachen, Germany, 8–10 September 2014, pp. 20–30 (2014)
13. Reuss, P., Althoff, K.: Dependencies between knowledge for the case factory maintenance approach. In: Proceedings of the LWA 2015 Workshops: KDML, FGWM, IR, and FGDB, Trier, Germany, 7–9 October 2015, pp. 256–263 (2015)
14. Richter, M.M.: The knowledge contained in similarity measure. In: Invited Talk at the First International Conference on Case-Based Reasoning, ICCBR 1995 (1995)
15. Richter, M.M.: Fallbasiertes Schlieen. In: Handbuch der künstlichen Intelligenz, pp. 407–430. Oldenbourg Wissenschaftsverlag (2003)
16. Roth-Berghofer, T.: Knowledge Maintenance of Case-based Reasoning Systems: The SIAM Methodology. Akademische Verlagsgesellschaft Aka GmbH, Berlin (2003)
17. Sangal, N., Jordan, E., Sinha, V., Jackson, D.: Using dependency models to manage complex software architecture. In: Proceedings of the 20th Annual ACM SIGPLAN Conference on Object-Oriented Programming, Systems, Languages, and Applications, OOPSLA 2005, pp. 167–176. ACM, New York (2005)

18. Sell, C., Winkler, M., Springer, T., Schill, A.: Two dependency modeling approaches for business process adaptation. In: Karagiannis, D., Jin, Z. (eds.) KSEM 2009. LNCS (LNAI), vol. 5914, pp. 418–429. Springer, Heidelberg (2009). doi:10.1007/978-3-642-10488-6_40
19. Smyth, B., Keane, M.: Remembering to forget: A competence-preserving case deletion policy for case-based reasoning systems. In: Proceedings of the 13th International Joint Conference on Artificial Intelligence, pp. 377–382 (1995)
20. Stahl, A.: Learning feature weights from case order feedback. In: Aha, D.W., Watson, I. (eds.) ICCBR 2001. LNCS, vol. 2080, pp. 502–516. Springer, Heidelberg (2001). doi:10.1007/3-540-44593-5_35
21. Stahl, A., Roth-Berghofer, T.R.: Rapid prototyping of CBR applications with the open source tool myCBR. In: Althoff, K.-D., Bergmann, R., Minor, M., Hanft, A. (eds.) ECCBR 2008. LNCS, vol. 5239, pp. 615–629. Springer, Heidelberg (2008). doi:10.1007/978-3-540-85502-6_42
22. Stram, R., Reuss, P., Althoff, K.-D., Henkel, W., Fischer, D.: Relevance matrix generation using sensitivity analysis in a case-based reasoning environment. In: Goel, A., Díaz-Agudo, M.B., Roth-Berghofer, T. (eds.) ICCBR 2016. LNCS, vol. 9969, pp. 402–412. Springer, Cham (2016). doi:10.1007/978-3-319-47096-2_27

Case-Based Recommendation for Online Judges Using Learning Itineraries

Antonio A. Sánchez-Ruiz(✉), Guillermo Jimenez-Diaz, Pedro P. Gómez-Martín, and Marco A. Gómez-Martín

Department of Software Engineering and Artificial Intelligence, Universidad Complutense de Madrid, Madrid, Spain
{antsanch,gjimenez}@ucm.es, {pedrop,marcoa}@fdi.ucm.es

Abstract. Online judges are online repositories with hundreds or thousands of programming exercises or *problems*. They are very interesting tools for learning programming concepts, but novice users tend to feel overwhelmed by the large number of problems available. Traditional recommendation techniques based on content or collaborative filtering do not work well in these systems due to the lack of user ratings or semantic descriptions of the problems. In this work, we propose a recommendation approach based on *learning itineraries*, i.e., the sequences of problems that the users tried to solve. Our experiments reveal that interesting learning paths can emerge from previous user experiences and we can use those learning paths to recommend interesting problems to new users. We also show that the recommendation can be improved if we consider not only the problems but also the order in which they were solved.

1 Introduction

Online judges are online repositories with hundreds or even thousands of programming exercises [9]. Each programming exercise is made of a public statement describing the problem to solve and a private set of test cases that will be used to automatically validate the solutions submitted to the system. This way, online judges can automatically compile and execute the code submitted by the users and check its correctness. Programming exercises usually impose certain restrictions regarding to execution time and memory usage so that solutions must be efficient as well as correct. Examples of such systems are the UVa Online Judge[1] and Codeforces[2], to mention just two of them.

These systems are usually used for training on-site programming contests such as the ACM-ICPC International Collegiate Programming Contest[3].

Supported by UCM (Group 910494) and Spanish Committee of Economy and Competitiveness (TIN2014-55006-R).

1 https://uva.onlinejudge.org.
2 http://codeforces.com/.
3 https://icpc.baylor.edu/.

D.W. Aha and J. Lieber (Eds.): ICCBR 2017, LNAI 10339, pp. 315–329, 2017.
DOI: 10.1007/978-3-319-61030-6_22

Some of them also hold online contests that coincide with the publication of new problems. This way, the users may practice with fresh problems and compete with others to climb positions in the system ranking.

Online judges are also valuable resources to teach and practice different programming skills. That is the case of *Acepta el reto*[4] (Spanish translation of *Take on the challenge*), an online judge developed by some of the authors of this paper, which is used in different subjects in Computer Science at the Universidad Complutense de Madrid. Teachers usually select a small set of problems with an appropriate level of difficulty, in which the solutions require putting into practice the concepts learned in the subject. This way, when the students try to solve those problems, they are actually involved in a *learning by doing* educational experience.

The role of the teacher proposing collections of suitable problems is necessary because, unfortunately, online judges pay little or no attention to the newbies who are not biased by programming contests but just want to practice algorithms or data structures. Usually, these users are overwhelmed by the large number of problems available in the repository and do not know how to choose which one they should try to solve next. This problem could be mitigated with the presence of some type of recommendation mechanism to guide those users. Unfortunately, recommender systems are quite uncommon in these systems and the existing ones are usually based on the *Global Ranking Method*, which just recommends the problem with more correct solutions in the system that the user has not resolved yet.

The lack of more sophisticated problem recommender systems in online judges can be explained by the fact that users hardly ever rate problems, they do not express their preferences on their profiles and the information about the problems is nonexistent or, at most, consists on a few tags with programming concepts that should be used in the solutions. The absence of user ratings and the very shallow description of the problems make very difficult and ineffective the use of classical approaches like collaborative filtering or content-based techniques.

In our previous work [8], we represented user-problem interactions as an implicit social network and, then, we used similarity-based link prediction techniques to recommend specific problems to each user. In this paper, we propose an alternative case-based approach based on *learning itineraries*. A learning itinerary comprises the sequence of problems that a user has attempted to solve (successfully or not), and it is a simplified view of the user interactions with the system. In this work, we evaluate different knowledge-light similarity measures for sequences to retrieve similar learning itineraries and recommend interesting problems to the user. Our hypothesis is that interesting learning paths can emerge from previous user experiences and constitute a collective source of expert knowledge that can be used to recommend interesting problems to new users. The results of our first experiments seem to support this hypothesis and

[4] https://www.aceptaelreto.com (Spanish only).

stress that, in addition to the set of problems attempted by each user, it is also important to consider the order in which they were solved.

The rest of the paper is organized as follows. Next section describes the online judge and the dataset used in our experiments. Section 3 details the concept of using learning itineraries to model the users and shows an example. Next section describes the process to recommend specific problems to the user using similar learning itineraries and the different similarity measures employed to compare them. Section 5 depicts the experimental setup and discusses the results obtained. Finally, the paper closes with some related work, conclusions and future lines of research.

2 *Acepta el reto* Online Judge

Acepta el reto (ACR) is an online judge created by two of the authors in 2014. It focuses on Spanish students, who find it hard to use other judges (with English statements) because of the language barrier. Problems are tagged according to the programming concepts needed to solve them, the kind of data structures required and some other aspects.

Users select the next problem to confront with and then try to solve it submitting code solutions in one of the accepted languages (currently C, C++ and Java). The system compiles the source code and runs it against many *test cases* whose solutions are known by the judge. The output generated by the submitted code is compared with the official solution and a verdict is provided.

From the system's point of view, a submission can be seen as a tuple (d, p, u, c, v) where d is the submission date, p and u are the problem and user respectively, c is the source code sent by the user, and v is the verdict emitted by the judge. As in many other online judges, the verdicts and their meanings are the following:

AC (*Accepted*): The submitted solution was correct because it produced the right answer, and it did not exceed the time and memory usage thresholds.

PE (*Presentation Error*): The solution was almost correct, though it failed to write the output in the exact required format (having an excess of blanks or line endings, for example).

CE (*Compile Error*): The solution did not even compile.

WA (*Wrong Answer*): The program failed to write the correct answer for one or more test cases.

RTE (*Runtime Error*): The program crashed during the execution (because of segmentation fault, floating point exception...).

TLE (*Time Limit Exceeded*): The execution took too much time and was cancelled.

MLE (*Memory Limit Exceeded*): The solution consumed too much memory and was aborted.

OLE (*Output Limit Exceeded*): The program tried to write too much information. This usually occurs if it runs into an infinite loop.

Table 1. Descriptive analysis of the ACR submissions: the original dataset (*Raw*) and the filtered dataset (*Solved*), where only the first AC-PE verdict for a user-problem submission is considered.

Metric	*Raw*	*Solved*
# Submissions	110,364	18,067
# Problems	289	289
# Users	3,678	2,892
Density	0.10	0.02
Earliest submission	2014/02/17	2014/02/17
Latest submission	2017/02/13	2017/02/13
Time span	1092 days	1092 days
Problems		
Maximum # submissions per problem	5,613	1,157
Median # submissions per problem	232	33
Average # submissions per problem	381.88	87.03
Minimum # submissions per problem	8	1
# Problems with at least 10 submissions	276	229
# Problems with at least 50 submissions	216	29
# Problems with at least 100 submissions	146	96
Users		
Maximum # submissions per user	2,576	249
Median # submissions per user	10	3
Average # submissions per user	30.01	6.84
Minimum # submissions per user	1	1
# Users with at least 5 submissions	2,415	801
# Users with at least 10 submissions	1,790	407
# Users with at least 20 submissions	1,198	145
Verdicts		
# Submissions with AC-PE	36,824	18,067
# Submissions with CE	7,061	-
# Submissions with runtime-limit error	31,924	-
# Submissions with wrong answer	33,443	-

Generally, users suffering a negative verdict try to fix their code and they then resubmit it. Sometimes, users resubmit *accepted* code with changes, in order to optimize the solution and improve their ranking position, but these improvements can lead into a negative verdict. However, from the system's point of view, the problem has been already solved by that user, despite the non-AC verdict in their last submission.

We carried out an exploratory analysis of the ACR database in order to familiarize ourselves with the data contained in it and to find relevant information for our recommendation purposes. Although the ACR system does allow all these resubmissions described above with no restrictions, for the sake of simplicity we filter the submissions in order to make easier to model the relationship between users and problems. After all, from a *user's* point of view, a problem can be:

- *Unattempted*: the user did not submit any solution to the problem yet.
- *Attempted*: the user submitted one or more solutions to the problem, but all of them were invalid.
- *Solved*: the user submitted several solutions for a problem and at least one was correct. In this category, we consider both AC and PE verdicts, since PE verdicts are close to being correct.

At the time of this writing (February 2017), ACR has 3,678 registered users, 289 problems and around 110,000 submissions (including resubmissions). The recommendation approach described in this paper will work with sequences of solved problems, ignoring the attempts that pursue to enhance their accepted submission. For this reason, we filter the dataset removing all the submissions representing attempts and we keep only the first accepted submission (AC or PE) for a user in a problem. This way, the number of submissions drops from the original 110,000 to 18,000. Table 1 provides a catalogue of descriptive statistics, as proposed in [3], about both datasets: the original (*Raw*) submission dataset and the filtered (*Solved*) dataset.

Figure 1 shows the number of submissions per month before and after filtering the dataset. In this analysis, we group together runtime errors (RTE) and all the verdicts related to limits exceeded (TLE, MLE and OLE). The high number of submissions in March 2014–2016 and December 2016 is due to a local contest organized by the authors on those months. ACR contains the problems of past editions and, the weeks before those contests, the contestants use the judge for training purposes. Figure 2 shows the cumulative number of submissions in both datasets.

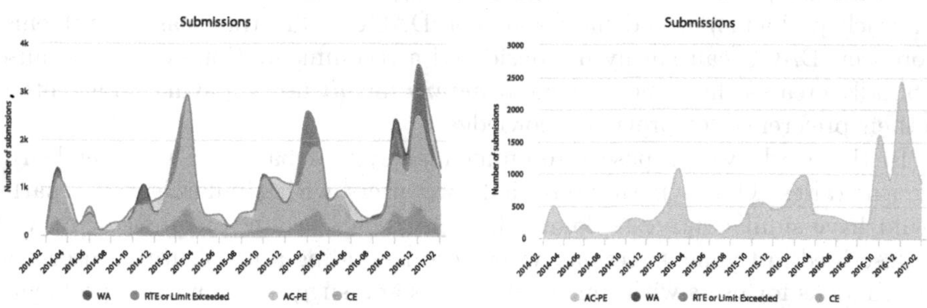

Fig. 1. Number of submissions per month in the raw (left) and filtered (right) dataset.

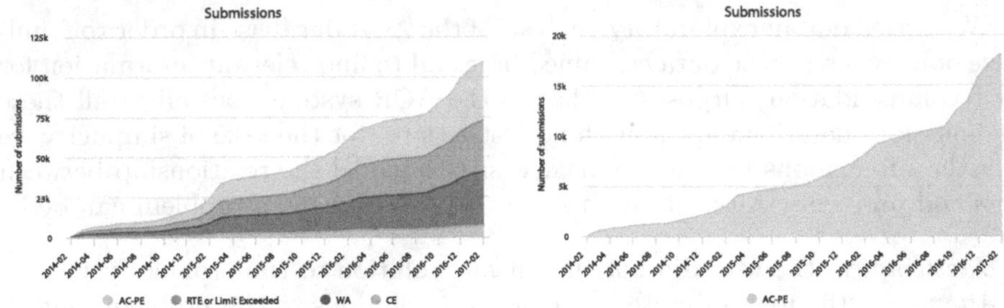

Fig. 2. Cummulative number of submissions per month in the raw dataset, categorized by verdict (left) and the cummulative number of accepted submissions in the filtered dataset (right).

3 Learning Itineraries

The submissions made by a specific user to an online judge constitute a *directed graph*, where nodes represent problems, and an edge between node A and node B exists if the user made a submission to problem A, followed by a submission to problem B. A self-loop in this graph is an indication of a problem that was tried twice in a row (Fig. 3a).

Users who are submitting solutions in an online judge are usually *learning* about programming and algorithms. The graph of attempted problems becomes their *learning itineraries* through these subjects. A recommendation regarding the next problem to resolve is, in this way, a hint about the path the user should follow in their own learning.

Nevertheless, online judges usually lack recommenders for suggesting users the next problem to solve. At best, they just list the problems unattempted by the user, in descending order of DACU (*Distinct ACcepted Users*). This is considered good enough under the hypothesis that a problem solved by many different users is easier than those solved by just a few.

Unfortunately, that assumption is not always true. It is common that problems that are prototypical for a family of algorithms (for example 8-queen or knapsack problems) would have a higher DACU value than easier problems. Moreover, DACU can hardly be considered a recommendation system, because it blindly creates the same *learning itinerary* for *all* users, paying no attention to their preferences or previous knowledge.

In this work, we propose a recommender system based on the user learning itineraries. Our hypothesis is that two users with similar learning paths would have similar interests, knowledge and even learning preferences. As will be described in the next sections, when recommending a problem to a user, the system looks for users with similar itineraries and suggest the untried problems.

Learning itineraries become our *cases*. An interesting aspect that makes this approach suitable for online judges is that these cases can be acquired automatically from the normal interactions between the users and the system, something important in a context where users are not used to rate problems or explicitly express their preferences.

Currently, each case is just a shortened version of a complete learning itinerary. Specifically, instead of considering all the submissions, we only use the first accepted solution for each problem, that is, the *filtered dataset* described in Sect. 2. This has the effect of *flattening* the learning itineraries to just a sequence without repetition (Fig. 3b).

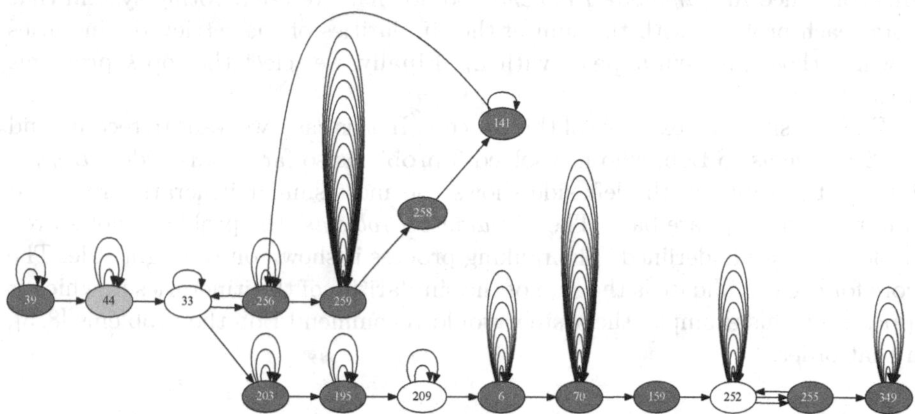

(a) Using the raw dataset. Dark nodes are problems with at least an AC submission, and shaded nodes are problems without AC but at least a PE submission.

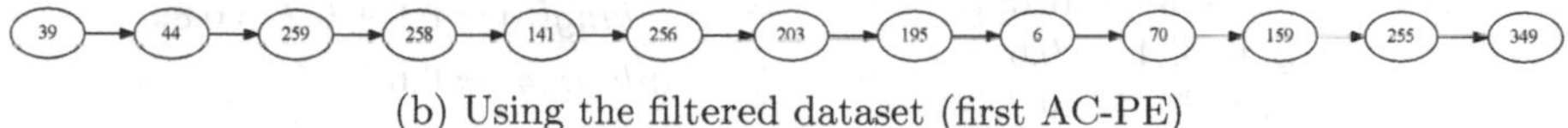

(b) Using the filtered dataset (first AC-PE)

Fig. 3. Learning itinerary of a particular user. Nodes represent problems (numbers are just ids with no meaning) and edges represent consecutive submissions.

4 Recommending Problems to Users

We propose a case-based recommendation approach based on the sequences of problems collected from the use of the online judge. These sequences represent previous experiences that can be exploited to recommend problems to the users of the online judge.

Our recommendation approach is based on a case base that stores learning itineraries or sequences of solved problems. A case is represented by a tuple (u_i, L_i), where u_i is a user and $L_i = [p_i, \ldots, p_n]$ is a learning itinerary containing the problems solved by u_i. The learning itinerary is ordered according to when u_i solved each problem, so if p_r appears before p_s in L_i is because u_i solved p_r before p_s.

The process to recommend k problems to a specific user is quite simple. Given a target user u_q, we use their itinerary L_q as the query to perform a recommendation. We retrieve from the case-base the most similar itinerary to L_q, according to a sequence similarity measure. Then, we create a list of *Candidate Problems* incorporating the problems from the retrieved itinerary that has not been solved by the user u_q yet. In other words, we only select the problems that do not belong to L_q. The construction of the *Candidate Problems* of size k follows an iterative approach, retrieving the next most similar itinerary from the case base until we collect k or more candidates.

Next, the adaptation phase creates the recommendation list ranking the problems contained in *Candidate Problems.* To do that, we use a voting system that scores each problem with the sum of the similarities of the retrieved itineraries in which those problem appears with L_q. Finally, we select the top k problems in the ranking.

Figure 4 shows an example of this process. In this case, we want to recommend $k = 2$ problems to Bob, who has solved 3 problems so far in this order: $L_{Bob} = [1, 6, 5]$. The table on the left side shows the most similar itineraries retrieved from the itinerary case base. The *Candidate Problems* (the problems not solved by Bob yet) are underlined. The ranking process is shown on the right side. The score for each candidate is the sum of the similarities of the itineraries in which it appears. In this example, the system would recommend Bob the problems [8, 3], in that order.

$L_{Bob} = [1, 6, 5]$

Itinerary	Similarity
6, 1, <u>8</u>, 5	0.75
1, <u>3</u>, 6, 5, <u>4</u>	0.6
5, <u>8</u>, <u>3</u>	0.2

(a) Most similar itineraries retrieved

$$ranking(8) = 0.75 + 0.2 = 0.95$$
$$ranking(3) = 0.6 + 0.2 = 0.8$$
$$ranking(4) = 0.6$$

(b) Candidate ranking

Fig. 4. Example to recommend 2 problems to Bob. The system would recommend the problems 8 and 3, in that order.

4.1 Similarity Measures for Learning Itineraries

Our case-based approach relies on finding similar learning itineraries or sequences of problems. Regarding the calculation of this similarity, learning itineraries are sequences that contain the identifiers (an integer) of the problems solved by the users.

The similarity measure, therefore, consists on the evaluation of the similarity between plain list of numbers. Moreover, we will consider the problem id's as numbers for simplicity, but they are just id's that may be seen as labels, not as numbers. In that sense, two different integers that are closed in $\mathbb{N}$ does not necessarily mean similar problems.

Let $L_1 = [p_{11}, p_{12}, \ldots]$ and $L_2 = [p_{21}, p_{22}, \ldots]$ be two different learning itineraries. We have considered four different similarity measures:

- Jaccard index [10]: The similarity between two lists is the ratio between the number of common elements and the total number of distinct elements. Let $E(L)$ be the set of elements in L, then:

$$J(L_1, L_2) = \frac{|E(L_1) \cap E(L_2)|}{|E(L_1) \cup E(L_2)|}$$

- Edit distance similarity [11]: It is based on the number of operations required to transform one sequence into the other. In our case we have considered as operations insertion, deletion and substitution. The number of operations is divided by the length of the longest sequence.
- Normalized compression distance (NCD) [6]: It uses the relationship between the lengths of the compressed sequences. Let x and y be string representations of L_1 and L_2, and let $Z(x)$ be the length of x compressed, then:

$$NCD(x, y) = \frac{Z(xy) - min\{Z(x), Z(y)\}}{max\{Z(x), Z(y)\}}$$

- Order constraints similarity: It is an extension of the Jaccard index that takes into consideration the number of common elements and ordering constraints in both sequences. Let $OC(L)$ be the set of constraints $a \prec b$ indicating that a appears before b in L. Then, the order constraints similarity is computed as follows:

$$OCS(L_1, L_2) = \frac{|E(L_1) \cap E(L_2)| + |OC(L_1) \cap OC(L_2)|}{|E(L_1) \cup E(L_2)| + |OC(L_1) \cup OC(L_2)|}$$

Note that our learning itineraries do not contain duplicated elements because we only consider the first time each problem is solved. In order to work with duplicates, the Jaccard and Order Constraints similarities should be extended to use multisets.

It is also worth mentioning that Jaccard similarity was not designed to work with sequences and, therefore, it does not take into account the particular order followed by the problems that appear in the sequences. Furthermore, all the similarity measures proposed are knowledge-light, in the sense that they do not take into account the semantic information of the items/problems in order to fine tune the similarity between paths. All the strategies are mainly binary regarding the distance between two problems: they are just either equals or distinct. In the future we plan to study how to incorporate this semantic information in the similarity assessment process.

5 Experimental Setup and Evaluation Results

In this section we evaluate the performance of different similarity measures to recommend new problems to ACR users. In order to evaluate the recommendation approach, we split users' itineraries into two subsequences. The former subsequence represents the problems that the user has already solved and we use it as a query to retrieve similar itineraries. Its length will be Q_L. The latter subsequence represents the problems that the user will solve in the future and we use it to validate the recommendations. The quality of the recommendations is evaluated using the following standard metrics:

- *Precision*, *Recall* and *F-Score* in top k recommendations [12].
- *At least one hit* (1-hit): ratio of recommendations in which at least one recommended problem was solved by the user. It corresponds to the metric *Success@k* with a success condition of guessing right at least one problem.
- *Mean Reciprocal Rank* (MRR): it evaluates the quality of a ranked list of recommendations based on the position of the first correct item [14]. Since we only provide one list of recommendations per user, the MRR can be computed as $MRR = 1/_{rank_i}$, where $rank_i$ is the position of the first attempted problem in the recommendation list.

It is worth noting that, in these experiments, we populate the case base with itineraries with length 5 or higher (i.e. users that have solved at least 5 problems). We made this decision for two main reasons. The first one is that most of the ACR users are transient users who only solve one or two problems before they stop using the system. We think that these users add noise to the case base and, anyway, they are probably not good examples for future recommendations. The second reason is that our recommendations are based on the similarity between sequences of problems, so we require the sequences to have a minimum length in order to be interesting. With this restriction, the case base contained 964 itineraries with a suitable length.

In the first experiment, we used a leave-one-out evaluation with all the itineraries available in the case base. Note that this experiment is not very realistic, in the sense that we do not consider the temporal constraints present in the recommender system. In other words, in order to recommend new problems to each user, we use all the other itineraries, including those that extend after the query's timestamp. Despite this unrealistic scenario, the leave-one-out approach evaluates the quality of the recommendations for every itinerary in the case base and can provide some insights.

There are two parameters to configure in this experiment: the number of recommended problems (k) and the length of the query (Q_L). As described above, Q_L is used to split the test itinerary in 2 parts: the query and the validation sequences (problems already solved and problems that will be solved in the future). Although we tried different configurations, the best results were obtained recommending only the best problem to the user ($k = 1$) and using query sequences with length 3 (results appear in Table 2).

It is interesting to highlight that the highest precision result corresponds to Jaccard (followed closely by Order Constraints) because this similarity measure does not take into account the relative order of the problems. Initially, we thought that the reason was that the query sequences were really short and, therefore, the order was not so important. However, these results remained consistent using longer queries. In fact, using longer queries or recommending more

Table 2. Results of the leave-one-out evaluation with $k = 1$ and $Q_L = 3$.

Similarity	Precision	Recall	F-Score	1-Hit	MRR
Jaccard	**0.4160**	0.0705	0.1109	**0.4160**	**0.7080**
Edit distance	0.3890	0.0660	0.1041	0.3890	0.6945
NCD	0.3444	0.0566	0.0894	0.3444	0.6722
Order constraints	0.4077	**0.0710**	**0.1113**	0.4077	0.7038

problems only reduces the precision of all the recommendations. Now, we suppose that these strange results can be explained because most of the ACR users are students from our faculty that only solve the problems that the teacher recommends during the course. This way, there are several students solving the same set of problems and, since we do not consider the temporal constraints in this experiment, the order of the problems is not very important. Recall (and therefore F-Score) values are really small because we only recommend one problem and the validation sequences usually contains several more problems, so the recommendation coverage is very small.

In the second experiment, we wanted to test a more realistic scenario. Now we only consider the information available at the instant of the recommendation. In order to do that, we selected a particular timestamp t to split the itineraries in two sets. The *test set* was built with itineraries of users who solved at least 5 problems before t and at least 5 problems after t. The sequence of problems solved before t was used as the query and the sequence of problems solved after t, for validation. The case base of itineraries was built with the remaining itineraries but removing any problem solved after t. The date selected to split the itineraries was 2016/10/20 because that timestamp allowed to build the largest test set with 117 itineraries.

In this experiment, the only parameter to configure is the number of recommendations (k) because the length of the queries (Q_L) is determined by the timestamp and might be different for each query (but at least 5). Moreover, this time we obtained the best results when we recommended only the best problem to each user (as shown in Table 3).

As we expected, the precision results are smaller than in the previous experiment because now we only use the information available at the time of the recommendation. This time, the recommendations based on Order Constraints and Edit Distance performed much better than Jaccard so we conclude that, in a realistic scenario, the order in which the problems are solved does seem to be important. Order Constraints is the clear winner in all the evaluation metrics with a precision of 29.06% and MRR of 0.6453. Like in the previous experiment, the small values of Recall and F-Score are due to the fact that we only recommend one problem. The NCD, which works very well in some domains, is very ineffective in our system maybe because the compressed sequences are really small.

Table 3. Evaluation results splitting the dataset at $t = 2016/10/20$ with $k = 1$.

Similarity	Precision	Recall	F-Score	1-Hit	MRR
Jaccard	0.1795	0.0115	0.0213	0.1795	0.5897
Edit distance	0.2308	0.0205	0.0367	0.2308	0.6154
NCD	0.1795	0.0173	0.0307	0.1795	0.5897
Order constraints	**0.2906**	**0.0233**	**0.0425**	**0.2906**	**0.6453**

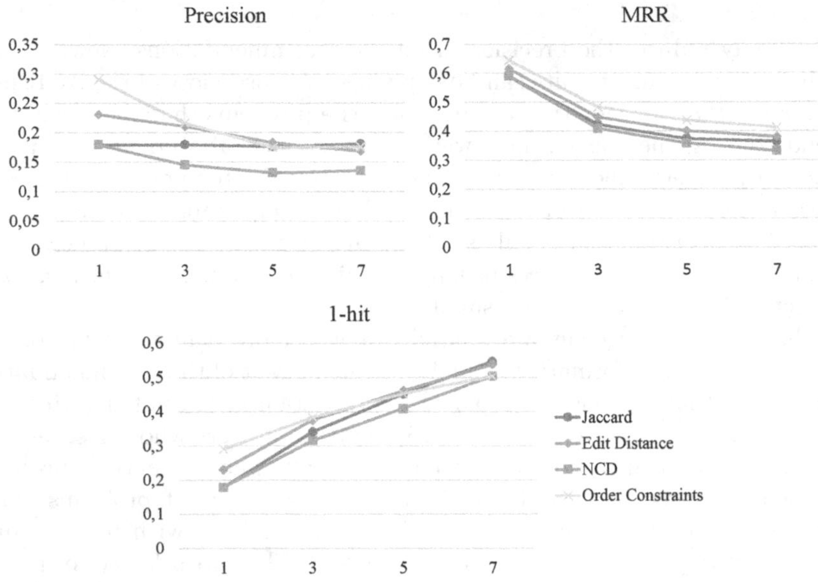

Fig. 5. Precision, 1-hit and MRR evolution when we increase the number of recommended problems (k).

Finally, Fig. 5 shows the evolution of the precision, 1-hit and MRR metrics when we increase the number of recommendations (k) from 1 to 7. Precision and MRR values tend to decrease as we recommend more problems, probably because those problems come from itineraries that are less similar and, therefore, less relevant to the user. On the other hand, the 1-hit metric increases with the number of recommendations because the more recommendations provided, the easier to guess right one of the future problems that the user will solve.

6 Related Work

An online judge like ACR can be seen as an online repository that stores a large amount of educational resources. These repositories traditionally suffer from the problem of how the student will find learning activities that best match their situational circumstances, prior knowledge, or preferences. Technology Enhanced

Learning (TEL) is the application domain that covers technologies that provide solutions to these problems when dealing with teaching and learning activities. Nowadays, recommender systems have become extremely interesting in TEL research, specially for recommending learning resources, peer students and sequences of resources or learning paths [5].

Most of the recommendation approaches that aim to suggest learning resources apply collaborative filtering or content-based techniques. Systems like CYCLADES [2] or ISIS [4] use collaborative filtering engines, which impose the rating of the learning resources by the students in order to find similar preferences. Our approach does not need explicit ratings but it uses the implicit results (attempted-accepted) in user submissions, which represents the interactions among users and problems in the online judge.

On the other hand, content-based techniques require an explicit description of the learning resources and, commonly, the student has to query the repository in terms of keywords or topics. This way, the recommender imposes the student a knowledge about the domain of the stored resources. ACR problems are indexed with metadata about the type of the previous knowledge needed to solve it. However, our previous work [8] highlighted that using the implicit similarity between problems according to the students who solved it can perform better results than using a content-based approach based in this problem metadata.

Case-based Reasoning techniques have also been considered for recommending learning activities. Alves et al. [1] proposes a case-based agent that helps the student in the current learning activity providing new learning resources using a Fuzzy approach. The work in [13] uses a simplified CBR model without the revision and adaptation phase for recommending learning objects in the computer programming domain. It follows a knowledge-intensive approach, where the educational resources are tagged and indexed using an ontology that provides knowledge about the similarity between the concepts that represent the domain topics. Additionally, the ontology defines successful learning paths through the domain concepts. Those paths are exploited by the recommendation strategy, which filters out the learning objects that cannot be covered by the user's current knowledge. Our approach does not describe successful learning paths but it supposes that these paths emerge from previous user experiences with the repository.

Case-based planning has been employed for suggesting learning paths, a sequence of recommended resources that guides a student towards achieving a learning goal [7]. In this work, the system retrieves abstract plans from a case-base and the chosen plan is incrementally personalized, selecting the learning resources that instantiate the plan as long as the student interacts with other resources. As stated before, we cannot define learning paths in ACR. However, a deep analysis of the problem resolution sequences could abstract prototypical sequences in order to help novel users. The identification of behavioural patterns in navigation history has been also employed before for recommendation purposes [15] so this could be an interesting research work to follow in the future.

7 Conclusions and Future Work

Online judges are online repositories with hundreds or even thousands of programming exercises. They can be very interesting tools to teach and train different programming concepts but they require a teacher or instructor to select subsets of problems with an appropriate difficulty for their students. This issue could be alleviated including recommendation methods, but the lack of user ratings and the shallow semantic description of the problems in these systems make difficult to use classical recommendation techniques based on content or collaborative filtering.

In our search for alternative recommendation approaches, we have previously tried to represent the user interactions as a social network and use link prediction techniques [8]. In this work, we propose a different approach based on learning itineraries, i.e., the sequences of problems the users tried to solve. Learning itineraries are abstract representations of the user interactions that consider the order in which the problems were solved.

In this work we have described a case-based recommender that leverages the implicit learning itineraries that emerge from the online judge submissions. The recommender provides a user with a catalogue of problems finding similar itineraries to the one followed by the target user and selecting the problems that she did not try to solve yet. Several sequence similarity metrics have been studied in order to select the most similar itineraries.

Our experiments reveal that interesting learning paths can emerge from previous user experiences and that we can use those learning paths to recommend problems to new users. The similarity measure that achieves better results in our experiments to retrieve relevant learning itineraries is based on the number of common ordering constraints between the problems of both itineraries.

There are several ways to continue our research. Currently our itineraries only consider the problems solved by a user. We plan to enrich this representation with information about the problems unsuccessfully attempted. From a pedagogical point of view, knowledge about failure can be as important as knowledge about success. Another line of improvement is related to the similarity measures used in this work because they do not contemplate different degrees of similarity between problems (they are the same problem or they are different). Although there is not much semantic knowledge about the problems available in the system, exercises are tagged with some programming concepts and the online judge stores statistics about how many users attempted to solve each problem and how many finally solved it. We will try to use this information to compare problems based on the programming techniques involved in their solutions and their difficulty.

The results presented in this paper are theoretical because the recommendation module has not been incorporated into the online judge yet. Every recommender system introduces some bias in the way users interact with the system so, for now, we would like to keep trying different approaches before we make a final decision.

References

1. Alves, P., Amaral, L., Pires, J.: Case-based reasoning approach to adaptive web-based educational systems. In: Eighth IEEE International Conference on Advanced Learning Technologies, ICALT 2008, pp. 260–261. IEEE (2008)
2. Avancini, H., Straccia, U.: User recommendation for collaborative and personalised digital archives. Int. J. Web Based Communities **1**(2), 163–175 (2005)
3. Dooms, S., Bellogín, A., De Pessemier, T., Martens, L.: A framework for dataset benchmarking and its application to a new movie rating dataset. ACM Trans. Intell. Syst. Technol. **7**(3), 41:1–41:28 (2016)
4. Drachsler, H., Hummel, H.G.K., Van den Berg, B., Eshuis, J., Waterink, W., Nadolski, R., Berlanga, A.J., Boers, N., Koper, R.: Effects of the ISIS recommender system for navigation support in self-organised learning networks. Educ. Technol. Soc. **12**(3), 115–126 (2009)
5. Drachsler, H., Verbert, K., Santos, O.C., Manouselis, N.: Panorama of recommender systems to support learning. In: Ricci, F., Rokach, L., Shapira, B. (eds.) Recommender Systems Handbook, pp. 421–451. Springer, Heidelberg (2015). doi:10.1007/978-1-4899-7637-6_12
6. Goldman, R.P., Kuter, U.: Measuring plan diversity: pathologies in existing approaches and a new plan distance metric. In: Proceedings of the Twenty-Ninth AAAI Conference on Artificial Intelligence, AAAI 2015, pp. 3275–3282. AAAI Press (2015)
7. Hulpuş, I., Hayes, C., Fradinho, M.O.: A framework for personalised learning-plan recommendations in game-based learning. In: Manouselis, N., Drachsler, H., Verbert, K., Santos, O.C. (eds.) Recommender Systems for Technology Enhanced Learning. Springer, New York (2014). doi:10.1007/978-1-4939-0530-0_5
8. Jimenez-Diaz, G., Gómez Martín, P.P., Gómez Martín, M.A., Sánchez-Ruiz, A.A.: Similarity metrics from social network analysis for content recommender systems. In: Goel, A., Díaz-Agudo, M.B., Roth-Berghofer, T. (eds.) ICCBR 2016. LNCS, vol. 9969, pp. 203–217. Springer, Cham (2016). doi:10.1007/978-3-319-47096-2_14
9. Kurnia, A., Lim, A., Cheang, B.: Online judge. Comput. Educ. **36**(4), 299–315 (2001)
10. Levandowsky, M., Winter, D.: Distance between sets. Nature **234**(5323), 34–35 (1971)
11. Levenshtein, V.I.: Binary codes capable of correcting deletions, insertions and reversals. In: Soviet Physics Doklady, vol. 10, pp. 707 (1966)
12. Pazzani, M.J.: A framework for collaborative, content-based and demographic filtering. Artif. Intell. Rev. **13**(5–6), 393–408 (1999)
13. Ruiz-Iniesta, A., Jimenez-Diaz, G., Gomez-Albarran, M.: A semantically enriched context-aware OER recommendation strategy and its application to a computer science OER repository. IEEE Trans. Educ. **57**(4), 255–260 (2014)
14. Said, A., Bellogín, A.: Comparative recommender system evaluation. In: Proceedings of the 8th ACM Conference on Recommender systems - RecSys 2014, pp. 129–136 (2014)
15. Zaíane, O.R.: Building a recommender agent for e-learning systems. In: Proceedings of the International Conference on Computers in Education, pp. 55–59. IEEE (2002)

kNN Sampling for Personalised Human Activity Recognition

Sadiq Sani[1(✉)], Nirmalie Wiratunga[1], Stewart Massie[1], and Kay Cooper[2]

[1] School of Computing Science and Digital Media, Robert Gordon University, Aberdeen AB10 7GJ, Scotland, UK
{s.a.sani,n.wiratunga,s.massie}@rgu.ac.uk

[2] School of Health Sciences, Robert Gordon University, Aberdeen AB10 7GJ, Scotland, UK
k.cooper@rgu.ac.uk

Abstract. The need to adhere to recommended physical activity guidelines for a variety of chronic disorders calls for high precision Human Activity Recognition (HAR) systems. In the SELFBACK system, HAR is used to monitor activity types and intensities to enable self-management of low back pain (LBP). HAR is typically modelled as a classification task where sensor data associated with activity labels are used to train a classifier to predict future occurrences of those activities. An important consideration in HAR is whether to use training data from a general population (subject-independent), or personalised training data from the target user (subject-dependent). Previous evaluations have shown that using personalised data results in more accurate predictions. However, from a practical perspective, collecting sufficient training data from the end user may not be feasible. This has made using subject-independent data by far the more common approach in commercial HAR systems. In this paper, we introduce a novel approach which uses nearest neighbour similarity to identify examples from a subject-independent training set that are most similar to sample data obtained from the target user and uses these examples to generate a personalised model for the user. This nearest neighbour sampling approach enables us to avoid much of the practical limitations associated with training a classifier exclusively with user data, while still achieving the benefit of personalisation. Evaluations show our approach to significantly out perform a general subject-independent model by up to 5%.

1 Introduction

Human Activity Recognition (HAR) is the computational discovery of human activity from sensor data and is increasingly being adopted in health, security, entertainment and defense applications [9]. An example of the application of HAR in healthcare is SELFBACK [2], a system designed to assist users with low back pain (LBP) by monitoring their level of physical activity in order to provide advice and guidance on how best to adhere to recommended physical activity

D.W. Aha and J. Lieber (Eds.): ICCBR 2017, LNAI 10339, pp. 330–344, 2017.
DOI: 10.1007/978-3-319-61030-6_23

guidelines. Guidelines for LBP recommend that patients should not be sedentary for long periods of time and should maintain moderate physical activity. SELFBACK continuously reads sensor data from a wearable device worn on the user's wrist, and recognises user activities in real time. This allows SELFBACK to compare the user's activity profile to the recommended guidelines for physical activity and produce feedback to inform the user on how well they are adhering to these guidelines. Other information in the user's activity profile include the durations of activities and, for walking, the counts of steps taken, as well as intensity e.g. slow, normal or fast. The categorisation of walking into slow, normal and fast allows us to better match the activity intensity (i.e. low, moderate or high) recommended in the guidelines. HAR is typically modelled as a classification task where sensor data associated with activity labels are used to train a classifier to predict future occurrences of those activities. This introduces two important considerations, representation and personalisation.

Many different representation approaches have been proposed for HAR. In this paper, we broadly classify these approaches into three: hand-crafted, transformational and deep representations. Previous works have not provided a definitive answer as to which feature extraction approach is best due to the often mixed or contradictory results reported in different works [14]. This may be attributed to the differences in the configurations (e.g. sensor types, sensor locations, types of activities etc.) used in different works. For this reason, we conduct a comparative study of five different representation approaches from the three representation classes in order to determine which representation works best for our particular configuration (single wrist-mounted accelerometer) and our choice of activity classes.

The second consideration for HAR is personalisation, where training examples can either be acquired from a general population (subject-independent), or from the target user of the system (subject-dependent). Previous works have shown using subject-dependent data to result in superior performance [3,8,16]. The relatively poorer performance of subject-independent models can be attributed to variations in activity patterns, gait or posture between different individuals [11]. However, training a classifier exclusively with user provided data is not practical in a real-world configuration as this places significant cognitive burden on the user to provide sufficient amounts of training data required to build a personalised model. In this paper, we introduce a nearest neighbour sampling approach for subject-independent training example selection. In doing so, we achieve personalisation by ensuring only those examples that best match a user's activity pattern influence the generation of the HAR model. Our approach uses nearest neighbour to identify subject-independent examples that are most similar to a small number of labelled examples provided by the user. In this way, our approach avoids the practical limitations of subject-dependent training. Our work draws inspiration from selective sampling in CBR where useful cases are sampled from the set of available cases for building effective case-bases [7,17].

The rest of this paper is organised as follows: in Sect. 2, we discuss important related work on personalised HAR and selective sampling of examples. Section 3 discusses the different feature representation approaches considered in this work, while our kNN sampling approach is described in Sect. 4. A description of our dataset is presented in Sect. 5, evaluations are presented in Sect. 6 and conclusions follow in Sect. 7.

2 Related Work

The common approach to classifier training in HAR is to use subject-independent examples to create a general classification model. However, comparative evaluation with personalised models, trained using subject-dependent examples, show this to produce more accurate predictions [3,8,16]. In [16], a general model was compared with a personalised model using a c4.5 decision tree classifier. The general model produced an accuracy of 56.3% while the personalised model produced an accuracy of 94.6% using the same classification algorithm, which is an increase of 39.3%. Similarly, [3,8] reported increases of 19.0% and 9.7% between personalised and general models respectively. However, all rely on access to subject-dependent training dataset. Such an approach has limited practical use for real-world applications because of the burden it places on users to provide sufficient training data.

Different types of semi-supervised learning approaches have been explored for personalised HAR e.g. Self-learning, Co-learning and Active learning, which bootstrap a general model with examples acquired from the user [11]. Both Self-learning and Co-learning attempt to infer accurate activity labels for unlabelled examples without querying the user. This way, both approaches manage to avoid placing any labelling burden on the user. In contrast, Active learning selectively chooses the most useful examples to present to the user for labelling. Hence, while Active learning does not avoid user labelling, it attempts to reduce it to a minimum using techniques such as uncertainty sampling which consistently outperform random sampling [12]. Our work does not focus on uncertainty, but instead uses similarity as the focus.

While semi-supervised learning approaches address the data acquisition bottleneck of subject-dependent training, they do not address the presence of noisy or inconsistent examples in the general model. It is our view that part of the reason why general models do not perform very well is that some examples are sufficiently distinct from the activity pattern of the current user that they contribute more to noise in the training set. Therefore, an attempt at selecting only the most useful examples from the training set for classifier training is likely to improve classification performance.

In CBR, sampling methods have been employed for casebase maintenance. Here the aim is to delete cases that fail to contribute to competence such that edited case-bases consistently lead to retrieval gains [15]. Case selection heuristics commonly exploit neighbourhood properties as a cue to identify areas of uncertainty and in doing so, active sampling approaches are adopted to inform case selection [4,17]. For our intended application, the criterion for example selection is very well defined. We seek to select examples from the available training set that are similar to examples supplied by the user, in order to personalise our classifier to the user's activity pattern. Accordingly, we use a k Nearest Neighbour sampling approach where the k most similar examples to the user's data are selected.

3 Feature Representation

Feature representation approaches for accelerometer data for the purpose of HAR can be divided into three categories: handcrafted features, frequency transform features and deep features.

3.1 Hand-Crafted Features

This is the most common representation approach for HAR and involves the computation of a number of defined measures on either the raw accelerometer data (time-domain) or the frequency transformation of the data (frequency domain) obtained using Fast Fourier Transforms (FFTs). These measures are designed to capture the characteristics of the signal e.g. average acceleration, variation in acceleration, dominant frequency etc. that are useful for distinguishing different classes of activities. For both time and frequency domain hand-crafted features, the input is a vector of real values $\overrightarrow{v} = v_1, v_2, \ldots, v_n$ for each axis x, y and z. A function θ_i (e.g. mean) is then applied to each vector $\overrightarrow{v}$ to compute a single feature value f_i. The final representation is a vector of length l comprised of these computed features $\overrightarrow{f} = f_1, f_2, \ldots, f_l$. The time-domain and frequency domain features used in this work are presented in Table 1. Further information on these features can be found in [5,20] respectively.

Table 1. Hand-crafted features for both time and frequency domains.

Time-domain features	*Frequency domain features*
Mean	Dominant frequency
Standard deviation	Spectral centroid
Inter-quartile range	Maximum
Lag-one-autocorrelation	Mean
Percentiles (10, 25, 50, 75, 90)	Median
Peak-to-peak amplitude	Standard deviation
Power	
Skewness	
Kurtosis	
Log-energy	
Zero crossings	
Root squared mean	

While hand-crafted features have worked well for HAR [9], a significant disadvantage is that they are domain specific. A different set of features need to be defined for each different type of input data i.e. accelerometer, gyroscope, time-domain or frequency domain values. Hence, some understanding of the characteristics of the data is required. Also, it is not always clear which features are likely to work best. Choice of features is usually made through empirical evaluation of different combinations of features or with the aid of feature selection algorithms [19].

3.2 Frequency Transform Features

Frequency transform feature extraction involves applying a single function ϕ on the vectors of raw accelerometer data to transform these into the frequency domain where it is expected that distinctions between different activities are better emphasised. Common transformations that have been applied include FFTs and Discrete Cosine Transforms (DCTs) [6,13]. FFT is an efficient algorithm optimised for computing the discrete Fourier transform of a digital input by decomposing the input into its constituent sine waves. DCT is a similar algorithm to FFT which decomposes an input into it's constituent cosine waves. Also, DCT returns an ordered sequence of coefficients such that the most significant information is concentrated at the lower indices of the sequence. This means that higher DCT coefficients can be discarded without losing information, making DCT better for compression. The main difference between frequency transform and frequency-domain hand-crafted features is that here, the coefficients of the transformation are directly used for feature representation without further feature computations. An overview of transform feature representation is presented in Fig. 1.

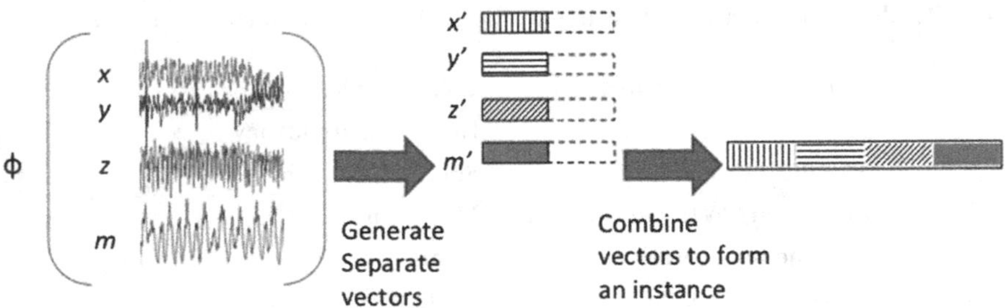

Fig. 1. Feature extraction and vector generation using frequency transforms.

A transformation function (DCT or FFT) ϕ is applied to the time-series accelerometer vector $\overrightarrow{v}$ of each axis $\mathbf{x}' = \phi(\mathbf{x})$, $\mathbf{y}' = \phi(\mathbf{y})$ and $\mathbf{z}' = \phi(\mathbf{z})$, as well as for the magnitude vector $\mathbf{m} = \{m_{i1}, \ldots, m_{il}\}$. The output of ϕ is a vector of coefficients which describe the sinusoidal wave forms that constitute the original signal. The final feature representation is obtained by concatenating the absolute values of the first l coefficients of $\mathbf{x}'$, $\mathbf{y}'$, $\mathbf{z}'$ and $\mathbf{m}'$ to produce a single feature vector of length $4 \times l$. The value $l = 48$ is used in this work, which is determined empirically.

3.3 Deep Features

Recently, deep learning approaches have been applied to the task of HAR due to their ability to extract features in an unsupervised manner. Deep approaches are able to stack multiple layers of operations to create a hierarchy of increasingly more abstract features [10]. Early work using Restricted Boltzmann Machines for HAR have only shown comparative performance to FFT and Principal Component Analysis [13]. More recent applications have used more of Convolutional Neural Networks (CNNs) due to their ability to model local dependencies that may exist between adjacent data points in the accelerometer data [18]. CNNs are a type of Deep Neural Network that have the ability for feature extraction by stacking multiple convolutional operators [10]. An example of a CNN is shown in Fig. 2.

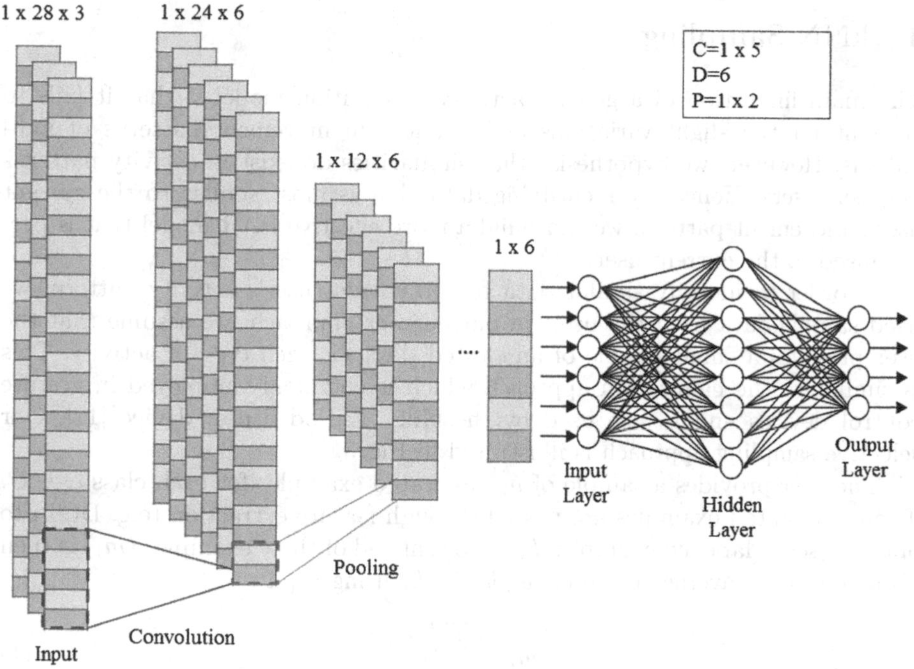

Fig. 2. Illustration of CNN

The input into the CNN in Fig. 2 is a 3-dimensional matrix representation with dimensions $1 \times 28 \times 3$ representing the width, length and depth respectively.

Tri-axial acceleromter data typically have a width of 1, a length l and a depth of 3 representing the x, y and z axes. A convolution operation is then applied by passing a convolution filter over the input which exploits local relationships between adjacent data points. This operation is defined by two parameters, D representing the number of convolution filters to apply and C, the dimensions of each filter. For this example, $D = 6$ and $C = 1 \times 5$. The output of the convolution operation is a matrix with dimensions $1 \times 24 \times 6$, these dimensions being determined by the dimension of the input and the parameters of the convolution operation applied. This output is then passed through a Pooling operation which basically performs dimensionality reduction. The parameter P determines the dimensions of the pooling operator which in this example is 1×2, which results in a reduction of the width of its input by half. The output of the pooling layer can be passed through additional Convolution and Pooling layers. The output of the final Pooling layer is then flattened into a 1-dimensional representation and then fed into a fully connected neural network. The entire network (including convolution layers) is trained through back propagation over a number of generations until some convergence criteria is reached. Detailed description of CNNs can be obtained in [10].

4 kNN Sampling

The main limitation of a general activity recognition model is that it fails to account for the slight variations and nuances in movement patterns of individuals. However, we hypothesise that similarities do exist in activity patterns between users. Hence, by identifying data that is most similar to the current user's movement pattern, we can build a more effective HAR model that is personalised to the current user.

In order to identify similar data to the current user's activity pattern, we need sample data from the user. In our current approach, we assume that the user provides a small sample of annotated data for each type of activity. This is similar to the calibration approach which is commonly employed in gesture control devices and is also used by the Nike + iPod fitness device [11]. Our selective sampling approach is illustrated in Fig. 3.

The user provides a sample of n_i annotated examples for each class $c_i \in C$. These annotated examples are passed through feature extraction (e.g. DCT) to obtain a set of labelled examples L_i. The centroid of these examples (m_i) is then obtained as the average of all examples in L_i using Eq. 1.

$$m_i = \frac{\sum_j^{n_i} l_{ij}}{n_i} \tag{1}$$

Where $l_{ij} \in L_i$. The centroid m_i is used along with kNN to obtain the k most similar training examples S_i from the set of training examples T_i that belong to class c_i. The selected examples S_i are then combined with the user labelled examples L_i to form a new training set T_i' which is used for training a personalised classifier.

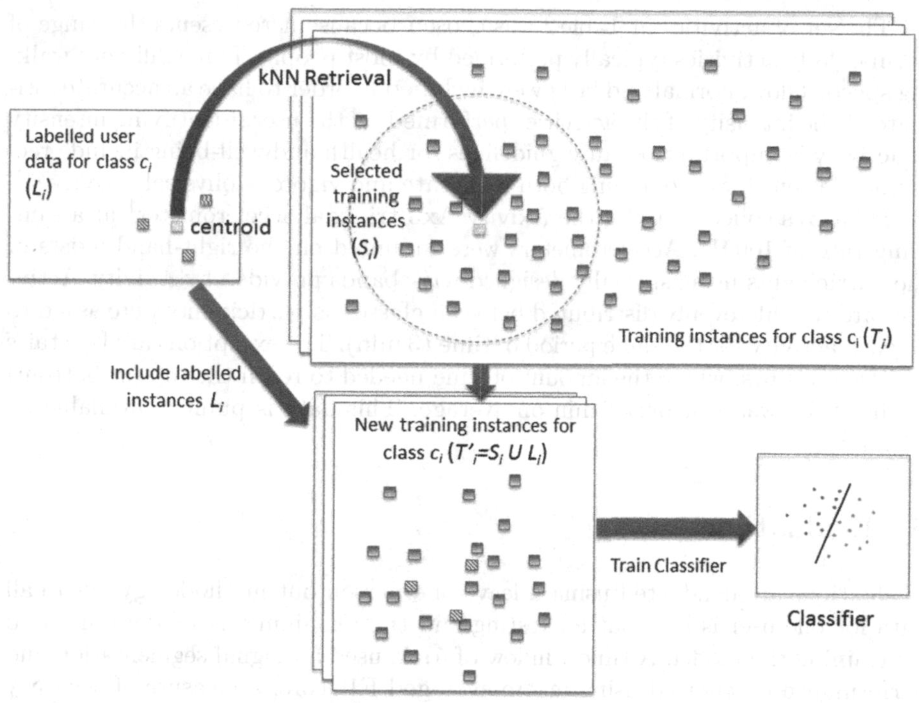

Fig. 3. Nearest neighbour sampling approach.

5 Dataset

A group of 50 volunteer participants was used for data collection. The age range of participants is 18–54 years and the gender distribution is 52% Female and 48% Male. Data collection concentrated on the activities provided in Table 2.

Table 2. Details of activities classes.

Activity	Description
Lying	Lying down relatively still on a plinth
Sitting	Sitting still with hands on desk or thighs
Standing	Standing relatively still
Walking slow	Walking at slow pace
Walking normal	Walking at normal pace
Walking fast	Walking at fast pace
Up stairs	Walking up 4–6 flights of stairs
Down stairs	Walking down 4–6 a flights of stairs
Jogging	Jogging on a treadmill at moderate speed

The set of activities in Table 2 was chosen because it represents the range of normal daily activities typically performed by most people. Three different walking speeds (slow, normal and fast) were included in order to have an accurate estimate of the intensity of the activities performed by the user. Identifying intensity of activity is important because guidelines for health and well-being include recommendations for encouraging both moderate and vigorous physical activity [1].

Data was collected using the Axivity Ax3 tri-axial accelerometer[1] at a sampling rate of 100 Hz. Accelerometers were mounted on the right-hand wrists of the participants using specially designed wristbands provided by Axivity. Activities are roughly evenly distributed between classes as participants were asked to do each activity for the same period of time (3 min). The exceptions are Up stairs and Down stairs, where the amount of time needed to reach the top (or bottom) of the stairs was just over 2 min on average. This data is publicly available on Github[2].

6 Evaluation

Evaluations are conducted using a leave-one-person out methodology where all data for one user is held out for testing and the remaining users' data are used for training the model. A time window of 5 s is used for signal segmentation and performance is reported using macro-averaged F1 score, a measure of accuracy that considers both precision (the fraction of examples predicted as class c_i that correctly belong to c_i) and recall (the fraction of examples truly belonging to class c_i that are predicted as c_i) for each class.

Our evaluation is composed of two parts. Firstly, we compare the different representations discussed in Sect. 3 using 2 classifiers: kNN and SVM. In the second section, we use the best representation/classifier combination to compare different selection approaches for generating personalised HAR models.

6.1 Feature Representations

In this section, we compare the feature representation appraoches presented in Sect. 3 as follows:

- Time: Time domain features
- Freq: Frequency domain features
- FFT: Frequency transform features using FFT coefficients
- DCT: Frequency transform features using DCT coefficients.

Each representation is evaluated with both a kNN and SVM classifier. In addition, we include a CNN classifier. The architecture of our CNN uses 3 convolution layers with convolution filter numbers D, set to 40, 20 and 10 respectively. Dimensions of each convolution filter C, are set to $1 \times 10 \times 3$. Each convolution

[1] http://axivity.com/product/ax3.

[2] https://github.com/selfback/activity-recognition/tree/master/activity_data.

layer is followed by a pooling layer with dimension P, set to 1×2. The output of the convolution is fed into a fully connected network with 2 hidden layers with 900 and 300 units respectively and an output layer with soft-max regression. Training of the CNN is performed for a maximum of 300 generations as longer training generations did not improve performance. The inclusion of CNN allows us to compare the performance of a state-of-the-art approach against conventional HAR approaches.

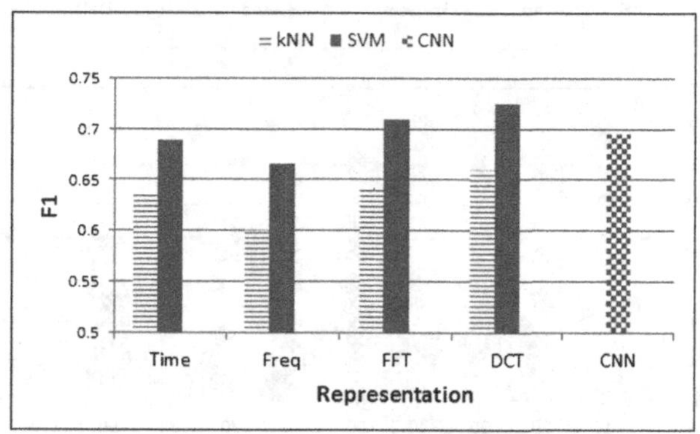

Fig. 4. Evaluation of different representations and classifiers.

Note from Fig. 4 that the best result is achieved using DCT representation with SVM classifier, while second best is FFT with SVM. In general, SVM out performed kNN on all representation types. The poor performance of kNN might be because the dataset does not provide clearly separable neighbourhoods. Indeed it is intuitive to think many examples from similar classes e.g. sitting and lying, as well as slow, normal and fast walking would be within close proximity in the feature space and might not be easily distinguishable using nearest neighbour similarity. CNN came in third best in the comparison. This indicates the potential of CNN for HAR, however, in our evaluation, it did not beat the much simpler frequency transform approaches. Our results are consistent with the findings of [14] where CNNs did not out perform conventional approaches. Also, the high cost of retraining a CNN makes this approach impractical for personalisation using our approach.

6.2 Selective Sampling

The seconds part of the evaluation uses the best representation/classifier combination, i.e. DCT+SVM to compare different sampling approaches for generating a personalised HAR model. The sampling approaches included in the comparison are as follows:

- All-Data: uses entire training set T without sampling;
- knnSamp: uses the kNN sampling approach presented in Sect. 4, but uses only the selected training examples S for classifier training;

- knnSamp$^+$: uses the kNN sampling approach presented in Sect. 4 and uses the combined set $T' = S \cup L$ for model generation; and
- Random: selects training examples at random for classifier training.

For any given user, 30% of test data is held-out to simulate user provided annotated data for personalisation. The remaining 70% forms the test data.

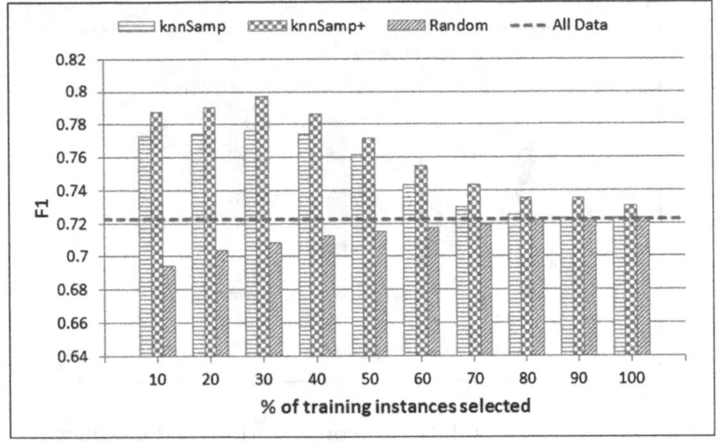

Fig. 5. Results for personalised model generation strategies.

Figure 5 shows the results of the different sampling approaches where the x-axis shows the percentage of training examples selected while the y-axis shows the F1 score. The horizontal line shows the result for All-Data. Observe that both knnSamp and knnSamp$^+$ significantly outperform all other approaches when no more than 50% of the training set is used, with the best result achieved using only 30% of the training set. F1 score declines after 50% as more of the noise from dissimilar examples in the training set are introduced into the model. The high accuracy of knnSamp compared to the other approaches indicates that the nearest neighbour selection strategy effectively selects useful similar examples for activity recognition. The best improvements of both knnSamp and knnSamp$^+$ compared to the other approaches are statistically significant at 99% using a paired T-test. Unsurprisingly, no improvement is achieved through random selection of training examples. Note that adding the user data to the entire training set without sampling produces only marginal improvement (+0.008 F1 Score).

6.3 Discussion

To further understand the performance gain of our personalisation approach, we present the break down of the performance (precision, recall and F1 score) by class for the best performing sampling method knnSamp$^+$ (with 30% sampling of training data) in Table 3. Here, we can see that personalisation had produced considerable improvement in the F1 scores of lying, sitting, walk_slow, walk_normal and walk_fast. From the confusion matrix in Fig. 6, we can observe that without personalisation, about 50% (547) of lying examples are predicted as sitting. However, personalisation produces better separation between lying and sitting which is evidenced by the higher recall score of lying (0.91) and higher precision of sitting (0.90) after personalisation.

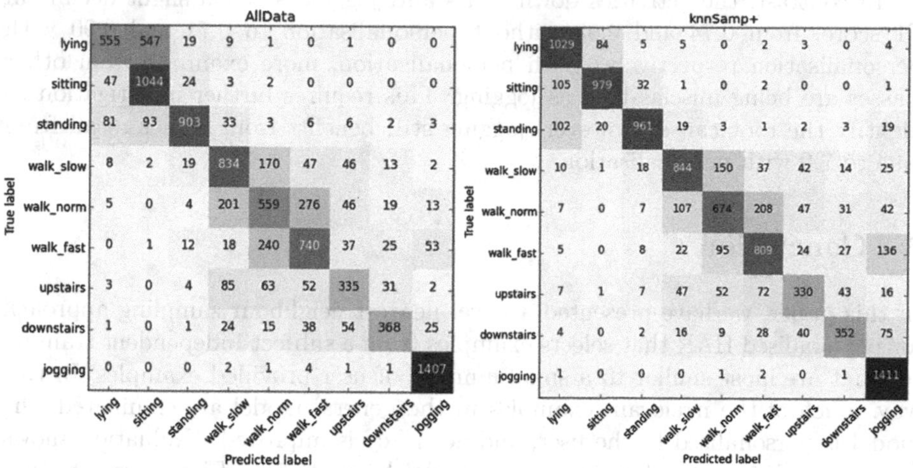

Fig. 6. Confusion matrices for All-Data (left) and knnSamp^{+} (right).

Table 3. Precision, recall and F1 scores by class for All-Data and knnSamp^{+}.

	All-Data			knnSamp^{+}		
	Precision	Recall	F1 Score	Precision	Recall	F1 Score
Lying	0.80	0.49	0.61	0.81	0.91	0.86
Sitting	0.62	0.92	0.74	0.90	0.87	0.89
Standing	0.91	0.82	0.86	0.92	0.85	0.89
Up stairs	0.63	0.57	0.60	0.68	0.57	0.62
Down stairs	0.79	0.70	0.74	0.75	0.67	0.71
Walk_fast	0.63	0.66	0.65	0.70	0.72	0.71
Walk_normal	0.53	0.51	0.52	0.68	0.60	0.64
Walk_slow	0.69	0.72	0.70	0.80	0.74	0.77
Jogging	0.93	0.98	0.95	0.82	1.00	0.90

A similar pattern can also be observed with the three different walking speeds. From Fig. 6, it can be observed that without personalisation, only about half (559 examples) of walking normal are predicted correctly, with most of the other half split between walking slow (201 examples) and walking fast (276 examples) giving a low recall score of 0.51. In addition, 240 walking fast examples are miss-classified as walking normal which results in a low precision score for walking normal of 0.53. However, with personalisation, the number of walking normal examples predicted correctly increases to 674 while the number of misclassified walking fast examples reduces to 95 which improves the recall and precision scores to 0.68 and 0.60 respectively.

In contrast, the activities down stairs and jogging suffer a slight decline in F1 score, from 0.74 and 0.95 without personalisation to 0.71 and 0.90 with personalisation respectively. With personalisation, more examples from other classes are being misclassified as jogging. This requires further investigation to identify the root cause. However, jogging still benefits from higher recall from 0.98 to 1.0 with personalisation.

7 Conclusion

In this paper, we have presented a novel nearest neighbour sampling approach for personalised HAR that selects examples from a subject-independent training set that are most similar to a small number of user provided examples. In this way, much of the irrelevant examples in the general model are eliminated, the model is personalised to the user, and accuracy is improved. Evaluation shows our approach to outperform a general model by up to 5% of F1 score. Another advantage of our approach is that it avoids the practical limitation of subject-dependent training by reducing the data collection burden on the user.

Many different representation approaches have been proposed for HAR without a definitive best approach, partly due to the differences in configurations (e.g. sensor types, sensor locations etc.) and partly due to different mix of activity classes. Therefore, it is important to determine which representation approach is best suited for the configuration used in SELFBACK i.e., a single wrist-mounted accelerometer, as well as the types of activities. Accordingly, another contribution of this paper is a comparative study of five representation approaches including state-of-the-art CNNs on our dataset. Results show a frequency transform approach using DCT coefficients to outperform the rest.

A number of considerations have been identified for future work. Firstly, a method that further reduces if not eliminates the need for user annotated data will further improve the user experience of our system. Secondly, evaluations in this paper have only been applied on short time durations that immediately follow the user examples. Test data covering longer durations are needed in order to evaluate the performance of the personalised model over longer periods of time. If the accuracy of the model drops due to long term changes in context, it would be interesting to be able to identify these context changes in order to initiate further rounds of personalisation. Note that success in automatic acquisition of labelled examples should significantly aid this process of continuous personalisation with minimal impact on user experience.

Acknowledgment. This work was fully sponsored by the collaborative project SELFBACK under contract with the European Commission (# 689043) in the Horizon 2020 framework. The authors would also like to thank all students and colleagues who volunteered as subjects for data collection.

References

1. Abel, M., Hannon, J., Mullineaux, D., Beighle, A.: Determination of step rate thresholds corresponding to physical activity intensity classifications in adults. J. Phys. Act. Health **8**(1), 45–51 (2011)
2. Bach, K., Szczepanski, T., Aamodt, A., Gundersen, O.E., Mork, P.J.: Case representation and similarity assessment in the SELFBACK decision support system. In: Goel, A., Díaz-Agudo, M.B., Roth-Berghofer, T. (eds.) ICCBR 2016. LNCS (LNAI), vol. 9969, pp. 32–46. Springer, Cham (2016). doi:10.1007/978-3-319-47096-2_3
3. Berchtold, M., Budde, M., Gordon, D., Schmidtke, H.R., Beigl, M.: ActiServ: activity recognition service for mobile phones. In: Proceedings of International Symposium on Wearable Computers, ISWC 2010, pp. 1–8 (2010)
4. Craw, S., Massie, S., Wiratunga, N.: Informed case base maintenance: a complexity profiling approach. In: Proceedings of the Twenty-Second AAAI Conference on Artificial Intelligence, 22–26 July 2007, Vancouver, British Columbia, Canada, p. 1618. AAAI Press (2007)
5. Figo, D., Diniz, P.C., Ferreira, D.R., Cardoso, J.M.: Preprocessing techniques for context recognition from accelerometer data. Pers. Ubiquitous Comput. **14**(7), 645–662 (2010)
6. He, Z., Jin, L.: Activity recognition from acceleration data based on discrete consine transform and SVM. In: Proceedings of IEEE International Conference on Systems, Man and Cybernetics, SMC 2009, pp. 5041–5044. IEEE (2009)
7. Hu, R., Delany, S., MacNamee, B.: Sampling with confidence: using k-NN confidence measures in active learning. In: Proceedings of the UKDS Workshop at 8th International Conference on Case-Based Reasoning, ICCBR 2009, p. 50 (2009)
8. Jatoba, L.C., Grossmann, U., Kunze, C., Ottenbacher, J., Stork, W.: Context-aware mobile health monitoring: evaluation of different pattern recognition methods for classification of physical activity. In: Proceedings of 30th Annual International Conference of the IEEE Engineering in Medicine and Biology Society, pp. 5250–5253, August 2008
9. Lara, O.D., Labrador, M.A.: A survey on human activity recognition using wearable sensors. Commun. Surv. Tutor. IEEE **15**(3), 1192–1209 (2013)
10. LeCun, Y., Bengio, Y.: Convolutional networks for images, speech, and time series. In: Arbib, M.A. (ed.) The Handbook of Brain Theory and Neural Networks, pp. 255–258. MIT Press, Cambridge (1998)
11. Longstaff, B., Reddy, S., Estrin, D.: Improving activity classification for health applications on mobile devices using active and semi-supervised learning. In: Proceedings of 4th International Conference on Pervasive Computing Technologies for Healthcare, pp. 1–7, March 2010
12. Miu, T., Missier, P., Plötz, T.: Bootstrapping personalised human activity recognition models using online active learning. In: Proceedings of IEEE International Conference on Computer and Information Technology; Ubiquitous Computing and Communications; Dependable, Autonomic and Secure Computing; Pervasive Intelligence and Computing (CIT/IUCC/DASC/PICOM) 2015, pp. 1138–1147. IEEE (2015)

13. Plötz, T., Hammerla, N.Y., Olivier, P.: Feature learning for activity recognition in ubiquitous computing. In: Proceedings of the Twenty-Second International Joint Conference on Artificial Intelligence, IJCAI 2011, pp. 1729–1734. AAAI Press (2011)
14. Ronao, C.A., Cho, S.-B.: Deep convolutional neural networks for human activity recognition with smartphone sensors. In: Arik, S., Huang, T., Lai, W.K., Liu, Q. (eds.) ICONIP 2015. LNCS, vol. 9492, pp. 46–53. Springer, Cham (2015). doi:10.1007/978-3-319-26561-2_6
15. Smyth, B., Keane, M.T.: Remembering to forget: a competence-preserving case deletion policy for case-based reasoning systems. In: Proceedings of the 14th International Joint Conference on Artificial Intelligence, IJCAI 1995 vol. 1, pp. 377–382. Morgan Kaufmann Publishers Inc., San Francisco (1995)
16. Tapia, E.M., Intille, S.S., Haskell, W., Larson, K., Wright, J., King, A., Friedman, R.: Real-time recognition of physical activities and their intensities using wireless accelerometers and a heart rate monitor. In: Proceedings of 11th IEEE International Symposium on Wearable Computers, pp. 37–40. IEEE (2007)
17. Wiratunga, N., Craw, S., Massie, S.: Index driven selective sampling for CBR. In: Ashley, K.D., Bridge, D.G. (eds.) ICCBR 2003. LNCS (LNAI), vol. 2689, pp. 637–651. Springer, Heidelberg (2003). doi:10.1007/3-540-45006-8_48
18. Zeng, M., Nguyen, L.T., Yu, B., Mengshoel, O.J., Zhu, J., Wu, P., Zhang, J.: Convolutional neural networks for human activity recognition using mobile sensors. In: Proceedings of 6th International Conference on Mobile Computing, Applications and Services, pp. 197–205 (2014)
19. Zhang, S., Mccullagh, P., Callaghan, V.: An efficient feature selection method for activity classification. In: Proceedings of IEEE International Conference on Intelligent Environments, pp. 16–22 (2014)
20. Zheng, Y., Wong, W.K., Guan, X., Trost, S.: Physical activity recognition from accelerometer data using a multi-scale ensemble method. In: IAAI (2013)

Case-Based Interpretation of Best Medical Coding Practices—Application to Data Collection for Cancer Registries

Michael Schnell[1,2(✉)], Sophie Couffignal[1], Jean Lieber[2], Stéphanie Saleh[1], and Nicolas Jay[2,3]

[1] Epidemiology and Public Health Research Unit, Department of Population Health, Luxembourg Institute of Health, 1A-B, rue Thomas Edison, 1445 Strassen, Luxembourg
{michael.schnell,sophie.couffignal,stephanie.saleh}@lih.lu

[2] UL, CNRS, Inria, Loria, 54000 Nancy, France
{michael.schnell,jean.lieber,nicolas.jay}@loria.fr

[3] Service d'évaluation et d'information médicales, Centre Hospitalier Régional Universitaire de Nancy, Nancy, France
n.jay@chru-nancy.fr

Abstract. Cancer registries are important tools in the fight against cancer. At the heart of these registries is the data collection and coding process. Ruled by complex international standards and numerous best practices, operators are easily overwhelmed. In this paper, a system is presented to assist operators in the interpretation of best medical coding practices. By leveraging the arguments used by the coding experts to determine the best coding option, the proposed system is designed to answer the coding questions from operators and provide an answer associated with a partial explanation for the proposed solution.

Keywords: Interpretation of best practices · Interpretive case-based reasoning · Coding standards · Cancer registries · User assistance · Decision support

1 Introduction

The Luxembourg National Cancer Registry (NCR) is a systematic, continuous, exhaustive and non redundant collection of data about cancers diagnosed and/or treated in Luxembourg. For every case matching the inclusion criteria of the NCR, data about the patient, the tumor, the treatment and the follow up are collected. The main objectives of the NCR are cancer monitoring (incidence rates, survival rates, comparisons on an international level, ...) and the evaluation of cancer case management (diagnosis, treatment, ...) in Luxembourg.

There are numerous cancer registries around the world (over 700 according to the Union for International Cancer Control[1]), with varying means and goals.

[1] http://www.uicc.org/sites/main/files/private/UICCCancerRegistries-whywhathow.pdf.

D.W. Aha and J. Lieber (Eds.): ICCBR 2017, LNAI 10339, pp. 345–359, 2017.
DOI: 10.1007/978-3-319-61030-6_24

In order for the collected data to be comparable, it is necessary to have a common definition of the collected data and the coding practices. This lead to the creation of various international coding standards, providing both common terminologies (e.g. the International Classification of Diseases (ICD)) and coding best practices [9]. It is essential to follow these standards in order to obtain standardized and reliable data. However, the broadness and complexity of the standards can make the work of the operators difficult. The operators are the people in charge of collecting and coding cancer cases. It takes months of time to attain excellence. Time and practice are essential. Complex cases add an extra level of difficulty.

The aim of this research is to address this complexity, by assisting both operators and coding experts in the interpretation of coding best practices.

As an illustrating example, let us consider the case of a particular male patient from the NCR. In 2013, he suffered from lasting pains in his side and a sudden loss of appetite. On January 12th, 2014, a CT scan of his left kidney revealed nothing out of the ordinary. As the patient's condition continued to deteriorate, a second scan was made on February 15th, 2014. This time, two suspicious neoplasms were found and the clinicians suspected cancer. Another CT scan made on March 10th, 2014 showed signs of multiple renal adenopathy, which reinforced the cancer suspicion. On June 2nd, 2014, a renal biopsy was carried out and the following histological findings pointed to a renal cell carcinoma. The operator, after reading the complete file and carefully selecting the important facts, determined that this type of cancer meets the inclusion criteria of the NCR and has to be coded into the database according to international standards. The most important values collected by the registry for this tumor are the incidence date (February 15th, 2014), the topography (C64.9 – Kidney) and the morphology (M-8312/3 – renal cell carcinoma). The majority of questions concerns these values and, thus they are primary focus of this research project.

This example was rather easy to code. For the operator, the task is more complex as the data are contained within the various letters and free text reports that constitute the medical record. These documents have to be evaluated and summarized. It is possible for two reports to provide conflicting data. Here, the first CT scan showed nothing, unlike the following ones. Sometimes, important data are simply missing from the patient record. This can be the case if the patient has continued his treatment abroad or in a different hospital, if the patient died from an unrelated cause (e.g. car accident) or if the patient refused further treatment. Another possible explanation for the missing information is the difference in objectives between treatment and coding. Some aspects are assumed implicitly by the clinicians. In the case of breast cancer, no mention of a palpation usually means that no tumor is palpable, though a palpation was actually performed. However, in the case of the NCR, both exam and result must be explicitly documented. As such, aspects deemed unimportant by the clinicians might actually be very important for the registry and vice-versa. Furthermore, most medical reports do not structure their data beyond simple sectioning or identifying information (type of report, clinician, patient, ...). The important information (e.g. the description and the conclusion) is found in the free text sections. In addition, this text can be very ambiguous (vague conclusions, inconsistencies between factual description and medical conclusion).

In order to solve these conflicts, which require not only a deep knowledge of the coding standards, but also a solid medical background, coding experts are consulted. The coding experts need to determine the coding practices which should be applied to the problematic patient record. However, as consistency is a key requirement for cancer registries (needed for temporal analysis and to track tendencies), experts have to ensure that two identical cases receive the same coding. If the standards clearly state how to solve such an instance, it is only a matter of finding the proper practices and interpreting them accordingly. This is not always possible, as the coding standards do not (and cannot) cover all possible aspects of a cancer patient. Should such a situation occur, a new practice is designed to complement existing ones. For any future identical patient, this new practice should then be applied (in order to guarantee the consistency of the registry). It is therefore crucial to remember these particular coding questions and how they were solved (e.g. what practices were eventually used).

Context and motivation are discussed in Sect. 2. Section 3 introduces some definitions and notations. Section 4 describes an approach to assist the data collection process for cancer registries and how case-based reasoning (CBR [1]) was applied. In Sect. 5, a prototype of the proposed method is described. The proposed method is discussed and compared with related work in Sect. 6. Section 7 presents a conclusion and points out what further efforts need to be undertaken in the future.

2 Context and Motivation

For the Luxembourg National Cancer Registry, the operators can ask questions at any time using a ticketing system. The operator provides a free text description of their question with the minimum amount of required data about the patient, the tumor and, if relevant, the treatment. However, for the most part, the operator chooses what is worth providing. Of course, should anything important be missing, the experts will ask additional data or provide a tentative answer taking into account the missing data (e.g. if the missing value is A, then solution B, else solution C).

While providing a very individualized response, this approach complicates the sharing process. As the operators can only see the questions asked by other operators from the same hospital, a common question will be asked and answered several times. This repetition can lead to inconsistent answers for the same question. As consistency is an important quality measure for cancer registries, this issue needs to be addressed. As of today, this issue is remedied partially with continuous training sessions for the operators, during which the most important questions are discussed with coding experts.

Answering all the problems encountered by the operators is very time consuming for coding experts. The aim of this research is to decrease this workload.

To achieve that goal, a shared tool for both operators and coding experts is implemented. It allows the operators to ask questions, tries to answer them as best as possible and provides experts with an interface to answer the remaining questions.

Given the similarity between the working process of the experts and case-based reasoning, we have chosen to base our approach on CBR. Nevertheless, other reasoning or optimization algorithms were also considered for this task. Very popular methods are black box learning algorithms (like neural networks). Indeed, given enough representative data, this approach would yield good results. Some papers explored this in a related domain, automatic data collection or annotation. This research area focuses on the creation of solutions for the annotation and coding of electronic medical patient records. In a workshop of the 2007 BioNLP conference, a shared task focused on the assignment of ICD-9-CM codes to radiology reports [15]. Several methods were proposed with very interesting results (see [8]). However, those good results are due to two factors specific to radiology. The classification only used around 40 diagnosis codes from ICD-9-CM (out of over 14 000) and a representative data set (with proportionate representations for every possible code) was provided. While there are considerably fewer codes for cancer registries, there is no comprehensive data set available for the learning and evaluation process of any of the proposed methods. This is probably one of the major problems for this kind of method. Another weakness is the explanation. By contrast to automatic coding, for which explaining the reason why the system has chosen to code a patient record in a given way may be slightly less important, it is essential for a decision support system.

3 Preliminaries

Case-Based Reasoning. In a given application domain, a *case* is the representation of a problem-solving episode frequently represented by a pair (`pb`, `sol(pb)`) where `pb` is a problem related to the application domain and `sol(pb)` is a solution of `pb`. Given a new problem `tgt`—the target problem—, case-based reasoning aims at solving `tgt` by reusing a case base. A *source case* is an element of the case base. A classical way to do so consists in selecting a source case judged similar to `tgt` (retrieval step) and to reuse it to solve `tgt`.

RDFS is a knowledge representation language of the semantic web [5]. An RDFS formula is a triple `(s p o)` that can be understood as a sentence in which `s` is the subject, `p` (the predicate) is a verbal group and `o` is an object. Thus `(romeo loves juliet)` is a triple stating that mister Montague has strong feelings for miss Capulet. An RDFS base is a set of triples and is generally assimilated to an RDFS graph where nodes are subjects and objects, and where edges are labeled by properties. E.g., the graph

$$\text{romeo} \underset{\text{loves}}{\overset{\text{loves}}{\rightleftarrows}} \text{juliet} \xrightarrow{\text{age}} 13$$

states that Romeo and Juliet love each other and that Juliet is 13.

Some properties are associated with semantics, in particular `rdf:type`, abbreviated as `a` and meaning "is an instance of", and `rfds:subClassOf`, abbreviated as `subc` and meaning "is a subclass of". For example, from

$$\mathtt{G} = \text{ romeo} \xrightarrow{\text{a}} \text{Man} \xrightarrow{\text{subc}} \text{Human}$$

it can be inferred that romeo $\overset{\text{a}}{\to}$ Human.

SPARQL is a query language for RDFS. In this paper, the only type of SPARQL query used is `ASK`. This query tests the existence of a subgraph in a given graph, using variables. In SPARQL, variable names start with `?`, e.g., `?x`, `?tumor`. For example, the following query tests if someone (`?x`) (in the queried graph) loves a human (`?human`): `ASK {?x loves ?human . ?human a Human}`.

RDFS was chosen for its status as a recognized knowledge representation language, with numerous available tools. It also provides access to the Linked Open Data, which are open knowledge bases. This enables the usage of previously coded medical knowledge for the resoning tool presented in this paper.

4 Case-Base Interpretation of Best Practices

This section describes the proposed approach to assist operators in their coding task. This research project has been elaborated after discussing actual coding problems with operators and experts from the Luxembourg National Cancer Registry. First, the running example is introduced, followed by an overview of the global architecture of the system. Finally, the representation of the cases and the steps of the proposed approach are detailed.

4.1 Introduction of the Running Example

For the following sections, the same example will be used to explain and demonstrate the proposed approach. In the descriptions below, important patient features are in ***bold italics***.

Target problem (`tgt`). The question concerns the nature (primary, metastasis, ...) of a lung tumor. This is a recurring question, as the lung is an organ that very easily develops metastases. As the coding of the tumor varies heavily based on its nature, it is an important question for the operator. The nature of the tumor depends on its localization and where the cancer initially developed. There are essentially two possibilities: primary or secondary. The tumor at the initial localization is the primary tumor. From that tumor, cells may detach themselves and, traveling through the body using the cardiovascular system, develop new tumors in other body parts. These new tumors are called metastases and are of secondary nature.

The target problem concerns a woman, born on December 5th, 1950. In ***2006***, ***breast cancer*** was diagnosed and treated. In ***2016***, a ***lung tumor*** was discovered within the right lower lobe. A CT scan indicated ***no mediastinal adenopathy***[2]. A histological analysis of a sample identified the morphology[3] of the cancer as ***adenocarcinoma***. The ***TTF1*** marker test was ***negative***[4]. After further testing, ***no other tumor site*** was found. In the patient record, it was noted that the ***oncologist considered*** the lung tumor to be of ***primary nature***.

For our example, three source cases are described hereafter. A case is a representation of a coding episode based on best coding practices. For the sake of simplicity, all the source cases concern the same subject, i.e. the nature of a lung tumor. For each case, the patient record is described, followed by the answer and a description of the arguments in favor of and against the proposed answer.

Source 1 ($\texttt{srce}_1$) concerns a woman, born on July 23rd, 1946. In ***2012***, she was diagnosed with ***breast cancer*** (***adenocarcinoma***) and treated. In ***2015***, a ***lung tumor*** was discovered within the middle left lobe. A histological analysis identified the morphology of the cancer as ***small-cell carcinoma***.

In this case, the answer to the question of the nature of the lung tumor was primary tumor. As for the argumentation, there was ***one strong argument in favor***, namely the morphology of the tumor. Indeed, small-cell carcinoma most commonly arise within the lung.

Source 2 ($\texttt{srce}_2$) concerns a woman, born on March 14th, 1930. In ***2011***, she was diagnosed with ***colorectal cancer*** and treated. In ***2013***, a ***lung tumor*** was discovered within the left middle lobe. A CT scan indicated ***no mediastinal adenopathy*** and showed ***multiple pulmonary opacities*** indicative of a lung metastasis. The patient was already very fragile, thus no further tests were performed. The ***oncologist concludes*** that the lung tumor was a ***metastasis*** of the previous cancer.

In this case, the answer to the question of the nature of the lung tumor was a metastasis (of the colorectal cancer). There were ***four weak arguments in favor*** in this case: the close antecedent, the absence of mediastinal adenopathy, the oncologist's opinion and the multiple pulmonary opacities.

Source 3 ($\texttt{srce}_3$) concerns a man, born on August 14th, 1953. In ***2000***, ***prostate cancer was*** diagnosed and treated. In ***2014***, a ***lung tumor*** is discovered within the upper left lobe. A CT scan indicated ***no mediastinal adenopathy***. A histological analysis of a sample identified the morphology of the cancer as ***adenocarcinoma***. After further testing, ***no other tumor site*** is found. In the patient record, it is noted that the ***oncologist considered*** the tumor to be of ***primary nature***.

In this case, the tumor was primary. There were ***three weak arguments in favor***: the oncologist's opinion, the fact that no other synchronous tumor was

---

[2] An adenopathy is an enlargement of lymph nodes, likely due to the cancer.

[3] The morphology describes the type and behavior of the cells that compose the tumor.

[4] For primary lung adenocarcinoma, TTF1 marker is usually positive.

found and the long time span between the previous cancer and the current one (a shorter time span would have been in favor of a metastasis). There was also ***one weak argument against***, namely the absence of mediastinal adenopathy.

4.2 Global Architecture

Figure 1 summarizes the main process for our approach. It uses a 4-R cycle adapted from [1] and four knowledge containers [16]. To solve a new problem, first a description must be provided. That description is then used with the domain knowledge (`DK`) and the retrieval knowledge (`RK`) to find a suitable source case `srce` and its solution `sol(srce)` within the case base. Then, in the reuse step, the solution `sol(tgt)` for `tgt` is produced from (`srce`, `sol(srce)`) together with the domain knowledge and adaptation knowledge (`AK`). The pair (`tgt`, `sol(tgt)`) may then be revised by an expert, leading to the pair (`tgt′`, `sol(tgt′)`). Finally, this pair may be retained by adding it into the case base.

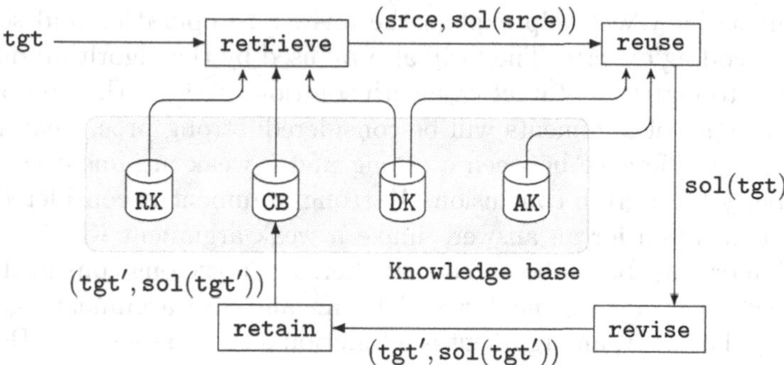

Fig. 1. Adapted 4-R cycle and knowledge containers for the proposed approach.

4.3 Case-Based Interpretation of Best Practices

A case (`srce`, `sol(srce)`) is composed of three parts: the question, the patient record and the solution.

The question part indicates the subject (incidence date, topography, tumor nature, ...) as well as the focused entity from the patient record. In the running example, the question is about the tumor's nature and focuses on the lung tumor.

The patient record represents the data from the hospital patient record (patient features, tumors, exams, treatments, ...) needed to answer the question. The relevant data depends on the subject and is defined by coding experts. For the source cases, only the required information is provided. For the target problems, this assumption cannot be made. The operator may simply not know what is needed and thus is encouraged to provide as much information as possible.

The solution contains the answer to the question, an optional textual explanation and the most important arguments in favor of (**pros**) and against (**cons**)

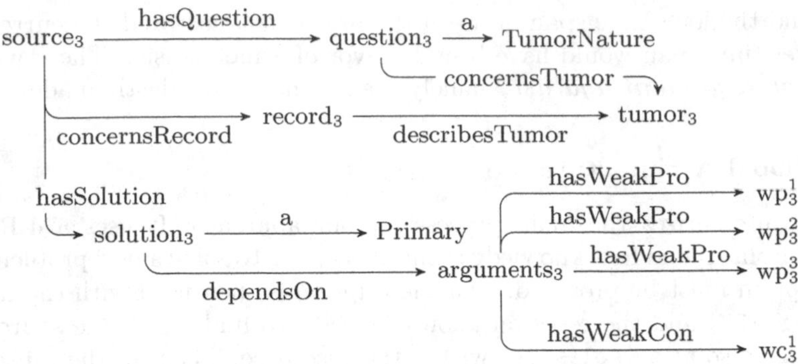

Fig. 2. Partial RDFS graph for source case $\texttt{srce}_3$ (patient record details in Fig. 3).

the given answer. The optional explanation is provided by the coding experts, and may point out key features or best practices for operators. The arguments have two uses. They will help explain the answer to operators and serve as a reminder for coding experts. They will also be used by the algorithm during the retrieval step to match the target case with solution cases. In the proposed approach, three types of arguments will be considered: strong pros, weak pros and weak cons. The difference between a strong and a weak argument comes from their reliability for a given conclusion. A strong argument is considered to be a sufficient justification for an answer, unlike a weak argument which is more of an indication or clue. It can be noted that there are no strong cons in the source cases. Indeed, such an argument would be an absolute argument against the given answer. Formally, an argument is a function a that associates a Boolean to a case and is stored as a SPARQL ASK query. The argument type (i.e., strong or weak and pro or con) is defined by the coding experts, in accordance with the coding standards and best practices.

A partial RDFS graph for $\texttt{srce}_3$ is shown in Figs. 2 and 3. One of the pros is that no other synchronous tumor is found. This argument $\texttt{wp}_3^1$ is formalized as follows:

$\texttt{wp}_3^1\texttt{(case)} =$

```
ASK {
  case hasQuestion ?question .
  case concernsRecord ?record .
  ?question concernsTumor ?tumor .
  FILTER NOT EXISTS {
    ?record describesTumor ?other_tumor .
    ?tumor != ?other_tumor .
    ?tumor isSynchronousWith ?other_tumor
  }
}
```

$\texttt{wp}_3^1$ applies to $\texttt{srce}_3$ means that $\texttt{wp}_3^1(\texttt{srce}_3) = \texttt{true}$.

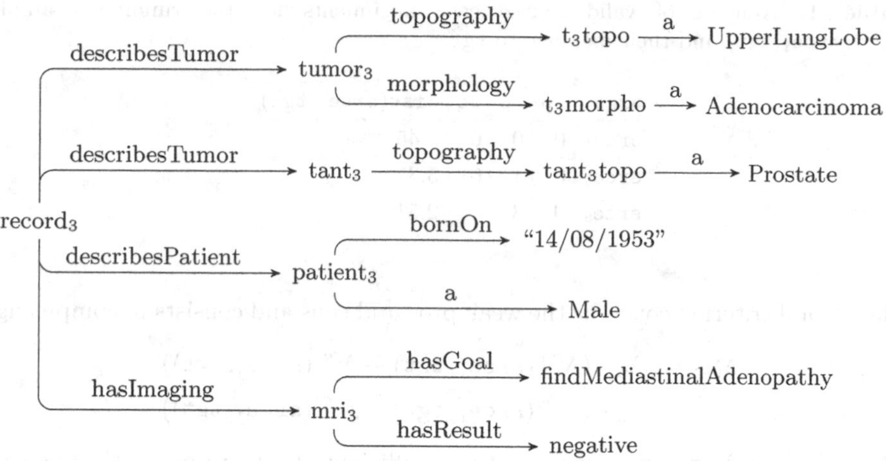

Fig. 3. Partial RDFS graph for the patient record of $\mathtt{srce}_3$.

4.4 Retrieve

The retrieval of source cases is limited to cases concerning the same subject as the target problem. For the running example, this means that only source cases concerning the nature of the tumor will be taken into account.

To find the most appropriate source case among the selected cases, the arguments will be considered. The arguments are part of the reasoning process which leads the coding experts to the final solution. As such, they can be used to identify similar cases.

Knowing the target problem `tgt`, retrieval knowledge consists in preferring one source case to another, the preferred source case being the retrieved one. This preference relation is denoted by the preorder $\preccurlyeq_{\mathtt{tgt}}$.

For a given source case `srce`, let `sp(srce)` be the set of its strong pros, `wp(srce)` the set of its weak pros and `wc(srce)` the set of its weak cons. For $\mathtt{srce}_3$, $\mathtt{sp}(\mathtt{srce}_3) = \emptyset$, $\mathtt{wp}(\mathtt{srce}_3) = \{\mathtt{wp}_3^1, \mathtt{wp}_3^2, \mathtt{wp}_3^3\}$ and $\mathtt{wc}(\mathtt{srce}_3) = \{\mathtt{wc}_3^1\}$.

Let $\mathtt{args} \in \{\mathtt{sp}, \mathtt{wp}, \mathtt{wc}\}$ be an argument type, $\mathcal{N}^{\mathtt{args}}(\mathtt{srce}, \mathtt{tgt})$ denotes the number of arguments of type `args` of a the source case `srce` which are valid for a case `tgt`.

$$\mathcal{N}^{\mathtt{args}}(\mathtt{srce}, \mathtt{tgt}) = |\{\mathsf{a} \in \mathtt{args}(\mathtt{srce}) \mid \mathsf{a}(\mathtt{tgt}) = \mathtt{true}\}|$$

Table 1 presents the different values of $\mathcal{N}^{\mathtt{args}}(\mathtt{srce}_\mathtt{i}, \mathtt{tgt})$ for `tgt` and the possible source cases $\mathtt{srce}_1$, $\mathtt{srce}_2$ and $\mathtt{srce}_3$. For example, out of the four weak pros of $\mathtt{srce}_2$, only one can be applied to `tgt`, thus $\mathcal{N}^{\mathtt{wp}}(\mathtt{srce}_2, \mathtt{tgt}) = 1$.

To compare two source cases $\mathtt{srce}_\mathtt{i}$ and $\mathtt{srce}_\mathtt{j}$, three criteria are combined. The first criterion concerns the strong pros and consists in computing:

$$\Delta^{\mathtt{s}}_{\mathtt{i},\mathtt{j}} = \mathcal{N}^{\mathtt{sp}}(\mathtt{srce}_\mathtt{i}, \mathtt{tgt}) - \mathcal{N}^{\mathtt{sp}}(\mathtt{srce}_\mathtt{j}, \mathtt{tgt})$$

Table 1. Number of valid source case arguments for the running example $\mathcal{N}^{\text{args}}(\texttt{srce}_\texttt{i}, \texttt{tgt})$ and their distance to tgt.

args	sp	wp	wc	$\texttt{dist}(\texttt{srce}_\texttt{i}, \texttt{tgt})$
$\texttt{srce}_1$	0	0	0	3.45
$\texttt{srce}_2$	0	1	0	5.37
$\texttt{srce}_3$	0	3	1	2.51

The second criterion concerns the weak pros and cons and consists in computing:

$$\begin{aligned}\Delta^{\texttt{w}}_{\texttt{i,j}} = {} & \lambda_{\texttt{p}} * (\mathcal{N}^{\texttt{wp}}(\texttt{srce}_\texttt{i}, \texttt{tgt}) - \mathcal{N}^{\texttt{wp}}(\texttt{srce}_\texttt{j}, \texttt{tgt})) \\ & - \lambda_{\texttt{c}} * (\mathcal{N}^{\texttt{wc}}(\texttt{srce}_\texttt{i}, \texttt{tgt}) - \mathcal{N}^{\texttt{wc}}(\texttt{srce}_\texttt{j}, \texttt{tgt}))\end{aligned}$$

where $\lambda_{\texttt{p}}$ and $\lambda_{\texttt{c}}$ are two non-negative coefficients that are currently fixed to $\lambda_{\texttt{p}} = 2$ and $\lambda_{\texttt{c}} = 1$. When more data are available, these parameter values will be reevaluated. The third criterion concerns the patient record similarity and consists in computing:

$$\Delta^{\texttt{dist}}_{\texttt{i,j}} = \texttt{dist}(\texttt{srce}_\texttt{j}, \texttt{tgt}) - \texttt{dist}(\texttt{srce}_\texttt{i}, \texttt{tgt})$$

where dist is a distance function between patient records. dist has been implemented using an edit distance between graphs [6].

These criteria are considered lexicographically, first $\Delta^{\texttt{s}}_{\texttt{i,j}}$, then $\Delta^{\texttt{w}}_{\texttt{i,j}}$ and finally $\Delta^{\texttt{dist}}_{\texttt{i,j}}$, that is $\texttt{srce}_\texttt{i} \preccurlyeq_{\texttt{tgt}} \texttt{srce}_\texttt{j}$ if

$$\Delta^{\texttt{s}}_{\texttt{i,j}} > 0 \text{ or } (\Delta^{\texttt{s}}_{\texttt{i,j}} = 0 \text{ and } (\Delta^{\texttt{w}}_{\texttt{i,j}} > 0 \text{ or } (\Delta^{\texttt{w}}_{\texttt{i,j}} = 0 \text{ and } \Delta^{\texttt{dist}}_{\texttt{i,j}} \geq 0)))$$

This means that, for our approach, the criterion based on the strong pros outweighs the one based on the weak pros and cons, which in turn outweighs the criteria based on the patient record similarities. This order has been chosen to match the coding experts' reasoning. For the implemented prototype, several source cases are retrieved, ordered by $\preccurlyeq_{\texttt{tgt}}$ and according to a threshold (which remains to be accurately fixed).

Table 2 shows the values of the various helpers for the running example. None of the strong arguments of the source cases are valid for tgt, thus the weak arguments are considered. The comparison shows that $\texttt{srce}_3$ is preferred to $\texttt{srce}_2$ and that both are preferred to $\texttt{srce}_1$.

Table 2. Comparing source cases with respect to the target problem of the running example case (with $\lambda_{\texttt{p}} = 2$ and $\lambda_{\texttt{c}} = 1$).

i	j	$\Delta^{\texttt{s}}_{\texttt{i,j}}$	$\Delta^{\texttt{w}}_{\texttt{i,j}}$	$\Delta^{\texttt{dist}}_{\texttt{i,j}}$	$\preccurlyeq_{\texttt{tgt}}$
1	2	0	−2	1.88	$\texttt{srce}_2 \preccurlyeq_{\texttt{tgt}} \texttt{srce}_1$
1	3	0	−5	−0.94	$\texttt{srce}_3 \preccurlyeq_{\texttt{tgt}} \texttt{srce}_1$
2	3	0	−3	−2.86	$\texttt{srce}_3 \preccurlyeq_{\texttt{tgt}} \texttt{srce}_2$

4.5 Reuse

Once an appropriate source case has been found, the solution associated to the source case is copied and then the arguments that do not apply to the target problem, if any, are simply removed. This step is repeated for every retrieved source case. For the running example, the most appropriate source case is $\texttt{srce}_3$. The answer for $\texttt{srce}_3$ is to consider the tumor to be of primary nature and thus, for $\texttt{tgt}$, the answer to the question is also a primary tumor. All the arguments of $(\texttt{srce}_3, \texttt{sol}(\texttt{srce}_3))$ apply, therefore $\texttt{sp}(\texttt{tgt}) = \texttt{sp}(\texttt{srce}_3)$, $\texttt{wp}(\texttt{tgt}) = \texttt{wp}(\texttt{srce}_3)$ and $\texttt{wc}(\texttt{tgt}) = \texttt{wc}(\texttt{srce}_3)$.

4.6 Revise and Retain

Currently, the retrieve and reuse steps have been implemented in a prototype described in Sect. 5. This section presents first thoughts about the revise and retain steps.

Let $(\texttt{tgt}, \texttt{sol}(\texttt{tgt}))$ be the reused case. It may be revised by a coding expert, to modify the answer and/or add, remove or modify some arguments. The expert may also want to remove information from $\texttt{tgt}$ as to keep only the relevant information with regard to the problem-solving process (i.e., the reuse of the arguments). In such a situation, $(\texttt{tgt}, \texttt{sol}(\texttt{tgt}))$ is substituted by $(\texttt{tgt}', \texttt{sol}(\texttt{tgt}'))$, where $\texttt{tgt}'$ is more general than $\texttt{tgt}$. $(\texttt{tgt}', \texttt{sol}(\texttt{tgt}'))$ is a generalized case that has a larger coverage than $(\texttt{tgt}, \texttt{sol}(\texttt{tgt}))$ [12].

When the system will be in use, the revise step is going to be triggered systematically, at least for the very beginning. Nevertheless, this should unburden the experts, since, hopefully, revising a case will be less time-consuming than solving it.

For now, it is planned to retain all the revised cases. Currently, between 100 and 200 cases per year require expert help. If the case base happens to be too large, a case base management process may be considered [17].

It may occur that the retrieve step fails, if some thresholds are chosen for the retrieval step. For example, it can be considered that for the source case to be retrieved, at least one of its pros has to be applicable to the target problem. In such a situation, the target problem is solved by the coding experts, and thus, the revise and retain steps enrich the case base. This constitutes a case authoring process.

5 A Prototype

The prototype is a web application, allowing an operator to ask questions. It tries to solve these questions, using the approach previously described (Sects. 4.4 and 4.5). The web application is composed of two parts: a form for the description of the target problem (see Fig. 4) and a presentation the proposed solutions accompanied by a summary of the target problem (see Fig. 5). In this first implementation, the number of items that can be provided by the user are fixed (e.g., there can only be one tumor, one antecedent and one synchronous tumor) and only a single question subject is possible, namely the tumor's nature.

Coding questions

New question

Subjet: Tumor nature — Information or sujet of the question

Patient

Birthdate: 05/12/1950

Gender: Female / Male — Medical gender of the patient

Tumor

Details of the tumor whose nature is uncertain

Incidence date: 24/03/2016

Topography: C34.3 - Lower lobe, lung

Side: Right

Morphology: 8140/3 - Adenocarcinoma

Medistinal adenopathy: Yes / No / Unknown — Is there any exam who showed signs of medistinal adenopathy ?

TTF1 marker: Positive / Negative / Unknown — Results of the TTF1 marker test, if any.

Pulmonary opacities: Yes / No / Unknown — Did any imaging results indicate multiple

Opinion clincian: Primary / Secondary / Unknown

Fig. 4. Form used to describe the target problem of the running example.

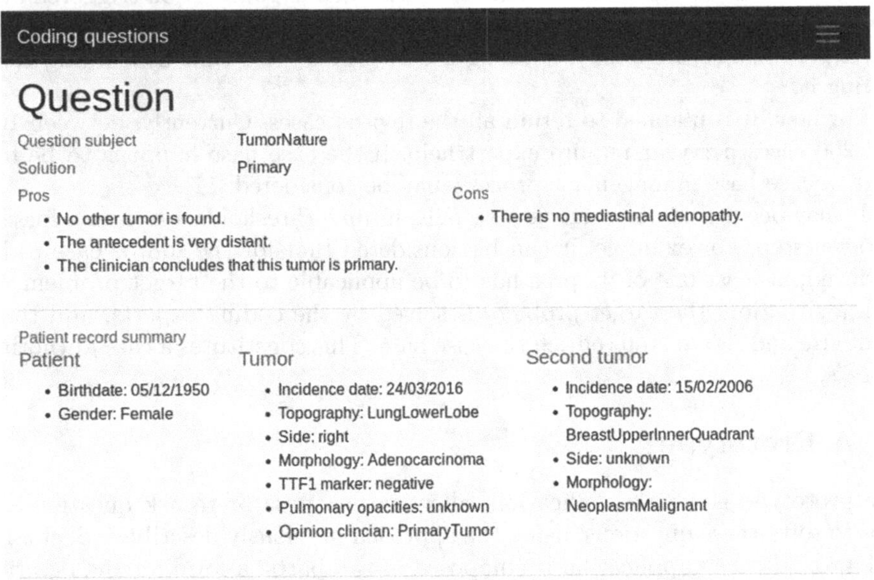

Coding questions

Question

Question subject: TumorNature

Solution: Primary

Pros
- No other tumor is found.
- The antecedent is very distant.
- The clinician concludes that this tumor is primary.

Cons
- There is no mediastinal adenopathy.

Patient record summary

Patient
- Birthdate: 05/12/1950
- Gender: Female

Tumor
- Incidence date: 24/03/2016
- Topography: LungLowerLobe
- Side: right
- Morphology: Adenocarcinoma
- TTF1 marker: negative
- Pulmonary opacities: unknown
- Opinion clincian: PrimaryTumor

Second tumor
- Incidence date: 15/02/2006
- Topography: BreastUpperInnerQuadrant
- Side: unknown
- Morphology: NeoplasmMalignant

Fig. 5. Summary of the described target problem and the proposed solution. The most appropriate source cases are shown similarly to the target problem (not visible in this screenshot).

6 Discussion and Related Work

The system described in this paper can be seen as an example of interpretative case-based reasoning. Other approaches in this area include Murdock et al. [14]. Their approach focuses on assisting intelligence analysts in evaluating hypotheses of hostile activities such as take over attempts by criminal groups. The hypothesis (target problem) is matched to a model (source case), which represents a general sequence of events for the given hypothesized event. Then, their system compares the facts from their target hypothesis with those from the model. If a successful match is found, their system relies on this match to generate arguments to justify or discredit the hypothesis. It is left to the user to decide whether or not the target hypothesis is valid. Contrarily to our approach, the arguments are used solely to explain the proposed solution.

Case-based reasoning has been used a lot in the legal domain (HYPO [3], CATO [2]). Here, source cases are old court cases. The argumentation focuses on the reuse of these precedents, on how similarities can be highlighted and differences downplayed, in order to justify the desired outcome for the target court case. This marks a difference with the approach described in this paper, where arguments are described per case and implicitly linked to the source case.

Particularly in the context of assisting users, explanations are essential, as they provide a measure of understanding for the user and promote the trustworthiness of the system. Similarly to this research, pros and cons are considered by McSherry in [13]. He describes a system for binary classification which uses the closest source case to provide the conclusion and the closest source case with the opposite conclusion to compute which attributes favor the conclusion (pros) and which attributes do not (cons). Unlike our approach, each argument is linked to a single attribute. Thus they cannot show how the combination of attributes might influence a given outcome.

In health sciences, case-based reasoning is not the only area that is currently very actively researched [4]. Automatic annotation of medical documents is another such area [15]. While our approach focuses on assisting operators in their tasks, these approaches seek to replace the need for operators in their current capacity. They focus on analyzing and annotating medical reports [10,11].

7 Conclusion and Future Work

In this paper, an approach to assist operators in the interpretation of best medical coding practices has been proposed. This approach is based on discussions with operators and coding experts on actual coding problems. A dozen tricky problems were discussed in detail, among a hundred simpler problems. The coding questions asked by the operators are compared to previous questions and solved by reusing the pros and cons of previously given solutions.

This approach has modeled the reasoning processes of the coding experts that were observed. A first prototype has been developed for this purpose and has to be deployed and evaluated (does the system decrease the experts' workload while maintaining the coding quality?).

Currently, the arguments used by our approach remain very simple. As such, they only cover a part of the problem-solving process and resemble hints or highlights. To better represent the solving process, more complex arguments are required. Complex arguments could be combined from simpler arguments using a few operators (e.g. and, or). This should allow for the inclusion of other arguments which, by themselves, do not favor or disfavor a given outcome, but might do so when combined. Furthermore, arguments of a source case are presently reused as such for the target problem. It is planned to examine how these arguments could be adapted to take into account the differences between the source case and the target problem.

Another crucial aspect for the cancer registries is the evolution of the coding practices. Any change in the coding practices will provoke changes in the case base and the associated knowledge containers. It might be interesting to consider methods to detect the needed changes and to help maintain the represented knowledge [7].

When the system is tested, validated, improved and routinely used by operators and experts, a second version of it will be designed that is less domain-dependent. The objective is to build a generic system for argumentative case-based reasoning using semantic web standards.

Acknowledgments. The authors wish to thank the anonymous reviewers for their remarks which have helped in improving the quality of the paper. The first author would also like to thank the Fondation Cancer for their financial support.

References

1. Aamodt, A., Plaza, E.: Case-based reasoning: foundational issues, methodological variations, and system approaches. AI Commun. **7**, 39–59 (1994)
2. Aleven, V., Ashley, K.D.: Teaching case-based argumentation through a model and examples empirical evaluation of an intelligent learning environment. In: Artificial intelligence in education, vol. 39, pp. 87–94 (1997)
3. Ashley, K.D.: Modeling Legal Arguments: Reasoning with Cases and Hypotheticals. MIT Press, Cambridge (1991)
4. Bichindaritz, I., Marling, C., Montani, S.: Case-based reasoning in the health sciences. In: Workshop Proceedings of ICCBR (2015)
5. Brickley, D., Guha, R.V.: RDF Schema 1.1, W3C recommendation (2014). https://www.w3.org/TR/rdf-schema/. Accessed Mar 2017
6. Bunke, H., Messmer, B.T.: Similarity measures for structured representations. In: Wess, S., Althoff, K.-D., Richter, M.M. (eds.) EWCBR 1993. LNCS, vol. 837, pp. 106–118. Springer, Heidelberg (1994). doi:10.1007/3-540-58330-0_80
7. Cardoso, S.D., Pruski, C., Silveira, M., Lin, Y.-C., Groß, A., Rahm, E., Reynaud-Delaître, C.: Leveraging the impact of ontology evolution on semantic annotations. In: Blomqvist, E., Ciancarini, P., Poggi, F., Vitali, F. (eds.) EKAW 2016. LNCS, vol. 10024, pp. 68–82. Springer, Cham (2016). doi:10.1007/978-3-319-49004-5_5
8. Crammer, K., Dredze, M., Ganchev, K., Talukdar, P.P., Carroll, S.: Automatic code assignment to medical text. In: Proceedings of the Workshop on BioNLP 2007: Biological, Translational, and Clinical Language Processing, pp. 129–136. Association for Computational Linguistics (2007)

9. European Network of Cancer Registries, Tyczynski, J.E., Démaret, D., Parkin, D.M.: Standards and guidelines for cancer registration in Europe: the ENCR recommendations. International Agency for Research on Cancer (2003)
10. Kavuluru, R., Han, S., Harris, D.: Unsupervised extraction of diagnosis codes from EMRs using knowledge-based and extractive text summarization techniques. In: Zaïane, O.R., Zilles, S. (eds.) AI 2013. LNCS, vol. 7884, pp. 77–88. Springer, Heidelberg (2013). doi:10.1007/978-3-642-38457-8_7
11. Kavuluru, R., Hands, I., Durbin, E.B., Witt, L.: Automatic Extraction of ICD-O-3 Primary Sites from Cancer Pathology Reports (2013). http://www.ncbi.nlm.nih.gov/pmc/papers/PMC3845766/
12. Maximini, K., Maximini, R., Bergmann, R.: An investigation of generalized cases. In: Ashley, K.D., Bridge, D.G. (eds.) ICCBR 2003. LNCS, vol. 2689, pp. 261–275. Springer, Heidelberg (2003). doi:10.1007/3-540-45006-8_22
13. McSherry, D.: Explaining the pros and cons of conclusions in CBR. In: Funk, P., González Calero, P.A. (eds.) ECCBR 2004. LNCS, vol. 3155, pp. 317–330. Springer, Heidelberg (2004). doi:10.1007/978-3-540-28631-8_24
14. Murdock, J.W., Aha, D.W., Breslow, L.A.: Assessing elaborated hypotheses: an interpretive case-based reasoning approach. In: Ashley, K.D., Bridge, D.G. (eds.) ICCBR 2003. LNCS, vol. 2689, pp. 332–346. Springer, Heidelberg (2003). doi:10.1007/3-540-45006-8_27
15. Pestian, J.P., Brew, C., Matykiewicz, P., Hovermale, D.J., Johnson, N., Cohen, K.B., Duch, W.: A shared task involving multi-label classification of clinical free text. In: Proceedings of the Workshop on BioNLP 2007: Biological, Translational, and Clinical Language Processing, pp. 97–104. Association for Computational Linguistics (2007). 1572411
16. Richter, M.M., Weber, R.O.: Case-Based Reasoning: A Textbook. Springer Science & Business Media, Berlin (2013)
17. Smyth, B., Keane, M.T.: Remembering to forget. In: Proceedings of the 14th International Joint Conference on Artificial Intelligence, pp. 377–382. Citeseer (1995)

Running with Cases: A CBR Approach to Running Your Best Marathon

Barry Smyth(✉) and Pádraig Cunningham

Insight Centre for Data Analytics, School of Computer Science,
University College Dublin, Dublin, Ireland
barry.smyth@ucd.ie

Abstract. Every year millions of people around the world train for, and compete in, marathons. When race-day approaches, and training schedules begin to wind down, many participants will turn their attention to their race strategy, as they strive to achieve their best time. To help with this, in this paper we describe a novel application of case-based reasoning to address the dual task of: (1) predicting a challenging, but achievable, personal best race-time for a marathon runner; and (2) recommending a race-plan to achieve this time. We describe how suitable cases can be generated from the past races of runners, and how we can predict a personal best race-time and produce a tailored race-plan by reusing the race histories of similar runners. This work is evaluated using data from the last six years of the London Marathon.

Keywords: Case-based reasoning · Recommender systems · Sports analytics

1 Introduction

Running a marathon is hard. It depends on months of dedicated training and the discipline to follow a time-consuming and exhausting schedule. To complete a successful marathon a runner also needs an appropriate race-plan for the event itself. The runner must select a suitable target finish-time and carefully plan the pacing of their race, stage by stage, kilometer by kilometer. The aim of this work is to help marathon runners to achieve a new personal best during their next race. We do this by using case-based reasoning to solve these dual problems by: (1) *predicting* a best achievable finish-time; and (2) *recommending* a tailored pacing plan to help the runner achieve this time on race-day.

This work sits at the intersection between personal sensing, machine learning, and connected health. An explosion of wearable sensors and mobile devices has created a tsunami of personal data [1], and the promise that it can be harnessed to help people to make better decisions and live healthier and more productive lives [2,3]. Indeed, within the case-based reasoning community there has been a long history of applying case-based, data-driven methods to a wide range of healthcare problems [4]. Recently, the world of sports and fitness has similarly

D.W. Aha and J. Lieber (Eds.): ICCBR 2017, LNAI 10339, pp. 360–374, 2017.
DOI: 10.1007/978-3-319-61030-6_25

embraced this data-centric vision, as teams and athletes embrace the power of data to optimise the business of sports and the training of athletes [5,6].

In this work we focus on recreational marathon running, helping runners to plan their race strategy for optimal performance. A key concept is that of *pace*, measured in minutes per mile/km, so that higher pace means slower speed. There is a growing body of research that explores the various factors that influence pacing during the marathon and other endurance events. For example, the work of [7] looks at the effect of age, gender, and ability on marathon pacing to conclude that female runners are typically more disciplined (running with less pace variation) than male runners; see also [8,9]. Then there is the runner's pacing *strategy*: how the runner plans to adjust their pace during the race [10]. There are 3 basic pacing strategies, based on how the pace of a runner varies between the first and second-half of a race. For example, we say that a runner completes an *even-split* if their pace is even throughout the race. Running a *positive-split* means the second-half of the race is *slower* (higher pace) than the first-half of the race, whereas a *negative-split* means the runner speeds-up in the second-half of the race, running it *faster* than the first-half. Many elites and disciplined runners will aim for even or slightly negative-splits [11]. Recreational runners typically run positive-splits, slowing during the second-half of the race, sometimes significantly. And for some this means *hitting the wall*, referring to the sudden onset of debilitating fatigue and a near-complete loss of energy that can occur late in a race [12]. At best, this temporarily slows even the swiftest of runners; at worst it reduces them to a shuffling gait for the rest of the race.

All this is to say that running the marathon is a challenge, and the difference between a good day and a terrible day, training aside, may well come down to how carefully a runner plans their race: what finish-time to aim for; positive vs. negative vs. even splits; avoiding hitting the wall etc. This is where we believe there is a significant opportunity to support marathon runners, by advising them on a suitable a target finish-time and providing them with a concrete race-plan, one that is personalized to their ability and tailored to the marathon course.

2 Problem Definition

The problem addressed by this work is how to help a marathon runner achieve a new personal best in their next race. We will assume we are dealing with a runner who has completed at least one previous marathon and so has a race record to serve as a starting point. As a reminder, this problem involves two related tasks: the *prediction* of a suitable target finish-time and the *recommendation* of an appropriate pacing plan.

2.1 Best Achievable Finish-Time

We start by assuming our runner wants to beat their current best-time, but by how much? If they are too conservative they will chose a finish-time that does not fully test them, and may leave them disappointed if they finish too comfortably on race-day. If they are too ambitious they may select finish-time that is beyond their ability and risk sabotaging their race; aiming for an overly ambitious target time is one sure way to end up hitting the wall later in the race. The point is that selecting a *best achievable* finish-time is non-trivial and getting it wrong can have a disastrous effect on race-day.

2.2 Race Plans and Pacing Profiles

Given a best achievable finish-time, the next task is to devise a suitable race-plan. We will assume the marathon is divided up into 8×5 km stages or *segments* (0–5 km, 5 km–10 km, ..., 35 km–40 km), plus a final 2.2 km segment (40 km–42.2 km); many big-city marathons measure times in 5 km segments and so race records typically provide these *split-times*. For the purpose of this work a race-plan, or *pacing plan*, will consist of a sequence of average paces (measured in kms per minute) for each of these race segments. For the avoidance of doubt we will refer to a pacing *plan* as a *proposed* set of paces for the runner and a pacing *profile* as the *actual* set of paces achieved by the runner in their race. For example, Fig. 1(a) illustrates a race-record for a runner who completed a marathon in 4 h and 13 min; that is an average pace of 6 mins/km. The corresponding pacing profile shows they ran a positive-split, starting their race 10% faster than their average pace during the first 5K and 10K segments of the race, and slowing in the second-half, to finish 7% slower than their average pace in the final segment.

Most conventional pacing plans are fairly simple by encouraging the runner to run an even-split. For example, if our runner wishes to run a 4-hour (240 min) marathon then this suggests an average pace of 5 min and 41 s per km for the duration of the race. Some plans may account for positive or negative splits, by advising the runner to run the first half of the race more quickly or slowly, respectively. However, in general, conventional plans are not especially tailored for the runner or the course, and so they leave considerable room for improvement, especially when it comes to achieving a new personal best. Thus, in this work we argue for a more *personalized* and *customized* approach to pacing, so that runners can benefit from a pacing plan that is suitable for their goal time, their personal fitness level and ability, and that is tailored for the peculiarities of a given marathon course.

3 Using CBR to Achieve a Personal Best in the Marathon

The key insight of this work is that we can use a CBR approach to both predict suitable target finish-times and recommend tailored pacing plans, based on the runner's own race experience and the experiences of similar runners who have achieved personal bests on the target course. To do this we will rely on a case base of *race pairs*, representing a pair of race records for a single runner. Each race record contains a pacing profile and a finish-time for a completed race; see Fig. 1(b). One of these race records corresponds to a *non personal best* (*nPB*) race, the other to a *personal best* (*PB*) race; the *nPB* plays the part of the *case description* while the *PB* is the *case solution*. Given a target/query runner (q), and their own recent race record (finish-time and pacing profile), we generate a finish-time prediction and pacing plan recommendation based on the *PB*'s of cases that have a similar *nPB* to q, as summarised in Fig. 1(c).

3.1 Case Generation

Each case in the case base corresponds to a single runner, r, with a *nPB* part and a *PB* part; see Eq. 1. To be represented in the case base, r must have at least two race records, and in general may have $n > 2$ race records if they have run many races; for example, in Fig. 1(b) we highlight 3 race records for r (for marathons m_1, m_2, m_3).

$$c_{ij}(r, m_i, m_j) = \Big\langle nPB_i(r, m_i), PB(r, m_j) \Big\rangle \tag{1}$$

The race record with the fastest finish-time is considered to be the personal best, and it is paired with the remaining $n-1$ non personal best records, producing $n-1$ cases. As per Fig. 1(b), r's best race is m_2, with a finish-time of 236 min. This is paired with the two *nPB* records (m_1 and m_3) to produce two cases, $c(r, m_1, m_2)$ and $c(r, m_3, m_2)$ as shown.

3.2 Case Retrieval

Retrieval is a three-step process, as shown in Algorithm 1. Given a query race record (q) — that is a runner, a finish-time, and a *nPB* pacing profile — we first filter the available cases (CB) based on their finish-times, so that we only consider cases for retrieval if their finish-times are within t minutes of the query finish-time. This ensures that we are basing our reasoning on a set of cases that are somewhat comparable in terms of performance and ability.

Next, we filter on the basis of gender, only considering cases for retrieval if they have the same gender as the query runner. The reason for this is that physiological differences between men and women have a material impact on marathon performance. On average women tend to finish about 12% slower than men [7], and thus men and women with the same approximate finish-times will be racing at very different levels of maximum effort; in relative terms, a 3-hour female finisher will be performing at a higher level of effort than a 3-hour male finisher, for example. Hence we separate male and female cases for more effective matching, during retrieval.

Finally, we perform a standard, distance-weighted kNN retrieval over the remaining candidate cases C, comparing q's pacing profile to their *nPB* profiles. These pacing profiles are real-valued vectors and, for now, we use a simple Euclidean-based similarity metric for similarity assessment. We select the top k most similar as the retrieved cases, R.

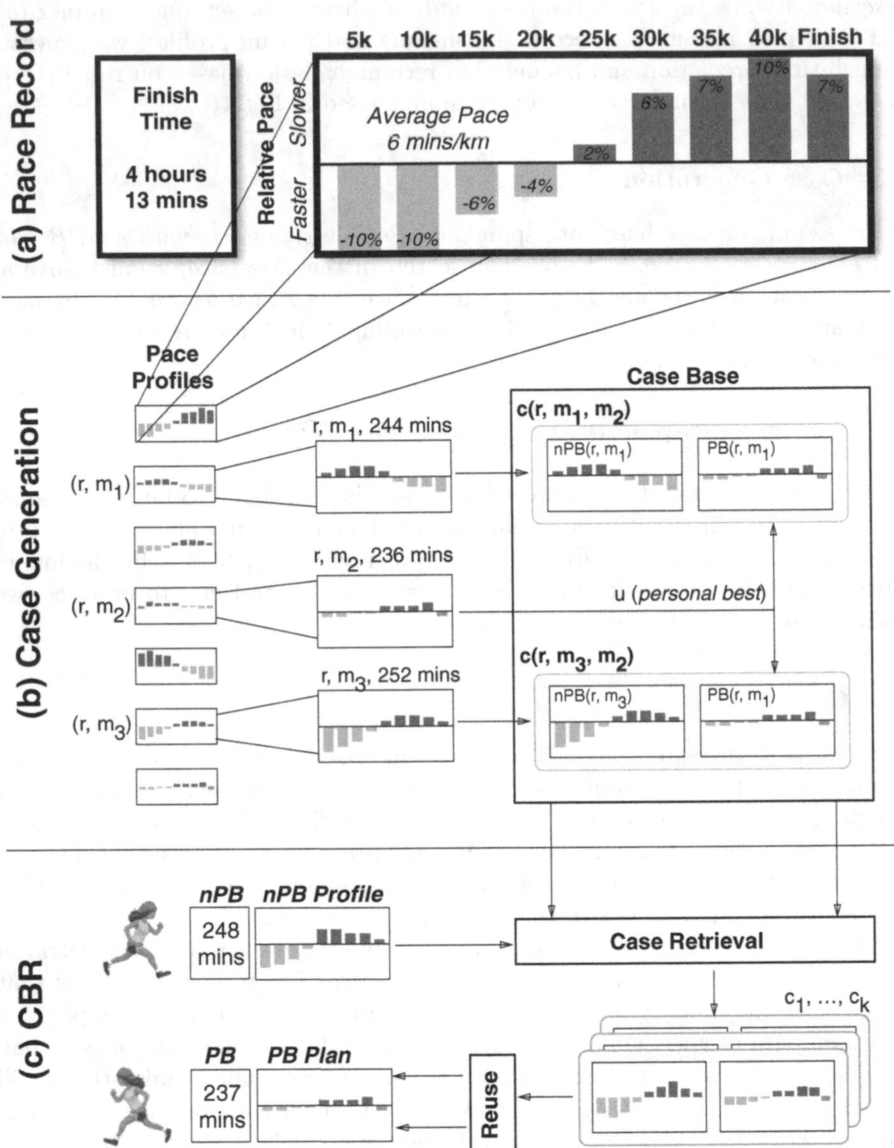

Fig. 1. Races, cases, predictions, and recommendations. (a) An example race record for a runner, showing a finish-time and a pacing profile containing pacing data for each of the 5 km race segments; (b) Converting race records into cases; (c) An overview of the CBR process: given an nPB race record as a query, the system retrieves a set of k cases with *similar* nPB parts, and combines these to generate a personal best finish-time prediction and a pacing plan to achieve this finish-time.

Algorithm 1. Outline CBR Algorithm.

```
Data: Given: q, query race record; CB, case base; k, number of cases to be
      retrieved; t, finish-time threshold.
Result: pb, predicted finish-time; pn, recommended pacing profile.
begin
    C = {c ∈ CB : Time(q) − t < Time(c) < Time(q) + t}
    C = {c ∈ C : c.gender == q.gender}
    if len(C) ≥ k then
        R = sort_k(sim(q, c) ∀ c ∈ C)
        pb = predict(q, R)
        pn = recommend(q, R)
        return pb, pn
    else
        return None
    end
end
```

3.3 Personal Best Finish-Time Prediction

Given a set of similar cases, R, we need to estimate the best achievable finish-time for q. Each case in R represents another runner with a similar *nPB* to q, but who has gone on to achieve a faster personal best on the same marathon course. The intuition is that since these *PBs* were achievable by these similar runners, then a similar *PB* should be achievable by the query runner.

For the purpose of this work we test three straightforward prediction approaches in which the predicted *PB* time of the query runner is a simple function of the personal best times of the retrieved cases. With each approach the predicted *PB* finish-times are weighted based on the relative difference between the query runner's finish-time and the corresponding *nPB* finish-time of a retrieved case; see Eq. 2. This adapts the *PB* times based on whether they were achieved by similar runners with slightly slower or faster *nPB* times.

$$w(q, c) = \frac{q(nPB).finish}{c(nPB).finish} \tag{2}$$

Best *PB*. In Eq. 3 the predicted finish-time is the weighted *PB* finish-time of the single fastest retrieved case, C_{best}; the case with the minimum *PB* time. For example, if the query runner has a finish-time of 245 min, and the nPB time for the best retrieved case is a slightly faster 240 min, then the weighting factor will be 1.02 (245/240). If the *PB* time of this retrieved case is 232 min then the predicted *PB* time for q will be just over 236 min.

$$PB_{best}(q, C) = w(q, C_{best}) \bullet Time(C_{best}(PB)) \tag{3}$$

Mean *PB*. Our second prediction approach calculates the *weighted mean* of the *PB* finish-times of the retrieved cases, as in Eq. 4. This will obviously tend to provide a slower predicted time than the *Best PB* approach above, but it will be more representative of the personal bests achieved by the similar runners.

$$PB_{mean}(q, C) = \frac{\sum_{\forall i \in 1...k} w(q, C_i) \bullet Time(C_i(PB))}{k} \tag{4}$$

Even *PB*. Finally, our third approach (*Even PB*) is based on the idea that more consistent or even pacing is better than more varied or erratic pacing. By measuring the coefficient of variation (*CoV*) of the relative paces in the *PB* pacing profile for each retrieved case, we can estimate their pacing variation, and select the one (C_{even}) with the lowest *CoV* value. The *PB* time of this C_{even} case is used as the predicted time for q.

$$PB_{even}(q, C) = w(q, C_{even}) \bullet Time(C_{even}(PB)) \tag{5}$$

3.4 Pacing Recommendation

Given a predicted personal best finish-time it is straightforward to turn this into an average pace (mins/km) for q to run her race. However, as discussed previously, such an approach is unlikely to be successful, due to a variety of factors including ability, discipline, and course conditions. Therefore, in this section we describe how to use CBR to fulfill the requirement for a more personalized and tailored pacing plan with which to achieve this new personal best finish-time. In fact, a key advantage of CBR in this regard, compared to a more global, eager learning approaches, is precisely that it retains access to the local cases and their pacing profiles. It means that we can use the *PB* profiles of retrieved cases as the basis for a pacing plan for q and, in what follows, we describe 3 different approaches as *companions* to our 3 prediction approaches.

Best Profile. The first recommendation approach (*Best Profile*) reuses the *PB* pacing profile of the case with the best *PB* time, C_{best}. In other words, we take the relative pacing from C_{best} and map its relative paces to the average pace for predicted *PB* time for the query runner. For example, if the predicted *PB* time is for 232 min, indicating an average pace of 5 min 30 s per km, and the *PB* profile in C_{best} calls for a first 5 km that is 5% faster than average pace, then the generated pacing profile for q will advise running the first 5 Km at just over 5 min and 13 s per km.

Mean Profile. The second approach (*Mean Profile*) generates a new pacing plan based on the *mean* relative segment paces of the *PB* profiles from the k retrieved cases. So, if, on average, the *PB* profiles of the retrieved cases have their runners starting 6% faster than their average pace then our *Mean Profile* plan for a 232-minute finish will suggest that q begins her race at 5 min and 10 s per km.

Even Profile. Finally, we can generate our pacing plan based on the *PB* profile of C_{even}, the case with the most even pacing profile. If its profile calls for the first 5 km to be run at a pace that is just 2% faster than average race pace then the pacing profile for q will suggest a pace of 5 min and 23 s per km.

3.5 A Worked Example

Up to now we have described an approach to generating cases from pairs of race records. We have shown how these cases can be reused (according to three different strategies, *Best*, *Mean*, *Even*) as the basis for finish-time predictions and pacing recommendations, to help q achieve a new personal best in their marathon. By design we have kept our case representation, retrieval, and reuse strategies straightforward, to make it easy to embed this approach in a practical application for runners.

A summary example of such an application is depicted in Fig. 1(c) in which a (query) runner submits a (*nPB*) race record for a 248-minute finish-time with a pace profile that indicates a strong positive-split; they started out fast and completed the first-half of the race quite a bit faster than the second-half of the race. In terms of providing this information to the CBR system, it may be feasible for the runner to provide their race number and for the appropriate record to be pulled from the public race records in real-time. Large marathons such as London or Boston could readily accommodate this by integrating the service as part of their site offering.

In any event, having submitted a race record, the CBR system retrieves a set of k cases, $c_1 \dots c_k$ and, using one of the strategies above, generates a predicted *PB* time and suitable race plan for the runner. In this example the predicted *PB* time indicates to the runner that she should be able to achieve a personal best finish-time of 237 min, a fairly significant 11-minute improvement over her previous time. The recommended pacing plan suggests running the race with a slight positive-split, starting out a few percent faster than their target average pace of 5 min and 37 s per km, while running a slightly slower second-half of the race; it is a much more evenly paced race plan than their *nPB* race record. Importantly, this pacing plan provides our runner with a set of concrete paces for each of the race segments based on how other runners (with similar *nPB* records to the query runner) have run to a PB on the same course; in short, it provides the runner with a more personalized and tailored race plan than might otherwise be available.

4 Evaluation

To evaluate our approach we will focus separately on the tasks of finish-time prediction and profile recommendation using data from the last six years of the London Marathon (2011–2016).

4.1 Dataset

Briefly, our London dataset includes 215,575 race records for 185,143 unique runners. Each race record contains, among other features, the 5 km split-times and finish-times that we need as the basis for our case base. 37% of runners are female and just over 20% (or 37,704 unique runners) appear more than once in the dataset. This means we can construct at least 37,704 cases. However, since many of these repeat runners have only run 2 races this does limit the variation available for the identification of personal best times and profiles. Consequently, in this study we limit our data to runners who have run 3 or more races; this way we can choose a personal best race from at least 3 races and every runner will be represented by at least 2 cases. There are 5,390 such runners and on average they have run 3.4 races each, leading to a case base of 12,968 individual cases with 5,390 personal bests.

4.2 Methodology

For the purpose of this evaluation we use a standard 10-fold cross-validation methodology to test prediction accuracy and recommendation quality. Briefly, we randomly hold-out 10% of cases to act as a test-set and use the remaining 90% of cases as the basis for prediction and recommendation, repeating this 10 times and averaging the results. Thus, the *nPB* part of each test case is used as a query and the *PB* part is held back to evaluate the finish-time prediction and pacing plan recommendation.

Evaluation Metrics. To evaluate prediction accuracy we calculate the average percentage error between the predicted personal best time and the actual personal best time held back from the test case. To evaluate the quality of the recommended pacing plan we estimate the similarity between this recommended plan and actual pacing profile that was also held back; our measure of profile similarity is based on the relative difference between the segment paces of the recommended plan and the segment paces of the actual pace profile.

Test Algorithms. For each prediction/recommendation session we generate predictions/recommendations using the 3 CBR approaches (*Best, Mean, Even*) as described earlier. While in theory we could possibly *mix-and-match* between prediction and recommendation strategies, for the purpose of this study, we use a *like-with-like* approach and pair each prediction strategy with the corresponding recommendation strategy. Thus when we refer to the *Best* strategy, for example, we mean the *Best* prediction strategy combined with the *Best* recommendation strategy.

4.3 Prediction Error and Profile Similarity vs. k

To begin with we will look at prediction accuracy and recommendation quality versus k, the number of cases retrieved; see Figs. 2 (a) and (b). Both graphs contain *ribbon-plots* for each of the 3 test algorithms. Each ribbon-plot shows the prediction error (or profile similarity) for all runners, as the marked central lines, and separately for male and female runners, as the boundary lines. In this way we can get a sense of how gender influences prediction error and pacing profile similarity, as k varies.

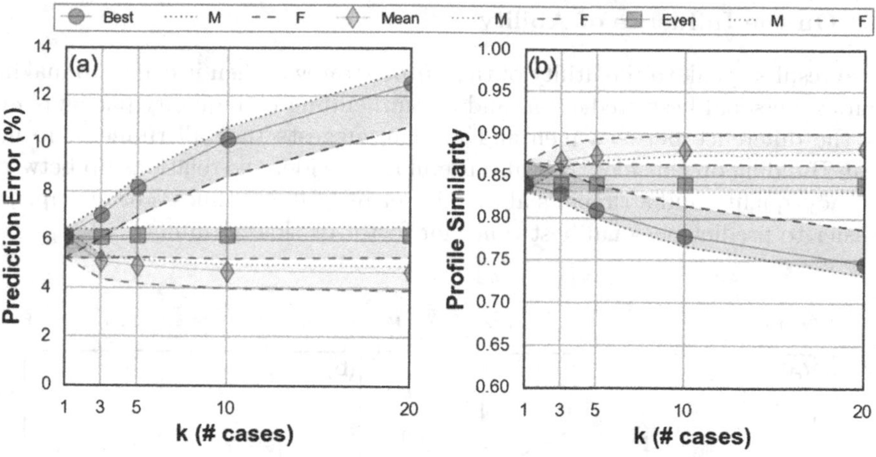

Fig. 2. Prediction error (a) and pacing profile similarity (b) versus k for *Best*, *Mean*, and *Even* strategies, for all runners and men and women.

With respect to prediction error, we can see how our strategies behave quite differently for increasing k. *Mean* produces the lowest errors and benefits from increasing values of k, before stabilising for $k \geq 10$; *Mean* achieves an error of $\approx$4.5%, for all runners, and as low as 4% for women. In contrast, *Even* produces predictions with an average error of $\approx$6% regardless of k, while the accuracy of *Best* deteriorates steadily with increasing k. The problem for *Best* is that more retrieved cases generally lead to faster best finish-times to use as a prediction. So *Best* tends to predict more and more ambitious personal best times as k increases, times that have not been achieved by our test runners.

When it comes to pacing profile similarity — our proxy for recommendation quality, and with higher similarity equating to better quality — we see a similar, albeit inverted pattern in Fig. 2(b). The *Mean* strategy out-performs the others, recommending pacing plans that are increasingly similar to actual *PB* pacing profiles (before stabilising at $k \approx 5$). The profile similarity of *Even* recommendations is largely unaffected by k, while the *Best* recommendations become less and less similar to the actual *PB* pacing profiles as k increases; once again more cases mean faster best C_{Best}'s with pacing profiles that are increasingly different from the personal best race records of the test users.

It is also interesting to note how the prediction accuracy (and profile similarity) for all strategies is better for women than for men, regardless of k. This is something that is not so surprising in marathon running research, where recent studies have shown how female runners tend to run in a more disciplined manner than their male counterparts [7,8]. In short, women run more evenly paced races, they are less likely to hit the wall, and so their races unfold in a more predictable fashion, compared to men's races. Our results suggest that this also extends to the matter of predicting a personal best time and recommending a bespoke pacing plan.

4.4 On the Influence of Ability

These results speak to the utility of the *Mean* strategy when it comes to making accurate personal best predictions and recommending high quality pacing plans. But the difference between men and women suggests that all runners are not equal. Building on this idea, it is also useful to consider the relationship between accuracy/quality and a runner's ability, in terms of finish-times. For example, is it easier to predict personal best times for faster or slower runners?

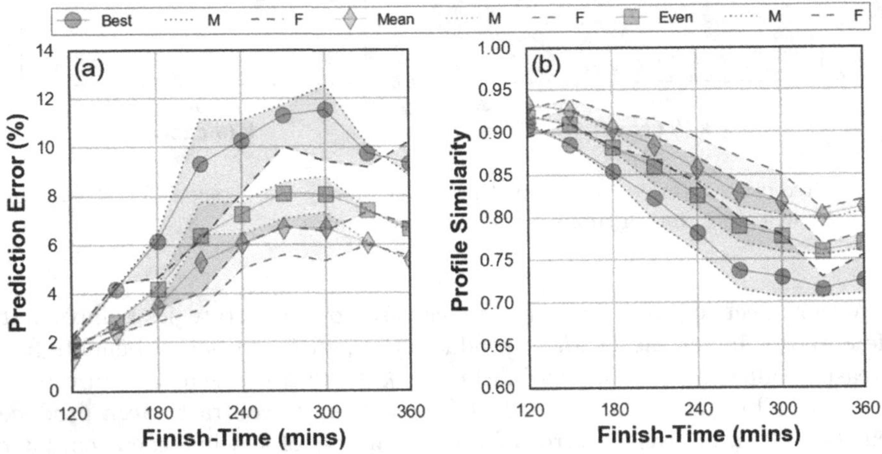

Fig. 3. Prediction error (a) and pacing profile similarity (b) vs. *nPB* finish-time for *Best*, *Mean*, and *Even* strategies, for all runners and men and women.

These prediction error and profile similarity results, for different finish-times, are presented in Figs. 3(a) and (b); for simplicity the results have been averaged over all values of k. Clearly, finish-times have a significant impact on both prediction error and profile similarity. For example, the fastest (elite) runners benefit from very accurate personal best predictions by all three strategies, but as finish-times increase so too do prediction errors, and at different rates for different strategies. Once again the *Mean* strategy benefits from the most accurate predictions across all finish-times, and the difference between men and women generally persists. It is interesting too to note that error rates begin to fall again after the 300-minute mark, suggesting that the *PBs* of slower finishers are more predictable.

The similarity of recommended pacing profiles to the actual personal best profiles falls as finish-times increase; see Fig. 3. The *Mean* strategy continues to perform better than *Even* and *Best* and women enjoy more similar pacing profiles than men.

4.5 Personal Best Differences

One more factor that may influence prediction accuracy and recommendation quality is the ambitiousness of a personal best. Are very ambitious personal bests more or less difficult to predict and plan for? To test this we define the *PB Difference* as the relative difference between a runner's non personal best time and their personal best time; for example, a PB difference of 0.1 indicates that the *PB* time of a case is 10% faster than its *nPB* time.

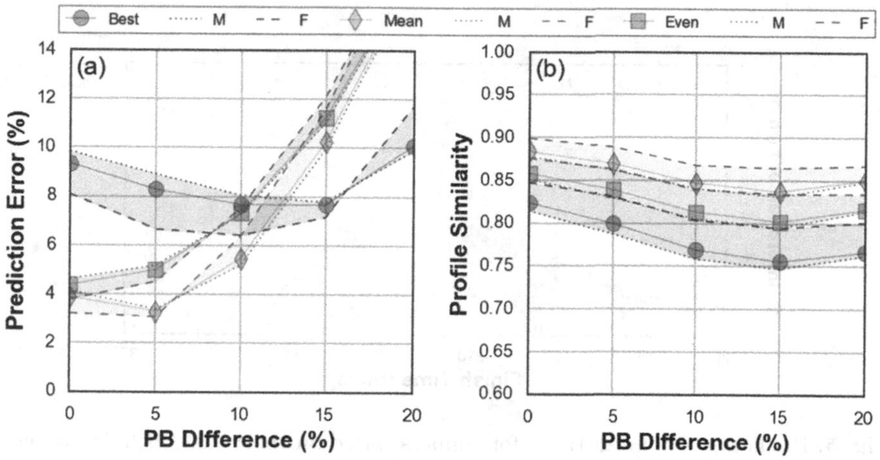

Fig. 4. Prediction error (a) and pacing profile similarity (b) vs. *PB Difference* for *Best*, *Mean*, and *Even* strategies, and for all runners and men and women.

Figures 4(a) and (b) shows how prediction error and profile similarity are influenced by *PB Difference*. As we might expect, runners whose *PB* times are less ambitious (lower *PB Difference* values) enjoy more accurate predictions for *Mean* and *Even*) than runners with larger PB differences; see Fig. 4(a). Likewise, profile similarity is also highest for runners with lower *PB Difference* values; see Fig. 4(b).

Again, the *Best* strategy performs poorly in general but with one notable exception. If *PB* times are more than 12% faster than their *nPB* times, then the inherently ambitious *Best* strategy out-performs *Mean* and *Even*. That said, it is worth noting that a 12% improvement in marathon finish-time represents a very significant improvement, and is not so common in practice; in our dataset less than 20% of runners achieve such an improvement.

5 Discussion

These results point to *Mean* as the best strategy to use in practice, because it offers more accurate *PB* finish-time predictions and more similar *PB* pacing profiles when compared with the actual *PBs* of the test runners. *Even* also performs well but *Best* appears to be far too ambitious.

5.1 Callibrating Personal Best Ambitions

With all this discussion about error rates and similarities it is easy to lose sight of the practical reason for the proposed system: to help marathon runners plan for, and achieve, new personal best finish-times. And so, as we draw to a conclusion, it is worth forming a more concrete sense of what these new *PB* times are likely to be for runners' *nPB* times. These results are presented in Fig. 5 for our most accurate strategy, *Mean*, as a graph of the predicted *PB* improvement in minutes versus the *nPB* time.

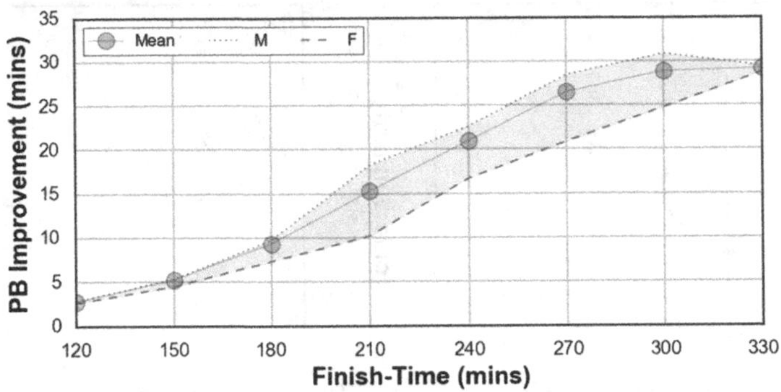

Fig. 5. Prediction *PB* finish-times for runners based on their non *PB* finish-times.

As we might expect, the predicted PB improvement increases with finish-time; it is 'easier' for a 5-hour marathoner to increase their time by 20 min than it is for a 3-hour marathoner. What is particularly encouraging here is the scale of the improvements on offer. For example, a 4-hour marathoner can expect to improve their PB time by just over 20 min on average (±75 s based on the error associated with predictions for such runners). This should be very encouraging for most recreational marathoners, and the combination of these accurate predictions and sensible, tailored pacing profiles may help many enthusiastic marathoners to improve significantly in their next race.

5.2 Pushing Beyond the *PB*

Our various strategies have been evaluated using the actual *PB* races of test runners and it should be pointed out that it remains unclear whether these runners *could* have produced even better *PBs* had they received the right pacing advice when they ran these races; no doubt some did but most will have adopted fairly simply pacing strategies. Perhaps with the right pacing advice the more ambitious *Even* or *Best* predictions might be achievable on the day.

Certainly conventional marathon wisdom suggests that a more even pacing profile is better and thus in practice we may find an additional benefit to presenting runners with the *Even* pacing plan even though its corresponding finish-time prediction is marginally less accurate (faster) than the *mean*. By following this *Even* profile the runner may achieve a more ambitious finish-time. Testing this hypothesis requires further work but it speaks to the potential benefits of presenting *Mean* and *Even* predictions and recommendations to runners, rather than focusing exclusively on just one strategy, such as *Mean*.

5.3 PB Quality

Another factor worth considering is the method by which *PBs* are identified. Currently they are the fastest race record for a runner. They may or may not reflect a genuine *PB*. As part of future work we plan to explore this by using different techniques for selecting more reliable *PBs*. Limiting our case base to runners who have run many races is one way to do this. Another way might be to limit *PBs* to those race records that have more even pacing profiles. Either way these approaches will provide a way to at least partially evaluate the likely impact of *PB* quality.

6 Conclusions

We have proposed a case-based reasoning solution to help marathon runners achieve a personal best at their next race. It represents a novel application of CBR to an important, open problem for marathon runners, namely how to select a best achievable target-time and how to pace the race to achieve it.

We have described and evaluated three CBR variations using a large-scale, real-world dataset from the London Marathon. The results indicate that the *Mean* CBR approach is capable of accurately predicting personal best finish-times and of recommending high-quality pacing plans. This may prove to be extremely valuable in helping runners to calibrate their expectations and plan effectively for their next race-day. By using training data from other marathons it will be possible to produce similarly personalised predictions and tailored recommendations for different marathon courses.

As part of future work we plan to explore other factors that likely impact prediction and recommendation performance, *PB* quality for example, as discussed, but also the length of time between a *nPB* and a *PB* in a case for instance; if the race-pair represent races 5 years apart, for example, will this affect prediction or recommendation? There is an exciting opportunity to also include this, and related ideas, into a new generation of exercise apps to help provide users with sensible training advice and real-time race feedback. Moreover, there is considerable opportunity to further enrich the case representation, either with additional race histories and/or context information, which has the potential to further improve prediction accuracy. Finally, there is no reason why this should be limited to marathon running or even running. Similar techniques may prove valuable for shorter races, for example, or other types of events, cycling, race-walking, triathalons etc., all of which are obvious targets for future research.

Acknowledgments. Supported by Science Foundation Ireland through the Insight Centre for Data Analytics under grant number SFI/12/RC/2289 and by Accenture Labs, Dublin.

References

1. Campbell, A.T., Eisenman, S.B., Lane, N.D., Miluzzo, E., Peterson, R.A., Lu, H., Zheng, X., Musolesi, M., Fodor, K., Ahn, G.-S.: The rise of people-centric sensing. IEEE Internet Comput. **12**(4), 12–21 (2008)
2. Mayer-Schönberger, V., Cukier, K.: Big Data: A Revolution That Will Transform How We Live, Work, and Think. Houghton Mifflin Harcourt, Boston (2013)
3. Ellaway, R.H., Pusic, M.V., Galbraith, R.M., Cameron, T.: Developing the role of big data and analytics in health professional education. Med. Teach. **36**(3), 216–222 (2014)
4. Bichindaritz, I., Montani, S., Portinale, L.: Special issue on case-based reasoning in the health sciences. Appl. Intell. **28**(3), 207–209 (2008)
5. Lewis, M.: Moneyball: The Art of Winning an Unfair Game. WW Norton & Company, New York City (2004)
6. Kelly, D., Coughlan, G.F., Green, B.S., Caulfield, B.: Automatic detection of collisions in elite level rugby union using a wearable sensing device. Sports Eng. **15**(2), 81–92 (2012)
7. Trubee, N.W.: The effects of age, sex, heat stress, and finish time on pacing in the marathon. Ph.D. thesis, University of Dayton (2011)
8. Deaner, R.O.: More males run fast: a stable sex difference in competitiveness in us distance runners. Evol. Hum. Behav. **27**(1), 63–84 (2006)
9. Haney Jr., T.A.: Variability of pacing in marathon distance running. Ph.D. thesis, University of Nevada, Las Vegas (2010)
10. Foster, C., Schrager, M., Snyder, A.C., Thompson, N.N.: Pacing strategy and athletic performance. Sports Med. **17**(2), 77–85 (1994)
11. Abbiss, C.R., Laursen, P.B.: Describing and understanding pacing strategies during athletic competition. Sports Med. **38**(3), 239–252 (2008)
12. Stevinson, C.D., Biddle, S.J.: Cognitive orientations in marathon running and "hitting the wall". Br. J. Sports Med. **32**(3), 229–234 (1998)

Weighted One Mode Projection of a Bipartite Graph as a Local Similarity Measure

Rotem Stram[1(✉)], Pascal Reuss[1,2], and Klaus-Dieter Althoff[1,2]

[1] Smart Data and Knowledge Services Group,
German Research Center for Artificial Intelligence, Kaiserslautern, Germany
{rotem.stram,pascal.reuss,klaus-dieter.althoff}@dfki.de
[2] Institute of Computer Science, Intelligent Information Systems Lab,
University of Hildesheim, Hildesheim, Germany

Abstract. Bipartite graphs are a common structure to model relationships between two populations. Many times a compression of the graph to one population, namely a one mode projection (OMP), is needed in order to gain insight into one of the populations. Since this compression leads to loss of information, several works in the past attempted to quantify the connection quality between the items from the population that is being projected, but have ignored the edge weights in the bipartite graph. This paper presents a novel method to create a weighted OMP (WOMP) by taking edge weights of the bipartite graph into account. The usefulness of the method is then displayed in a case-based reasoning (CBR) environment as a local similarity measure between unordered symbols, in an attempt to solve the long-tail problem of infrequently used but significant symbols of textual CBR. It is shown that our method is superior to other similarity options.

Keywords: Bipartite graph · One-mode projection · Textual case-based reasoning · Local similarity · Weights · Long-tail

1 Introduction

Complex network analysis is a field that is currently being vastly researched both under theoretical models and for practical use. The bipartite graph is a special type of network where nodes belong to two distinct populations, and includes only connections between population, but not within them. An example of such a graph can be seen in Fig. 1(a).

Bipartite graphs can model many real world systems, such as economic networks where countries are connected to the products they export [10], or collaboration networks of scientific coauthoring of papers where each author is connected to the paper they (co)authored [13,18]. Even human preferences can be modeled and studied using bipartite graphs [14,29].

Many times the goal of researching this type of networks is to model the relationships between items of only one population based on their connections to the

D.W. Aha and J. Lieber (Eds.): ICCBR 2017, LNAI 10339, pp. 375–389, 2017.
DOI: 10.1007/978-3-319-61030-6_26

other, for instance the economic relations between countries, or coauthorships. To this end the network is many times projected onto the population we want to focus on, in a process called one-mode projection (OMP). Here nodes from one population are connected to each other if they share at least one neighbor in the bipartite graph.

Looking at the example of coauthorships, many times authors collaborate on more than one paper, some more than others. If we look at authors l_1, l_2, and l_3, authors l_1 and l_2 could have coauthored five papers together, while authors l_1 and l_3 only one. Clearly the relationship between l_1 and l_2 is different from the relationship between l_1 and l_3. This information is lost if we disregard the number of neighbors r_i two items share in the bipartite graph.

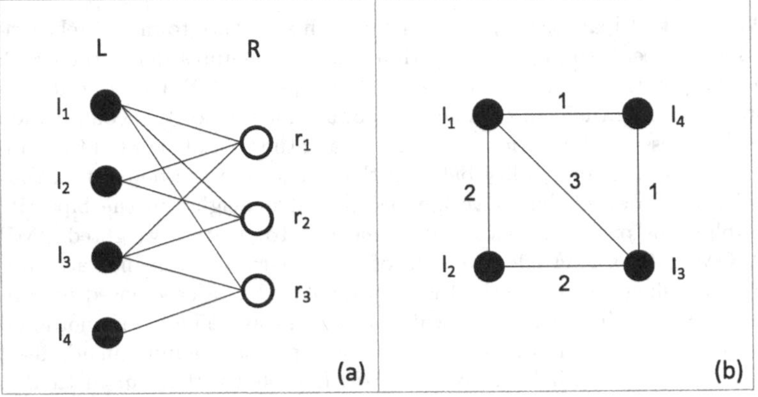

Fig. 1. (a) A bipartite graph consisting of two populations, L and R. (b) The OMP of the L population with simple weights counting the number of common neighbors

To reduce information loss, several methods have been proposed to take the number of shared neighbors into account and introduce edge weight in the projected network. The simplest form of weighting is to count the number of common neighbors two nodes share [19] (see Fig. 1(b)). When looking at nodes from one population in the bipartite graph, a higher degree may cause a lower impact of the nodes in the other population on each other. As an example we consider again a collaboration network. Two authors who collaborated on a paper might have a stronger connection if they were the sole authors, as opposed to a paper with many other authors. In order to include this information, Newman introduced a factor of $1/(n_k - 1)$ to each weight, where n_k is the number of authors, or the degree, of paper k [16,17]. Another problem that might arise with such projections is that adding another connection for authors who already collaborated on many papers before should not have the same impact as a new connection between authors who collaborated on only one or two papers in the past. To add a saturation effect, Li et al. suggested using a hyperbolic tangent function [15].

Another method to evaluate the relationship between two nodes l_1, l_2 from the same population using a OMP is described by Zweig et al. [30]. Here random graph models are used to find the expected occurrence of a connection motif between two nodes l_1 and l_2, namely $M(l_1, r, l_2)$, where r is a common neighbor in the bipartite graph, and use it to quantify the *interestingness* of this motif. Only the pairs with the highest interestingness are connected in the OMP. Although the resulting one-mode graph does not contain weights, each edge describes a strong relationship.

A problem that all these methods share is that all weights on the OMP, if they exist, are symmetrical. Going back to our collaboration network example, a new author with very little published papers would likely give a higher weight to his relationship with a new coauthor, than an author who already has many publications. All the methods described 'till now would give the same weight to a connection between these two authors. Moreover, many papers are written by a single author, and this information will be lost in the projection since only collaborations are taken into account. Zhou et al. proposed looking at each connection in a bipartite graph as a resource that is being allocated from nodes of population L to nodes from population R, and vice versa [29]. This means that each node $l \in L$ equally distributes its resource to all nodes $r \in R$ it is connected to, and then all nodes $r \in R$ distribute their resources to all nodes $l \in L$ they are connected to. This creates a path between each two nodes $l_1, l_2 \in L$ that share at least one neighbor $r \in R$, with a weight corresponding to the resource allocation between all members of this path. As a result, walking this path from two different directions would result in two different weights.

The methods described above assume that the connections between the populations have an equal weight, and disregard the possibility that the edges in the bipartite graph may be weighted. This work presents a new method to find the weights between two items from the same population that are connected by at least one neighbor in a bipartite graph, while taking into account the edge weights of the bipartite graph, thus creating a weighted OMP (WOMP).

We will first describe our method for WOMP in Sect. 2, then we will discuss its usefulness in modeling similarities between keywords in a textual case-based reasoning (CBR) system in Sect. 3. Section 4 describes the area of application of this work. Experiment results will be shown and analyzed in Sect. 5, demonstrating the superiority of the WOMP over other methods in determining similarities in CBR systems. Section 6 will talk about other works in the CBR field, among others, that are related to this work, while the conclusions and future work will be discussed in Sect. 7.

2 Method

We turn to look at a bipartite graph with two populations of nodes L and R, where each edge between nodes $l_i \in L$ and $r_j \in R$ holds a weight w_{ij}. Our goal is to find the weight $w_{ab}^{L \to L}$ between each $l_a, l_b \in L$ that share at least one common neighbor in R. To derive this weight we expand the resource allocation method described in [29] to include weights in the original bipartite graph.

The idea behind the resource allocation method is that each node in the graph holds a certain amount of resources, that is then distributed to its neighbors. The weight of an edge then describes part of the resources that is passed along the edge. To find the weight between two nodes from the same population we need to follow the distribution path of the resources.

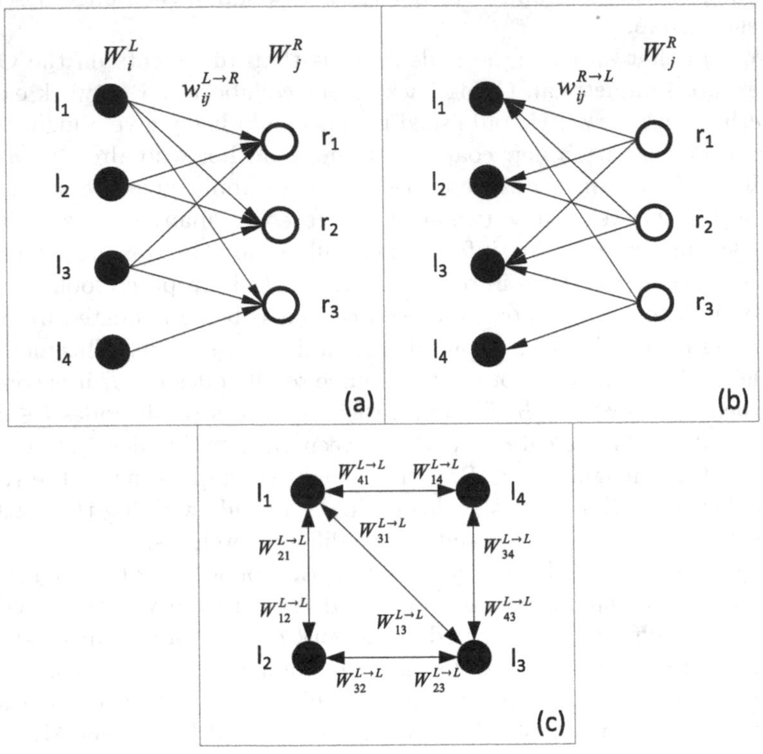

Fig. 2. (a) The flow of resources from population L to R in a bipartite graph. (b) The flow of resources from population R to L in a bipartite graph. (c) The WOMP of the bipartite graph.

Consider a bipartite graph $G(L, R, E)$ where E is the edge list containing tuples (l_i, r_j, w_{ij}), where w_{ij} is the weight between nodes $l_i \in L$ and $r_j \in R$, and $|L| = n$, $|R| = m$. Let's say we want to find the WOMP of population L. First we define the amount of resources that each node $l_i \in L$ has as:

$$W_i^L = \sum_{j=1}^{m} w_{ij} \tag{1}$$

In case there is no edge between l_i and r_j we consider $w_{ij} = 0$. Next, we define the resource that l_i allocates to r_j as the ratio between the amount of resources that this edge contributed to l_i and the total amount of resources that l_i possesses:

$$w_{ij}^{L \to R} = \frac{w_{ij}}{W_i^L} \tag{2}$$

It is clear to see that $w_{ij}^{L \to R} \in [0, 1]$, and represents the portion of resources that flow through this edge. From here we can conclude that the resources that node r_j accumulates is the sum of those that have been allocated to it from all its neighbors:

$$W_j^R = \sum_{i=1}^{n} w_{ij}^{L \to R} \tag{3}$$

This flow of resources is visualized in Fig. 2(a). Now we switch directions and distribute resources from R to L. Nodes r_j allocate the following to their neighbors l_i:

$$w_{ij}^{R \to L} = \frac{w_{ij}^{L \to R}}{W_j^R} \tag{4}$$

The change in flow direction can be seen in Fig. 2(b). Please note that $w_{ij}^{R \to L}$ is calculated analogously to $w_{ij}^{L \to R}$, as the ratio between the amount of resources that this edge contributed to r_j and the total amount of resources that r_j possesses.

To find the weight between two nodes $l_a, l_b \in L$ one must follow the flow of resources from l_a to l_b:

$$w_{ab}^{L \to L} = \sum_{j=1}^{m} p_{aj} \cdot p_{bj} \cdot (w_{aj}^{L \to R} + w_{bj}^{R \to L}) \tag{5}$$

Where $p_{ij} \in \{0, 1\}$ indicates whether or not there is an edge between l_i and r_j. To make this notion concrete we give an example of how to find the weight $w_{12}^{L \to L}$ between nodes l_1 and l_2 from Fig. 1(a). One can see that their shared neighbors are r_1 and r_2. First we follow the flow of resources from left to right, and then we follow the flow from right to left:

$$w_{12}^{L \to L} = w_{11}^{L \to R} + w_{21}^{R \to L} + w_{12}^{L \to R} + w_{22}^{R \to L}$$

In order to find the weight $w_{21}^{L \to L}$, the same links are used but the flow direction of resources is switched:

$$w_{21}^{L \to L} = w_{21}^{L \to R} + w_{11}^{R \to L} + w_{22}^{L \to R} + w_{12}^{R \to L}$$

One should note that at this stage $w_{ab}^{L \to L} \geq 0$, and allows values greater than 1. To illustrate this we look at another specific case, namely $w_{41}^{L \to L}$, we have:

$$w_{41}^{L \to L} = w_{43}^{L \to R} + w_{13}^{R \to L}$$

It is clear to see that $w_{41}^{L\to L} \geq 1$, since $w_{43}^{L\to R} = \frac{w_{43}}{W_4^L} = \frac{w_{43}}{w_{43}} = 1$ and $w_{13}^{R\to L} \geq 0$. The next step is then to normalize the weights to values in $[0, 1]$, and to do that the following normalization is used:

$$W_{ab}^{L\to L} = \frac{w_{ab}^{L\to L}}{w_{bb}^{L\to L}} \tag{6}$$

Where $w_{bb}^{L\to L}$ describes the highest possible portion of resources that can flow to l_b.

This method produces asymmetrical weights for the projection onto population L, creating a directed graph where each connection is bi-directional. An illustration of a WOMP can be seen in Fig. 2(c).

3 Similarities in Textual Case-Based Reasoning

In the world of expert systems, an attempt is made to mimic the responses of experts of a given field to certain situations, and possibly to surpass the experts based on some performance measure. Case-based reasoning (CBR) is a paradigm that can be used to implement an expert system. Under CBR, situations may be described in many different ways, from attribute-value pairs, to object-oriented (OO) classes, to graphs.

The idea behind CBR is that similar problems have similar solutions. In order to solve a problem that is described by a situation, an attempt is made to find past situations that are similar to the current one and adapt their solutions to fit the problem [21]. A perfect CBR system would be able to evaluate the a-posteriori utility of each case c_i in the case base to a new problem. This utility function is, however, unknown, and so an approximation attempt is made using heuristics [25]. This means that a CBR system depends heavily on the similarity measure between two situations to perform well.

Two types of similarities are used in CBR, local and global. If we focus on an attribute-value type case description, the local similarity can be defined as the similarity between the values of each attribute. Attributes with numerical values may use a distance measure to define this similarity, while symbolic attributes may utilize taxonomies or similarity tables to model the relationships between the different symbols. The global similarity describes the similarity of whole cases by amalgamating the local similarities. We define $sim_{local}(v, w)$ as the local similarity function between two values v, w of a given attribute, and $sim_{global}(c_1, c_2)$ as the global similarity of two cases c_1, c_2.

Many times the sources for the situation descriptions are in the form of free-text, and a popular method to tackle this is to transform the text into an attribute-value form by extracting wanted features from it [6,7,27,28]. Usually the values are an unordered set of symbols describing keywords and phrases, meaning that the next step is to model the similarity between them. Many times the extracted terms are presented to the experts in the field, and those experts then provide insight into the local similarity. Unfortunately, descriptions in free-text form can cause an explosion of keywords for each attribute, many of which are informative and descriptive of the situation but are used very rarely. There is only so much information experts can provide a developer about the situation descriptions, and so to best utilize their support experts may be asked to model the relationships only between the most frequently used symbols. This creates a long tail of rarely used attribute values that are informative to the case description, but are excluded from the similarity modeling process. A possible solution to this problem is to simply define $sim_{local}(v, w) = equal(v, w)$ where $equal(v, w)$ is the equality function, if either v or w is unmodeled. This solution is not informative and could affect the quality of the retrieval. In order to prevent this and make full utilization of these values, we propose to use WOMP to supplement our knowledge about the relationships between all values of an attribute.

4 Application Area

This work is a contribution to the OMAHA project [1], which is a joint project with Airbus and Lufthansa System to assist aircraft technicians in diagnosing faults using CBR methods. It is a step in the toolchain that was developed in order to tackle the challenges presented by this project [20]. The problem descriptions of past experiences are given in free-text form, and following the toolchain are transformed into an attribute-value form by extracting keywords from the text and assigning them to features. The toolchain was also developed to extract knowledge from the dataset, such as completion rules for the queries, or the importance of each attribute [26]. Although the most common keywords were modeled by the experts in the field, i.e. the experts explicitly quantified their similarity values, many others were disregarded due to time and labor constraints. Our goal is to quantify the similarities of these keywords using WOMP as follows:

1. For each attribute create a bipartite graph where L is the set of all keywords that appear under the given attribute, and R is the set of all possible diagnoses. A keyword $k \in L$ in connected to a diagnosis $d \in R$ if it appeared in a case with diagnosis d. The weight of each edge is the number of cases with diagnosis d that k appeared in.
2. Find the WOMP of the keywords L according to the method described in Sect. 2.
3. Use the weights of the edges between the keywords as their similarity value for the given attribute.

Unfortunately the Airbus fault description dataset does not contain well defined diagnoses yet, so in order to test our hypothesis we used a different dataset with similar conditions, namely the internet movie database[1] (IMDb). A casebase was built using the MyCBR tool [5], where each case describes a movie with only one attribute, namely the keywords related to the movie as reported by IMDb, and the diagnosis for each case is the genre of the movie. This means that the system receives a set of keywords as a query, and tries to diagnose the genre by retrieving movies with similar sets of keywords.

[1] http://www.imdb.com/.

5 Experimental Results

Two disjoint sets of movie descriptions were constructed by randomly choosing movies that were released between 2005 and 2015, contained a set of keywords, and belonged to one of the following genres: horror, action, romance, and comedy. Short films were ignored. One set contained 6,000 items, namely 1,500 movies from each genre and was used as a training set, while the other contained 500 movies from each genre, 2,000 in total, and was used as the test set.

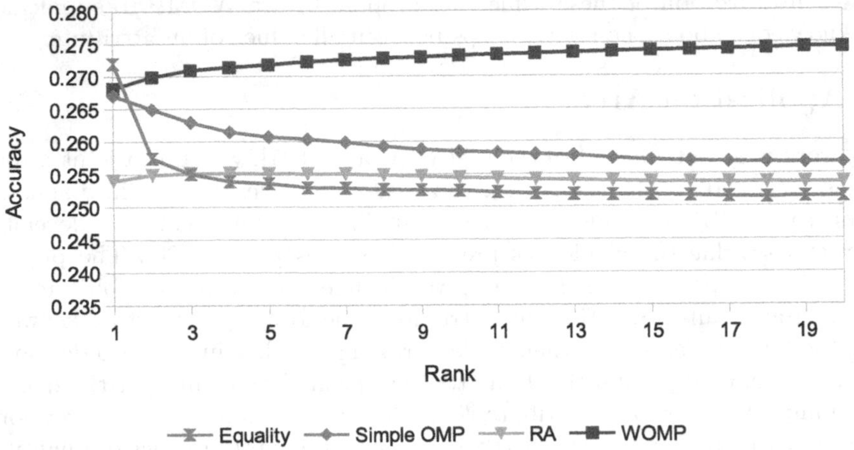

Fig. 3. The accuracy results of the four methods that were tested: the equality function, the simple OMP representing the number of common neighbors, the resource allocation (RA) method suggested by Zhou et al., and the WOMP method proposed in this paper.

Four weight functions were used to define the local similarity between the keywords. First, the equality function was used. Then, a bipartite graph was built where population L described the keywords, while R contained the genres. The edge weights described the number of movies each keyword appeared in that belong to the given genre. The second similarity function described the edge weights of the simple OMP, counting the number of neighbors each two keywords shared, disregarding the edge weights in the bipartite graph, and normalizing this number by the maximal degree of the nodes in L. The third function was the resource allocation (RA) method described by Zhou et al. [29], where again edge weights of the bipartite graph are disregarded. Lastly, the WOMP was used to define the similarity function while utilizing the information in the edges. Only movies from the training set were used to model each similarity. To evaluate how well these similarities performed, four case bases were built from the test set, one for each similarity function, and a retrieval test was performed on each movie in the test set. A case was deemed correctly retrieved if it belonged to the same genre as the query case. To quantify how well each similarity function performed confusion matrices were constructed for the highest ranked retrieved results in the first 1–20 positions. The retrieval accuracy as described by Eq. 7 was then calculated for each matrix.

$$accuracy = \frac{CorrectDiagnoses}{AllDiagnoses} \tag{7}$$

Figure 3 shows the results of the evaluation. While all four similarity functions performed above the random accuracy level (25%), WOMP produced the best results. The equality function started off as the best method for the first rank, but then quickly decline and became the worst method starting from the third rank. The graph for the simple OMP is similar in shape to the equality function, however its decline is smoother, and the overall score is higher. The results for the RA method performed the worse, and then the second worse, however its shape is interesting. For the first 5 ranks the accuracy is increased, and then it starts to slowly decline.

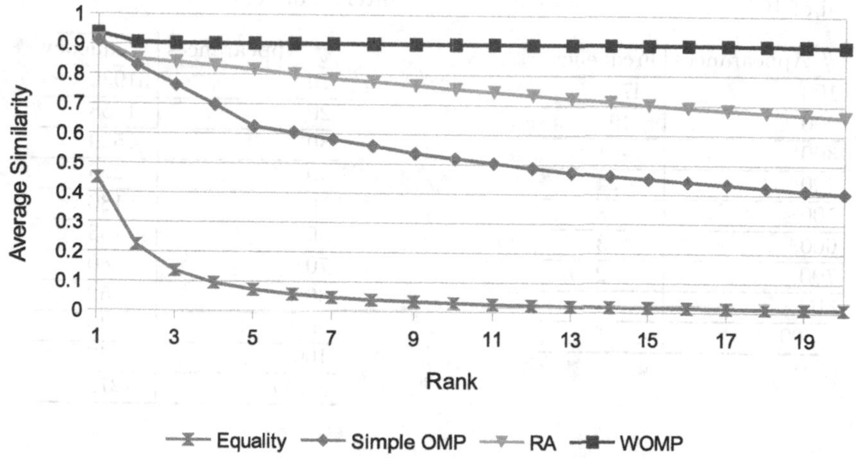

Fig. 4. The average similarity value by rank of the three methods that were tested: the equality function, the simple OMP representing the number of common neighbors, and the WOMP method discussed in this paper.

It is clear that the proposed method, namely WOMP, received the best results starting from rank 2 by a comparably large margin. Another difference that can be seen is the rise in accuracy in the WOMP when considering more ranks, which can be compared to RA, as opposed to the decrease thereof in the other two methods. This can be explained by the differences in average similarity values for different ranks, as shown in Fig. 4. The equality function and the simple OMP show a rapid decline for higher ranks. This is probably due to the limited similarity values two keywords can take ($\{0, 1\}$ for the equality function and $\{0, 0.25, 0.5, 0.75, 1\}$ for the simple OMP), and the relatively big step between each value (1 for the equality function and 0.25 for the simple OMP). This leads to low confidence in the lower ranks, and a lower accuracy. The WOMP and RA, on the other hand, produce more finely grained similarity values, creating a continuous and smoother transition between ranks. Even well after rank 10 the average similarity value for WOMP is above 90%, creating a ripe condition for a confidence vote, leading to the rise in accuracy the further we get through the ranks. Even though it is not yet visible at rank 20, logically one can assume that accuracy will decrease after a certain point for WOMP, and the shape of its graph should be comparable to the RA. One can also see that the similarity values for RA are also quite high, even though the accuracy for this method is quite low. This leads to the conclusion that RA may not be a suitable weighting method when the bipartite graph is weighted.

Table 1. The frequency of the keywords for different number of appearances in the dataset with bucket size interval of 100.

# Appearances	Frequency
100	34771
200	191
300	41
400	24
500	8
600	3
700	2
800	2
>800	1

Table 2. The frequency of the keywords for different number of appearances in the dataset with bucket size interval of 10.

# Appearances	Frequency
10	31930
20	1458
30	559
40	286
50	186
60	133
70	80
80	52
90	45
100	42
>100	272

In order to demonstrate the long tail abilities of the WOMP, we turn to look at term frequency in the dataset. Table 1 shows the frequency of the keyword under the different buckets of number of appearances with an interval of 100. This table supports the long tail assumption that many keywords are infrequently used in the dataset. To make matters more precise, Table 2 focuses on the first bucket, and divides it into even smaller buckets with an interval of 10. When considering that the dataset contains 35,043 keywords in total, 91% of them appear less than 10 times. Our assumption is that experts do not have the resources to model the similarities of infrequent terms, and in order to demonstrate that WOMP can help with this long tail, we constructed another case base where the 9% most frequent terms remain with the similarity values as calculated by WOMP to simulate the experts' input (since it was shown to produce the best results), and the remaining 91% were modeled with the equality function.

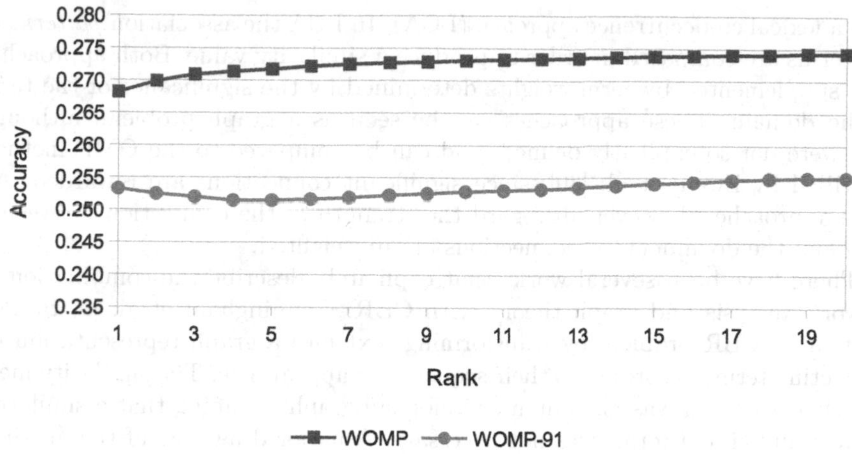

Fig. 5. The accuracy results of the WOMP and a WOMP version where the 91% least frequent terms had their similarity values replaced by those produced by the equality function

Figure 5 compares the retrieval accuracy of the both versions of the WOMP. One can clearly see the boost in accuracy that WOMP provides when it is used to model the similarities of the least frequent terms.

6 Related Work

Textual CBR is a well researched field, where problem descriptions and solutions in textual form are processed and transformed into cases that can be compared to each other. Usually, cases are represented in an attribute-value form. One of the first examples of this is PRUDENTIA, a system that transforms legal texts into cases and allows the retrieval of similar cases. This system as described by Weber et al. [27] follows experts guidelines and the strict structure of legal texts to extract terms and assign them to the correct attribute. The similarity between terms is completely modeled by experts.

A more recent example is described by Bach et al. [4], where cases in attribute-value form were extracted from textual service reports of vehicle problems. Here natural language processing (NLP) methods were used to extract terms from the text. The relationship between these terms was completely modeled by the experts, organizing them in taxonomies and similarity tables. This work is particularly similar to OMAHA, however the long-tail problem of infrequent terms was solved by disregarded anything the experts did not model.

The task of modeling similarities of terms without the help of an expert was tackled by Chakraborti et al. [8]. Here the 1st, 2nd, and 3rd co-occurrence degrees of terms in documents were explored and combined using a weighted sum to find the similarity value between two terms. A similar approach was further explored by Sani et al. [22] who compared the 1st degree co-occurrence with a lexical co-occurrence approach (LCA). In LCA the association *patterns* of two terms are compared in order to produce a similarity value. Both approaches were supplemented by term weights determined by the significance of the term to the domain. These approaches can be seen as a graph problem, although they were not so explicitly defined, and can be compared to the OMP method described by Zweig et al. [30], since significant connections are rewarded. All these approaches, however, disregard the strength of the connection between a term and the document, as connections are unweighted.

There have been several works that explicitely describe the combination of network analysis and graph theory with CBR. Cunningham et al. [9] tackled the textual CBR problem by transforming text into a graph representation by connecting terms according to their sequence of appearance. The similarity measure that was used was maximum common subgraph, meaning that a similarity between individual terms was not necessary. A major drawback of this method is that the complexity of the similarity assessment is polynomial, as opposed to linear when using attribute-value form.

Another work that combines CBR and graph theory is the Text Reasoning Graph (TRG) as described by Sizov et al. [23,24]. The TRG models causal relationships with textual entailments and paraphrase relations. In their first attempt, the TRG required that the solution of each case contain an analysis part, from which the TRG was extracted. Case similarity was calculated based on the vector space model with TF-IDF weights, while the graph was used only in the reuse step of the CBR cycle [2]. The TRG was later expanded to include the problem description. Two cases are then compared by looking at the problem description part of the graph and finding the so called longest common paraphrase (LCP). Combining the LCP with an informativeness measure of the phrases creates a ranked list of useful cases. As stated before, this method requires an analysis description of how each case was solved, something that may not be readily available in many applications, including our own.

An approach that was similar to ours is described by Jimenes-Diaz et al. [12]. Here OMP was used for link prediction in a recommender system. The idea here was to create a system that recommends programming tasks for students to practice on, according to previously solved tasks. The authors created a unweighted bipartite graph of tasks and students who solved them and derived the simple OMP, with number of common neighbors in the bipartite graph as weight. These weights were then used as a similarity measure between two tasks with the goal of predicting new links between tasks and users. A comparison was made between several weighting methods, and the simple OMP was found to produce the best predictions. This comparison is closely related to Zhou et al. [29], on which our WOMP method is based, who used resource allocation instead of simple OMP to evaluate the relationships between two nodes from the same population in a bipartite graph. The usefulness of this method was demonstrated on a movie recommendation system, where a user was recommended movies according the ones he liked in the past. Resource allocation was shown to be a powerful method compared to others.

When looking outside the scope of CBR there have been other attempts at estimating the similarities between object, most notably SimRank [11]. Here a PageRank-like algorithm was utilized to iteratively find similarities between nodes in a graph, with an extension to bipartite graphs. The main idea behind this algorithm is that "two objects are similar if they are related to similar objects." This work was later expanded with SimRank++ to take edge weights into account [3]. Both SimRank and SimRank++, however, produce symmetrical weights and have a relatively high time complexity.

7 Conclusions and Future Work

In this paper we presented a novel method to employ edge weights of bipartite graphs when building a OMP of a single population, namely the WOMP. This method is a generalization of the resource allocation based OMP presented by Zhou et al. [29]. The resulting OMP is a directed graph where all edges are bidirectional and are differently weighted in each direction, creating an asymmetrical similarity value between each two nodes in the graph.

This method was then used as a similarity measure between keywords extracted from free text, and evaluated as a supplementary similarity function for textual CBR. The idea here was to use WOMP as a similarity function between keywords that are infrequent but informative and have not been modeled by the experts in the field due to various constraints. An evaluation of the accuracy of WOMP weights, as opposed to the equality function, the simple OMP, and the unweighted resource allocation method was made and it was shown that WOMP produced superior results. A simulation of experts evaluation was also compared to WOMP, and has shown the contribution of this method when weighing infrequent keywords.

The WOMP uses resource allocation to model the relationship between two items from a single population in a bipartite graph. The edge weights of the bipartite graph are regarded as partial resources that make a whole, while each node contains the same amount of resources. This means that weights with different scales but a similar ratio produce the same $w_{ab}^{L\to L}$ values. In the future we plan on integrating the actual edge weight into the resource, thus allowing different amounts of resources to produce different results even if the scales are the same.

References

1. German aerospace center - dlr, lufo-projekt omaha gestartet. http://www.dlr.de/lk/desktopdefault.aspx/tabid-4472/15942_read-45359
2. Aamodt, A., Plaza, E.: Case-based reasoning: foundational issues, methodological variations and system approaches. AI Commun. **7**(1), 39–59 (1994)
3. Antonellis, I., Molina, H.G., Chang, C.C.: Simrank++: query rewriting through link analysis of the click graph. Proc. VLDB Endow. **1**(1), 408–421 (2008)

4. Bach, K., Althoff, K.-D., Newo, R., Stahl, A.: A case-based reasoning approach for providing machine diagnosis from service reports. In: Ram, A., Wiratunga, N. (eds.) ICCBR 2011. LNCS, vol. 6880, pp. 363–377. Springer, Heidelberg (2011). doi:10.1007/978-3-642-23291-6_27
5. Bach, K., Sauer, C., Althoff, K.D., Roth-Berghofer, T.: Knowledge modelling with the open source tool myCBR. In: CEUR Workshop Proceedings (2014)
6. Baudin, C., Waterman, S.: From text to cases: machine aided text categorization for capturing business reengineering cases. In: Proceedings of the AAAI 1998 Workshop on Textual Case-Based Reasoning, pp. 51–57 (1998)
7. Brüninghaus, S., Ashley, K.D.: Bootstrapping case base development with annotated case summaries. In: Althoff, K.-D., Bergmann, R., Branting, L.K. (eds.) ICCBR 1999. LNCS, vol. 1650, pp. 59–73. Springer, Heidelberg (1999). doi:10.1007/3-540-48508-2_5
8. Chakraborti, S., Wiratunga, N., Lothian, R., Watt, S.: Acquiring word similarities with higher order association mining. In: Weber, R.O., Richter, M.M. (eds.) ICCBR 2007. LNCS, vol. 4626, pp. 61–76. Springer, Heidelberg (2007). doi:10.1007/978-3-540-74141-1_5
9. Cunningham, C., Weber, R., Proctor, J.M., Fowler, C., Murphy, M.: Investigating graphs in textual case-based reasoning. In: Funk, P., González Calero, P.A. (eds.) ECCBR 2004. LNCS, vol. 3155, pp. 573–586. Springer, Heidelberg (2004). doi:10.1007/978-3-540-28631-8_42
10. Hidalgo, C.A., Hausmann, R.: The building blocks of economic complexity. Proc. Natl. Acad. Sci. **106**(26), 10570–10575 (2009)
11. Jeh, G., Widom, J.: Simrank: a measure of structural-context similarity. In: Proceedings of the Eighth ACM SIGKDD International Conference on Knowledge Discovery and Data Mining (2002)
12. Jimenez-Diaz, G., Gómez Martín, P.P., Gómez Martín, M.A., Sánchez-Ruiz, A.A.: Similarity metrics from social network analysis for content recommender systems. In: Goel, A., Díaz-Agudo, M.B., Roth-Berghofer, T. (eds.) ICCBR 2016. LNCS, vol. 9969, pp. 203–217. Springer, Cham (2016). doi:10.1007/978-3-319-47096-2_14
13. Lambiotte, R., Ausloos, M.: N-body decomposition of bipartite author networks. Phys. Rev. E **72**(6), 066117 (2005)
14. Lambiotte, R., Ausloos, M.: Uncovering collective listening habits and music genres in bipartite networks. Phys. Rev. E **72**(6), 066107 (2005)
15. Li, M., Fan, Y., Chen, J., Gao, L., Di, Z., Wu, J.: Weighted networks of scientific communication: the measurement and topological role of weight. Phys. A: Stat. Mech. Appl. **350**(2), 643–656 (2005)
16. Newman, M.E.: Scientific collaboration networks. I. Network construction and fundamental results. Phys. Rev. E **64**(1), 016131 (2001)
17. Newman, M.E.: Scientific collaboration networks. II. Shortest paths, weighted networks, and centrality. Phys. Rev. E **64**(1), 016132 (2001)
18. Newman, M.E.: The structure of scientific collaboration networks. Proc. Nat. Acad. Sci. **98**(2), 404–409 (2001)
19. Ramasco, J.J., Morris, S.A.: Social inertia in collaboration networks. Phys. Rev. E **73**(1), 016122 (2006)
20. Reuss, P., Stram, R., Juckenack, C., Althoff, K.-D., Henkel, W., Fischer, D., Henning, F.: FEATURE-TAK - framework for extraction, analysis, and transformation of unstructured textual aircraft knowledge. In: Goel, A., Díaz-Agudo, M.B., Roth-Berghofer, T. (eds.) ICCBR 2016. LNCS (LNAI), vol. 9969, pp. 327–341. Springer, Cham (2016). doi:10.1007/978-3-319-47096-2_22

21. Richter, M., Weber, R.: Case-Based Reasoning: A Textbook. Springer Science & Business Media, Heidelberg (2013)
22. Sani, S., Wiratunga, N., Massie, S., Lothian, R.: Term similarity and weighting framework for text representation. In: Ram, A., Wiratunga, N. (eds.) ICCBR 2011. LNCS, vol. 6880, pp. 304–318. Springer, Heidelberg (2011). doi:10.1007/978-3-642-23291-6_23
23. Sizov, G., Öztürk, P., Aamodt, A.: Evidence-driven retrieval in textual CBR: bridging the gap between retrieval and reuse. In: Hüllermeier, E., Minor, M. (eds.) ICCBR 2015. LNCS, vol. 9343, pp. 351–365. Springer, Cham (2015). doi:10.1007/978-3-319-24586-7_24
24. Sizov, G., Öztürk, P., Štyrák, J.: Acquisition and reuse of reasoning knowledge from textual cases for automated analysis. In: Lamontagne, L., Plaza, E. (eds.) ICCBR 2014. LNCS, vol. 8765, pp. 465–479. Springer, Cham (2014). doi:10.1007/978-3-319-11209-1_33
25. Stahl, A.: Learning similarity measures: a formal view based on a generalized CBR model. In: Muñoz-Ávila, H., Ricci, F. (eds.) ICCBR 2005. LNCS, vol. 3620, pp. 507–521. Springer, Heidelberg (2005). doi:10.1007/11536406_39
26. Stram, R., Reuss, P., Althoff, K.-D., Henkel, W., Fischer, D.: Relevance matrix generation using sensitivity analysis in a case-based reasoning environment. In: Goel, A., Díaz-Agudo, M.B., Roth-Berghofer, T. (eds.) ICCBR 2016. LNCS (LNAI), vol. 9969, pp. 402–412. Springer, Cham (2016). doi:10.1007/978-3-319-47096-2_27
27. Weber, R., Martins, A., Barcia, R.: On legal texts and cases. In: Textual Case-Based Reasoning: Papers from the AAAI 1998 Workshop (1998)
28. Yang, C., Orchard, R., Farley, B., Zaluski, M.: Automated case base creation and management. In: Chung, P.W.H., Hinde, C., Ali, M. (eds.) IEA/AIE 2003. LNCS, vol. 2718, pp. 123–133. Springer, Heidelberg (2003). doi:10.1007/3-540-45034-3_13
29. Zhou, T., Ren, J., Medo, M., Zhang, T.C.: Bipartite network projection and personal recommendation. Phys. Rev. E **76**(4), 046115 (2007)
30. Zweig, K.A., Kaufmann, M.: A systematic approach to the one-mode projection of bipartite graphs. Soc. Netw. Anal. Min. **1**(3), 187–218 (2011)

SCOUT: A Case-Based Reasoning Agent for Playing Race for the Galaxy

Michael Woolford and Ian Watson(✉)

Department of Computer Science, University of Auckland, Auckland, New Zealand
mwoo119@aucklanduni.ac.nz, ian@cs.auckland.ac.nz
https://www.cs.auckland.ac.nz/research/gameai/

Abstract. Game AI is a well-established area of research. Classic strategy board games such as Chess and Go have been the subject of AI research for several decades, and more recently modern computer games have come to be seen as a valuable test-bed for AI methods and technologies. Modern board games, in particular those known as German-Style Board Games or Eurogames, are an interesting mid-point between these fields in terms of domain complexity, but AI research in this area is more sparse. This paper discusses the design, development and performance of a game-playing agent, called SCOUT, that uses the Case-Based Reasoning methodology as a means to reason and make decisions about game states in the Eurogame Race for the Galaxy. The purpose of this research is to explore the possibilities and limitations of Case-Based Reasoning within the domain of Race for the Galaxy and Eurogames in general.

1 Introduction

Historically, the most prominent examples of game AI research have focused on achieving and exceeding human skill levels of performance in classic board games [4, 16, 17], while others have used those games as a test bed for experimenting with specific technologies and methodologies, or within the bounds of various limitations such as avoiding using domain knowledge [5, 15, 18].

Laird and van Lent [8] argued that despite impressive successes in their specific domains, this research had done little to progress the field towards development of a general human-level AI, and that modern computer games of many different genres, including computer strategy games, provided a superior test bed for human-level AI.

Race for the Galaxy (RftG) falls into a category of modern board games known as Eurogames. These games typically involve more complex rule-sets than traditional card and board games, have mixtures of hidden and open information and deterministic and stochastic elements, and are less abstract. Because of this, they bear more similarities to computer strategy games than do traditional board games. In recent years several agents for playing various Eurogames have been developed [6, 7, 19]. In general the approach to creating these agents has been more in keeping with the approaches taken in classic

D.W. Aha and J. Lieber (Eds.): ICCBR 2017, LNAI 10339, pp. 390–402, 2017.
DOI: 10.1007/978-3-319-61030-6_27

strategy board game AI systems. In contrast, a key aim of this project is to train SCOUT from examples of games played by a human player, using the Case-Based Reasoning (CBR) methodology [2]. Our hope is that if SCOUT can successfully mimic the decisions made by a human player, then it can implicitly gain some of the benefit of the human's strategic reasoning, and demonstrate a style of playing the game that resembles that of the human. Of course the primary goal of playing a game like RftG is to win, so that remains our main focus in terms of results, but we would also like to observe the way in which SCOUT goes about winning, and try to encourage diverse and adaptive play styles. Additionally, where possible we aim to limit our use of RftG domain knowledge in developing the system, with an eye toward exploring methods that could be generalised to other Eurogames.

2 Race for the Galaxy

Race for the Galaxy [9] is a popular Eurogame in which players attempt to build the best empire by constructing a tableau of cards. The game is highly stochastic as the game progresses as cards are randomly drawn from the deck, lending a high degree of variety and unpredictability to gameplay, but player actions are resolved deterministically, resulting in a richly strategic and skillful game. RftG has several expansions which increase the complexity of the game further, and can be played by up to six players. Currently, SCOUT is designed to be played only in a two-player game without expansions.

The rules are significantly more complex than classic board games such as chess; fortunately, it is not necessary for the purposes of this paper to understand how to play the game. A complete description of the game and its rules is available at the publisher's website [14].

Multiple computer implementations of RftG have been developed for online play. Currently, the most commonly used game engine is that hosted by the site boardgamearena.com, while our work was done with an offline open-source game engine developed by Keldon Jones [7]. In terms of its reasoning processes, SCOUT is designed to function largely independently of the game engine with which it is playing RftG, and could be implemented to work with any game engine. Henceforth we will use "RftG game engine" when referring to this part of the system in general, and "Keldon game engine" when referring to the specific engine used in our implementation. The Keldon AI is sophisticated by the standards of popular game AI and plays the game competently at an intermediate level. It is generally outplayed by a skilled human player but is able to win with favourable draws, and it will regularly beat a novice human player (Fig. 1).

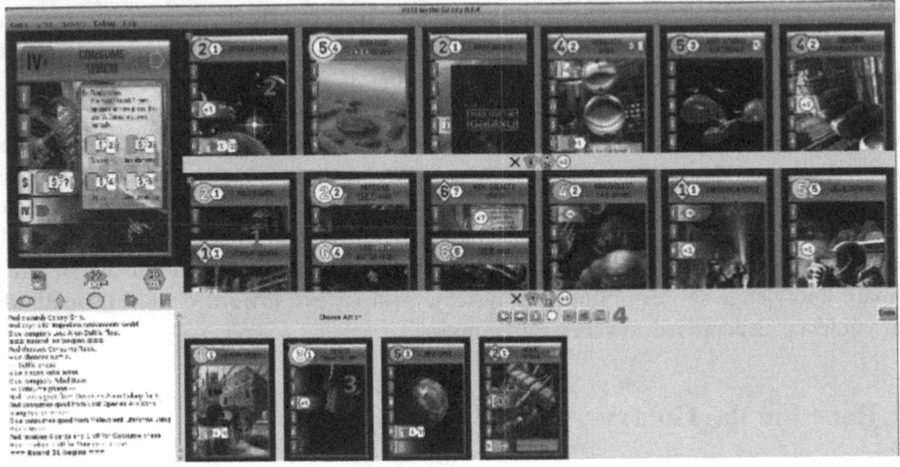

Fig. 1. A game of RftG in progress in the Keldon engine [7], with the Human player close to defeating the Keldon AI with a military strategy.

3 SCOUT's Design

This section aims to give an overview of each of the functional elements of the latest version of SCOUT. The next section will detail the specific design choices and developments which lead to this structure.

SCOUT consists of a group of independent modules, each of which handles one aspect of its functionality, along with a multipart case-base. The aim of this approach is to be flexible and to facilitate easy experimentation with, and comparison of, different approaches to developing an AI system for RftG or potentially other Eurogames. This was inspired by Molineaux and Aha's TIELT system for integrating AI systems with RTS games [1, 10].

There are 6 modules in the current iteration of SCOUT:

1. The Head module
2. The Interface module
3. The Case Controller module
4. The Placement Reasoning module
5. The Phase Reasoning module
6. The Payment Reasoning module

In brief, the Head module determines how to process incoming requests from the RftG game engine and facilitates communication between separate modules; The Interface receives game information and decision requests from the game engine and translates them into the specification used by SCOUT; The Case Controller module organises and maintains a registry of the case-bases; The Placement, Phase, and Payment modules each reason about the game state and make decisions when a relevant request is made by the game engine.

This model is very flexible, for example: In order to work with a different implementation of RftG, only the Interface module would need to be modified. If we wished to try a completely new reasoning process for card placement, we could swap out the Placement module. Alternatively, if we wish to disable any part of SCOUT's reasoning system and defer back to the Keldon AI, this can be achieved with a simple switch in the Head module. Meanwhile, all other parts of the system function unchanged. This is particularly useful during testing, as it allows us to measure the influence of another part of the system on SCOUT's overall performance in isolation. Modules make requests of one another via the Head but their internal processes are irrelevant to each other, particularly with regards to the reasoning modules. The Case Controller is of course specific to a CBR approach, but the Placement system, for example, could be reworked to classify a game state using a neural network while the Phase module continued to use the CBR system and each would still work in tandem.

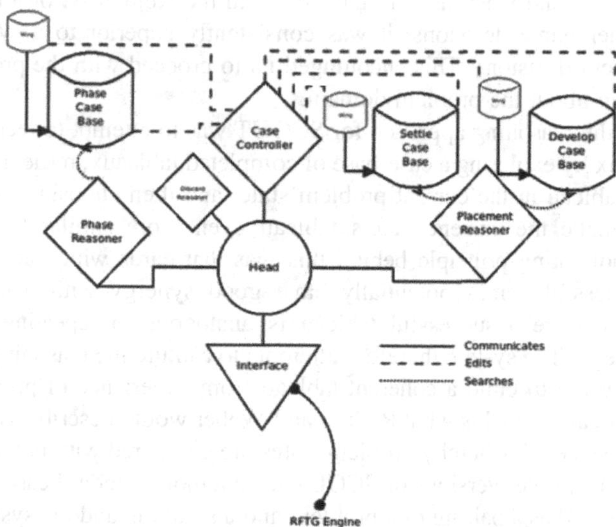

Fig. 2. Visibility between the modules that constitute SCOUT.

4 SCOUT's Development

The basis of SCOUT's reasoning faculties was an initial case-base generated by a human player playing 1,000 games against the Keldon AI. Every game that the human player won (748 games) was stored, and cases were extracted from it. By using these cases, we hoped that SCOUT would be able to take advantage of the human player's superior skill by mimicking their play style. We also experimented with cases generated by running the Keldon AI against itself, in order to generate more cases than a human player could do in a reasonable amount of time.

SCOUT maintains three distinct case-bases that are interrelated but organised independently of one another. These are:

1. The Phase Case Base
2. The Settlement Case Base
3. The Development Case Base

Each case base is used for a particular decision, clearly indicated by its name: The Settlement and Development Case-Bases are used by the Placement Reasoner to make Settle and Develop decisions respectively, and the same case-bases are used by the Payment Reasoner to make Payment decisions, while the Phase Case-Base is used by the Phase Reasoner to make Phase decisions.

4.1 SCOUT Prototype

The initial prototype of SCOUT, programmed in Python, was capable of making placement decisions using a k-NN algorithm on a case-base with simplified cases with only two indexed features. Despite it's simplicity, it was capable of performing its task with some success, and when used in tandem, with the Keldon AI or a human player making the other game decisions, it was consistently superior to a system making random placement decisions. This encouraged us to proceed with the project and was also illuminating about the problem domain.

In essence, the reasoning approach for SCOUT was to attempt to recreate previous winning tableaux by exploring a case-base of completed tableaux, retrieving those most similar to the tableau in the current problem state, and then choosing to place a card which would make the current state's tableau even more similar to the retrieved tableau. The motivating principle behind this was that cards which are together in a tableau in successful games potentially have good synergy with one another and attempting to recreate a successful tableau is analogous to repeating a successful placement strategy. The system therefore attempts to capture the reasoning process of a human player trying to build a coherent tableau from experience of prior games.

This type of case model is what Richter and Weber would describe as "an extended view of a CBR system", whereby problem states are compared with potential solutions directly [13, p. 41]. Later versions of SCOUT used a more standard case model, where a case is represented as a pairing of a problem and a solution, and the system compares problem states to other problems in the case-base, as opposed to comparing potential solutions.

A major factor in beginning with this approach was that completed game tableaux were able to be exported from a completed game within the Keldon engine by default, and thus we were able to prototype the system before beginning the challenging task of reverse-engineering the Keldon game engine to produce more sophisticated cases.

4.2 Retrieval

The reasoning approach for SCOUT's placement system was essentially to attempt to find the game states from previous successful games most similar to the current game state and adapt the decision made in that case to the current situation. This was a more standard CBR approach than the system used by the prototype. The rationale behind this was that if a case was similar enough to the current state in the relevant features then SCOUT could take advantage of all of the reasoning and planning that the player used to make a decision in the original case.

A k-NN algorithm was used to retrieve cases. Each case in the case-base now represented a specific game state, defined in the same terms as the problem situation, and with a single solution. Deciding what card to place in the current game state essentially became a classification problem, where each of the 95 cards were represented by a class, and cases with a particular card as a solution belonged to the class representing that card. By correctly classifying the problem case, the system determined the best card to place.

As is typical, the algorithm passed through the entire case-base and evaluated the similarity of a case to the problem case by summing the weighted similarities of each indexed case feature. From each of the *k* best matching cases an appropriate solution for the problem was adapted and added to a multiset of solutions, and finally, a single element of the multiset was randomly selected as the solution. The elements in the multiset were frequently homogeneous because SCOUT's retrieval algorithms were effective in classifying the cases consistently. Therefore the random element was much less pronounced and often completely deterministic, as all retrieved cases yielded the same solution. When the stochastic element did come into effect this was generally heavily biased toward a single good possibility, with an improbable secondary possibility providing some desirable variation.

Indexing

SCOUT processes cases into an internal representation of the case with 23 features, representing a game state, paired with the decision that was made in response to that game state, which constitutes the case's solution. The features are as follow, and are unindexed where not specified:

- **game id.** A non-unique nominal id shared by all cases which were generated within the same game.
- **case name.** A unique nominal id for each case.
- **game round.** An integer representing the game round in which the game state occurs. Indexed with high importance for both decision types.
- **player chips.** An integer representing the number of victory point chips the player possesses.
- **player hand.** A set of nominal ids representing the cards in the player's hand. Indexed with low importance for both decision types.
- **player hand size.** An integer representing the number of cards in the player's hand. Indexed with high importance for Action/Phase Selection decisions.
- **player military.** An integer representing the player's military score. Indexed with high importance for Placement decisions.
- **player goods.** A set of nominal ids representing the player's goods. Indexed with high importance for Action/Phase Selection decisions.
- **player score.** An integer representing the player's total score. Indexed with low importance for both decision types.
- **player tableau.** A set of nominal ids representing the cards in the player's tableau. Indexed with very high importance for both decision types.

- **player tableau size.** An integer representing the number of cards in the player's tableau. Indexed with high importance for both decision types.
- **opponent chips.** An integer representing the number of victory point chips the opponent possesses.
- **opponent hand size.** An integer representing the number of cards in the opponent's hand. Indexed with high importance for Action/Phase Selection decisions.
- **opponent goods.** A set of nominal ids representing the opponent's goods.
- **opponent military.** An integer representing the opponent's military score
- **opponent score.** An integer representing the opponent's total score.
- **opponent tableau.** A set of nominal ids representing the cards in the opponent's tableau. Indexed with moderate importance for Action/Phase Selection decisions.
- **opponent tableau size.** An integer representing the number of cards in the opponent's tableau. Indexed with very low importance for both decision types.
- **score difference.** An integer representing the player's score minus the opponent's total score. Indexed with moderate importance for Action/Phase Selection decisions.
- **deck.** An integer representing the number of cards currently in the deck.
- **discard.** An integer representing the number of cards currently in the discard pile.
- **pool.** An integer representing the number of victory point chips currently available. Indexed with high importance for Action/Phase Selection decisions.

Two of these features, case name and game id, have no meaning within the game and are only used to identify cases, but the remaining features are all potentially indexed features. We judged this number to be too high; especially since player tableau and opponent tableau in particular are highly complex features in themselves. Thus, we aimed to identify which of these features were most relevant to Placement decisions. Reducing the number of features as much as possible was important because k-NN algorithms have a tendency to be sensitive to irrelevant, interacting and noisy features [12] Such identifications could be made with domain expertise, however since we wished to find a method by which this could be done naively, and also which had the potential to expose unexpected patterns, we used statistical analysis on the case-base to identify the most relevant features. For each numeric feature, the distribution of values across the entire case-base was compared to its distribution of values among cases from each class separately, and if these distributions were found to vary significantly across several classes and the general distribution then these features were determined to be of relevance in terms of case similarity. For example, the Development "Contact Specialist" is almost always played when the player has 0 or −1 military score, while across the entire case base cards are played with various military scores.

5 Results and Discussion

5.1 SCOUT's Performance vs. Keldon AI

SCOUT's overall performance does not reach the Keldon AI's level, let alone that of a human player, but it does demonstrate an ability to play reasonably and competitively. This section will cover the results of many runs of games against the Keldon AI, along with other benchmarks. This will be followed by discussion about SCOUT's strengths and weaknesses as indicated by these results.

For comparison we tested other agents controlling the placement decision: the Keldon AI, a Random agent, and a Human. The Random agent merely selects one of the possible options with equal probability. The purpose of this is to give an indication of the absolute minimum level of performance possible. The Human player is the same whose cases comprise SCOUT's case-base. Contrasting the Random agent, this is intended to give a rough indication of ideal performance.

Each test was comprised of 5,000 games against a player controlled by a pure Keldon AI agent, except for the Human, for which the test was only 1,000 games. The number of games was selected by running the Keldon AI against itself until it reached a stable victory rate. The victory rate includes tied matches, hence the Keldon AI's victory rate against itself being slightly greater than 50% (Table 1).

Table 1. Victory rate of four different agents controlling all strategically significant decisions vs. the Keldon AI agent

Full controlling agent	Victory rate
SCOUT2	30.2%
Keldon AI	51.0%
Random	0.04%
Human	74.8%

The Human's win rate is clearly the best, but the results show the total inability of the Random agent to win a game (barring very exceptional circumstances). This demonstrates the reasoning quality of the controlling agent. This is the most important result, as it clearly indicates that SCOUT, though not as strong as the Keldon AI in overall performance, is capable of playing and winning in a way that a non-reasoning agent cannot.

5.2 Score Distribution

Figure 2 shows the distribution of score ratios across 5,000 games between SCOUT and Keldon AI. Scores are best measured relative to the opponents score, as it is not useful to measure scores in absolute terms across multiple matches. Shorter matches typically have a lower winning score, but they are not necessarily indicative of inferior performance to a higher score from a different match, indeed the opposite is often the case. Within a single match, however, close scores typically give some indication that the performance of the competing players was also close.

Representing the score of a match as $S_i = S_{si}/S_{ki}$ where S_{si} is SCOUT's score in a match and S_{ki} is Keldon AI's score, gives log-normal distribution of scores. These scores show that although SCOUT loses the majority of matches against Keldon AI, it usually achieves a competitive and respectable score. The score ratios have a median of 0.85, indicating that although reaching a 50% win rate against Keldon AI would mean an 66% increase in win rate; it would take only an 18% increase in SCOUT's scoring to bring it to that level.

5.3 Phase Selection

Comparing the frequency of selected Actions/Phases of SCOUT to the Keldon AI and the human player whose cases trained it, it can be seen that SCOUT's choices follow the same general trend as both other players (Fig. 3). This is evidence of reasonable play compared to the random agent that would have an equal frequency distribution across all actions.

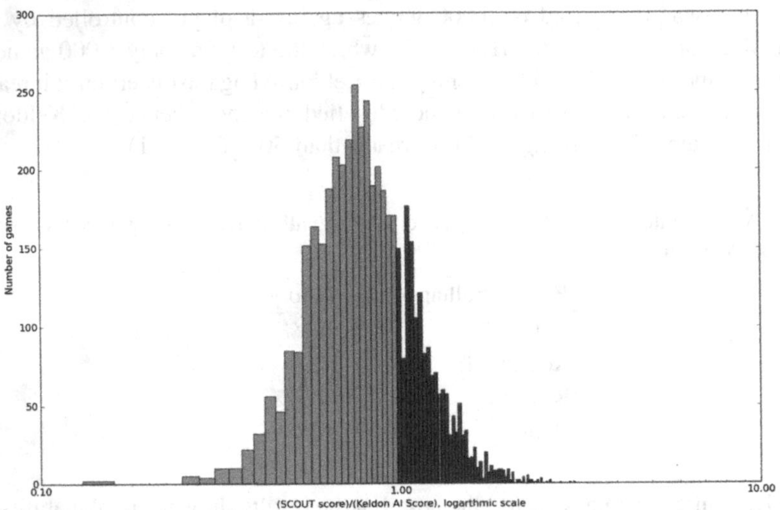

Fig. 3. Distribution of scores from 5000 games between SCOUT and Keldon AI, SCOUT won or drew the game when x $\geq$ 1.0 (31.7%). Note that the value of won or drawn games here is slightly higher as it includes those drawn matches that went against SCOUT in the tiebreaker.

A noticeable feature is that despite our observance that SCOUT follows a Consume-Produce strategy as frequently as possible, it in fact does not select Produce as frequently as either other player. This is likely explained by SCOUT's inability to perfectly manage its producer cards against its consumer cards. We regularly observed it producing many more goods than it could consume at once, and hence over three turns it would call Produce, then Consume, then Consume again, whereas a skilled human player, and to a lesser extent Keldon AI, tend to have more balanced numbers and thus call Produce and Consume on alternate turns.

A more promising observation is that across the first four action types, SCOUT's frequencies are more similar to the human player than to Keldon AI, indicating some success in mimicking the human's play style, at least in terms of selection frequency.

5.4 Directly Observed Games

Finally, during development we directly observed and analysed many hundreds of games played between SCOUT and Keldon AI. This section details results and observations of a random selection of games observed during the 4th era of a run. Game 1 was selected to be observed at random, and the rest are those that occurred in sequence thereafter. It represents better than average performance by SCOUT, which won 50% of these games. These observations are limited by their subjective nature, but they are useful in terms of gaining more insight into SCOUT's performance than simply raw scores and victory rates can provide (Fig. 4).

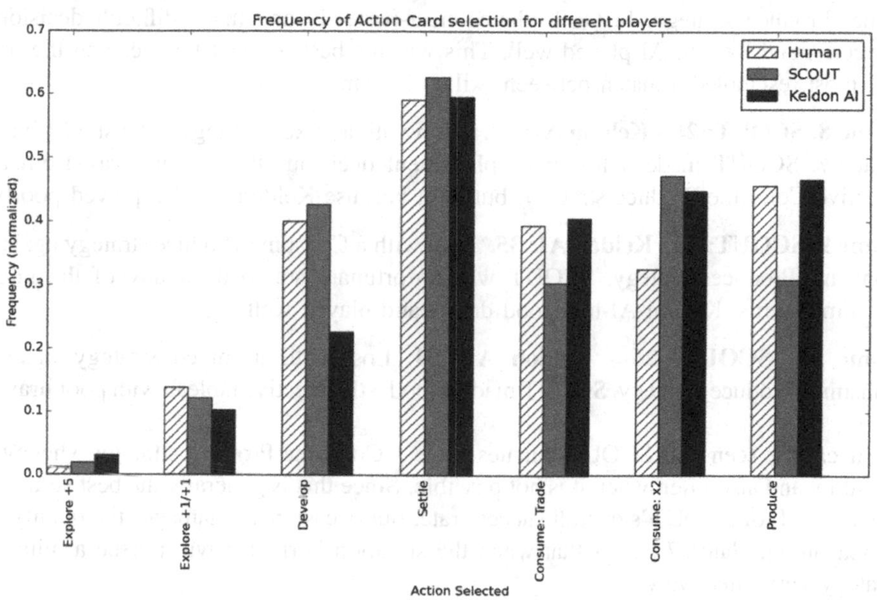

Fig. 4. Frequency of Action Card selection for different agents

Game 1, SCOUT: 26 - Keldon AI: 23. Won with Consume-Produce strategy against Consume-Produce Strategy. Made a questionable early decision to pass on a good placement, likely triggered by having another good card in hand that SCOUT played two turns later. Both cards could have been placed with correct hand management, however.

Game 2, SCOUT: 36 - Keldon AI: 33. Won with Consume-Produce strategy against a tableau with several large worlds and the key card Galactic Survey: SETI. SCOUT played near-perfectly.

Game 3, SCOUT: 17 Keldon AI: 25. Lost with an incoherent tableau against Consume-Produce strategy. SCOUT's draws were sufficient that it could have played a good Develop strategy. Keldon AI played well.

Game 4, SCOUT: 30 - Keldon AI: 25. Won with Consume-Produce strategy against an incoherent tableau. Made some poor placement decisions but won due to unfortunate draws for Keldon AI.

Game 5, SCOUT: 20 - Keldon AI: 35. Lost with an attempted Consume-Produce strategy against a Consume-Produce strategy. SCOUT made fatal errors, repeatedly wasting goods and actions, probably due to our feature generalisation indicating that the tableau was similar to other tableaux with subtle but important differences.

Game 6, SCOUT: 25 - Keldon AI: 26. Lost with Consume-Produce strategy against Military strategy. Both agents played reasonably.

Game 7, SCOUT: 46 - Keldon AI: 43. Won with Military strategy against a Consume- Produce strategy. SCOUT played near-perfectly and made difficult decisions correctly, and Keldon AI played well. This was the best of the 10 games and the one that most resembled a match between skilled humans.

Game 8, SCOUT: 26 - Keldon AI: 24. Won with a mixed strategy against a Military Strategy. SCOUT made a few poor placement decisions that led it away from an effective Consume-Produce strategy, but won because Keldon AI also played poorly.

Game 9, SCOUT: 20 - Keldon AI: 35. Lost with a Consume-Produce strategy against Consume-Produce Strategy. SCOUT was unfortunate not to draw any of the good consumer cards. Keldon AI had good draws and played well.

Game 10, SCOUT: 26 - Keldon AI: 31. Lost with a mixed strategy against Consume-Produce Strategy. SCOUT made a modestly effective tableau with poor draws.

It can be seen that SCOUT pursues a clear Consume-Produce strategy whenever possible, and also often when it is not possible. Since this is generally the best strategy, this is good for SCOUT's overall success rate, but occasionally damages the quality of its reasoning. Game 7 shows that when the situation is right it will pursue a military strategy very effectively.

6 Future Work and Conclusions

The aim of this research was to explore the use of CBR to play a modern board game, a domain that has received comparatively little attention in game AI research, despite offering many interesting challenges. SCOUT has demonstrated that a system using CBR and very limited domain knowledge can create a feasible agent for playing RftG. As of yet, however, it does not play at the same level as the current standard of RftG AI, the Keldon AI, which uses more conventional search and evaluation methods. The Keldon AI is a sophisticated system that has been developed and improved over many years, and reaching its level of performance is a high benchmark. Therefore while it is disappointing that SCOUT's performance is not up to this standard, we have had some success in creating a system which can make reasonable decisions for a complete and complex game.

In attempting to create a system from scratch, which includes the capacity to reason about various types of decisions and to evaluate, maintain, and improve its own case-bases, we have undertaken a large project with a broad scope. This may have come at the expense of focused optimisation of key elements. As a result, we do not believe SCOUT currently reaches the full potential of an agent using this methodology. This leaves the potential for future work in refining these aspects of the system, which could include systematically deriving feature weighting, or the development of more sophisticated retrieval algorithms. From a broader perspective, a hybrid approach which combines SCOUT's ability to recognise successful combinations of cards and make decisions in terms of a coherent strategy, combined with a system that can evaluate possible moves in the terms of game itself, such as that of the Keldon AI, may result in a system that is superior to both, and also closer to a human player's reasoning process. Combining CBR with other methodologies is a popular approach to such systems [3, 20]. SCOUT's architecture has the potential to be used as a basis for different AI agents for RftG, as could our fork of Keldon Jones' RftG engine, with the improved modularity of its control system.

While this paper focused on training SCOUT with human players' cases in an attempt to benefit from their reasoning, it may also be interesting to experiment with automatic case elicitation as per Powell's CHEBR system [11], beginning with a small or empty initial case-base. We have demonstrated, however, that a random agent is completely incapable of winning a game of RftG, so a different approach would need to be taken in the early stages of generating the case-base. In particular, an evaluation function that took more into account than the final result would be necessary. The overall performance of SCOUT's learning functionality proved to be limited, but there is potential to adjust its parameters and tweak its deletion policies.

Most importantly, future work that aims to improve upon SCOUT's performance would require access to a much larger case-base of games by skilled human players. This could open up the possibility for using data mining techniques to gain insight into feature weights, and of course give greater coverage in the initial case-bases. A case-base that includes negative cases to indicate potentially poor decisions to SCOUT, may also improve performance [2].

References

1. Aha, D.W., Molineaux, M., Ponsen, M.: Learning to win: case-based plan selection in a real-time strategy game. In: Muñoz-Ávila, H., Ricci, F. (eds.) ICCBR 2005. LNCS, vol. 3620, pp. 5–20. Springer, Heidelberg (2005). doi:10.1007/11536406_4
2. Aamodt, A., Plaza, E.: Case-based reasoning: foundational issues, methodological variations, and system approaches. AI Commun. **7**(1), 39–59 (1994)
3. Auslander, B., Lee-Urban, S., Hogg, C., Muñoz-Avila, H.: Recognizing the enemy: combining reinforcement learning with strategy selection using case-based reasoning. In: Althoff, K.-D., Bergmann, R., Minor, M., Hanft, A. (eds.) ECCBR 2008. LNCS, vol. 5239, pp. 59–73. Springer, Heidelberg (2008). doi:10.1007/978-3-540-85502-6_4
4. Campbell, M., Hoane, A.J., Hsu, F.: Deep blue. Artif. Intell. **134**(1), 57–83 (2002)
5. Fogel, D.: Blondie24: Playing at the Edge of AI. Morgan Kaufmann, Burlington (2001)
6. Heyden, C.: Implementing a computer player for Carcassonne. Master's thesis, Maastricht University (2009)
7. Jones, K.: Race for the Galaxy AI. www.keldon.net/rftg. Accessed 22 Oct 2016

8. Laird, J., VanLent, M.: Human-level AI's killer application: interactive computer games. AI Mag. **22**(2), 15 (2001)
9. Lehmann, T.: Game Preview: Race for the Galaxy, 26 September 2008. Boardgame News
10. Molineaux, M., Aha, D.W.: TIELT: a testbed for gaming environments. In: AAAI 2005, p. 1690. AAAI Press (2005)
11. Powell, J., Hauff, B., Hastings, J.: Utilizing case-based reasoning and automatic case elicitation to develop a self-taught knowledgeable agent. In: Challenges in Game Artificial Intelligence: Papers from the AAAI Workshop (2004)
12. Reyes, O., Morell, C., Ventura, S.: Evolutionary feature weighting to improve the performance of multi-label lazy algorithms. Integr. Comput.-Aided Eng. **21**(4), 339–354 (2014)
13. Richter, M.M., Weber, R.O.: Case-Based Reasoning. Springer, Heidelberg (2013)
14. Rio Grande Games. Race for the Galaxy. http://riograndegames.com/Game/240-Race-for-the-Galaxy. Accessed 22 Oct 2016
15. Rubin, J., Watson, I.: Investigating the effectiveness of applying case-based reasoning to the game of Texas Hold'em. In: FLAIRS Conference, pp. 417–422 (2007)
16. Schaeffer, J., Burch, N., Bjornsson, Y., Kishimoto, A., Muller, M., Lake, R., Lu, P., Sutphen, S.: Checkers is solved. Science **317**(5844), 1518–1522 (2007)
17. Silver, D., Huang, A., Maddison, C., Guez, A., Sifre, L., Van den Driessche, G., Schrittwieser, J., Antonoglou, I., Panneershelvam, V., Lanctot, M.: Mastering the game of Go with deep neural networks and tree search. Nature **529**(7587), 484–489 (2016)
18. Sinclair, D.: Using example-based reasoning for selective move generation in two player adversarial games. In: Smyth, B., Cunningham, P. (eds.) EWCBR 1998. LNCS, vol. 1488, pp. 126–135. Springer, Heidelberg (1998). doi:10.1007/BFb0056327
19. Szita, I., Chaslot, G., Spronck, P.: Monte-carlo tree search in settlers of catan. In: Herik, H.J., Spronck, P. (eds.) ACG 2009. LNCS, vol. 6048, pp. 21–32. Springer, Heidelberg (2010). doi:10.1007/978-3-642-12993-3_3
20. Wender, S., Watson, I.: Combining case-based reasoning and reinforcement learning for unit navigation in real-time strategy game AI. In: Lamontagne, L., Plaza, E. (eds.) ICCBR 2014. LNCS, vol. 8765, pp. 511–525. Springer, Cham (2014). doi:10.1007/978-3-319-11209-1_36

Conversational Process-Oriented Case-Based Reasoning

Christian Zeyen[(✉)], Gilbert Müller, and Ralph Bergmann

Business Information Systems II, University of Trier, 54286 Trier, Germany
{zeyen,muellerg,bergmann}@uni-trier.de
http://www.wi2.uni-trier.de

Abstract. Current approaches for retrieval and adaptation in process-oriented case-based reasoning (POCBR) assume a fully elaborated query given by the user. However, users may only have a vague idea of the workflow they desire or they lack the required domain knowledge. Conversational case-based reasoning (CCBR) particularly addresses this problem by proposing methods which incrementally elicit the relevant features of the target problem in an interactive dialog. However, no CCBR approaches exist that are capable of automatically creating questions from the case descriptions that go beyond attribute-value representations. In particular, no approaches exist that are applicable to workflow cases in graph representation. This paper closes this gap and presents a conversational POCBR approach (C-POCBR) in which questions related to structural properties of the workflow cases are generated automatically. An evaluation in the domain of cooking workflows reveals that C-POCBR can reduce the communication effort for users during retrieval.

Keywords: Process-oriented case-based reasoning · Workflow retrieval · Conversational case-based reasoning · Workflows

1 Introduction

Process-oriented case-based reasoning (POCBR) [18] addresses the integration of case-based reasoning (CBR) [5,24] with process-aware information systems [26]. A case in POCBR is usually a workflow or process description expressing procedural experiential knowledge. Among other things, POCBR aims at providing experience-based support for the modeling of workflows [11,14]. In particular, new workflows can be constructed by reuse of already available workflows that have to be adapted for new purposes and circumstances. In traditional POCBR, retrieval and adaptation are fully automatic and assume a fully elaborated query from the beginning [6,20,21]. However, in practice, users may only have a vague idea of the workflow they desire or they lack detailed domain knowledge and thus have serious difficulties to provide a precise query.

Conversational case-based reasoning (CCBR) [1,2,9] addresses this problem by focusing on the interactive nature of problem solving in particular.

D.W. Aha and J. Lieber (Eds.): ICCBR 2017, LNAI 10339, pp. 403–419, 2017.
DOI: 10.1007/978-3-319-61030-6_28

CCBR approaches include methods which incrementally elicit the relevant features of the target problem in an interactive dialog, often with the aim of minimizing the communication effort for the user. The basic assumption behind CCBR is that guided question answering requires less domain expertise than providing detailed queries from scratch. CCBR research so far focuses on methods for question selection and dialog inferencing and is mainly applied to diagnosis, help-desk support, and product recommendation [12,15–17]. Only very few approaches have been proposed that address synthetic applications [13,23,27]. Today, no CCBR approaches exist that automatically elicit questions from case descriptions that go beyond attribute-value representation to construct queries for retrieval. In particular, no such approach exists so far that is applicable for workflow representations as required for POCBR.

We present a new conversational POCBR approach, called C-POCBR. We consider graph-based workflow representations for cases and we propose an approach that considers the structural properties of workflows during the C-POCBR retrieval. Questions related to structural properties of cases are automatically constructed based on extracted workflow fragments and a respective question selection strategy is proposed. Thereby, we aim at reducing the effort and the required expertise for the definition of queries in POCBR. We illustrate and evaluate the approach in the cooking domain [19].

In the following, Sect. 2 briefly introduces POCBR and CCBR before Sect. 3 describes our C-POCBR approach. An experimental evaluation is presented in Sect. 4 while Sect. 5 summarizes our findings and discusses future work.

2 Foundations and Related Work

We now briefly describe relevant foundations and related work in the fields of POCBR and CCBR.

2.1 Process-Oriented CBR

POCBR [18] aims at supporting various tasks in process-aware information systems [26] such as process and workflow modeling, monitoring, analysis, or execution. In this paper, we focus on workflow modeling by reuse of best-practice workflows from a repository (case base). Thus, we aim at retrieving a workflow from a repository for reuse that is best suited to a specific situation.

In POCBR, cases are often represented as processes or workflows. Broadly speaking, a workflow describes a logical or chronological order (referred to as the control-flow) of tasks that are needed to reach a certain outcome – the workflow output [26]. Tasks exchange physical products or data, which is defined by the data-flow. In cooking workflows, tasks represent required cooking steps and exchange ingredients in order to produce a certain dish. We describe workflows as semantically labeled directed graphs by adopting the representation by Bergmann and Gil [6].

Definition 1. *A workflow is a directed graph $W = (N, E)$ with a set of nodes N and a set of edges $E \subseteq N \times N$. Nodes $N = N^D \cup N^T \cup N^C$ can be data nodes N^D, task nodes N^T, or control-flow nodes N^C. Each node $n \in (N^D \cup N^T)$ has a semantic label $S(n) \in \Sigma$, where Σ is a language for semantic annotations. Edges $E = E^C \cup E^D$ can be control-flow edges $E^C \subseteq (N^T \cup N^C) \times (N^T \cup N^C)$, which define the order of the tasks and control-flow nodes or data-flow edges $E^D \subseteq (N^D \times N^T) \cup (N^T \times N^D)$, which define how the data is shared between the tasks.*

A workflow $W' = (N', E')$ is a *partial workflow* (we write $W' \subseteq W$) of a workflow $W = (N, E)$ if W' is a subgraph of W with $N' \subseteq N$ and $E' \subseteq E$. Figure 1 gives an example of a purely sequential cooking workflow describing the preparation of a tomato sandwich.

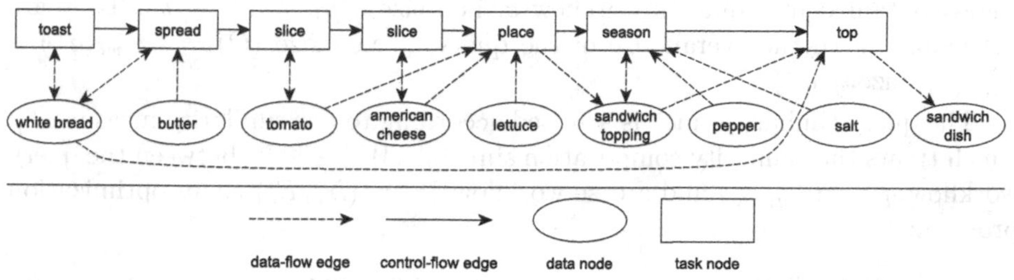

Fig. 1. Example of a cooking workflow

The language Σ for the semantic labels of nodes is structured hierarchically in two distinct domain-specific taxonomies, i.e., a data taxonomy of cooking ingredients and a task taxonomy of cooking steps. Thereby, workflows can be generalized regarding their semantic labels [20]. Generalized workflows provide a more general description and thus stand for a set of more specific workflows. For example, the workflow in Fig. 1 can be generalized by generalizing the ingredient *american cheese* to the more general ingredient *cheese* from the data taxonomy. A workflow W^* is a generalization of a workflow W (we write $W^* \sqsupseteq W$) if there exists a total mapping of data nodes and task nodes from W^* to W, in which each semantic label in W^* is more general (or equal) according to the taxonomies than the respective label in W.

In order to obtain a reusable workflow, similarity search or process model querying can be applied [8]. Outside of POCBR, various query languages have been proposed [3,4,25], which are used in visual query editors to formulate graph-based queries. Matching workflows from a repository are then obtained by applying graph edit measures [7] or graph/subgraph similarity measures [6,10]. We focus on similarity search as it is able to provide results even if exact matches are not available.

Queries in POCBR are used to describe the users' requirements for retrieving the most useful workflows [22]. In previous work [22], we proposed a process-oriented query language (POQL) to specify such queries. A POQL query

$Q = (Q^+, Q^-)$ consists of a query part $Q^+ = (q^+)$ with a single query workflow and a restriction part $Q^- = (q_1^-, \ldots, q_n^-)$ with several restriction workflows q_i^-. The query workflow represents properties the searched workflow should fulfill. Each restriction workflow represents one undesired situation that should be avoided. For a POQL query Q and a case workflow W_c, we define the following similarity measure:

$$\text{sim}(Q, W_c) = \frac{\text{sim}^+(q^+, W_c) \cdot \text{size}_{wf}(q^+)}{\text{size}_q(Q)} + \frac{\sum_{q^- \in Q^-} (\text{sim}^-(q^-, W_c) \cdot \text{size}_{wf}(q^-))}{\text{size}_q(Q)} \tag{1}$$

The similarities sim^+ and sim^- are weighted with the number of nodes and edges contained in a query's workflow q, i.e., $\text{size}_{wf}(q) = |N| + |E|$. They are normalized with the overall size of the query Q, i.e., $\text{size}_q(Q) = \text{size}_{wf}(q^+) + \sum_{q^- \in Q^-} \text{size}_{wf}(q^-)$.

The query similarity sim^+ is assessed according to our similarity measure [6], which treats the similarity computation $\text{sim}^+(q^+, W_c) \in [0, 1]$ between the query workflow $q^+ = (N_q, E_q)$ and a case workflow $W_c = (N_c, E_c)$ as an optimization problem:

$$\text{sim}^+(q^+, W_c) = \max\{\text{sim}_m(q^+, W_c) | \text{ admissible mapping } m\} \tag{2}$$

The similarity computation requires a search for the best possible admissible mapping $m : N_q \cup E_q \rightarrow N_c \cup E_c$ of nodes and edges of q^+ to those of W_c. A mapping is admissible, if it is type-preserving, partial, and injective. The core of the similarity model is a local similarity measure for semantic descriptions $\text{sim}_\Sigma : \Sigma^2 \rightarrow [0, 1]$. In our example domain, similarity values between semantic labels are derived from the data and task taxonomy that reflect the closeness of the concepts (refer to [5] for more details).

The restriction similarity sim^- is assessed with a binary measure that returns 1 if W_c does not fulfill the restriction q^-. If a restriction workflow q^- is a generalization of a partial workflow of W_c, the similarity is 0.

$$\text{sim}^-(q^-, W_c) = \begin{cases} 0.0 & \text{if } \exists W' \subseteq W_c : q^- \sqsupseteq W' \\ 1.0 & \text{otherwise} \end{cases} \tag{3}$$

2.2 Conversational CBR

While in many CBR applications a complete description of the target problem is assumed to be available in advance, CCBR [1,2,9,15] particularly addresses the interactive nature of problem solving. In CCBR, the user only has to answer posed questions, which presumably requires less domain expertise than providing queries from scratch. The CCBR dialog [1] often begins by asking the user to specify a brief textual description of her problem. Subsequently, a dialog is started consisting of a sequence of questions to be answered by the user. A goal in a conversation is to pose relevant questions, potentially suitable to elaborate the query and to determine the most useful case efficiently. The user interface consists of a question and solution display. If the user answers a question, the dialog component extends the query based on the given answer, performs a similarity-based retrieval, and updates the solution and question displays. Users can delete or alter answers to previously asked questions at any time. By selecting a solution, the conversation terminates.

To perform the dialog, the case representation in CCBR is enriched with an additional set of question-answer pairs stated in natural language. Thus, case authoring can become more demanding in CCBR, because suitable questions need to be formulated. Hence, the automatic creation of questions is desirable and often achieved by deriving questions from case attributes. CCBR research focuses on enhancing case representation to include knowledge relevant for the questioning strategy [9], methods for question selection, and methods for dialog inferencing and termination [2,9,12,15]. Our research is based on the similarity variance measure proposed by Kohlmaier et al. [12] which prefers questions, whose answers most probably have the highest influence on the similarity distribution of the most similar cases. CCBR finds its application mostly in analytical applications such as sequential diagnosis [17], customer help-desk support, or product recommendation [12,16]. Only very few approaches have been proposed that go beyond interactive query elicitation. Leake and Wilson [13] describe an approach for interactive case acquisition, retrieval, and adaptation for a specific design problem. Muñoz-Avila et al. [23] describe an interactive case-based planner which recursively applies a CCBR approach to guide the planning procedure of a hierarchical task network (HTN) planner. Weber et al. [27] propose a CCBR approach as part of their adaptive workflow system CBRflow. However, they use a traditional case representation consisting of manually defined question-answer pairs to explicitly acquire reasons and constraints for a specific workflow adaptation instance.

Today, no CCBR approaches exist that elicit questions to construct queries for cases represented as workflows as required for POCBR.

3 A Conversational POCBR Approach

Based on the generic CCBR approach, we now present a new approach that is particularly tailored to POCBR and thus named conversational POCBR (C-POCBR). In a nutshell, users are guided through the query process by a sequence of questions about their desired workflows. The more questions are answered, the more knowledge about desired and undesired properties is available, which is stored in an internal query for retrieval. A major focus is put on the automatic creation of questions to avoid that they need to be specified manually. For this purpose, we consider workflow fragments as characteristic properties of a workflow, which we refer to as features. The basic idea is to extract features from the workflows stored in the case base automatically, which are then used as the subject of questions. In order to conduct efficient conversations, we rank features by their ability to distinguish workflows from one another. Furthermore, identified relations between features enable to generate coherent follow-up questions and to infer irrelevant features based on already answered questions.

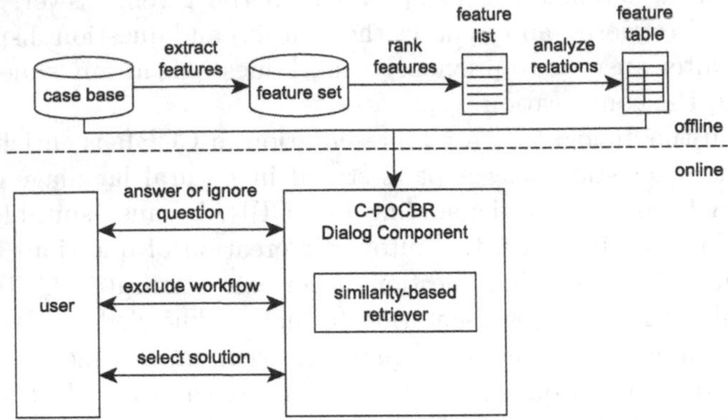

Fig. 2. Conversational POCBR process

Figure 2 illustrates the conversational POCBR process. The process is divided into two phases. The *offline phase* comprises pre-computations for the initial setup. During this phase, extraction, ranking, and analysis of features takes place and a feature table is created. Subsequently, the actual conversation is conducted in the *online phase*. In the following, we describe both phases in more detail.

3.1 Offline Phase

At first, features are extracted based on the graph-based representation of the workflows. As those features will occur in the questions posed to the user, they must be as simple and understandable as possible. For this purpose, we consider various design guidelines investigated in related work [1,12,24]. In principle, a feature can be any fragment of a workflow. In a workflow, the smallest possible feature consists of a single workflow item. This can be a single node such as a data or a task node. More complex features can be created by extracting partial workflows. To derive questions on a more general level of detail, we apply a generalization algorithm [20], which generalizes semantic labels based on the domain taxonomies. The generalization produces a generalized workflow W^* from the original workflow W, from which more general features can be extracted. We extract and annotate two different kinds of features for each workflow W in the case base:

- *specific feature nodes* and *generalized feature nodes*, i.e., single nodes from W and single nodes for all generalizations within the taxonomy up to the respective node in the generalized workflow W^*.
- *specific feature workflows* and *generalized feature workflows*, i.e., partial workflows (consisting of more than one node) from W and W^*, respectively.

A *feature workflow* describes structural properties of a workflow. Its definition is inspired by the idea of streamlets [21] proposed for compositional adaptation in POCBR:

Definition 2. *For a workflow* $W = (N, E)$ *and a data node* $d \in N^D$, *a feature workflow* W_d *of* W *is a partial workflow* $W_d = (N_d, E_d)$ *that consists of all task nodes* $N_d^T \subseteq N^T$ *connected to d and connected by control-flow edges. Moreover,* W_d *comprises all data nodes* $N_d^D \subseteq N^D$ *connected to* N_d^T *and the subset of edges* $E_d = E \cap ((N_d^T \times N_d^D) \cup (N_d^D \times N_d^T) \cup (N_d^T \times N_d^T))$ *connecting the nodes.*

A feature workflow W_d is a workflow according to Definition 1 and consists of at least one task and one data node, i.e., d.

Figure 3 exemplifies all features extracted from a cooking workflow (see dotted rectangles). The specific workflow is depicted in the middle of the figure. Related features (such as specific and generalized features) are arranged near one another. For instance, the specific feature node *pepper* is related to the generalized feature node *flavoring*. Based on the taxonomy, an additional generalized feature node *spice* laying inbetween those two is extracted as well.

With respect to the cooking domain, we applied some domain-specific restrictions. For the sake of simplicity, we assume a simplified workflow structure without control-flow nodes and with the control-flow restricted to a single sequence of tasks. Thus, parallel or alternative sequences as well as cycles are omitted. For the feature extraction, we omit single task nodes as they are mostly of no relevance when considered on their own. In addition, to obtain easy-to-understand feature workflows, we exclude tasks (marked with "*") that produce new data by consuming other data.

In the second step of the offline phase, features are sorted in descending order by their ability to distinguish workflows from one another. By this means, we reduce the length of a conversation. We adopt the simVar measure by Kohlmaier et al. [12], which utilizes the similarity variance as a ranking criterion. It estimates the variance of the similarity of the most similar cases assuming that the value of the respective feature in the query is known. Features with a higher simVar value are preferred.

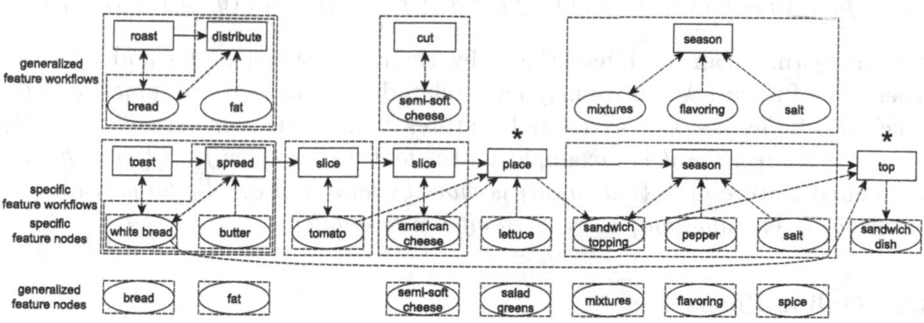

Fig. 3. Examples of a workflow's features

According to the POQL query (see Sect. 2.1), the user can either select a feature as desired or undesired during the conversation. Thus, the similarity variance is pre-computed for both situations. To calculate simVar, all similarities between the extracted features and the workflows stored in the case base must be computed. Each feature is added into the query part and the restriction part of an empty query, respectively. Then, for both queries, the similarities to each workflow from the case base are computed (according to Eq. 1) and cached. For a feature f and a case base CB, we define the similarity variance as follows:

$$\begin{aligned} \text{simVar}(f, CB) &= \frac{1}{|CB|} \sum_{W \in CB} (\text{sim}(Q_f, W) - \mu_f)^2 \\ \mu_f &= \frac{1}{|CB|} \sum_{W \in CB} \text{sim}(Q_f, W) \end{aligned} \tag{4}$$

Q_f denotes the query consisting of the feature f. $\text{sim}(Q_f, W)$ is the semantic similarity between the query Q_f and a workflow W. Moreover, μ_f is the arithmetic mean of the similarities between the query and each workflow W from the case base CB. The simVar value is computed in two ways for each feature f. The feature can either be added to the query part Q^+ of the query Q or it can be added to the restriction part Q^-. Thus, simVar^+ and simVar^- are computed separately and the average simVar is defined by the arithmetic mean:

$$\text{simVarMean}(f, CB) = \frac{\text{simVar}^+(f, CB) + \text{simVar}^-(f, CB)}{2} \tag{5}$$

Initially, features are ranked by their simVarMean value in descending order. Features with a value of 0 are ignored since they are not suitable to distinguish workflows from one another as they are part of every workflow in the case base.

In the next step, relations between features are analyzed. For each feature f all related features are determined and stored in a feature table FT. Formally, the set of related features $F_{rel}(f)$ of a feature $f \in FT$ contains those features $g \in FT$ that share a common partial workflow with f which is either a generalization of f or g:

$$F_{rel}(f) = \{g \in FT | \exists f' \subseteq f : (f' \sqsupseteq g \vee g \sqsupseteq f') \vee \exists g' \subseteq g : (g' \sqsupseteq f \vee f \sqsupseteq g')\}$$

Related features can be differentiated by their number of nodes and by their generality of nodes. A feature may have related features that are *larger*, *equally large*, or *smaller* as well as related features which are *more specific*, *equally specific*, or *more general*. For example, for the feature workflow $f_1 = \{$*slice, ham*$\}$, the related feature $g_1 = \{$*cut, meat*$\}$ is *more general* and *equally large* while the feature $g_2 = \{$*parma-ham*$\}$ is *more specific* and *smaller*.

3.2 Online Phase

The online phase of the C-POCBR dialog component is described in Algorithm 1. The dialog component iteratively creates and displays questions until the user

```
Input  : CB: Case Base, FT: Feature Table
Output: A solution workflow S
C-POCBR Dialog Algorithm(CB, FT)
    Q ← ∅, CW ← CB, CF ← FT
    repeat
        q ← questionSelection(Q, CF)
        displayQuestion(q)
        if userIgnoresQuestion then
            CF ← updateCandidateFeatures(CF, q)
        end
        if a ← userAnswersQuestion then
            Q ← extendQuery(Q, a)
            CW ← retrieveAndDisplayCandidateWorkflows(CW, Q)
            CF ← updateCandidateFeatures(CF, CW, a)
        end
        if W ← userExcludesWorkflow then
            CB ← CB \ {W}
            CW ← retrieveAndDisplayCandidateWorkflows(CB, Q)
            CF ← updateCandidateFeatures(CF, CW)
        end
    until S ← userSelectsSolution(CW) OR stoppingCriteria
    return S
```

Algorithm 1. C-POCBR Dialog Algorithm

selects a workflow or until stopping criteria are fulfilled. The set of candidate workflows is updated, if a question is answered or if a workflow is excluded by the user.

The dialog component is always initialized with the full case base CB as well as with the feature table FT. The dialog starts with an empty query Q[1]. Initially, the set of candidate features CF, i.e., relevant features to be asked in a question, is the full set of features from the feature table. The initial set of candidate workflows CW encompasses the whole case base.

In the main loop, the dialog component selects a question based on the candidate features CF considering the simVarMean scoring, the previously answered questions, as well as the feature relations (details are described below). Each question involves one or in certain cases several candidate features and is displayed to the user. Then, the user has four options to react:

1. *Ignoring a question:* In this case, the feature being subject of the question as well as larger related features and more specific related features are removed from the set of candidate features and the question selection determines the next best question.
2. *Answering a question:* If a question is answered by the user, the query Q is extended and a similarity-based retrieval with the extended query on the current set of candidate workflows CW is performed. After each retrieval, workflows that are less similar than the average of all workflow's similarities are removed from CW and thus are not included in subsequent retrievals. By this means, only the most suitable workflows are retained with respect to the current query. The workflow with the highest similarity from CW is displayed

[1] In principle an initial pre-modeled query could be used as well, but we have not yet investigated this option.

to the user. In addition, the table of candidate features CF is updated as well. Only features contained in the candidate workflows CW remain in CF, which ensures that only relevant questions are posed. Thus, with each retrieval performed, CW and CF are further reduced. In addition, the ranking of the remaining features CF is updated according to simVarMean (see Eq. 5) by using CW instead of CB.

3. *Excluding a suggested workflow:* The user may explicitly exclude a suggested workflow as possible solution, which removes the workflow from the case base CB (only temporary for this dialog) and triggers a new retrieval. As a consequence, it is likely that more candidate workflows CW than before exist because of the lower average similarity of all workflows to the current query. Consequently, more candidate features CF may become available.
4. *Selecting a solution:* If the user selects a workflow as the desired solution the retrieval terminates successfully.

We now describe in more detail the question selection method applied. We provide three major types of questions, which are depicted in Table 1. Based on the ranking of the candidate features, the subject matter of a question is determined. If the user answers that the suggested feature is desired, specific follow-up questions are selected in the subsequent iterations of the main loop. Those follow-up questions aim at further refining the previous question asked. Follow-up questions are derived from the set of related features stored in the feature table.

Table 1. Question sequence in a conversation

Order	Question type	Subject matter	Example
1.	Initial feature question (FQ)	Highest ranked feature	Q: Is $\{meat\}$ a desired feature? A: desired, undesired, irrelevant
2.	Follow-up specialization question (SQ)	More specific feature(s)	Q: Is there a suitable specialization for $\{meat\}$? $\{poultry\}, \{ham\}, \{chicken\}, \ldots$ A: apply, select undesired feature(s), irrelevant
3.	Follow-up enlargement question (EQ)	Larger feature(s)	Q: Is there a suitable enlargement for $\{chicken\}$? $\{shred, chicken\}, \{chop, chicken\}, \ldots$ A: apply, select undesired feature(s), irrelevant

At the beginning of a conversation the highest ranked feature from the candidate features is suggested in a feature question (FQ). This type of question is not related to previously suggested features and it will be asked as long as the user selects the suggested feature as irrelevant or undesired.

In case of a previously answered FQ as desired, a first follow-up question, i.e., a specialization question (SQ), is posed suggesting one or (if available) several equally large but more specific features. Again, the features are sorted by their simVarMean value. The user can choose a specialization, select features as undesired, or mark all specializations as irrelevant. This type of question is repeated as long as the user chooses specializations and as long as further specializations are available.

Following the SQs, an enlargement question (EQ) is displayed to the user that suggests, if available, larger and not more general features than the previously selected and/or specialized feature. More general features are omitted since they would widen the current context of the previous feature. Just as in SQ, the user has three different options: choose an enlargement, select an enlargement as undesired, or mark all enlargements as irrelevant. If no more EQs are available, the next initial FQ is selected, addressing a new and potentially unrelated subject matter.

When the set of candidate features CF is updated due to an ignored or answered question, irrelevant features can be inferred based on the relations between features. If a question is marked as irrelevant, all the related features (e.g., more specific and larger features) are marked as irrelevant, too. If suggested features are selected as undesired, they are added to the restriction part of the current query and related irrelevant features are no longer considered as candidate features, to prevent the system from repetitively asking the user what she does not like. If a feature is marked as desired, also related features such as more general features are removed from the candidates table. If a user chooses a specialization or an enlargement, the target feature that is already present in the query is replaced with the new feature. In this event, related features of the target feature without those that are still relevant for the new feature are removed from the feature table.

4 Evaluation

We now describe the evaluation comparing the presented C-POCBR approach with a traditional POCBR approach in which the user models a POQL query (see Sect. 2.1) manually using a query editor. The evaluation aims at testing three hypotheses and is conducted with a simulated user as well as with human users. Hypothesis H1 states that if the user's requirements can be fulfilled by a workflow from the case base, then this workflow must be retrievable by correctly answering all questions posed. However, as questions are created automatically, real users may give wrong answers due to misunderstood questions. Therefore, hypothesis H2 targets the basic utility from the user's perspective. Furthermore, hypothesis H3 relates to the user interaction effort by comparing the conversational approach with the query modeling approach. The following hypotheses are formulated under the assumption that the user's requirements can be fulfilled entirely by one workflow in the case base:

- H1. The desired workflow is retrieved with C-POCBR when all questions are answered correctly.
- H2. The C-POCBR dialog enables users to retrieve the desired workflow.
- H3. C-POCBR reduces the communication effort required to retrieve the desired workflow.

4.1 Evaluation Setup

For the experiments, we used the already existing CookingCAKE system [19], which is part of the CAKE framework[2]. It already includes a graphical POQL editor, which is used as implementation of the POCBR approach. In addition, we implemented the C-POCBR approach[3] as an extension of CookingCAKE. Thus, both systems to be compared use the same case base, similarity measures, and retrieval implementation[4].

In all experiments, we use a case base of 61 cooking workflows that describe the preparation of sandwich recipes. We created search scenarios for the evaluation that describe queries in plain text to be given to the users. According to the structure of POQL queries, a search scenario describes a required workflow together with several restriction workflows. Queries are constructed in a semi-automatic process in which each workflow from the case base is turned into a textual description of a search scenario that contains sufficient information to unambiguously specify it. In this process, feature workflows (see Definition 2) are added iteratively to the query either as requirement or as restriction until a unique specification is obtained. In a last step, textual descriptions are written by hand based on the constructed queries. In total for 60 workflows (out of 61) an appropriate search scenario description could be constructed.

4.2 Experimental Evaluation

Hypothesis H1 is tested using an experiment with a simulated user, which automatically answers the posed questions of the C-POCBR approach correctly. We adopt the methodology by Aha et al. [1], who evaluate a conversational retrieval with a leave-one-in cross validation. Consequently, in each of the 60 search scenarios the corresponding target workflow remains in the case base. During a conversation, the algorithm ignores questions that are not relevant in the specific search scenario; all other questions are answered according to the described search scenario. The conversation for a scenario is considered successful, if the proposed best fitting workflow that is displayed during the conversation is equal to the workflow from which the search scenario was derived. It turned out that the target workflow is retrieved in each of the 60 search scenarios, which fully confirms hypothesis H1. In average, 10.25 questions were asked in the dialog.

Hypotheses H2 and H3 are tested in experiments with eight human users who simultaneously performed the experiments on different computers while all interactions are being logged. After a familiarization phase in which the users are introduced to the usage of the POCBR and the C-POCBR approach, we randomly distributed the eight participants evenly to one of two groups. Furthermore, we randomly chose four textual search scenarios of similar size. Each user evaluated both approaches on the basis of the four scenarios. Thus, each

[2] See cake.wi2.uni-trier.de.

[3] Online demo available at cookingcake.wi2.uni-trier.de/conversation.

[4] During the experiments, the available adaptation methods of CookingCAKE are not used.

approach was used 16 times in total. The first group evaluated the POCBR approach with two scenarios and conducted the C-POCBR with the other scenarios afterwards. The second group evaluated both approaches in the opposite order. Finally, all users filled in a questionnaire capturing their subjective experience during the experiment.

Table 2. Experimental results: avg. values across all successful retrievals and users

	POCBR	C-POCBR
Number of successful conversations	15/16	14/16
Total conversation time	5:34 min	5:40 min (30 questions)
Required conversation time	4:46 min	2:16 min (9 questions)

Table 2 summarizes selected measures extracted from the logged experiment data. The values shown are average results over all successful queries and all users for the POCBR and the C-POCBR approach. The *number of successful conversations* shows that only a few of the 16 query runs were not successful as the target workflow was not identified and selected by a user. Thus, hypothesis H2 can be confirmed.

To assess the communication effort, the conversation time used in the POCBR and the C-POCBR approach were compared. In addition, the number of questions posed in the C-POCBR approach were determined. The *total conversation time* is the time span from the start of the conversation (in C-POCBR) or point in time when the user begins to enter the query in the POQL editor (in POCBR) until the desired workflow is retrieved and identified by the user. For POCBR and C-POCBR those time spans are quite comparable. We discovered that users following the POCBR approach tend to completely model the given query scenario, before they start the retrieval for the first time. Sometimes, the first retrieval does not lead to the desired workflow and modifications of the query have to be performed until the desired workflow is retrieved. In C-POCBR the users follow the dialog and investigate the presented workflow. In average 30 questions are answered before the user identified that the desired workflow is displayed. When analyzing these results in more detail, we found out that quite often the desired workflow is presented to the user but she did not recognize it as the desired result. In those cases, the dialog could have been terminated earlier if the user would have analyzed the displayed result more thoroughly. We analyzed this effect in detail and determined the *required conversation time*, i.e., the time until the desired workflow is displayed the first time in the dialog loop. We also determined the number of questions the user was asked during this period. We can see that if users would have checked the displayed workflows more thoroughly, the C-POCBR approach could have been more than twice as fast as the POCBR approach. The fact that this does not happen is an indication that the workflow presentation in the C-POCBR implementation needs additional explanation functions that better allow the user to identify how the presented workflow relates to the answers of her query. However, with respect to the dialog component, we consider hypothesis H3 at least partially confirmed.

Figure 4 shows the results obtained from the questionnaires that the users filled after they performed the conversation. The values are average ratings over 16 conversations with C-POCBR. Users consider the majority of the posed C-POCBR questions to be comprehensible and relevant. Moreover, the retrieval results were also rated to be reasonable with respect to the answered questions. The results indicate that the automatic creation of questions provides useful questions for the conversation. Users did state different opinions on whether the question sequences are sensible.

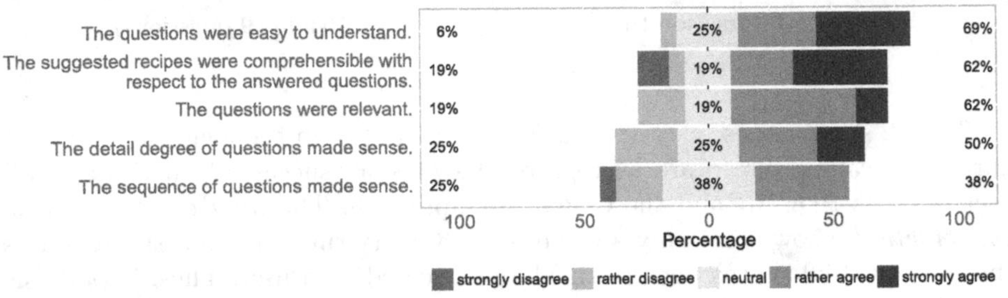

Fig. 4. Average user ratings of C-POCBR conversations on a five-point likert scale

In addition, users compared both approaches at the end of all experiments. Three out of eight users stated that the C-POCBR approach could be enhanced by allowing the user to model an initial query. Five out of eight participants stated the POCBR approach to be easier to use than the C-POCBR approach. Users indicated that sometimes general concepts in questions were difficult to understand as the subsumption of concepts was not always clear to them. Six out of eight participants claimed that POCBR is faster than C-POCBR, although this subjective assessment clearly conflicts with the measured values. One reason may be that the users are more actively involved during the modeling with the POCBR approach.

5 Conclusions and Future Work

We presented a novel approach to conversational POCBR that conducts an interactive dialog with users to facilitate the retrieval of workflows. To save effort for defining suitable questions, a method for the automatic creation of questions based on extracted features was described. Our work showed that those features are meaningful subjects of questions and that they are suitable to distinguish workflow cases from one another. The quality and performance of conversations was improved by ranking, analyzing, and selecting relevant features. We evaluated the approach with simulated and real users and showed that when questions are always answered correctly, the desired workflow is found in a straight-forward manner. Furthermore, our results indicate that the conversational query process has the potential to be faster than the traditional query approach and thus is able to reduce the communication effort for users.

The evaluation revealed some issues that should be investigated in future work. We discovered that users did not recognize the target workflow immediately and that more general questions caused problems of comprehension. Thus, the presentation and explanation of workflows and features should be improved. For the sake of simplicity, we omitted control-flow elements such as loops, parallel, and alternative sequences and restricted the description of nodes to their semantic label. Thus, it is desirable to evaluate the approach in domains with more complex workflows. In such domains, we assume that the conversation more strongly outperforms the modeling due to the increased complexity involved in modeling. In addition to POCBR domains, we assume the questioning strategy presented in this work to be also applicable more broadly in CCBR with complex case representations, provided that feature vocabularies are organized in a hierarchy and co-occurring features are identified. Also, future work should investigate how adaptability of workflows can be considered during a conversation. By this means, interactive retrieval could be combined with interactive adaptation to provide an even more powerful problem solver for users.

Acknowledgments. This work was funded by the German Research Foundation (DFG), project number BE 1373/3-3.

References

1. Aha, D.W., Breslow, L.A., Muñoz-Avila, H.: Conversational case-based reasoning. Appl. Intell. **14**(1), 9–32 (2001)
2. Aha, D.W., McSherry, D., Yang, Q.: Advances in conversational case-based reasoning. Knowl. Eng. Rev. **20**(3), 247–254 (2005)
3. Awad, A.: BPMN-Q: a language to query business processes. In: Enterprise Modelling and Information Systems Architectures - Concepts and Applications. LNI, vol. P-119, pp. 115–128. GI (2007)
4. Beeri, C., Eyal, A., Kamenkovich, S., Milo, T.: Querying business processes. In: Dayal, U., Whang, K., Lomet, D.B., Alonso, G., Lohman, G.M., Kersten, M.L., Cha, S.K., Kim, Y. (eds.) Proceedings of the 32nd International Conference on Very Large Data Bases, pp. 343–354. ACM (2006)
5. Bergmann, R.: Experience Management: Foundations, Development Methodology, and Internet-based Applications. Springer, Heidelberg (2002)
6. Bergmann, R., Gil, Y.: Similarity assessment and efficient retrieval of semantic workflows. Inf. Syst. **40**, 115–127 (2014)
7. Dijkman, R., Dumas, M., García-Bañuelos, L.: Graph matching algorithms for business process model similarity search. In: Dayal, U., Eder, J., Koehler, J., Reijers, H.A. (eds.) BPM 2009. LNCS, vol. 5701, pp. 48–63. Springer, Heidelberg (2009). doi:10.1007/978-3-642-03848-8_5
8. Dijkman, R.M., Rosa, M.L., Reijers, H.A.: Managing large collections of business process models - current techniques and challenges. Comput. Ind. **63**(2), 91–97 (2012)

9. Gu, M., Aamodt, A.: A knowledge-intensive method for conversational CBR. In: Muñoz-Ávila, H., Ricci, F. (eds.) ICCBR 2005. LNCS (LNAI), vol. 3620, pp. 296–311. Springer, Heidelberg (2005). doi:10.1007/11536406_24
10. Kapetanakis, S., Petridis, M., Knight, B., Ma, J., Bacon, L.: A case based reasoning approach for the monitoring of business workflows. In: Bichindaritz, I., Montani, S. (eds.) ICCBR 2010. LNCS (LNAI), vol. 6176, pp. 390–405. Springer, Heidelberg (2010). doi:10.1007/978-3-642-14274-1_29
11. Kim, J., Suh, W., Lee, H.: Document-based workflow modeling: a case-based reasoning approach. Expert Syst. Appl. **23**(2), 77–93 (2002)
12. Kohlmaier, A., Schmitt, S., Bergmann, R.: A similarity-based approach to attribute selection in user-adaptive sales dialogs. In: Aha, D.W., Watson, I. (eds.) ICCBR 2001. LNCS (LNAI), vol. 2080, pp. 306–320. Springer, Heidelberg (2001). doi:10.1007/3-540-44593-5_22
13. Leake, D.B., Wilson, D.C.: Combining CBR with interactive knowledge acquisition, manipulation and reuse. In: Althoff, K.-D., Bergmann, R., Branting, L.K. (eds.) ICCBR 1999. LNCS, vol. 1650, pp. 203–217. Springer, Heidelberg (1999). doi:10.1007/3-540-48508-2_15
14. Madhusudan, T., Zhao, J.L., Marshall, B.: A case-based reasoning framework for workflow model management. Data Knowl. Eng. **50**(1), 87–115 (2004)
15. Mcsherry, D.: Increasing dialogue efficiency in case-based reasoning without loss of solution quality. In: IJCAI 2003, pp. 121–126. Morgan Kaufmann (2003)
16. McSherry, D.: Explanation in recommender systems. Artif. Intell. Rev. **24**(2), 179–197 (2005)
17. McSherry, D.: Conversational case-based reasoning in medical decision making. Artif. Intell. Med. **52**(2), 59–66 (2011)
18. Minor, M., Montani, S., Recio-García, J.A.: Process-oriented case-based reasoning. Inf. Syst. **40**, 103–105 (2014)
19. Müller, G., Bergmann, R.: CookingCAKE: a framework for the adaptation of cooking recipes represented as workflows. In: Kendall-Morwick, J. (ed.) Workshop Proceedings from (ICCBR 2015). CEUR, vol. 1520, pp. 221–232. CEUR-WS.org (2015)
20. Müller, G., Bergmann, R.: Generalization of workflows in process-oriented case-based reasoning. In: Russell, I., Eberle, W. (eds.) Proceedings of the 28th International Florida Artificial Intelligence Research Society Conference, FLAIRS 2015, pp. 391–396. AAAI Press (2015)
21. Müller, G., Bergmann, R.: Learning and applying adaptation operators in process-oriented case-based reasoning. In: Hüllermeier, E., Minor, M. (eds.) ICCBR 2015. LNCS (LNAI), vol. 9343, pp. 259–274. Springer, Cham (2015). doi:10.1007/978-3-319-24586-7_18
22. Müller, G., Bergmann, R.: POQL: a new query language for process-oriented case-based reasoning. In: Bergmann, R., Görg, S., Müller, G. (eds.) Proceedings of the LWA 2015. CEUR Workshop Proceedings, vol. 1458, pp. 247–255. CEUR-WS.org (2015)
23. Muñoz-Avila, H., Aha, D.W., Breslow, L.A., Nau, D.S.: HICAP: an interactive case-based planning architecture and its application to noncombatant evacuation operations. In: Proceedings of the 16th National Conference on Artificial Intelligence AAAI/IAAI, pp. 870–875. AAAI Press (1999)
24. Richter, M.M., Weber, R.O.: Case-Based Reasoning - A Textbook. Springer, Heidelberg (2013)

25. Störrle, H., Acretoaie, V.: Querying business process models with VMQL. In: Proceedings of the 5th ACM SIGCHI Annual International Workshop on Behaviour Modelling - Foundations and Applications, BMFA 2013, pp. 4:1–4:10. ACM, New York (2013)
26. Van Der Aalst, W.M.P.: Business process management: a comprehensive survey. ISRN Softw. Eng. **2013**, 1–37 (2013). doi:10.1155/2013/507984. Hindawi Publishing Corporation
27. Weber, B., Wild, W., Breu, R.: CBRFlow: enabling adaptive workflow management through conversational case-based reasoning. In: Funk, P., González Calero, P.A. (eds.) ECCBR 2004. LNCS (LNAI), vol. 3155, pp. 434–448. Springer, Heidelberg (2004). doi:10.1007/978-3-540-28631-8_32

Maintenance for Case Streams: A Streaming Approach to Competence-Based Deletion

Yang Zhang(✉), Su Zhang, and David Leake

School of Informatics and Computing, Indiana University, Bloomington, IN 47405, USA
yz90@umail.iu.edu, zhangsu@indiana.edu, leake@cs.indiana.edu

Abstract. The case-based reasoning community has extensively studied competence-based methods for case base compression. This work has focused on compressing a case base at a single point in time, under the assumption that the current case base provides a representative sample of cases to be seen. Large-scale streaming case sources present a new challenge for competence-based case deletion. First, in such contexts, it may be infeasible or too expensive to maintain more than a very small fraction of the overall set of cases, and the current system snapshot of the cases may not be representative of future cases, especially for domains with concept drift. Second, the interruption of processing required to compress the full case base may not be practical for large case bases in real-time streaming contexts. Consequently, such settings require maintenance methods enabling continuous incremental updates and robust to limited information. This paper presents research on addressing these problems through the use of *sieve streaming*, a submodular data summarization method developed for streaming data. It demonstrates how the approach enables the maintenance process to trade off between maintenance cost and competence retention and assesses its performance compared to baseline competence-based deletion methods for maintenance. Results support the benefit of the approach for large-scale streaming data.

Keywords: Case-based maintenance · Case deletion · Sieve streaming · Streaming algorithm

1 Introduction

Case base maintenance has been extensively studied in case-based reasoning research, with particular focus on competence-preserving methods for controlling case-base growth (e.g., [9,13,14,19,21]). These commonly compress the case base periodically, based on criteria such as competence models generated from the full current case base, under the *representativeness assumption* that problems in the current case base are a good proxy for the entire problem space [15]. For standard CBR scenarios, such methods have been shown to provide good compression while limiting competence loss. However, the increasing prevalence

D.W. Aha and J. Lieber (Eds.): ICCBR 2017, LNAI 10339, pp. 420–434, 2017.
DOI: 10.1007/978-3-319-61030-6_29

of streaming data sources presents a new context for CBR maintenance. For example, in e-commerce applications, streams of orders can exceed millions of cases per day, and biomedical sensors may generate a stream of tens of millions of readings each day, at the rate of thousands of readings a second [11]. Some CBR research addresses streaming data issues such as temporal aspects and information aggregation across cases, but to our knowledge, not the maintenance methods required for large streaming sources under real-time constraints.

Handling large case streams in real time will require case-base maintenance. However, this presents challenges for traditional competence-based deletion. First, in the large-scale streaming context, a representative case sample may be difficult to obtain, or infeasible to store, due to data size; the system may need to be fielded with only a small fraction of potential cases and potential case coverage. Second, periodically compressing the entire case base may require interrupting system processing, with scale increasing the time required for full case-base compression. Consequently, there is a need for maintenance strategies explicitly targeting big data streaming scenarios. In addition, even a case base with good competence at a given time may be affected by later concept drift [12], requiring robustness to concept drift as well.

The need to handle streaming data is well known in the knowledge discovery community. This paper examines the applicability of a knowledge discovery method, Badanidiyuru et al.'s *sieve-streaming* [2] algorithm, to streaming CBR maintenance, and its tradeoffs with respect to standard CBR approaches. Sieve streaming is a data summarization algorithm for ongoing extraction of representative elements from a data stream in a single pass without access to the full dataset, using a fixed memory size, and with guarantees on level of approximation to the optimal set. We propose applying this to integrate ongoing maintenance into the case base, replacing the standard case base with a set of candidate case bases managed by sieve streaming, and between which the system can shift, based on ongoing quality estimates, every time a case is processed. In a previous small-scale study we did an initial test of this approach [20]. This paper presents a more extensive evaluation, including for concept drift settings. The experiments support the method's capability for continuous case-based maintenance robust to concept drift.

2 Previous Work on Case-Base Maintenance and Concept Drift

Case-base maintenance has been extensively studied in CBR, but the need for case-based maintenance to address concept drift has received less attention. Leake and Wilson [8] propose addressing domains where external environment is changing by using a trend-based maintenance approach, triggered by diachronic analysis of changes over time; they later proposed that patterns in the types of solved problems could also be treated as a form of trend information and exploited [18].

Lu and Zhang [10] propose the use of competence models to detect concept drift, which could in turn be used to guide case-base maintenance. Widmer and Kubat [17] present methods for storing concepts associated with recurring contexts to re-use when those contexts arise; such methods could be applied to CBR as well. Cunningham et al. [3] observe that a CBR system can address concept drift by selecting a new case base, composed only of recent cases, when accuracy falls below a threshold, and note the possibility of a CBR system addressing concept drift by incremental replacement of selected examples over time; this paper provides and evaluates such an approach. Delany et al. [4] present a two-stage maintenance protocol to address the concept drift in spam, including both ongoing case addition and periodic re-indexing.

3 Sieve Streaming for Case-Base Maintenance

3.1 Integrating Sieve-Streaming Maintenance into the CBR Cycle

Case-base maintenance is often seen as a process separate from core CBR processing, that is triggered outside of other steps in the CBR cycle. Our approach integrates maintenance directly into the case base. This integration is important because it enables the use of maintenance information—in the form of candidate case bases—to be applied flexibly, with problems always solved using the best candidate case base.

The approach builds on the sieve-streaming algorithm, introduced by Badanidiyuru et al. as a means to "summarize a massive dataset 'on the fly"' [2]. Sieve streaming selects a high utility subset of a data stream in real time, without requiring storing the entire dataset or multiple passes through the data. The utility of the selected subset is guaranteed to be within a constant factor of the optimum possible on the stream, regardless of stream size.

In sieve streaming, the utility of a subset as a whole is derived from the utility of its members, and each member's utility is evaluated based on its relation to other members of this subset. Those characteristics make the sieve streaming algorithm easily applicable to case-base maintenance: For case-base maintenance, utility can be instantiated as competence, with finding a high utility subset corresponding to finding a high competence compressed case base.

In the sieve-streaming algorithm, maintenance is done continuously, and a set of approximations of the optimal subset is used to help select and retain data points (cases) that provide a sufficient gain of utility. For each approximation, a "sieve" is constructed (in the CBR context, each sieve can be considered as a candidate case base). Corresponding to the changes of the utility characteristics of the sieves, those sieves will be created or deleted dynamically. The sieve with maximal utility is the desired result of the algorithm: the case base to use.

As illustrated in Fig. 1, in our sieve-streaming-based maintenance approach, the sieve streaming maintenance system replaces the standard case base. As problems are solved, the resulting case is provided to the sieve streaming mechanism, which determines whether to add the case to one or more sieves. When cases are retrieved, they are retrieved from the highest utility sieve.

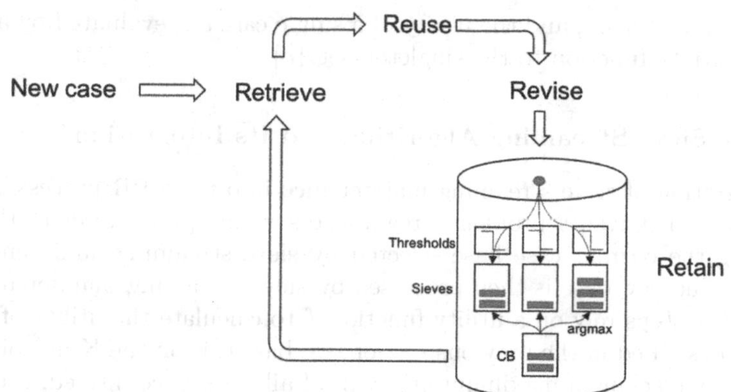

Fig. 1. Illustration of CBR framework with adapted sieve-streaming algorithm (CBR cycle adapted from [1] and Sieve-Streaming process from [2])

3.2 Defining Utility for Sieve Streaming Maintenance

Applying sieve streaming to generating competent compact case bases requires a utility function reflecting competence. However, issues arise for basing utility on commonly used competence notions such as coverage and reachability [14] and relative coverage (RC) [15]. Those models focus directly on the contributions of a particular case. This is useful to guide case-base editing operations(case deletion, addition, etc.), for example, for ordering cases presented to condensed nearest neighbor [5]. However, the sieve streaming utility function must reflect an estimate of overall competence. The choice to retain a case is made based on its potential utility contribution to a particular sieve; this depends on the dynamically changing contents of sieves.

Sieve streaming requires that the utility function be submodular, i.e., that case addition satisfies a "diminishing returns property:" the marginal gain of adding a case to a set must decrease as the set of cases becomes large. As the basis of our utility function, we follow Badanidiyuru et al. in using a function based on K-medoid clustering. K-medoid clustering aims to minimize a "loss function" the average dissimilarity between each point and the medoid of the cluster to which it is assigned. In the context of competence, if the medoid cluster points correspond to the cases retained in the case base, and other points in the cluster correspond to problems to be solved, this loss function can be seen as reflecting the average level of dissimilarity in retrieved cases for solving the other problems in the case set. We use a monotone submodular form of the K-medoid loss function [7]. Our function assesses utility for a set of cases, compared to a reference set; Sect. 3.4 describes our choices for reference sets. The K-medoid loss function can be used to estimate utility for both single cases and sets of cases for any non-negative similarity measures. The utility of sets of cases is estimated by averaging the pair-wise minimum dissimilarities between cases from that set and the evaluation set, and the utility of a single case c is evaluated by applying the same utility function to the singleton set $\{c\}$.

3.3 The Sieve Streaming Algorithm and Its Integration

The integration of sieve streaming maintenance into the CBR process is shown in Algorithm 1. Given a problem stream, the system applies CBR to the problem, using the current case base selected by sieve streaming, and generates a new case. The new case is then processed by sieve streaming maintenance. The maintenance steps rely on a utility function f to calculate the utility of a set of cases. As described in the previous section, we base this on the K-medoid utility function. By tracking a maximum utility m of all cases encountered, the system dynamically generates a new set of approximations of optimal case base choices, O_{new}, and deletes or adds sieves during processing. For approximations that first appeared in O_{new}, new empty sieves will be created in a case base candidate list l; and for approximations in O_{old} but not in O_{new}, the corresponding sieves and the cases they contain will be deleted from l permanently.

We denote sieves as S_v, where v is the threshold for the sieve. For each sieve S_v, the algorithm calculates the marginal gain of utility $\Delta_f(S_v \cup \{c\})$ for adding case c to a sieve S_v. If the current number of cases in sieve S_v does not reach the pre-determined upper boundary of sieve capacity k, and the calculated marginal gain is greater than the marginal value threshold $(\frac{v}{2} - f(S_v))/(k - |S_v|)$, the case is added to the sieve. Otherwise, the case is discarded. Finally, the system identifies the sieve with maximum utility as the case base, to be used for the next retrieval.

In the sieve streaming algorithm, once a sieve is full, the sieve is no longer changed. The sieve can be removed when the m value is updated (line 8) the corresponding approximation value no longer appears in the set O_{new}.

3.4 Evaluation Set Sampling for Utility Calculations

In streaming approach, retention decisions are based on the utility calculation, which estimates the competence contributions of cases. Because utility is used to choose between sieves, it must be calculated independently of the sieve contents. Consequently, a separate dataset must be generated for this purpose. One possibility is simply to sample from the first cases that have been seen. We call this the static sampling method.

The static method has limitations for a large scale stream with concept drift, as prior cases may no longer apply. The reservoir sampling algorithm [16] was introduced as a complement to static sampling, for situations such as concept drift. The evaluation set is first sampled uniformly from existing cases, then updated dynamically from incoming cases by substituting new items for old with a certain probability. This may increase the representativeness of cases.

Algorithm 1. *Streaming maintenance integrated with CBR stream processing. Maintenance algorithm adapted from Badanidiyuru et al. [2]*

$O_{new} \leftarrow \phi$
$O_{old} \leftarrow \phi$
$l \leftarrow \phi$
$m \leftarrow 0$

while problem stream not empty **do**
    read problem p from stream

CBR problem processing and case generation
    $c \leftarrow$ make-case(p, CBR(p, $C_{current}$))

Maintenance process
    m $\leftarrow \max(m, f(\{c\}))$
    $O_{old} \leftarrow O_{new}$
    $O_{new} \leftarrow \{(1+\epsilon)^i | i \in \mathbb{Z}, m < (1+\epsilon)^i < 2km\}$
    for each $v \in O_{new}$ and $\notin O_{old}$ **do**
        $S_v \leftarrow \phi$
        add S_v to l
    for each $v \in O_{old}$ and $\notin O_{new}$ **do**
        remove S_v from l
    for each sieve S_v in l **do**
        $\Delta_f(S_v \cup \{c\}) = f(S_v \cup \{c\}) - f(S_v)$
        if $|S_v| < k$ and $\Delta_f(S_v \cup \{c\}) \geq \frac{\frac{v}{2} - f(S_v)}{k - |S_v|}$ **then**
            $S_v \leftarrow S_v \cup \{c\}$
    $C_{current} \leftarrow \arg\max_{S_v \in l} f(S_v)$ ▷ Set current case base to best sieve

3.5 Complexity of Sieve Streaming vs. Conventional Maintenance

Given that a motivation for sieve streaming maintenance is handling large-scale case streams, computational complexity considerations are important. We first consider space complexity. By Badanidiyuru's analysis [2], when the system instantiates the sieves, the number of sieves is constrained by $O = \{(1+\epsilon)^i | i \in \mathbb{Z}, m < (1+\epsilon)^i < 2\,\mathrm{km}\}$, where k is the maximum capacity of each sieve, ϵ is a user-settable parameter determining the quality of the set of approximations O, and m is the maximum utility of the cases seen (smaller ϵ values increase memory requirements). From this, the number of sieves equals $log_{(1+\epsilon)}2k$. Because the maximum capacity for the sieves is restricted by k, with $|S_i| \leq k$, multiplying the number of sieves and maximum capacity establishes that the maximum number of cases stored in memory for the streaming approach is $klog_{(1+\epsilon)}2k$. Because k and ϵ are user-determined, this shows that the maximum memory cost is a constant, independent of the number of incoming cases n. Consequently, the space complexity of the streaming approach is $O(1)$. Because the maximum memory cost refers to the upper bound of space cost, while usually only some of sieves reach the maximum size k [2], in practice cost is expected to be below the maximum.

The time complexity of each application of the sieve streaming maintenance process follows the basic sieve-steaming algorithm cost $\frac{nlogk}{\epsilon}$ [2], where n is the number of input cases seen in the stream. As previously mentioned, k and ϵ are fixed user-set parameters, so $\frac{logk}{\epsilon}$ is constant. The desired level of approximation to an optimal compressed case base, reflected in the user's choice of parameter settings, will have an important effect in practice, as will the cost of utility calculations, determined by the utility strategy chosen. Because sieve streaming maintenance is a triggered for every new case, the overall time complexity of the streaming maintenance approach is O(n).

The baseline methods we consider are Smyth and McKenna's [15] CNN-FP and RC-FP. CNN-FP selects footprint cases based on Condensed Nearest Neighbor [5]; RC-FP also applies CNN but first sorts the candidate cases based on their Relative Coverage. When CNN-FP is applied periodically by adding cases until reaching a size threshold t, maximum space is t, so space complexity is $O(1)$. Maintenance will be triggered $\frac{n}{t}$ times for a n case input stream, and CNN requires a maximum of t^2 time, so for n cases processed the maximum time cost is nt, for time complexity $O(n)$.

Processing cost for the streaming approach, CNN-FP and RC-FP depends on the similarity method chosen. However, the similarity assessment cost is a function of the number of cases stored in the sieves (for the streaming method) or the compressed case base (for CNN-FP or RC-FP), independent of n. Thus the streaming approach and CNN-FP both have $O(n)$ time complexity and $O(1)$ space complexity. The primary benefit of the sieve approach lies in the ability to exploit the time-accuracy trade-off with different combinations of k and ϵ, and to have theoretical guarantees on the competence level achieved. For instance, with small k and large ϵ, there will be fewer sieves in the case base, and each sieve will have less capacity, reducing the maintenance processing time, but also the competence. Conversely, with large k and small ϵ, more sieves will be generated, and each sieve will have larger capacity, leading to better case base competence at the cost of higher time. While the ability to choose maximum case base size in CNN-FP also enables trading off time and accuracy, the CNN-FP offers no guarantees for quality level at a particular case base size.

4 Evaluation

We conducted experiments to address four questions:

1. How do parameter settings for the streaming approach enable tuning performance to trade off processing speed, case base size, competence, and retrieval quality?
2. How does the performance of the streaming approach compare to the baseline methods for speed, compressed case base size, competence, and retrieval quality?
3. How well does streaming retention handle concept drift, in terms of accuracy and speed of response to concept drift, compared to baseline methods?

4. How does the streaming approach perform for large-scale case streams, for competence and processing time?

Question two extends preliminary tests of Zhang et al. [20]. The other questions address new issues.

4.1 Experimental Design

Performance Criteria: Performance results consider four factors: (1) processing time, (2) case base size, (3) competence, and (4) retrieval quality. Competence is calculated for the whole case base; retrieval quality is calculated by summing the retrieval distance between test cases and retrieved cases for all cases processed in the input stream.

Datasets: Experiments were run using three test domains:

1. **Travel Case Base:**[1] This was selected as a widely used CBR benchmark. It contains 1470 cases, each with a unique identifier plus 6 categorical features and 3 numerical features.
2. **3D Spatial Network:**[2] This case base contains 434,874 cases, each with a unique ID and three numerical features. This was selected to provide a large scale test.
3. **Artificial Concept Drift Datasets**: This synthetic dataset was generated to enable controlled experiments on the effects of concept drift. We generated two variants, each with 2000 cases, each of which including four numerical attributes. The 2000 cases belong to five different groups. Each group includes 400 cases. Two types of concept drift data were generated:
   (a) **Sudden Concept Drift**: Here the groups have almost no overlap. The four features in the each group are randomly generated from the same Gaussian distributions, and the Gaussian distributions for the five case groups are constructed with standard deviation of 10 and means of 100, 200, 300, 400 and 500, respectively. We concatenate the sets of cases into an input stream with 2000 cases consisting of 5 groups, with four sudden concept drifts occurring, between the different groups.
   (b) **Gradual Concept Drift**: The dataset is generated as above, except for changing the standard deviation to 30 instead of 10. This results in approximately a 10% overlap between any two groups.

4.2 Question 1: Tuning Performance Characteristics

A potential benefit of the streaming approach is the ability to adjust parameter settings to trade off competence retention and speed. Our first experiments

---

[1] http://cbrwiki.fdi.ucm.es/mediawiki/index.php/Case_Bases.

[2] https://archive.ics.uci.edu/ml/datasets/3D+Road+Network+(North+Jutland,+Denmark) [6].

examine how parameter settings can affect processing speed, case base size, competence, and retrieval quality. We tested the system with varying ϵ and k on a 1000-case stream randomly selected from the 3D spatial network, with ϵ from 0.05 to 0.50, in increments of 0.05, and k from 25 to 200 in increments of 25.

As shown in Fig. 2a, running time decreases as ϵ increases, and running time increases as k increases. As shown in Fig. 2b, case base size increases when k increases, with ϵ having slightly more impact on case base size when k is large. Figure 2c shows competence improvement with larger k, but ϵ does not have the significant impact on the competence for this case base. In Fig. 2d, ϵ has less impact on accumulated retrieval distance than k. As k becomes small, distance increases significantly.

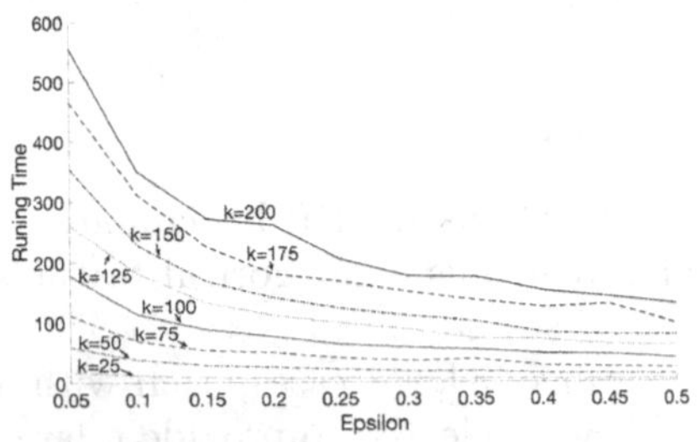

(a) Running time with different k, ϵ

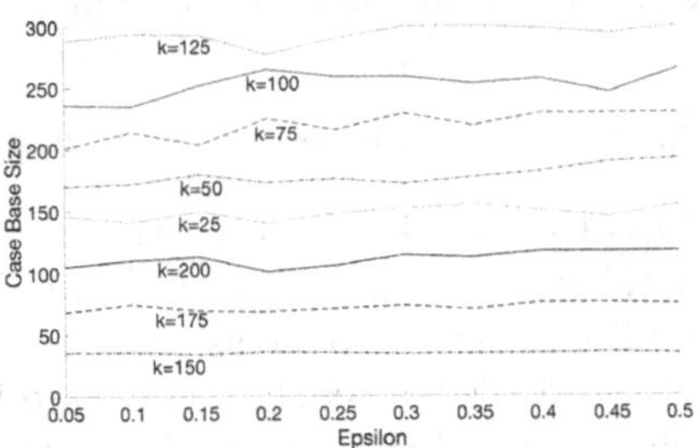

(b) Case Base Size with different k, ϵ

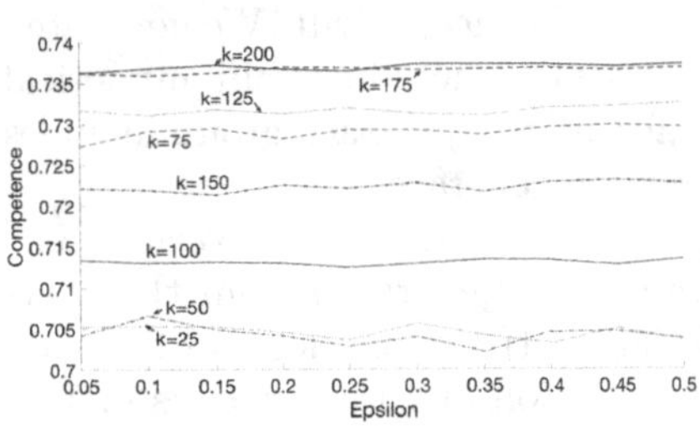

(c) Competence with different k, ϵ

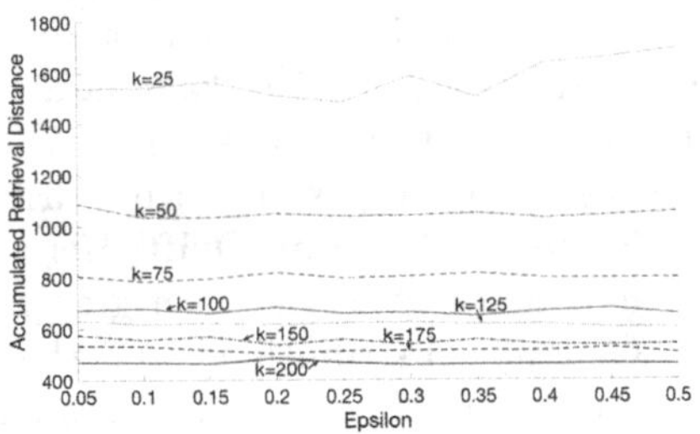

(d) Accumulated Retrieval Distance with different k, ϵ

Fig. 2. Effects of parameter changes

4.3 Question 2: Performance Comparison

To address question two, we evaluated the performance of the streaming approach along with baseline algorithms CNN-FP and RC-FP in four dimensions: *Speed*, *Case Base Size*, *Competence* and *Retrieval Quality*. Test results are averaged over five runs; each test run is conducted on a case stream of 1000 cases randomly selected from *Travel Agent Case Base*. For CNN-FP and RC-FP, the case base size limit was 100 cases; compression was triggered when that limit was reached, with the maximum size of the compressed case base limited to 50 cases. Based on tests of the relationship between those parameters, we selected an ϵ value of 0.1, at which the number of sieves is fairly stable for different ϵ values, and a maximum case base size k of 200 cases. Figure 3 shows the experimental results.

1. Maintenance Processing Time (Fig. 3a): CNN-FP and Sieve-Streaming have similar time performance. Both CNN-FP and Sieve-Streaming run much faster than RC-FP
2. Case Base Size (Fig. 3b): Compared to CNN-FP and RC-FP, Sieve-Streaming results in a more compact case base with more consistent size.
3. Competence (Fig. 3c): CNN-FP and RC-FP start with better competence. However, as more cases are processed, the competence of the streaming approach rises continuously, exceeding the baseline algorithms rapidly.
4. Retrieval Quality (Fig. 3d): RC-FP has the lowest accumulated retrieval distances and best retrieval quality. CNN-FP and the streaming approach have similar retrieval quality.

These results are generally similar to the sample result in Zhang et al. [20], and further support the promise of the approach. They also support the initially surprising result that the streaming approach does not necessarily increase processing speed. However, as the results for Question One show, speed depends on parameter settings, and it depends on similarity and stream characteristics as well. Our previous tests had the surprising result that CNN-FP ran faster than the streaming approach. For these experiments, we optimized the implementation of the streaming approach, which yielded faster running time, very close to CNN-FP. We note that in these experiments, the similarity metric is simple. Because CNN-FP relies on a larger number of similarity measurements, we expect that the streaming approach would outperform CNN-FP for similarity metrics that are more complex. We see this as a subject for further investigation.

4.4 Question 3: Concept Drift

To test the ability of the streaming approach to handle sudden or gradual concept drift, we tested it with the artificial concept drift dataset described in Sect. 4.1. The streaming approach was evaluated on the input case stream with two evaluation set sampling settings, static and dynamic, as introduced Sect. 3.4. In both settings measurements are averaged over 10 runs. Figure 4 shows the results.

Figures 4a and b show case base size during processing. For sudden concept drift in Fig. 4a, we noticed that with the static sampled evaluation set, the case base size dropped immediately when concept drift occurred, then rapidly recovered to a steady level. On the other hand, with the dynamically sampled evaluation set, slight fluctuations of case base size could be observed, but the overall trend was more steady and lower than with static sampling. For gradual concept drift in Fig. 4b, with static sampling, the case base size also dropped immediately and recovered to a higher level in a short period, then gradually

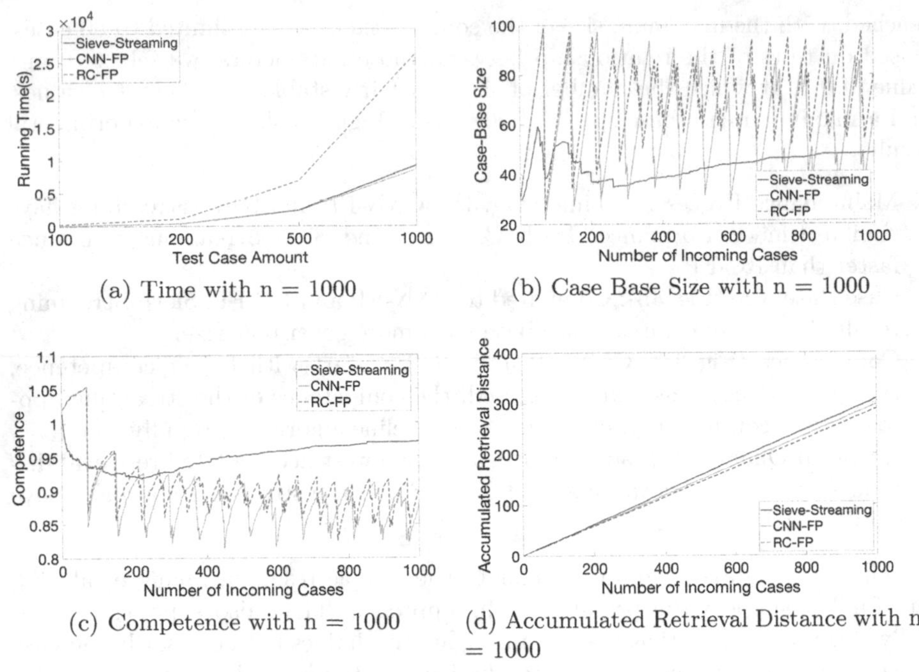

(a) Time with n = 1000

(b) Case Base Size with n = 1000

(c) Competence with n = 1000

(d) Accumulated Retrieval Distance with n = 1000

Fig. 3. Performance comparison

decreased; with dynamic sampling, the fluctuations could also be observed, and the overall level is similar to or slightly higher than static sampling.

Figure 4c shows that, for the static setting, the competence of case base drops first then recovers with case base size changes; and for the dynamic setting, the plot follows a stairstep pattern of increases, and the average competence level is better than the static setting. Figure 4d shows an upward trend for both static and dynamic settings, and static sampling showed more clear "stairs" and had a higher competence level than dynamic sampling.

In both Fig. 4e and f, the static setting shows better accuracy. We believe this is due to fluctuations in case base size changes with the dynamic setting. Also, with gradual concept drift, static sampling fits better with the overlaps in the data stream. Generally, with both static and dynamic sampling, the streaming maintenance approach is sensitive to sudden or gradual concept drift and then adapts rapidly with making use of previous knowledge.

We tested CNN-FP with same datasets. It also captured the concept changes but the streaming approach performed better in both competence and retrieval accuracy. Here the changes of case base size more reflected the periodic compressions than concept drift. Figures 5a and b show that with CNN-FP competence is lower than with the streaming method. For both sudden and gradual concept drifts, the competence first dropped then recovered slowly, and always to lower competence than streaming method. The competence plot of the streaming

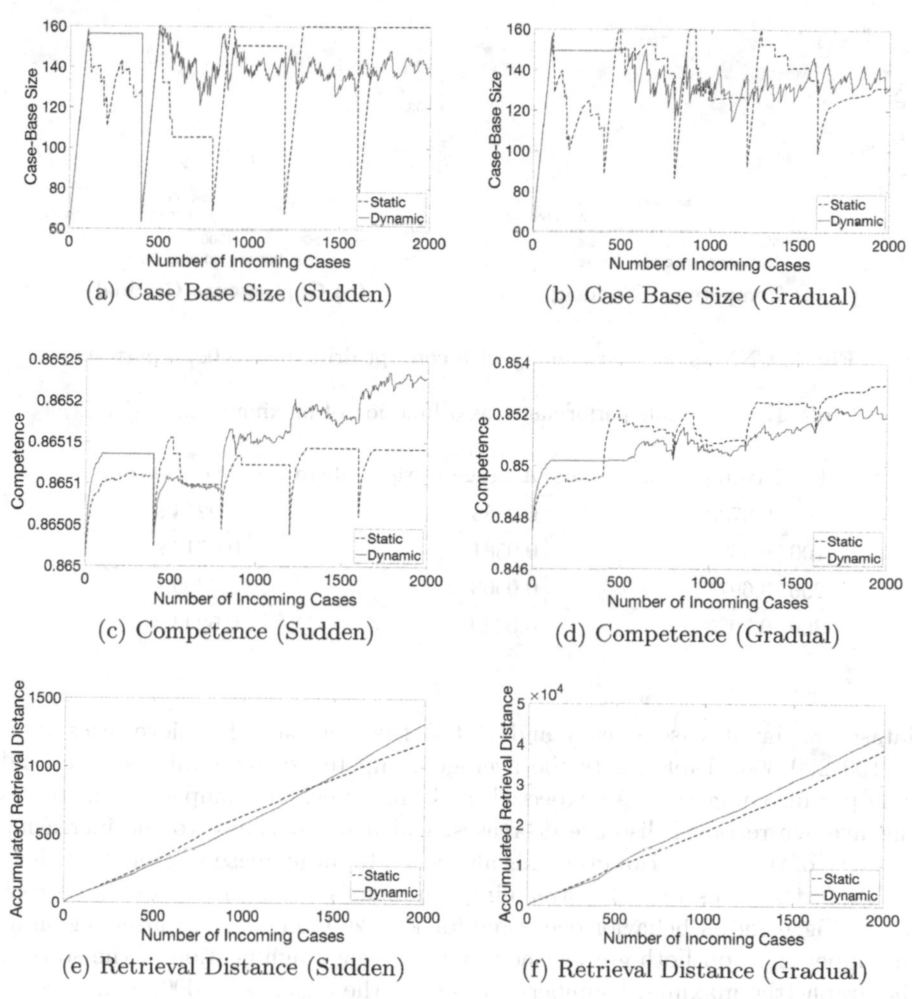

Fig. 4. Concept drift with n = 2000, 5 patterns

method here used same experimental data as Fig. 4c and d but with a different scale. The result supports the stability of the streaming method in this scenario.

When concept drift is detected, the system will delete old sieves together with their contents, which could lead to the immediate changes of case base size and competence when selecting a sieve to return as new case base. As cases pass through the maintenance component, new cases are accumulated in sieves again, and both the size and competence of case base recover.

4.5 Question 4: Scale-Up to Large Case Streams

To investigate streaming for large scale input streams, we randomly selected a stream with n = 400,000 from 434,874 examples from a 3D spatial network

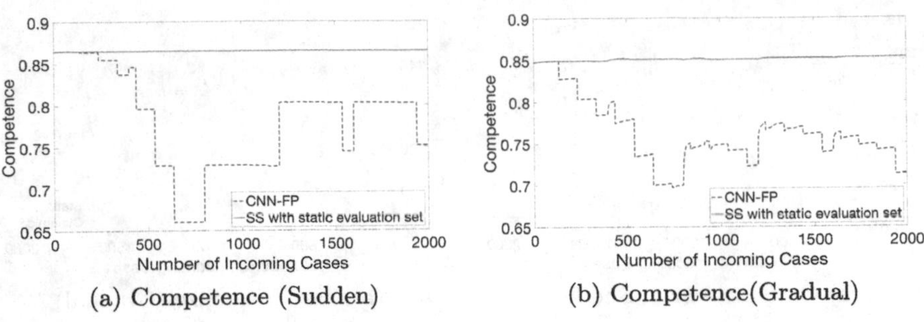

(a) Competence (Sudden)

(b) Competence(Gradual)

Fig. 5. CNN vs. sieve-streaming with concept drift, n = 2000, 5 patterns

Table 1. Large scale performance as a function of maximum sieve size (k)

k	Average competence	Average retrieval distance	Average time
50	0.6733	0.0883	0.0254 s
100	0.6681	0.0581	0.0712 s
200	0.6971	0.0503	0.3299 s
300	0.7023	0.0429	0.6244 s

dataset as input case stream and tested the approach for sieve sizes $k = 50, 100, 200, 300$. Table 1 lists the average competence, retrieval distance, and time per incoming case. As expected, as k increases, the competence increases and average retrieval distance decreases, which we attribute to the increasing capacity of the sieves. However, this also leads to an increase in execution time.

To further illustrate the streaming approach's performance for large case bases, Fig. 6 shows behavior over time for k = 200. The raised offset region at the upper right on both graphs shows an extreme magnification of the start of the graph (the maximum number of cases on the offset is 10,000, compared to 400,000 cases on the overall graph). Case base size and competence both follow a

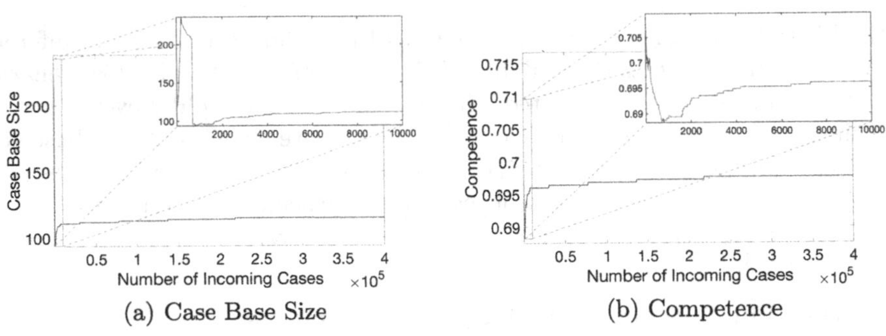

(a) Case Base Size

(b) Competence

Fig. 6. Large scale case base with n = 400,000

similar pattern; the value is variable in the beginning and then grows gradually. This suggests that the streaming approach is effective in capturing coverage of the large case base. On the other hand, the retrieval distance (tested but not shown in the figure, for reasons of space) is steady and effectively linear. The results suggest that the streaming approach is capable of handling large streams.

5 Conclusion

In this paper, we examined the properties of a case base maintenance approach based on sieve streaming. In contrast to traditional methods, the streaming approach can guarantee the competence level of the case base, given specific memory restrictions. We illustrated the trade-off between accuracy and efficiency for different parameter settings, which can be chosen based on user needs. With larger ϵ and smaller k, the streaming approach can react rapidly, with the drawback of losing accuracy. The streaming approach can also provide increased accuracy, with the drawback of reducing time efficiency. For the large scale input case streams, the streaming approach consistently maintains a substantial case base size and a steady competence level. The approach also demonstrated its effectiveness in handling concept drift. For both sudden and gradual concept drift the system is sensitive and responds rapidly. This supports that the streaming maintenance approach is promising for on-line or real-time environments, satisfactory for handling concept drift, and suitable for large-scale case streams. Interesting future issues include further study of processing speed compared to CNN-FP and further study of handling of concept drift in real-world scenarios.

References

1. Aamodt, A., Plaza, E.: Case-based reasoning: foundational issues, methodological variations, and system approaches. AI Commun. **7**(1), 39–52 (1994)
2. Badanidiyuru, A., Mirzasoleiman, B., Karbasi, A., Krause, A.: Streaming submodular maximization: massive data summarization on the fly. In: Proceedings of the 20th ACM SIGKDD International Conference on Knowledge Discovery and Data Mining, pp. 671–680. ACM (2014)
3. Cunningham, P., Nowlan, N., Delany, S., Haahr, M.: A case-based approach to spam filtering that can track concept drift. Technical report TCD-CS-2003-16, Computer Science Department, Trinity College Dublin (2003)
4. Delany, S.J., Cunningham, P., Coyle, L.: An assessment of case-based reasoning for spam filtering. Artif. Intell. Rev. **24**(3), 359–378 (2005)
5. Hart, P.E.: The condensed nearest neighbor rule. IEEE Trans. Inf. Theory **14**, 515–516 (1968)
6. Kaul, M., Yang, B., Jensen, C.S.: Building accurate 3D spatial networks to enable next generation intelligent transportation systems. In: 2013 IEEE 14th International Conference on Mobile Data Management (MDM), vol. 1, pp. 137–146. IEEE (2013)
7. Krause, A., Gomes, R.G.: Budgeted nonparametric learning from data streams. In: Proceedings of the 27th International Conference on Machine Learning (ICML-2010), pp. 391–398 (2010)

8. Leake, D.B., Wilson, D.C.: Categorizing case-base maintenance: dimensions and directions. In: Smyth, B., Cunningham, P. (eds.) EWCBR 1998. LNCS, vol. 1488, pp. 196–207. Springer, Heidelberg (1998). doi:10.1007/BFb0056333
9. Leake, D.B., Wilson, D.C.: Remembering why to remember: performance-guided case-base maintenance. In: Blanzieri, E., Portinale, L. (eds.) EWCBR 2000. LNCS, vol. 1898, pp. 161–172. Springer, Heidelberg (2000). doi:10.1007/3-540-44527-7_15
10. Lu, N., Zhang, G., Lu, J.: Concept drift detection via competence models. Artif. Intell. **209**, 11–28 (2014)
11. Redmond, S., Lovell, N., Yang, G., Horsch, A., Lukowicz, P., Murrugarra, L., Marschollek, M.: What does big data mean for wearable sensor systems? Yearb. Med. Inform. **9**(1), 135–142 (2014)
12. Schlimmer, J.C., Granger, R.H.: Incremental learning from noisy data. Mach. Learn. **1**(3), 317–354 (1986)
13. Smyth, B., Keane, M.: Remembering to forget: a competence-preserving case deletion policy for case-based reasoning systems. In: Proceedings of the Thirteenth International Joint Conference on Artificial Intelligence, pp. 377–382. Morgan Kaufmann, San Mateo (1995)
14. Smyth, B., McKenna, E.: Building compact competent case-bases. In: Althoff, K.-D., Bergmann, R., Branting, L.K. (eds.) ICCBR 1999. LNCS, vol. 1650, pp. 329–342. Springer, Heidelberg (1999). doi:10.1007/3-540-48508-2_24
15. Smyth, B., McKenna, E.: Competence models and the maintenance problem. Comput. Intell. **17**(2), 235–249 (2001)
16. Vitter, J.S.: Random sampling with a reservoir. ACM Trans. Math. Softw. (TOMS) **11**(1), 37–57 (1985)
17. Widmer, G., Kubat, M.: Learning in the presence of concept drift and hidden contexts. Mach. Learn. **23**(1), 69–101 (1996)
18. Wilson, D., Leake, D.: Maintaining case-based reasoners: dimensions and directions. Comput. Intell. **17**(2), 196–213 (2001)
19. Wilson, D., Martinez, T.: Reduction techniques for instance-based learning algorithms. Mach. Learn. **38**(3), 257–286 (2000). http://dx.doi.org/10.1023/A%3A1007626913721
20. Zhang, Y., Zhang, S., Leake, D.: Case-base maintenance: a streaming approach (2016)
21. Zhu, J., Yang, Q.: Remembering to add: competence-preserving case-addition policies for case base maintenance. In: Proceedings of the Fifteenth International Joint Conference on Artificial Intelligence, pp. 234–241. Morgan Kaufmann (1999)

Author Index

Printed in the United States
By Bookmasters